STUDENT ATLAS OF

World Geography

Sixth Edition

John L. Allen
University of Wyoming

Consulting Editor
Christopher J. Sutton
Western Illinois University

Mc Graw Hill **Higher Education**

Boston Burr Ridge, IL Dubuque, IA New York San Francisco St. Louis
Bangkok Bogotá Caracas Kuala Lumpur Lisbon London Madrid Mexico City
Milan Montreal New Delhi Santiago Seoul Singapore Sydney Taipei Toronto

The McGraw-Hill Companies

Higher Education

STUDENT ATLAS OF WORLD GEOGRAPHY, SIXTH EDITION

Published by McGraw-Hill, a business unit of The McGraw-Hill Companies, Inc., 1221 Avenue of the Americas, New York, NY 10020.
Student Atlas is published by the **Contemporary Learning Series** group within the McGraw-Hill Higher Education division.

1 2 3 4 5 6 7 8 9 0 QPD/QPD 0 9

MHID 0-07-352760-2
ISBN 978-0-07-352760-4
ISSN 1531-0221

Managing Editor: *Larry Loeppke*
Senior Managing Editor: *Faye Schilling*
Senior Developmental Editor: *Debra Henricks*
Editorial Coordinator: *Mary Foust*
Editorial Assistant: *Nancy Meissner*
Production Service Assistant: *Rita Hingtgen*
Permissions Coordinator: *Shirley Lanners*
Senior Marketing Manager: *Julie Keck*
Marketing Communications Specialist: *Mary Klein*
Marketing Coordinator: *Alice Link*
Senior Project Manager: *Jane Mohr*
Design Specialist: *Tara McDermott*

Compositor: Lachina Publishing Services
Cartography: Carto-Graphics, Hudson, WI
Cover Image: inset: ©Corbis/RF

We would like to thank Digital Wisdom Incorporated for allowing us to use their Mountain High Maps cartography software. This software was used to create maps 126, 128–131, 133, 135, 136, 138, 140–142, 144, 146, 147, 149, 151, 152a, 153, 154, 156, and 158–162.

www.mhhe.com

A Note to the Student

The study of geography has become an increasingly important part of the curriculum in secondary schools and institutions of higher education over the last decade. This trend, a most welcome one from the standpoint of geographers, has begun to address the massive problem of "geographic illiteracy" that has characterized the United States, almost alone among the world's developed nations. When a number of international comparative studies on world geography were undertaken, beginning in the 1970s, it became apparent that most American students fell far short of their counterparts in Europe, Russia, Canada, Australia, and Japan in their abilities to recognize geographic locations, to identify countries or regions on maps, or to explain the significance of such key geographic phenomena such as population distribution, economic or urban location, or the availability of natural resources. Indeed, many American students could not even locate the United States on world maps, let alone countries like France, or Indonesia, or Nigeria. This atlas, and the texts it is intended to accompany, is a small part of the process of attempting to increase the geographic literacy of American students. As the true meaning of "the global community" becomes more apparent, such an increase in geographic awareness is not only important but necessary. If the United States has learned any lesson from the tragic events at the World Trade Center and the Pentagon on September 11, 2001, these lessons would surely include the considerations that we are not isolated from events that transpire in other parts of the world; our boundaries do not make us secure; and we ignore the conditions of political, economic, cultural, and physical geography outside those boundaries at our great peril.

The maps in the *Student Atlas of World Geography* are designed to introduce you to the patterns or "spatial distribution" of the wide variety of human and physical features of the earth's surface and to help you understand the relationships between these patterns. We call such relationships "spatial correlation" and whenever you compare the patterns made by two or more phenomena that exist at or near the earth's surface—the distribution of human population and the types of climate, for example—you are engaging in spatial correlation. Like the maps, the data sets in the atlas are intended to enable you to make comparisons between the distributions of different geographic features (for example, population growth and literacy rates) and to understand the character of the geographic variation in a single geographic feature. In many instances, the data in the tables of your atlas are the same data that have been used to produce the maps. At the very outset of your study of this atlas, you should be aware of some limitations of the data tables. In some instances, there may be data missing from a table. In such cases, the cause may represent the failure of a country to report information to a central international body (like the United Nations or the World Bank), or it may mean that the shifting of political boundaries and changed responsibility for reporting data have caused some countries (for example, those countries that made up the former Soviet Union or the former Yugoslavia) to delay their reports. It is always our aim to use the most up-to-date data that is possible. Subsequent editions of this atlas will have increased data on countries like Serbia, Montenegro, or Kosovo when it becomes available. In the meantime, as events continue to restructure our world, it's an exciting time to be a student of world geography!

You will find your study of this atlas more productive if you study the maps and tables on the following pages in the context of the five distinct themes that have been developed as part of the increasing awareness of the importance of geographic education:

1. *Location: Where Is It?* This theme offers a starting point from which you discover the precise location of places in both absolute terms (the latitude and longitude of a place) and in relative terms (the location of a place in relation to the location of other places). When you think of location, you should automatically think of both forms. Knowing something about absolute location will help you to understand a variety of features of physical geography, since such key elements are so closely related to their position on the earth. But it is equally important to think of location in relative terms. The location of places in relation to other places is often more important as a determinant of social, economic, and cultural characteristics than the factors of physical geography.
2. *Place: What Is It Like?* This theme investigates the political, economic, cultural, environmental, and other characteristics that give a place its identity. You should seek to understand the similarities and differences of places by exploring their basic characteristics. Why are some places with similar environmental characteristics so very different in economic, cultural, social, and political ways? Why are other places with such different environmental characteristics so seemingly alike in terms of their institutions, their economies, and their cultures?
3. *Human/Environment Interactions: How Is the Landscape Shaped?* This theme illustrates the ways in which people respond to and modify their environments. Certainly the environment is an important factor in influencing human activities and behavior. But the characteristics of the environment do not exert a controlling influence over human activities; they only provide a set of alternatives from which different cultures, in different times, make their choices. Observe the relationship between the basic elements of physical geography such as climate and terrain and the host of ways in which humans have used the land surfaces of the world.
4. *Movement: How Do People Stay in Touch?* This theme examines the transportation and communications systems that link people and places. Movement or "spatial interaction" is the chief mechanism for the

spread of ideas and innovations from one place to another. It is spatial interaction that validates the old cliché, "the world is getting smaller." We find McDonald's restaurants in Tokyo and Honda automobiles in New York City because of spatial interaction. Advanced transportation and communications systems have transformed the world into which your parents were born. And the world your children will be born into will be very different from your world. None of this would happen without the force of movement or spatial interaction.

5. *Regions: Worlds Within a World.* This theme helps to organize knowledge about the land and its people. The world consists of a mosaic of "regions" or areas that are somehow different and distinctive from other areas. The region of Anglo-America (the United States and Canada) is, for example, different enough from the region of Western Europe that geographers clearly identify them as two unique and separate areas. Yet despite their differences, Anglo-Americans and Europeans share a number of similarities: common cultural backgrounds, comparable economic patterns, shared religious traditions, and even some shared physical environmental characteristics. Conversely, although the regions of Anglo-America and Eastern Asia are also easily distinguished as distinctive units of the earth's surface, they have a greater number of shared physical environmental characteristics. But those who live in Anglo-America and Eastern Asia have fewer similarities and more differences between them than is the case with Anglo-America and Western Europe: different cultural traditions, different institutions, different linguistic and religious patterns. An understanding of both the differences and similarities between regions like Anglo-America and Europe on the one hand, or Anglo-America and Eastern Asia on the other, will help you to understand the world around you. At the very least, an understanding of regional similarities and differences will help you to interpret what you read on the front page of your daily newspaper or view on the evening news report on your television set.

Not all of these themes will be immediately apparent on each of the maps and tables in this atlas. But if you study the contents of *Student Atlas of World Geography*, along with the reading of your text and think about the five themes, maps and tables and text will complement one another and improve your understanding of global geography.

John L. Allen

About the Authors

John L. Allen is professor and chair of geography, emeritus, at the University of Wyoming where he taught from 2000 to 2007, and emeritus professor at the University of Connecticut, where he taught from 1967 to 2000. He is a native of Wyoming. He received his bachelor's degree in 1963 and his M.A. in 1964 from the University of Wyoming, and in 1969 his Ph.D. from Clark University. His special areas of interest are perceptions of the environment and the impact of human societies on environmental systems. Dr. Allen is the author and editor of many books and articles as well as several other student atlases, including the best-selling *Student Atlas of World Politics.*

Christopher J. Sutton is professor of geography at Western Illinois University. Born in Virginia and raised in Illinois, he received his bachelor's degree (1988) and master's degree (1991) in Geography from Western Illinois University. In 1995 he earned his Ph.D. in Geography from the University of Denver. He is the author of numerous research articles and educational materials. A broadly trained geographer, his areas of interest include cartographic design, cultural geography, and urban transportation. After teaching at Northwestern State University of Louisiana for three years, Dr. Sutton returned to Western Illinois University in 1998, serving as chair of the Department of Geography from 2002 to 2007. Additionally, Dr. Sutton has served as president of the Illinois Geographical Society.

Acknowledgments

Nozar Alaolmolki
Hiram College

Barbara Batterson-Rossi
Palomar College

A. Steele Becker
University of Nebraska at Kearney

Koop Berry
Walsh University

Daniel A. Bunye
South Plains College

Winifred F. Caponigri
Holy Cross College

Femi Ferreira
Hutchinson Community College

Eric J. Fournier
Samford University

William J. Frazier
Columbus State College

Hari P. Garbharran
Middle Tennessee State University

Baher Gosheh
Edinboro University of Pennsylvania

Donald Hagan
Northwest Missouri State University

Robert Janiskee
University of South Carolina

David C. Johnson
University of Louisiana

Effie Jones
Crichton College

Cub Kahn
Marythurst University

Artimus Keiffer
Franklin College

Leonard E. Lancette
Mercer University

Donald W. Lovejoy
Palm Beach Atlantic College

Mark Maschhoff
Harris-Stowe State College

Richard Matthews
University of South Carolina

Madolia Mills
University of Colorado–Colorado Springs

Robert Mulcahy
Providence College

Otto H. Muller
Alfred University

J. Henry Owusu
University of Northern Iowa

Steven Parkansky
Morehead State University

William Preston
California Polytechnic State University, San Luis Obispo

Neil Reid
The University of Toledo

A. L. Rydant
Keene State College

Deborah Berman Santana
Mills College

Steven Slakey
University of La Verne

Rolf Sternberg
Montclair State University

Richard Ulack
University of Kentucky

David Woo
California State University, Haywood

Donald J. Zeigler
Old Dominion University

Table of Contents

Unit V Global Patterns of Environmental Disturbance 89

Unit VI Global Political Patterns 121

Introduction: How to Read an Atlas

An atlas is a book containing maps which are "models" of the real world. By the term "model" we mean exactly what you think of when you think of a model: a representation of reality that is generalized, usually considerably smaller than the original, and with certain features emphasized, depending on the purpose of the model. A model of a car does not contain all of the parts of the original but it may contain enough parts that it is recognizable as a car and can be used to study principles of automotive design or maintenance. A car model designed for racing, on the other hand, may contain fewer parts but would have the mobility of a real automobile. Car models come in a wide variety of types containing almost anything you can think of relative to automobiles that doesn't require the presence of a full-size car. Since geographers deal with the real world, virtually all of the printed or published studies of that world require models. Unlike a mechanic in an automotive shop, we can't roll our study subject into the shop, take it apart, put it back together. We must use models. In other words, we must generalize our subject, and the way we do that is by using maps. Some maps are designed to show specific geographic phenomena, such as the climates of the world or the relative rates of population growth for the world's countries. We call these maps "thematic maps" and Units I through VI of this atlas contain maps of this type. Other maps are designed to show the geographic location of towns and cities and rivers and lakes and mountain ranges and so on. These are called "reference maps" and they make up many of the maps in Unit VII. All of these maps, whether thematic or reference, are models of the real world that selectively emphasize the features that we want to show on the map.

In order to read maps effectively—in other words, in order to understand the models of the world presented in the following pages—it is important for you to know certain things about maps: how they are made using what are called *projections*; how the level of mathematical proportion of the map or what geographers call *scale* affects what you see; and how geographers use *generalization* techniques such as simplification and symbols where it would be impossible to draw a small version of the real world feature. In this brief introduction, then, we'll explain to you three of the most important elements of map interpretation: projection, scale, and generalization.

The Coordinate System

Map Projections

Perhaps the most basic problem in *cartography*, or the art and science of map-making, is the fact that the subject of maps—the earth's surface—is what is called by mathematicians "a non-developable surface." Since the world is a sphere (or nearly so—it's actually slightly flattened at the poles and bulges a tiny bit at the equator), it is impossible to flatten out the world or any part of its curved surface without producing some kind of distortion. This "near sphere" is represented by a geographic grid or coordinate system of lines of latitude or *parallels* that run east and west and are used to measure distance north and south on the globe, and lines of longitude or *meridians* that run north and south and are used to measure distance east and west. All the lines of longitude are half circles of equal length and they all converge at the poles. These meridians are numbered from 0 degrees (Prime or Greenwich Meridian) east and west to 180 degrees. The meridian of 0 degrees and the meridian of 180 degrees are halves of the same "great circle" or line representing a plane that bisects the globe into two equal hemispheres. All lines of longitude are halves of great circles. All the lines of latitude are complete circles that are parallel to one another and are spaced equidistant on the meridians. The circumference of these circles lessens as you move north or south from the equator. Parallels of latitude are numbered from 0 degrees at the equator north and south to 90 degrees at the North and South poles. The only line of latitude that is a great circle is the equator, which equally divides the world into a northern and southern hemisphere. In the real world, all these grid lines of latitude and longitude intersect at right angles. The problem for cartographers is to convert this spherical or curved grid into a geometrical shape that is "developable"; that is, it can be flattened (such as a cylinder or cone) or is already flat (a plane). The reason the results of the conversion process are called "projections" is that we imagine a world globe (or some part of it) that is made up of wires running north-south and east-west to represent the grid lines of latitude and longitude and other wires or even solid curved plates to represent the coastlines of continents or the continents themselves. We then imagine a light source at some location inside or outside the wire globe that can "project" or cast shadows of the wires representing grid lines onto a developable surface. Sometimes the basic geometric principles of projection may be modified by other mathematical principles to yield projections that are not truly geometric but have certain desirable features. We call these types of projections "arbitrary." The three most basic types of projections are named according to the type of developable surface: cylindrical, conic, or azimuthal (plane). Each type has certain characteristic features: they may be *equal area* projections in which the size of each area on the map is a direct proportional representation of that same area in the real world but shapes are distorted; they may be *conformal* projections in which area may be distorted but shapes are shown correctly; or they may be *compromise* projections in which both shape and area are distorted but the overall picture presented is fairly close to reality. It is important to remember that all maps distort the geographic grid and continental outlines in characteristic ways. The only representation of the world that does not distort either shape or area is a globe. You can see why we must use projections—can you imagine an atlas that you would have to carry back and forth across campus that would be made up entirely of globes?

Cylindrical Projections

Cylindrical projections are drawn as if the geographic grid were projected onto a cylinder. Cylindrical projections have the advantage of having all lines of latitude as true parallels or straight lines. This makes these projections quite useful for showing geographic relationships in which latitude or distance north-south is important (many physical features, such as climate, are influenced by latitude). Unfortunately, most cylindrical-type projections distort area significantly. One of the most famous is the Mercator projection shown on the next page.

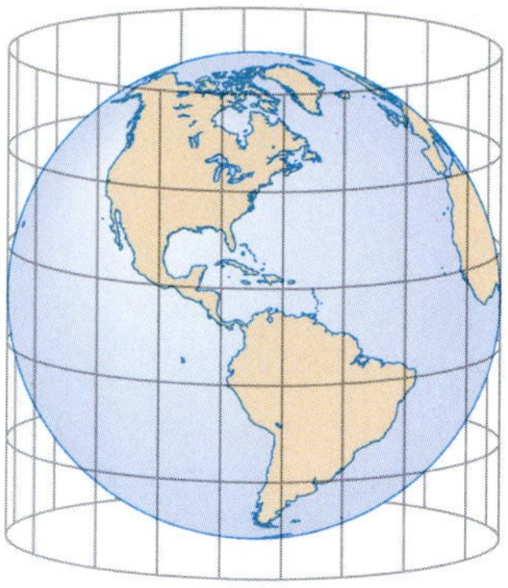

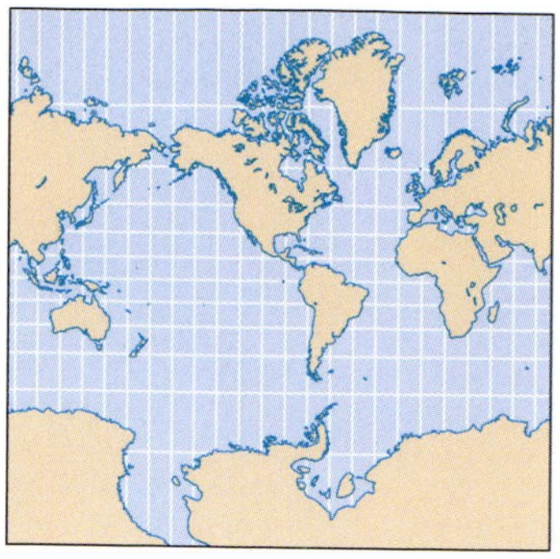

Cylindrical Projection: The Mercator Projection

This projection makes areas disproportionately large as you move toward the pole, making Greenland, which is actually about one-seventh the size of South America, appear to be as large as the southern continent. But the Mercator projection has the quality of conformality: landmasses on the map are true in shape and thus all coastlines on the map intersect lines of latitude and longitude at the proper angles. This makes the Mercator projection, named after its inventor, a sixteenth-century Dutch cartographer, ideal for its original purpose as a tool for navigation—but not a good projection for attempting to show some geographical feature in which areal relationship is important. Unfortunately, the Mercator projection has often been used for wall maps for schoolrooms and the consequence is that generations of American school children have been "tricked" into thinking that Greenland is actually larger than South America. Much better cylindrical-type projections are those like the Robinson projection used in this atlas that is neither equal area nor conformal but a compromise that portrays the real world much as it actually looks, enough so that we can use it for areal comparisons.

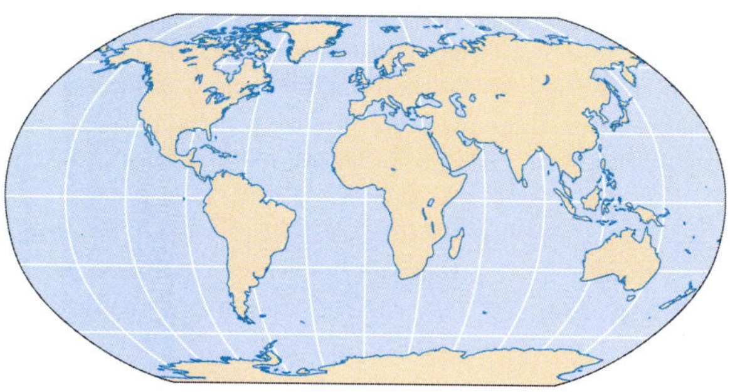

The Robinson Projection

Conic Projections

Conic projections are those that are imagined as being projected onto a cone that is tangent to the globe along a standard parallel, or a series of cones tangent along several parallels or even intersecting the globe. Conic projections usually show latitude as curved lines and longitude as straight lines. They are good projections for areas with north-south extent, like the preceding map of Europe, and may be either conformal, equal area, or compromise, depending on how they are constructed. Many of the regional maps in the last map section of this atlas are conic projections.

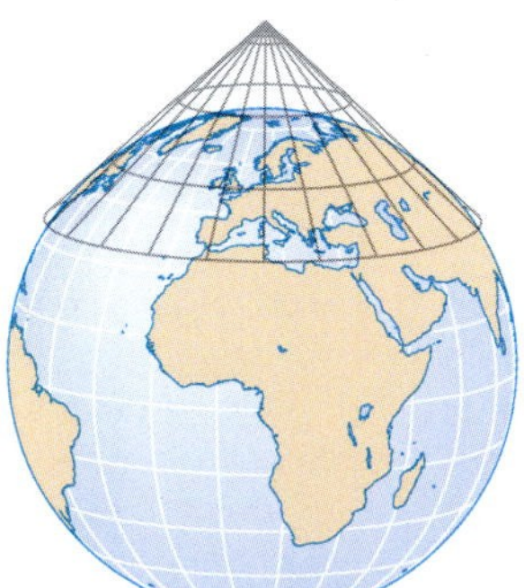

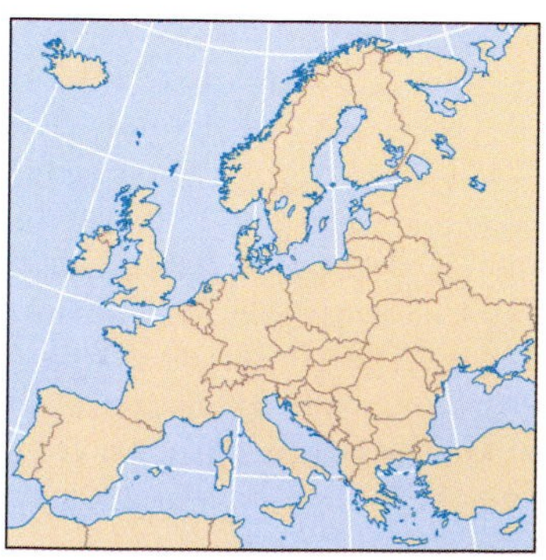

Conic Projection of Europe

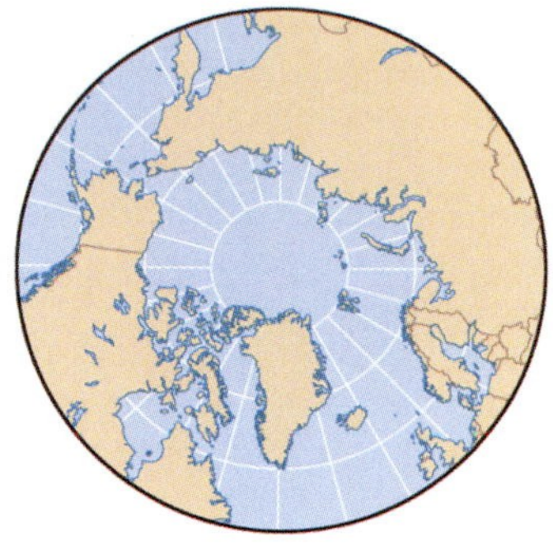

Azimuthal Projection of the North Polar Region

Azimuthal Projections

Azimuthal projections are those that are imagined as being projected onto a plane or flat surface. They are named for one of their essential properties. An "azimuth" is a line of compass bearing and azimuthal projections have the property of yielding true compass directions from the center of the map. This makes azimuthal maps useful for navigation purposes, particularly air navigation. But, because they distort area and shape so greatly, they are seldom used for maps designed to show geographic relationships. When they are used as illustrative rather than navigation maps, it is often in the "polar case" projection shown above where the plane has been made tangent to the globe at the North Pole.

Map Scale

Since maps are models of the real world, it follows that they are not the same size as the real world or any portion of it. Every map, then, is subject to generalization, which is another way of saying that maps are drawn to certain scales. The term *scale* refers to the mathematical quality of *proportional representation*, and is expressed as a ratio between an area of the real world or the distance between places on the real world and the same area or distance on the map. We show map scale on maps in three different ways. Sometimes we simply use the proportion and write what is called a *natural scale* or representative fraction: for example, we might show on a map the mathematical proportion of 1:62,500. A map at this scale is one that is one sixty-two thousand five-hundredth the size of the same area in the real world. Other times we convert the proportion to a written description that approximates the relationship between distance on the map and distance in the real world. Since there are nearly 62,500 inches in a mile, we would refer to a map having a natural scale of 1:62,500 as having an "inch-mile" scale of "1 inch represents 1 mile." If we draw a line one inch long on this map, that line represents a distance of approximately one mile in the real world. Finally, we usually use a graphic or linear scale: a bar or line, often graduated into miles or kilometers, that shows graphically the proportional representation. A graphic scale for our 1:62,500 map might be about five inches long, divided into five equal units clearly labeled as "1 mile," "2 miles," and so on. Our examples on the following page show all three kinds of scales.

The most important thing to keep in mind about scale, and the reason why knowing map scale is important to being able to read a map correctly, is the relationship between proportional representation and generalization. A map that fills a page but shows the whole world is

Map 1 Small Scale Map of the United States

Map 2 Map of the Northeast

Map 3 Map of Southeastern New England

Map 4 Large Scale Map of Boston, MA

much more highly generalized than a map that fills a page but shows a single city. On the world map, the city may appear as a dot. On the city map, streets and other features may be clearly seen. We call the first map, the world map, a *small scale* map because the proportional representation is a small number. A page-size map showing the whole world may be drawn at a scale of 1:150,000,000. That is a very small number indeed—hence the term *small scale* map even though the area shown is large. Conversely, the second map, a city map, may be drawn at a scale of 1:250,000. That is still a very small number but it is a great deal larger than 1:150,000,000! And so we'd refer to the city map as a *large scale* map, even though it shows only a small area. On our world map, geographical features are generalized greatly and many features can't even be shown at all. On the city map, much less generalization occurs—we can show specific features that we couldn't on the world map—but generalization still takes place. The general rule is that the smaller the map scale, the greater the degree of generalization; the larger the map scale, the less the degree of generalization. The only map that would not generalize would be a map at a scale of 1:1 and that map wouldn't be very handy to use. Examine the relationship between scale and generalization in the four maps on this page.

Generalization on Maps

A review of the four maps on this page should give you some indication of how cartographers generalize on maps. One thing that you should have noticed is that the first map, that of the United States, is much simpler than the other three and that the level of *simplification* decreases with each map. When a cartographer simplifies map data, information that is not important for the purposes of the map is just left off. For example, on Map 1 the objective may have been to show cities over 1 million in population. To do that clearly and effectively, it is not necessary to show and label rivers and lakes. The map has been simplified by leaving those items out. Map 4, on the other hand, is more complex and shows and labels geographic features that are important to the character of the city of Boston; therefore, the Charles River is clearly indicated on the map.

Another type of generalization is *classification*. Map 1 shows cities over 1 million in population. Map 2 shows cities of several different sizes and a different symbol is used for each size classification or category. Many of the thematic maps used in this atlas rely on classification to show data. A thematic map showing population growth rates (see Map 26) will use different colors to show growth rates in different classification levels or what are sometimes called *class intervals*. Thus, there will be one color applied to all countries with population growth rates between 1.0 percent and 1.4 percent, another color applied to all countries with population growth rates between 1.5 percent and 2.1 percent, and so on. Classification is necessary because it is impossible to find enough symbols or colors to represent precise values. Classification may also be used for qualitative data, such as the national or regional origin of migrating populations. Cartographers show both quantitative and qualitative classification levels or class intervals in important sections of maps called *legends*.

These legends, as in the samples that follow, make it possible for the reader of the map to interpret the patterns shown.

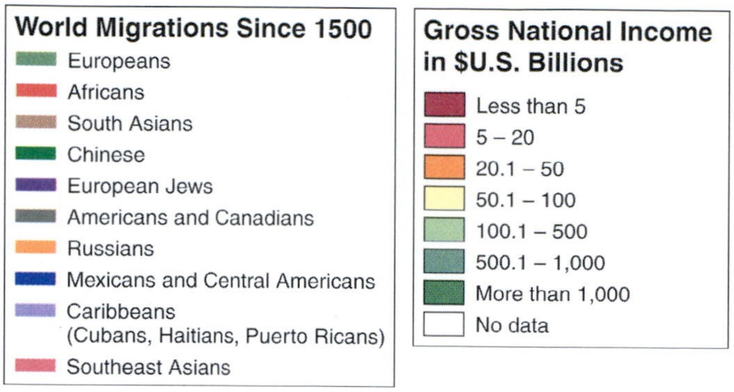

Map Legends from Maps 25 and 51

A third technique of generalization is *symbolization* and we've already noted several different kinds of symbols: those used to represent cities on the preceding maps, or the colors used to indicate population growth levels on Map 26. One general category of map symbols is quantitative in nature and this category can further be divided into a number of different types. For example, the symbols showing city size on Maps 1 and 2 on the preceding page can be categorized as *ordinal* in that they show relative differences in quantities (the size of cities). A cartographer might also use lines of different widths to express the quantities of movement of people or goods between two or more points as on Map 25.

The color symbols used to show rates of population growth can be categorized as *interval* in that they express certain levels of a mathematical quantity (the percentage of population growth). Interval symbols are often used to show physical geographic characteristics such as inches of precipitation, degrees of temperature, or elevation above sea level. The following sample, for example, shows precipitation (from Map 3a).

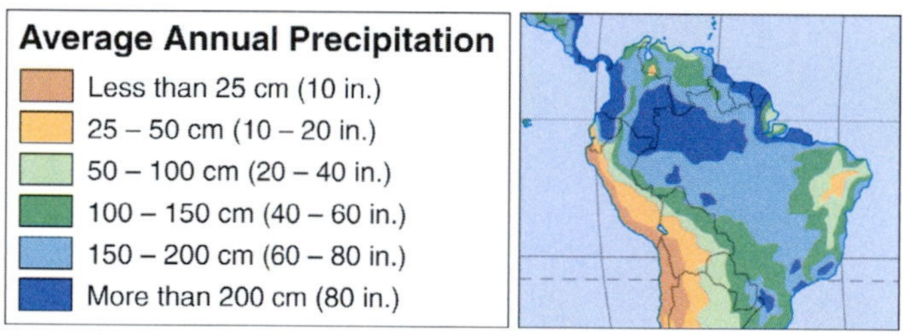

Interval Symbols

Still another type of mathematical symbolization is the *ratio* in which sets of mathematical quantities are compared: the number of persons per square mile (population density) or the growth in gross national product per capita (per person). The map below shows GDP change per capita (from Map 53).

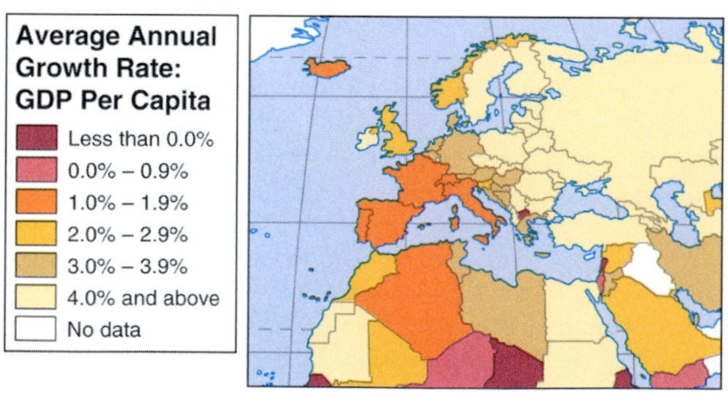

Ratio Symbols

Finally, there are a vast number of cartographic symbols that are not mathematical but show differences in the kind of information being portrayed. These symbols are called *nominal* and they range from the simplest differences such as land and water to more complex differences such as those between different types of vegetation. Shapes or patterns or colors or iconographic drawings may all be used as nominal symbols on maps. The following map (from Map 8) uses color to show the distribution of soil types.

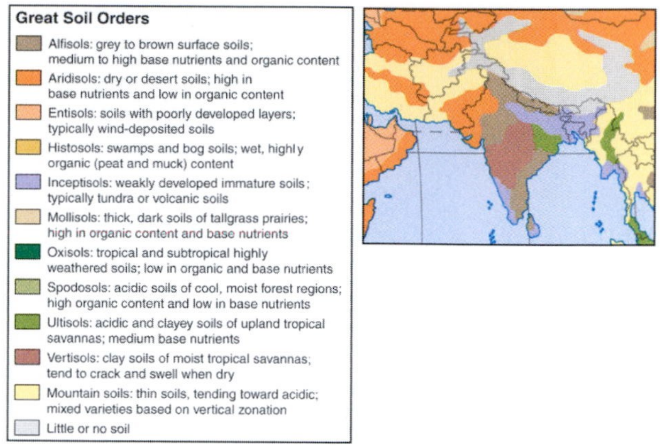

Nominal Symbols

The final technique of generalization is what cartographers refer to as *interpolation*. Here, the maker of a map may actually show more information on the map than is actually supplied by the original data. In understanding the process of interpolation it is necessary for you to visualize the quantitative data shown on maps as being three dimensional: *x* values provide geographic location along a north-south axis of the map; *y* values provide geographic location along the east-west axis of the map; and *z* values are those values of whatever data (for example, temperature) are being shown on the map at specific points. We all can imagine a real three-dimensional surface in which the *x* and y values are directions and the *z* values are the heights of mountains and the depths of valleys. On a topographic map showing a real three-dimensional surface, contour lines are used to connect points of equal elevation above sea level. These contour lines are not measured directly; they are estimated by interpolation on the basis of the elevation points that are provided.

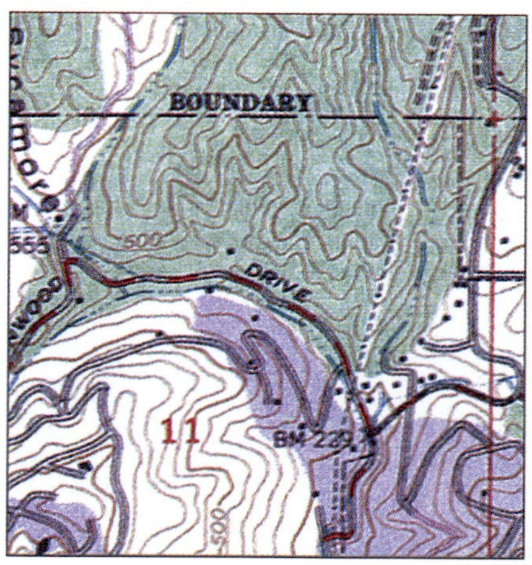

Interpolation

It is harder to imagine the statistical surface of a temperature map in which the x and y values are directions and the z values represent degrees of temperature at precise points. But that is just what cartographers do. And to obtain the values between two or more specific points where z values exist, they interpolate based on a class interval they have decided is appropriate and use *isolines* (which are statistical equivalents of a contour line) to show increases or decreases in value. The diagram to the right shows an example of an interpolation process. Occasionally interpolation is referred to as *induction*. By whatever name, it is one of the most difficult parts of the cartographic process.

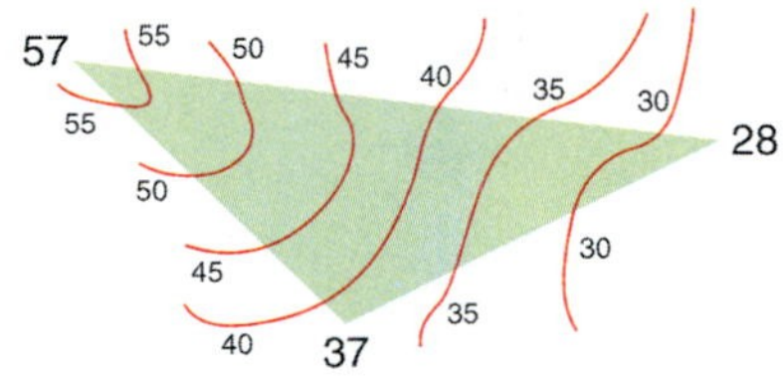

Degrees of Temperature (Celsius)
Interval = 5 degrees

And you thought all you had to do to read an atlas was look at the maps! You've now learned that it is a bit more involved than that. As you read and study this atlas, keep in mind the principles of projection and scale and generalization (including simplification, classification, symbolization, and interpolation) and you'll do just fine. Good luck and enjoy your study of the world of maps as well as maps of the world!

Unit I

Global Physical Patterns

Map 1 World Political Divisions

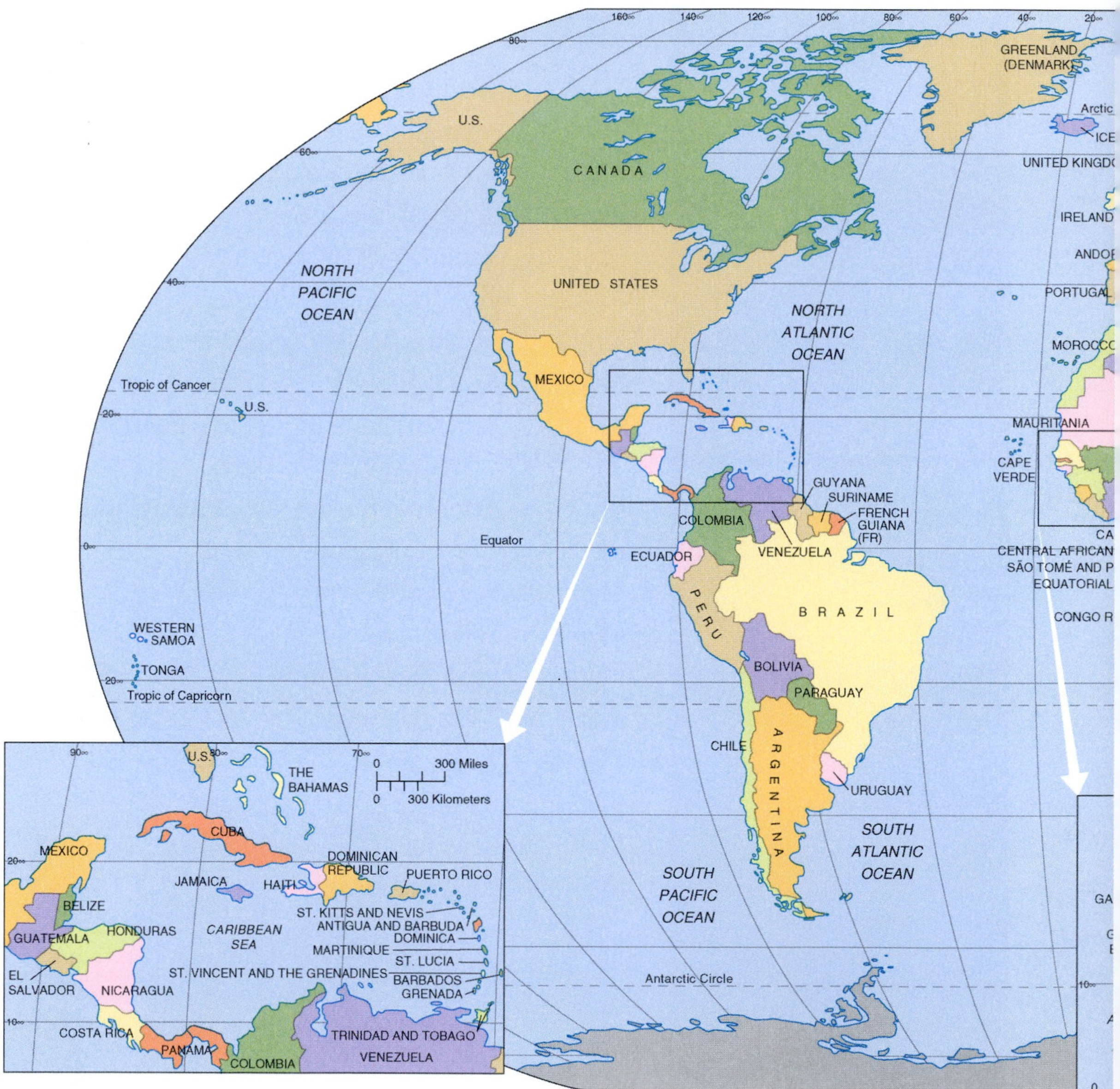

Scale: 1 to 125,000,000

0 1000 2000 Miles

0 1000 2000 3000 Kilometers

Note: All world maps are Robinson projection.

The international system includes the political units called "states" or countries as the most important component. The boundaries of countries are the primary source of political division in the world and for most people nationalism is the strongest source of political identity. State boundaries are an important indicator of cultural, linguistic, economic, and other geographic divisions as well, and the states themselves normally serve as the base level

for which most global statistics are available. The subfield of geography known as "political geography" has as its primary concern the geographic or spatial character of this international system and its components.

ARCTIC OCEAN
CANADA BASIN
12,468 ft.
BANKS I.
QUEEN ELIZABETH ISLANDS
GREENLAND
CHUKCHI SEA
Pt. Barrow
BEAUFORT SEA
VICTORIA ISLAND
7,008 ft.
Baffin Bay
Gunnbjørn Fjeld 12,139 ft.
BROOKS RANGE
BAFFIN ISLAND
Davis Strait
Arctic Circle
Bering Strait
Yukon R.
Mt. McKinley 20,320 ft.
Mackenzie R.
Great Bear Lake
NORTH AMERICA
BERING SEA
Mt. Logan 19,551 ft.
Great Slave Lake
Hudson Bay
988 ft.
LABRADOR SEA
Gulf of Alaska
ROCKY MTS.
9,869 ft.
ALEUTIAN ISLANDS
Mt. Fairweather 15,300 ft.
Mt. Robson 12,972 ft.
NORTHERN LOWLAND
LAURENTIAN PLATEAU
REYKJANES RIDGE
11,195 ft.
26,576 ft.
Mt. Waddington 13,163 ft.
Mt. Columbia 12,294 ft.
GREAT
L. Winnipeg
Strait of Belle Isle
509 ft.
ALEUTIAN TRENCH
Mt. Rainier 14,410 ft.
Borah Pk. 12,662 ft.
Granite Pk. 12,799 ft.
GREAT PLAINS
CENTRAL LOWLAND
Mississippi R.
St. Lawrence R.
NEWFOUNDLAND
17,248 ft.
Columbia R.
Gannett Pk. 13,804 ft.
Cloud Pk. 13,167 ft.
The Great Lakes
15,398 ft.
Kings Pk. 13,528 ft.
Cape Sable
Cape Cod
20,700 ft.
Uncompahgre Pk. 14,309 ft.
Grays Pk. 14,270 ft.
Ohio R.
APPALACHIAN MTS.
NORTH ATLANTIC OCEAN
20,664 ft.
Boundary Pk. 13,140 ft.
Colorado R.
Blanca Pk. 14,345 ft.
AZORES
Mt. Whitney 14,494 ft.
Wheeler Pk. 13,161 ft.
Cape Hatteras
MADEIRA
Jebel Toubkal 13,665 ft.
NORTH PACIFIC OCEAN
Mt. Elbert 14,433 ft.
Rio Grande
BERMUDA
Pico de Teide 12,188 ft.
MIDWAY IS.
5,758 ft.
Humphreys Pk. 12,633 ft.
SIERRA MADRE
Gulf of Mexico
17,921 ft.
22,950 ft.
CANARY ISLANDS
HAWAIIAN RIDGE
13,199 ft.
NORTH AMERICAN BASIN
MID-ATLANTIC RIDGE
HAWAIIAN ISLANDS
Tropic of Cancer
CAPE VERDE PLAIN
Mauna Kea 13,796 ft.
Cabo San Lucas
Pico de Orizaba 18,406 ft.
WEST INDIES
CAPE VERDE ISLANDS
3,468 ft.
25,218 ft.
Tajumulco (Vol.) 13,845 ft.
CARIBBEAN SEA
17,783 ft.
Cape Verde
15,778 ft.
23,925 ft.
18,767 ft.
Chirripó (Vol.) 12,530 ft.
Niger R.
Pico Bolívar 16,427 ft.
COCOS RIDGE
Nevado de Huila 18,865 ft.
LLANOS
GUIANA HIGHLANDS
Equator
13,783 ft.
Chimborazo 20,702 ft.
Cape Palmas
17,101 ft.
GALAPAGOS ISLANDS
Rio Negro
Amazon
PHOENIX ISLANDS
17,996 ft.
AMAZON BASIN
SELVAS
21,448 ft.
ASCENSION
15,877 ft.
ANDES MOUNTAINS
SOUTH AMERICA
CAMPOS
BRAZILIAN HIGHLANDS
Rio São Francisco
WESTERN SAMOA
Nevado Huascarán 22,133 ft.
TUAMOTU ARCHIPELAGO
19,850 ft.
TONGA IS.
TONGA
SOCIETY IS.
NAZCA RIDGE
PLATEAU OF MATO GROSSO
Rio Paraguay
Paraná
35,704 ft.
Tropic of Capricorn
ATACAMA TRENCH
26,465 ft.
1,079 ft.
18,879 ft.
RIO GRANDE RISE
KERMADEC
Nevado Ojos del Salado 22,615 ft.
33,292 ft.
KERMADEC IS.
TRENCH
SOUTHWEST PACIFIC BASIN
EAST PACIFIC RISE
Cerro Aconcagua 22,831 ft.
PAMPAS
CENTRAL LOWLAND
SOUTH PACIFIC OCEAN
16,963 ft.
Río de la Plata
SOUTH ATLANTIC OCEAN
17,278 ft.
CHATHAM ISLANDS
13,049 ft.
CHONOS ARCH.
20,382 ft.
15,998 ft.
Strait of Magellan
5,023 ft.
FALKLAND ISLANDS
SOUTH GEORGIA
PACIFIC-ANTARCTIC RIDGE
Cape Horn
SOUTH SANDWICH TRENCH
Drake Passage
BELLINGSHAUSEN PLAIN
17,192 ft.
1,641 ft.
WEDDELL PLAIN
Antarctic Circle
ANTARCTIC PENINSULA
15,847 ft.
ROSS SEA
WEDDELL SEA
Vinson Massif 16,066 ft.
160°
140°
120°
100°
80°
60°
40°
20°
0°

ZEMLYA FRANTSA IOSIFA
SEVERNAYA ZEMLYA
NOVOSIBIRSKIYE OSTROVA
ARCTIC OCEAN
SVALBARD
NOVAYA ZEMLYA
BARENTS SEA
KARA SEA
LAPTEV SEA
EAST SIBERIAN SEA
KOLYMA LOWLAND
Arctic Circle
CENTRAL SIBERIAN PLATEAU
WEST SIBERIAN PLAIN
URAL MTS.
Lake Onega
Lake Ladoga
BALTIC SEA
NORTH SEA
NORTH EUROPEAN PLAIN
EUROPE
Grossglockner 12,461 ft.
Dufourspitze 15,203 ft.
Gora Elbrus 18,510 ft.
CASPIAN DEPRESSION
ARAL SEA
CASPIAN SEA
BLACK SEA
CAUCASUS
ANATOLIAN PLATEAU
TURANIAN PLATEAU
Mt. Ararat 16,854 ft.
Pik Kommunizma 24,590 ft.
Nowshak 24,557 ft.
K2 28,250 ft.
Qolleh-ye Damavand 18,386 ft.
MEDITERRANEAN SEA
SYRIAN DESERT
Persian Gulf
ZAGROS MTS.
Lake Balkhash
ALTAI MTS.
ASIA
GOBI DESERT
TIEN SHAN
TARIM BASIN
HINDU KUSH
CENTRAL HIGHLANDS
PLATEAU OF TIBET
HIMALAYA
Mt. Everest 29,035 ft.
Kanchenjunga 28,208 ft.
Tirich Mir 25,230 ft.
Indus R.
Ganges R.
Lake Baikal
Amur R.
MANCHURIAN PLAIN
SEA OF OKHOTSK
SAKHALIN
BERING SEA
Klyuchevskaya (Vol.) 15,584 ft.
KURIL TRENCH 34,558 ft.
HOKKAIDO
JAPAN TRENCH
SEA OF JAPAN
HONSHU
Fuji Yama 12,388 ft.
KYUSHU
IZU TRENCH
NORTH PACIFIC OCEAN
NORTHWEST PACIFIC BASIN
EAST CHINA SEA
TAIWAN
Yu Shan 13,114 ft.
BONIN TR.
Tropic of Cancer
MARIANA ISLANDS
MARIANA TRENCH
36,203 ft.
LIBYAN DESERT
OASES OF FEZZAN
SAHARA
SAHEL
Lake Chad
AFRICA
RED SEA
ARABIAN PLATEAU
Hadur Shu'ayb 12,008 ft.
Ras Dashen Terara 15,158 ft.
Gulf of Aden
ARABIAN SEA
DECCAN PLATEAU
Bay of Bengal
ANDAMAN ISLANDS
NICOBAR ISLANDS
HAINAN
LUZON
PHILIPPINE SEA
SOUTH CHINA SEA
PHILIPPINE TRENCH
MINDANAO
Gulf of Thailand
Gunong Kinabalu 13,455 ft.
CAROLINE ISLANDS
MARSHALL ISLANDS
GILBERT ISLANDS
Equator
CENTRAL
ETHIOPIAN HIGHLANDS
ADAMAWA HIGHLANDS
Gulf of Guinea
Margherita Pk. 16,763 ft.
Lake Victoria
Mt. Kenya 17,058 ft.
Mt. Kilimanjaro 19,340 ft.
Lake Tanganyika
Lake Nyasa
Zambezi R.
Congo
PLATEAU
AMIRANTE IS.
MASCARENE PLAT.
MID-INDIAN RIDGE
CHAGOS-LACCADIVE PLATEAU
MALDIVE ISLANDS
CHAGOS ARCH.
DIEGO GARCIA
NINETYEAST RIDGE
SUMATRA
BORNEO
CELEBES
EAST INDIES
MELANESIA
Puncak Jaya 16,503 ft.
Mt. Wilhelm 14,793 ft.
NEW GUINEA
JAVA
JAVA TRENCH
Gunung Semeru 12,060 ft.
Gunung Rinjani 12,224 ft.
INDIAN OCEAN
Mozambique Channel
Cape Frio
KALAHARI DESERT
Orange R.
DRAKENSBERG
Cape of Good Hope
Cape Ste. Marie
MASCARENE IS.
SOUTHWEST INDIAN RIDGE
EAST INDIAMAN RIDGE
North West Cape
Tropic of Capricorn
GREAT SANDY DESERT
WESTERN PLATEAU
AUSTRALIA
GREAT VICTORIA DESERT
LAKE EYRE BASIN
GREAT DIVIDING RANGE
GREAT BARRIER REEF
CORAL SEA
NEW HEBRIDES
NEW CALEDONIA
SOUTH PACIFIC OCEAN
North Cape
PERTH BASIN
BROKEN RIDGE
GREAT AUSTRALIAN BIGHT
SOUTH AUSTRALIAN BASIN
TASMANIA
South East Cape
TASMAN SEA
Mt. Cook 12,316 ft.
AMSTERDAM I.
ST. PAUL I.
KERGUÉLEN ISLANDS
KERGUELEN PLATEAU
PRINCE EDWARD ISLAND
CROZET ISLANDS
SOUTHEAST INDIAN RIDGE
SOUTH INDIAN BASIN
TASMAN PLATEAU
MACQUARIE RIDGE
AUCKLAND ISLANDS
CAMPBELL PLATEAU
ENDERBY PLAIN
Antarctic Circle
BALLENY ISLANDS
Scale: 1 to 111,922,000
0 1000 2000 Miles
0 1000 2000 3000 Kilometers
ANTARCTICA

Map 3a Average Annual Precipitation

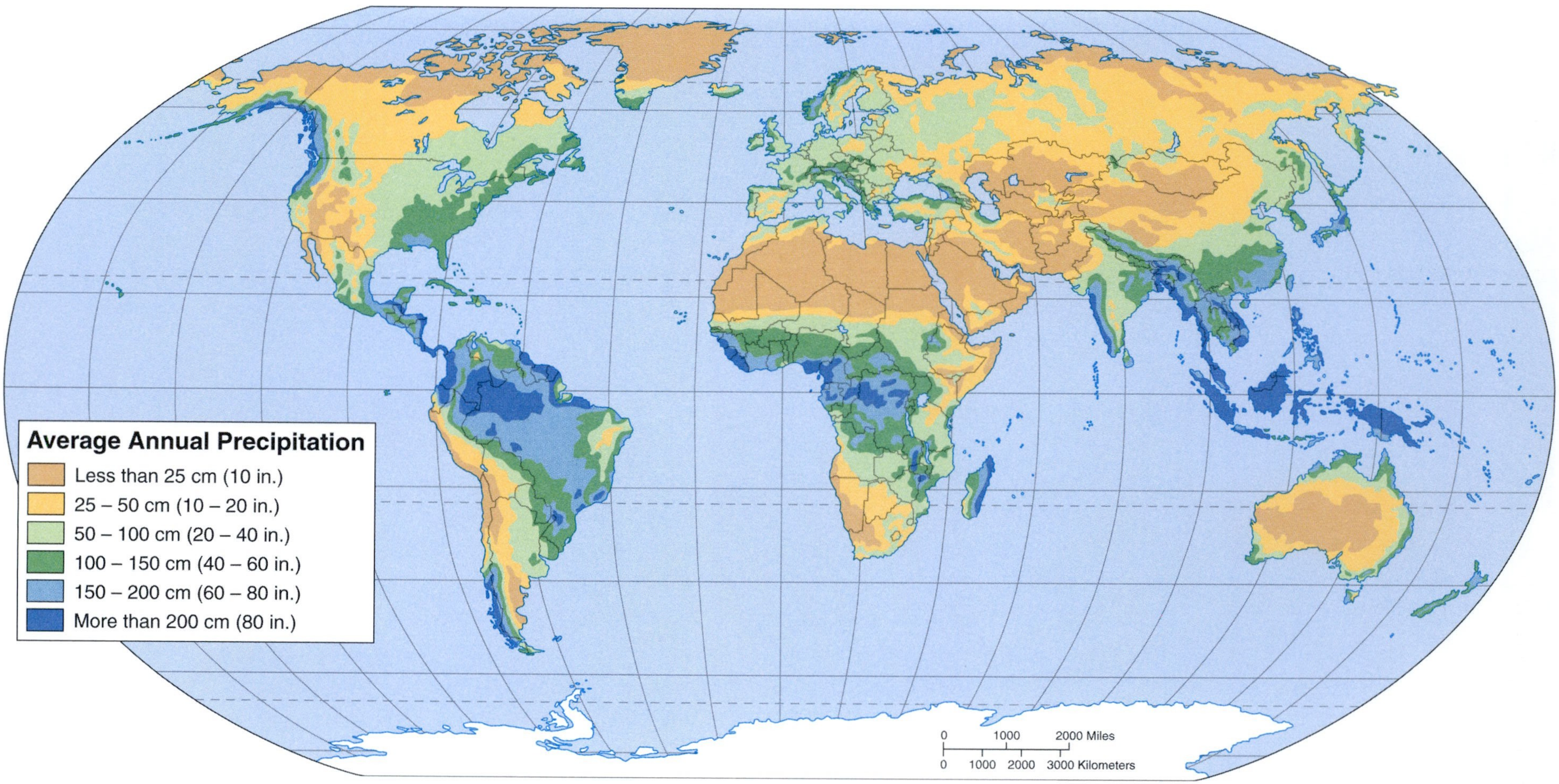

The two most important physical geographic variables are precipitation and temperature, the essential elements of weather and climate. Precipitation is a conditioner of both soil type and vegetation. More than any other single environmental element, it influences where people do or do not live. Water is the most precious resource available to humans, and water availability is largely a function of precipitation. Water availability is also a function of several precipitation variables that do not appear on this map: the seasonal distribution of precipitation (is precipitation or drought concentrated in a particular season?), the ratio between precipitation and temperature (how much of the water that comes to the earth in the form of precipitation is lost through mechanisms such as evaporation and transpiration that are a function of temperature?), and the annual variability of precipitation (how much do annual precipitation totals for a place or region tend to vary from the "normal" or average precipitation?). In order to obtain a complete understanding of precipitation, these variables should be examined along with the more general data presented on this map. The study of precipitation and other climatic elements is the concern of the branch of physical geography called "climatology."

Map 3b Seasonal Average Precipitation, November through April

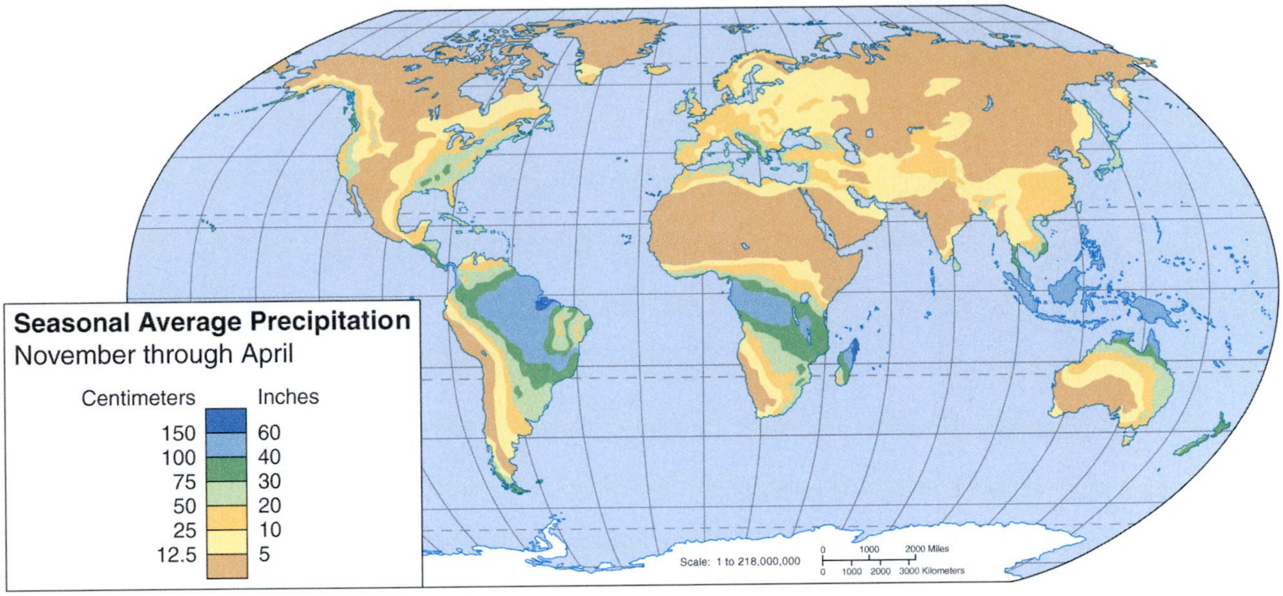

Map 3c Seasonal Average Precipitation, May through October

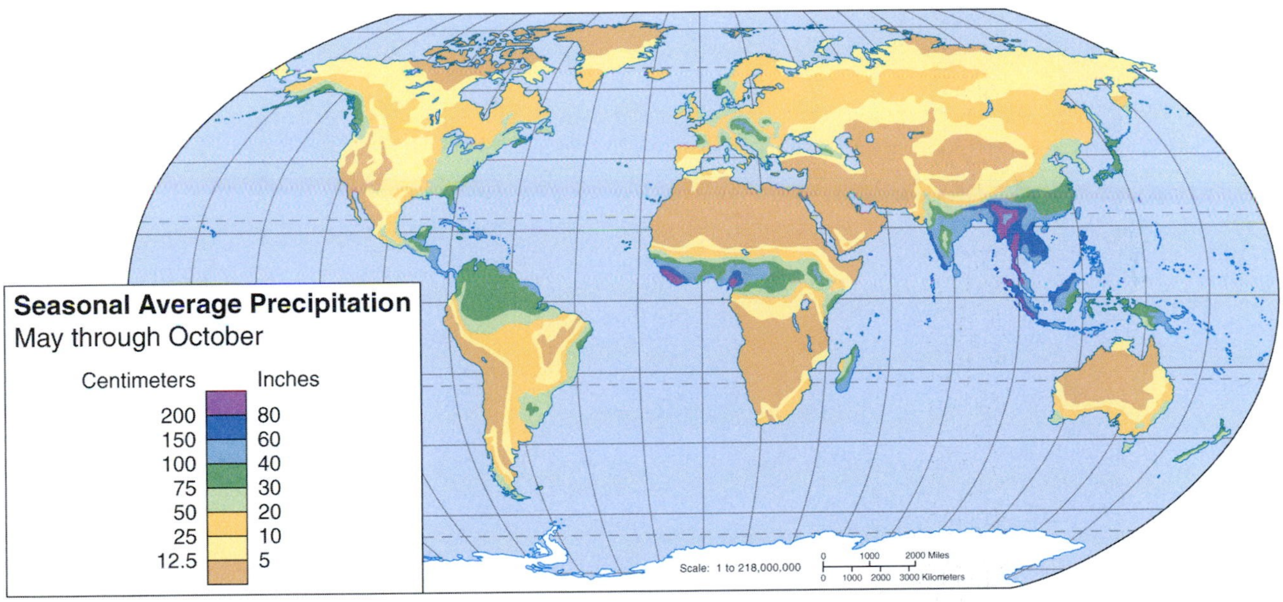

Seasonal average precipitation is nearly as important as annual precipitation totals in determining the habitability of an area. Critical factors are such things as whether precipitation coincides with the growing season and thus facilitates agriculture or during the winter when it is less effective in aiding plant growth, and whether precipitation occurs during summer with its higher water loss through evaporation and transpiration or during the winter when more of it can go into storage. Several of the world's great climate zones have pronounced seasonal precipitation rhythms. The tropical and subtropical savanna grasslands have a long winter dry season and abundant precipitation in the summer. The Mediterranean climate is the only major climate with a marked dry season during the summer, making agriculture possible only through irrigation or other adjustments to cope with drought during the period of plant growth. And the great monsoon climates of South and Southeast Asia have their winter dry season and summer rain that have conditioned the development of Asian agriculture and the rhythms of Asian life.

Map 3d Variation in Average Annual Precipitation

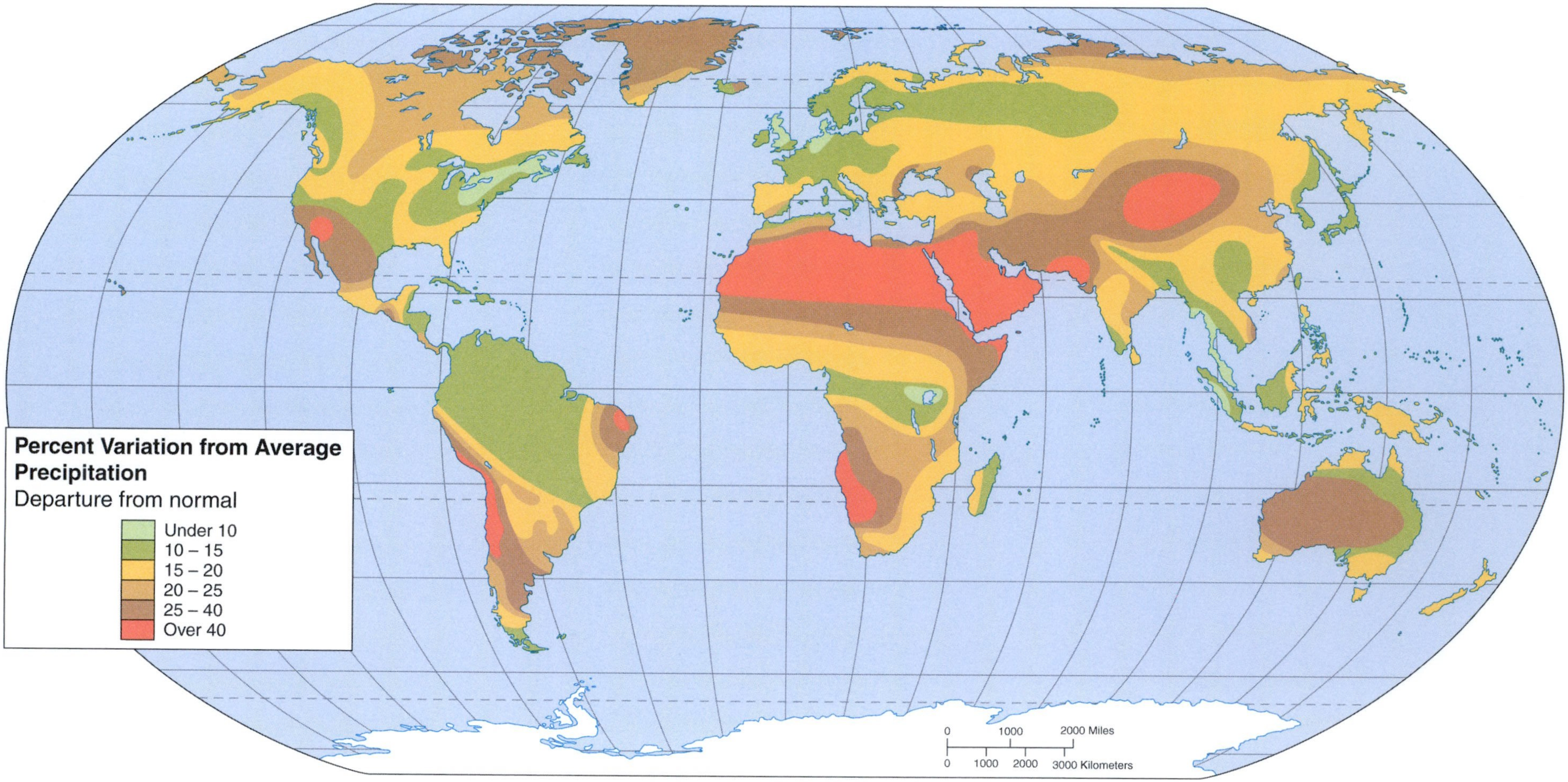

While annual precipitation totals and seasonal distribution of precipitation are important variables, the variability of precipitation from one year to the next may be even more critical. You will note from the map that there is a general spatial correlation between the world's drylands and the amount of annual variation in precipitation. Generally, the drier the climate, the more likely it is that there will be considerable differences in rainfall and/or snowfall from one year to the next. We might determine that the average precipitation of the mid-Sahara is 2 inches per year. What this really means is that a particular location in the Sahara during one year might receive .5", during the next year 3.5", and during a third year 2". If you add these together and divide by the number of years, the "average" precipitation is 2" per year. The significance of this is that much of the world's crucial agricultural output of cereals (grains) comes from dryland climates (the Great Plains of the United States, the Pampas of Argentina, the steppes of Ukraine and Russia, for example), and variations in annual rainfall totals can have significant impacts on levels of grain production and, therefore, important consequences for both economic and political processes.

Map 4a Temperature Regions and Ocean Currents

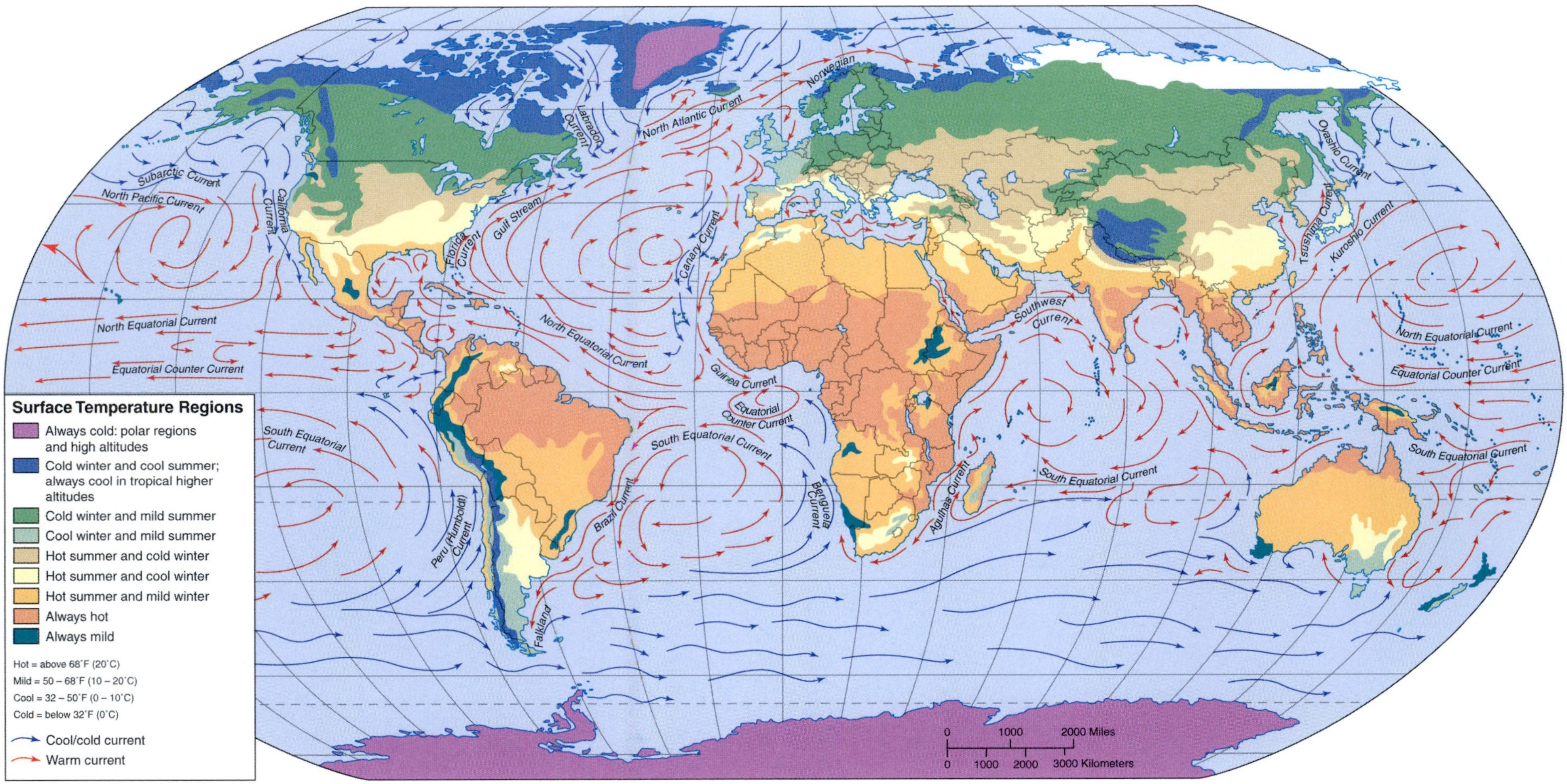

Along with precipitation, temperature is one of the two most important environmental variables, defining the climate conditions so essential for the distribution of such human activities as agriculture and the distribution of the human population. The seasonal rhythm of temperature, including such measures as the average annual temperature range (difference between the average temperature of the warmest month and that of the coldest month), is an additional variable not shown on the map but, like the seasonality of precipitation, should be a part of any comprehensive study of climate. The ocean currents illustrated exert a significant influence over the climate of adjacent regions and are the most important mechanism for redistributing surplus heat from the equatorial region into middle and high latitudes. Physical geographers known as "climatologists" study the phenomenon of temperature and related climatic characteristics.

Map 4b Average January Temperature

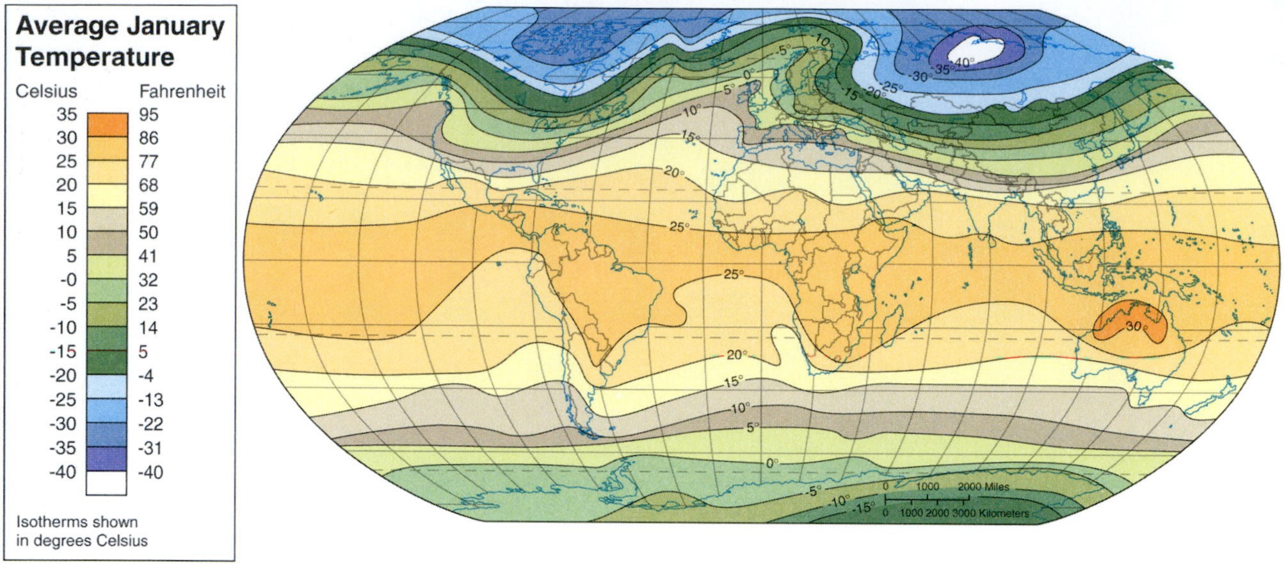

Map 4c Average July Temperature

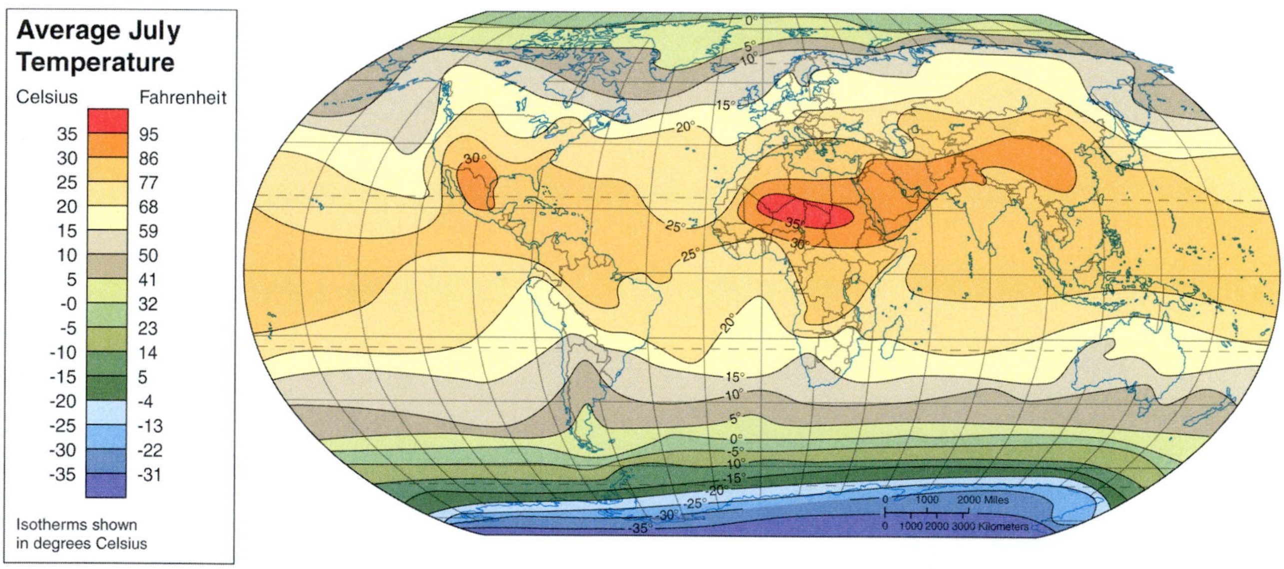

Where moisture availability tends to mark the seasons in the tropics and subtropics, in the mid-latitudes, seasons are marked by temperature. Temperature is determined by latitudinal transition, by altitude or elevation above sea level, and by location of a place relative to the world's landmasses and oceans. The most important of these controls is latitude, and temperatures generally become lower with increasing latitude. Proximity to water, however, tends to moderate temperature extremes, and "maritime" climates influenced by the oceans will be warmer in the winter and cooler in the summer than continental climates in the same general latitude. Maritime climates will also show smaller temperature ranges, the difference between January and July temperatures, while climates of the continental interiors, far from the moderating influences of the oceans, will tend to have greater temperature ranges. In the Northern Hemisphere, where there are both large landmasses and oceans, the range is great. But in the Southern Hemisphere, dominated by water and, hence, by the more moderate maritime air masses, the temperature range is comparatively small. Significant temperature departures from the "normal" produced by latitude may also be the result of elevation. With exceptions, lower temperatures produced by topography are difficult to see on maps of this scale.

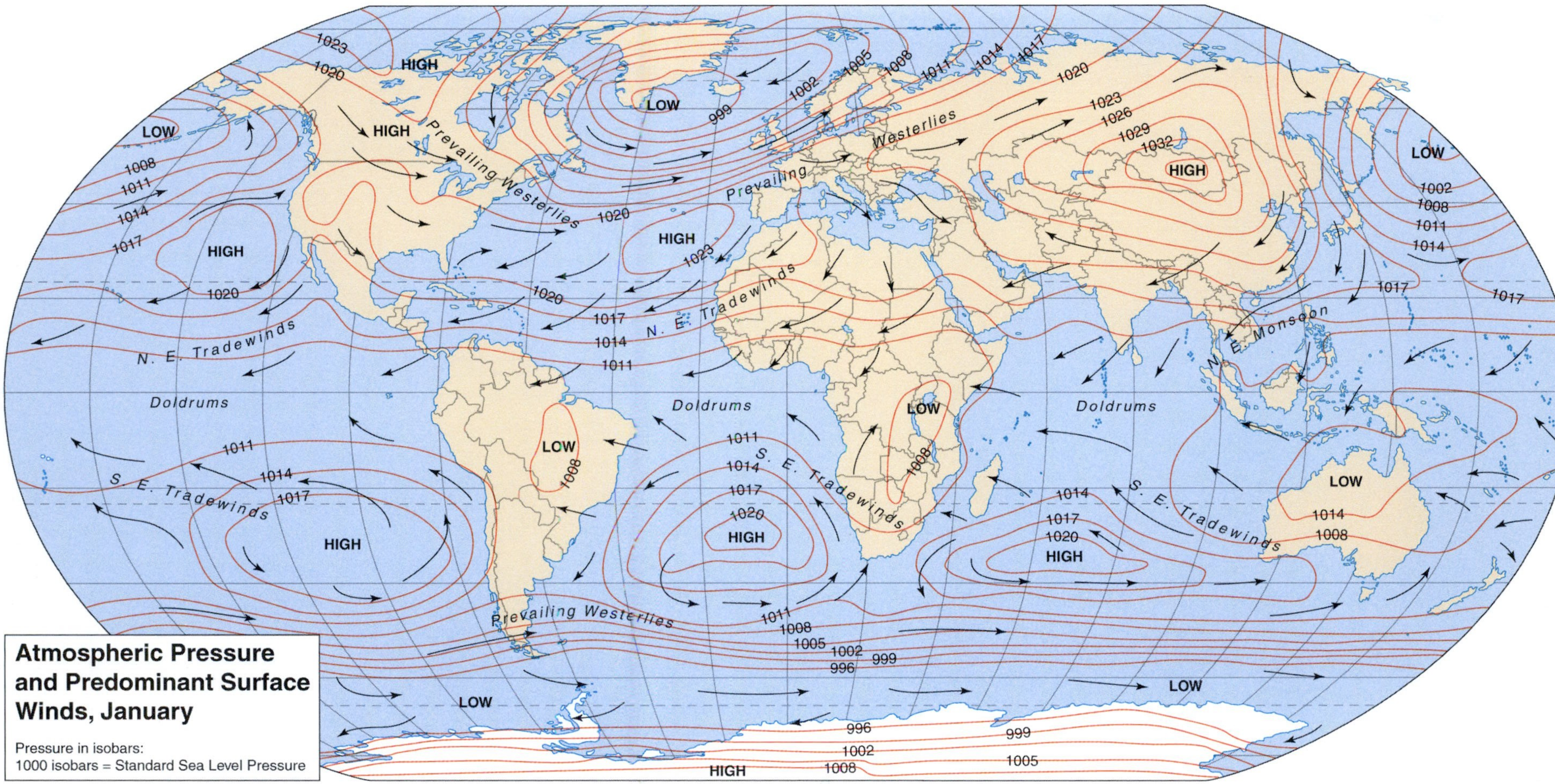

Atmospheric Pressure and Predominant Surface Winds, January

Pressure in isobars:
1000 isobars = Standard Sea Level Pressure

Map 5b Atmospheric Pressure and Predominant Surface Winds, July

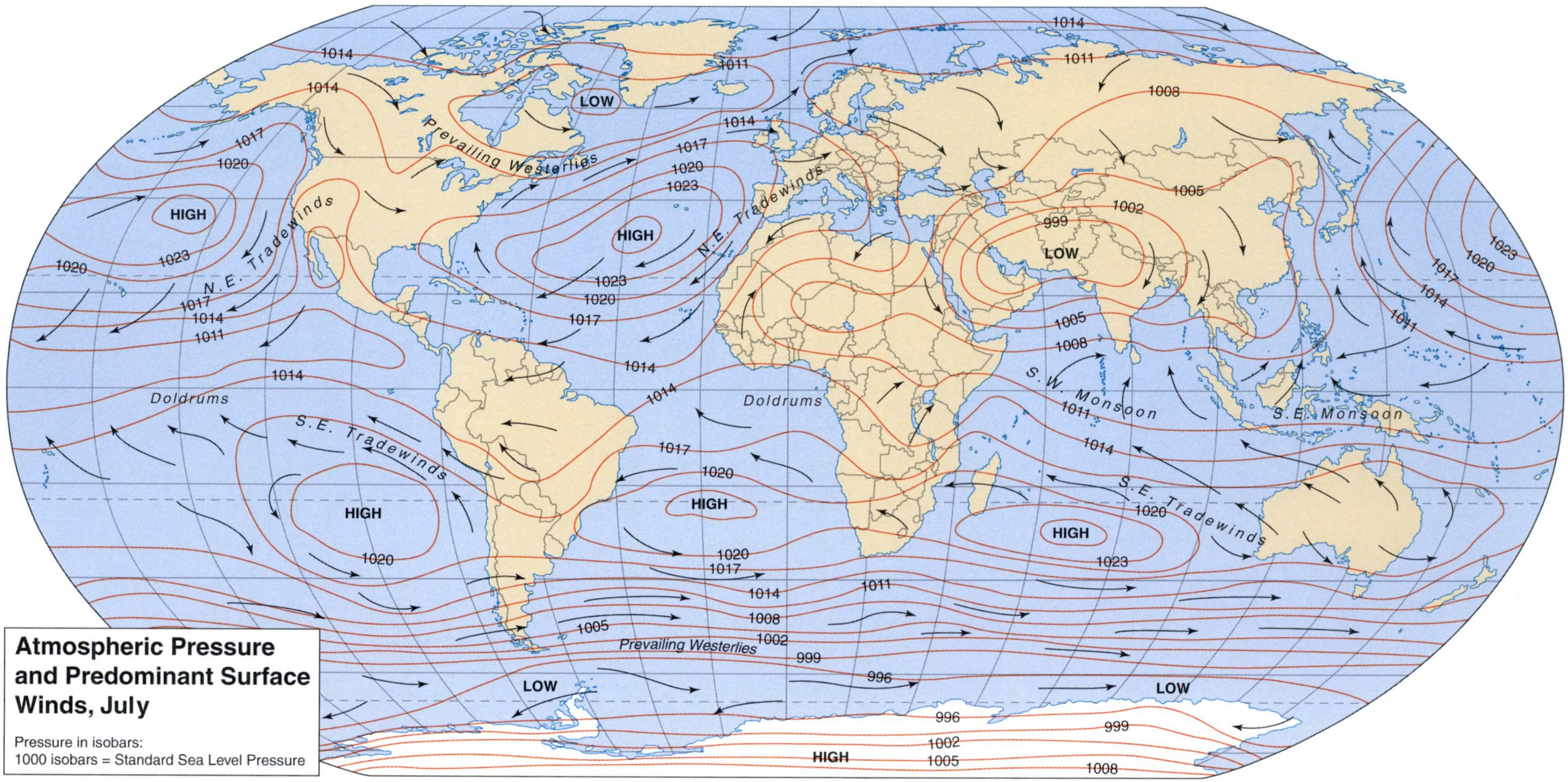

Atmospheric pressure, or the density of air, is a function largely of air temperature: the colder the air, the denser and heavier it is, hence the higher its pressure; the warmer the air, the lighter and less stable it is, hence the lower its pressure. Global pressure systems are the alternating low and high pressure systems that, from the equator north and south, include: the equatorial low (sometimes called the intertropical convergence) centered on the equator for much of the year; the subtropical highs with their centers near the 30th degrees of north and south latitude; the subpolar lows or polar fronts centered near the 60th parallel of north and south latitude; and the polar highs near the north and south poles. Air flows from high pressure to low pressure regions, and this air flow constitutes the earth's major surface winds such as the tropical tradewinds and the prevailing westerlies. This flow of air is one of the chief mechanisms by which surplus heat energy from the equatorial region is redistributed to higher latitudes. It is also the primary conditioner of the world's major precipitation belts, with rainfall and snowfall associated primarily with lower atmospheric pressure conditions.

Map 6 Climate Regions

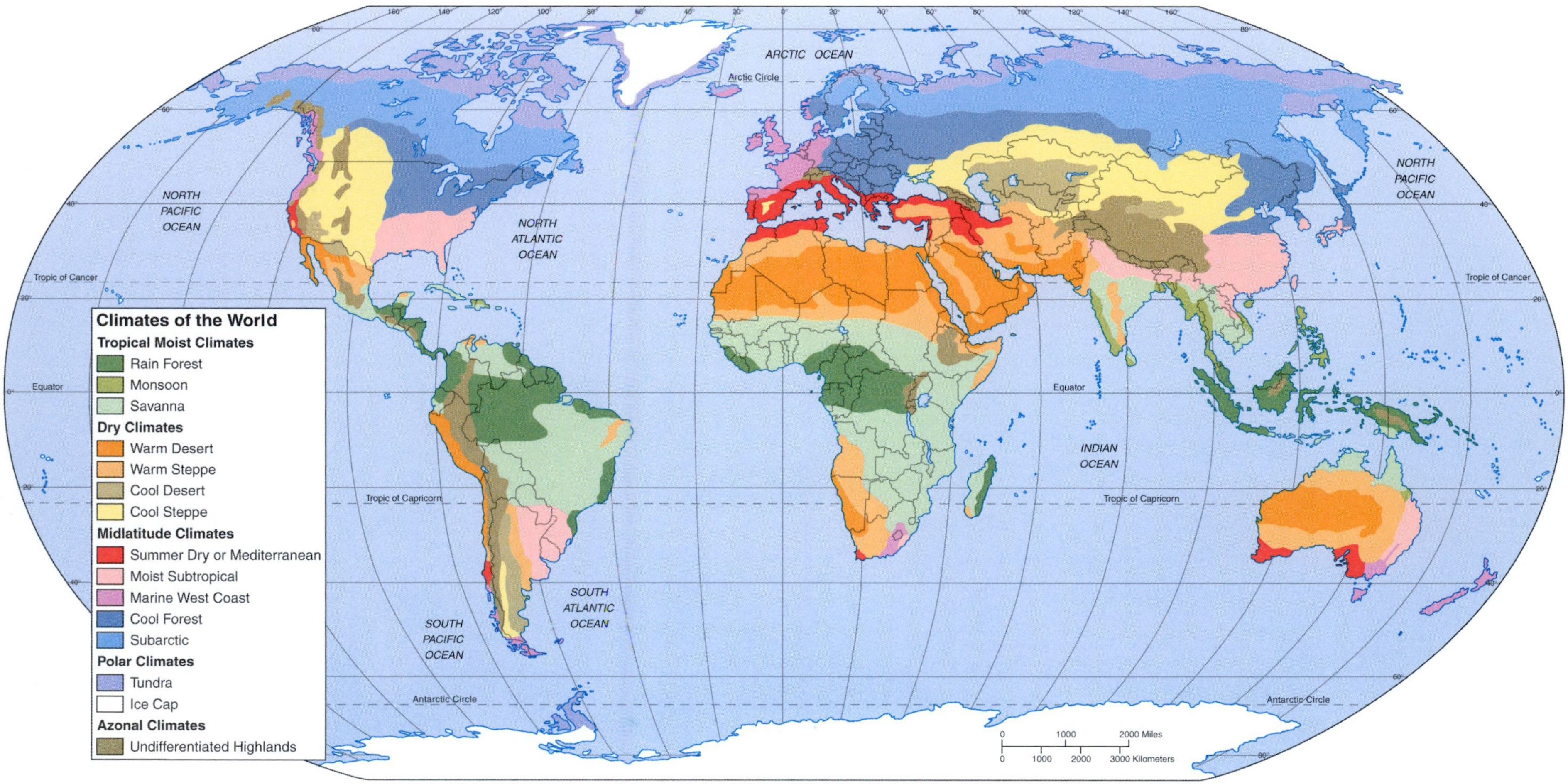

Of the world's many patterns of physical geography, climate or the long-term average of weather conditions such as temperature and precipitation is the most important. It is climate that conditions the distribution of natural vegetation and the types of soils that will exist in an area. Climate also influences the availability of our most crucial resource: water. From an economic standpoint, the world's most important activity is agriculture; no other element of physical geography is more important for agriculture than climate. Ultimately, it is agricultural production that determines where the bulk of human beings live, and therefore, climate is a basic determinant of the distribution of human populations as well. The study of climates or "climatology" is one of the most important branches of physical geography.

The climate classification system shown on this map is based on that developed by Wladimir Köppen. To establish his climate regions, Köppen used the climatic parameters of *precipitation, temperature,* and *evapotranspiration* as they impacted certain kinds of major vegetative associations. Hence the names for many of the climate regions are also the names of vegetative regions.

Map 7 Vegetation Types

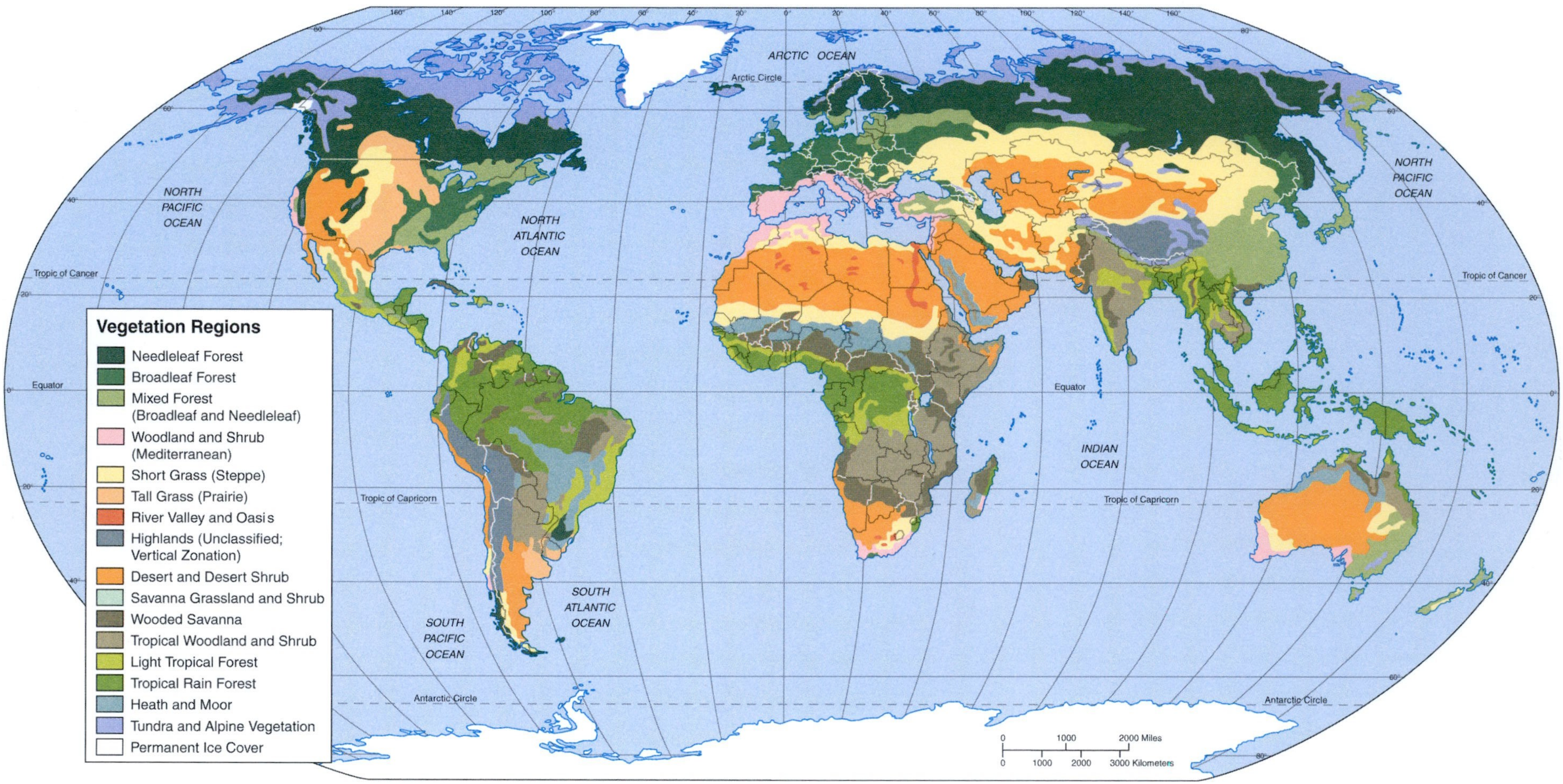

Vegetation is the most visible consequence of the distribution of temperature and precipitation. The global pattern of vegetative types or "habitat classes" and the global pattern of climate are closely related and make up one of the great global spatial correlations. But not all vegetation types are the consequence of temperature and precipitation or other climatic variables. Many types of vegetation in many areas of the world are the consequence of human activities, particularly the grazing of domesticated livestock, burning, and forest clearance. This map shows the pattern of natural or "potential" vegetation, or vegetation as it might be expected to exist without significant human influences, rather than the actual vegetation that results from a combination of environmental and human factors. Physical geographers who are interested in the distribution and geographic patterns of vegetation are "biogeographers."

Map 8 Soil Orders

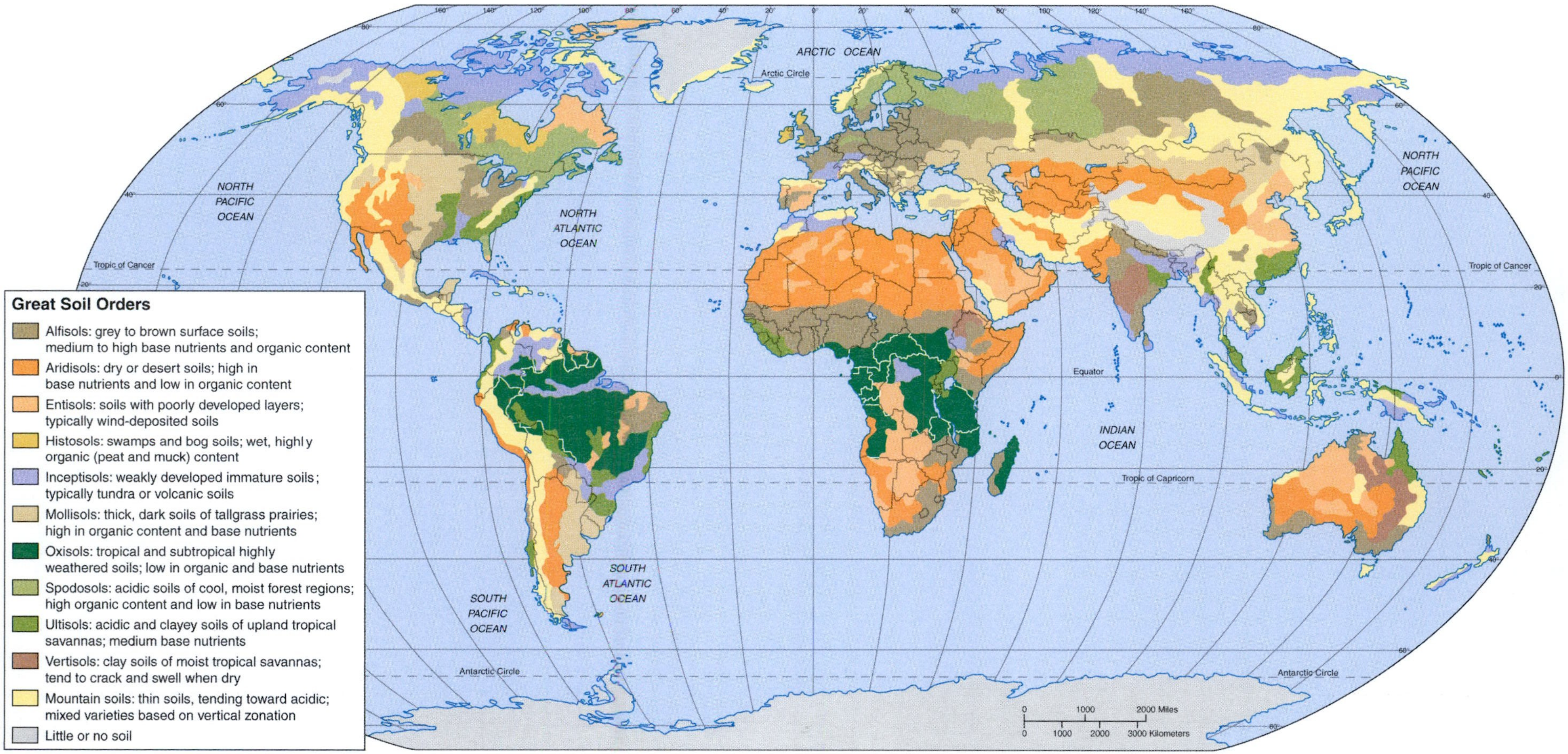

The characteristics of soil are one of the three primary physical geographic factors, along with climate and vegetation, that determine the habitability of regions for humans. In particular, soils influence the kinds of agricultural uses to which land is put. Since soils support the plants that are the primary producers of all food in the terrestrial food chain, their characteristics are crucial to the health and stability of ecosystems. Two types of soil are shown on this map: zonal soils, the characteristics of which are based on climatic patterns; and azonal soils, such as alluvial (water-deposited) or aeolian (wind-deposited) soils, the characteristics of which are derived from forces other than climate. However, many of the azonal soils, particularly those dependent upon drainage conditions, appear over areas too small to be readily shown on a map of this scale. Thus, almost none of the world's swamp or bog soils appear on this map. People who study the geographic characteristics of soils are most often "soil scientists," a discipline closely related to that branch of physical geography called "geomorphology."

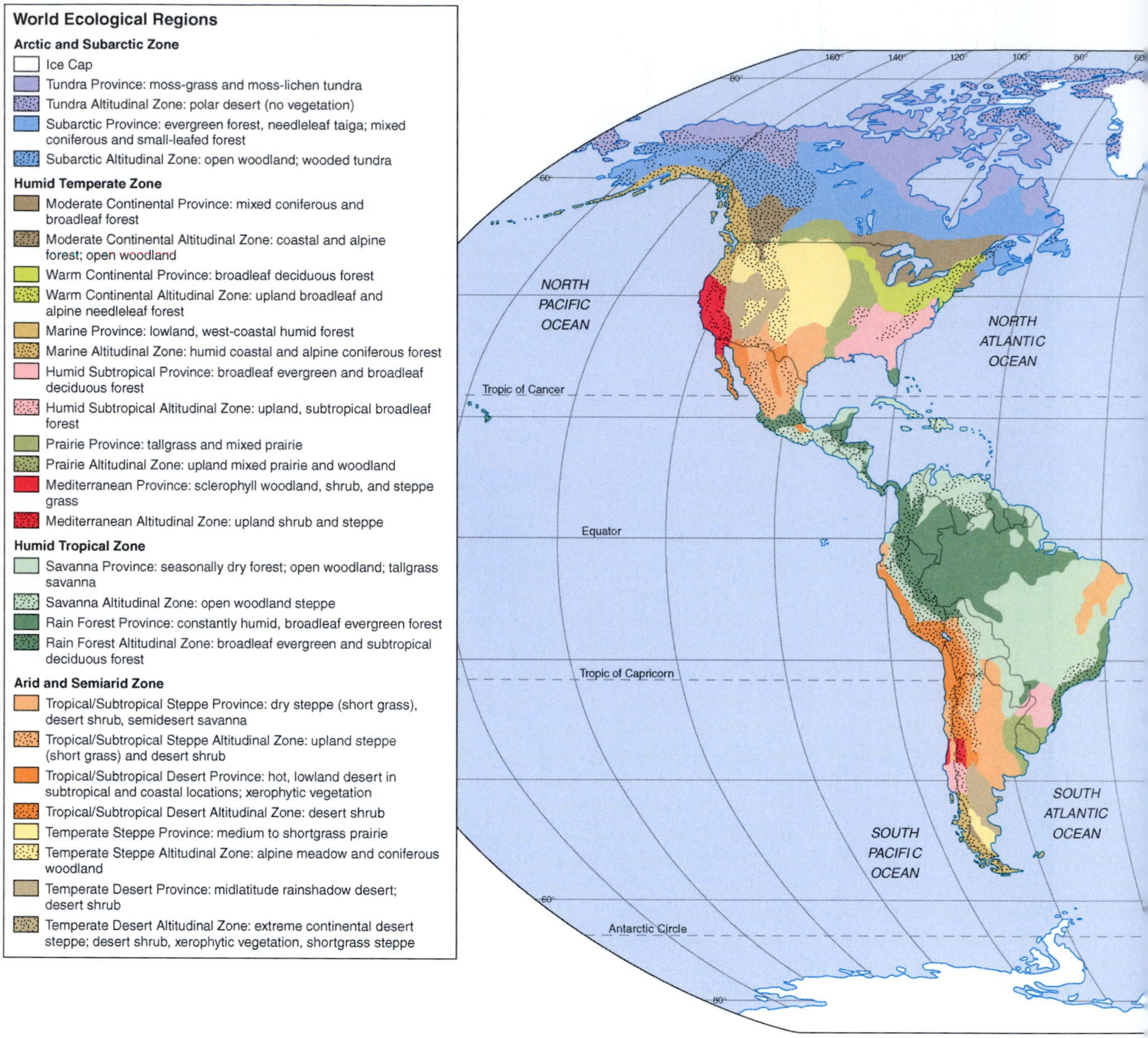

Ecological regions are distinctive areas within which unique sets of organisms and environments are found. We call the study of the relationships between organisms and their environmental surroundings "ecology." Within each of the ecological regions portrayed on the map, a particular combination of vegetation, wildlife, soil, water, climate, and terrain defines that region's habitability, or ability to support life, including human life.

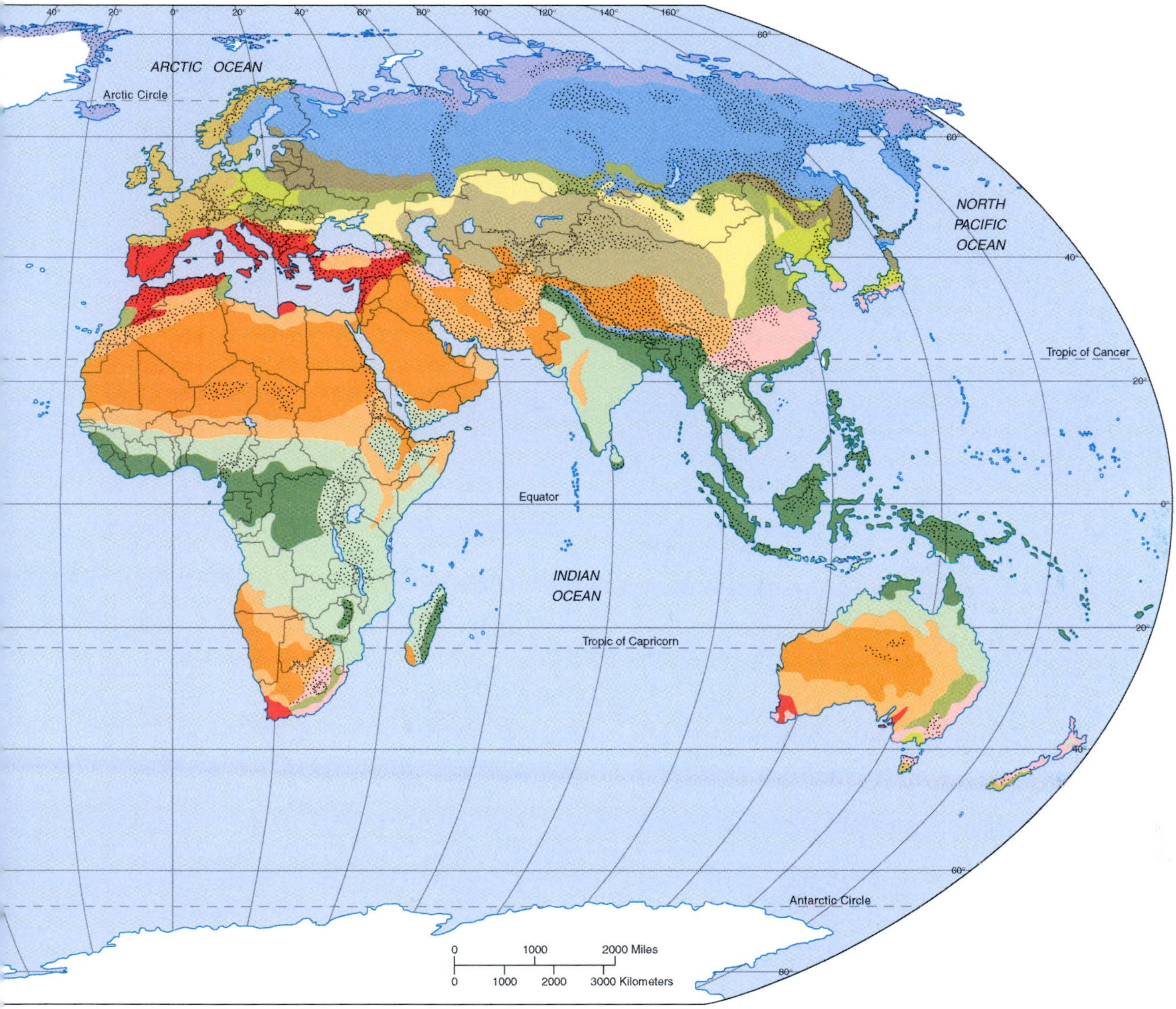

Like climate and landforms, ecological relationships are crucial to the existence of agriculture, the most basic of our economic activities, and important for many other kinds of economic activity as well. Biogeographers are especially concerned with the concept of ecological regions since such regions so clearly depend upon the geographic distribution of plants and animals in their environmental settings.

Map 10 Plate Tectonics

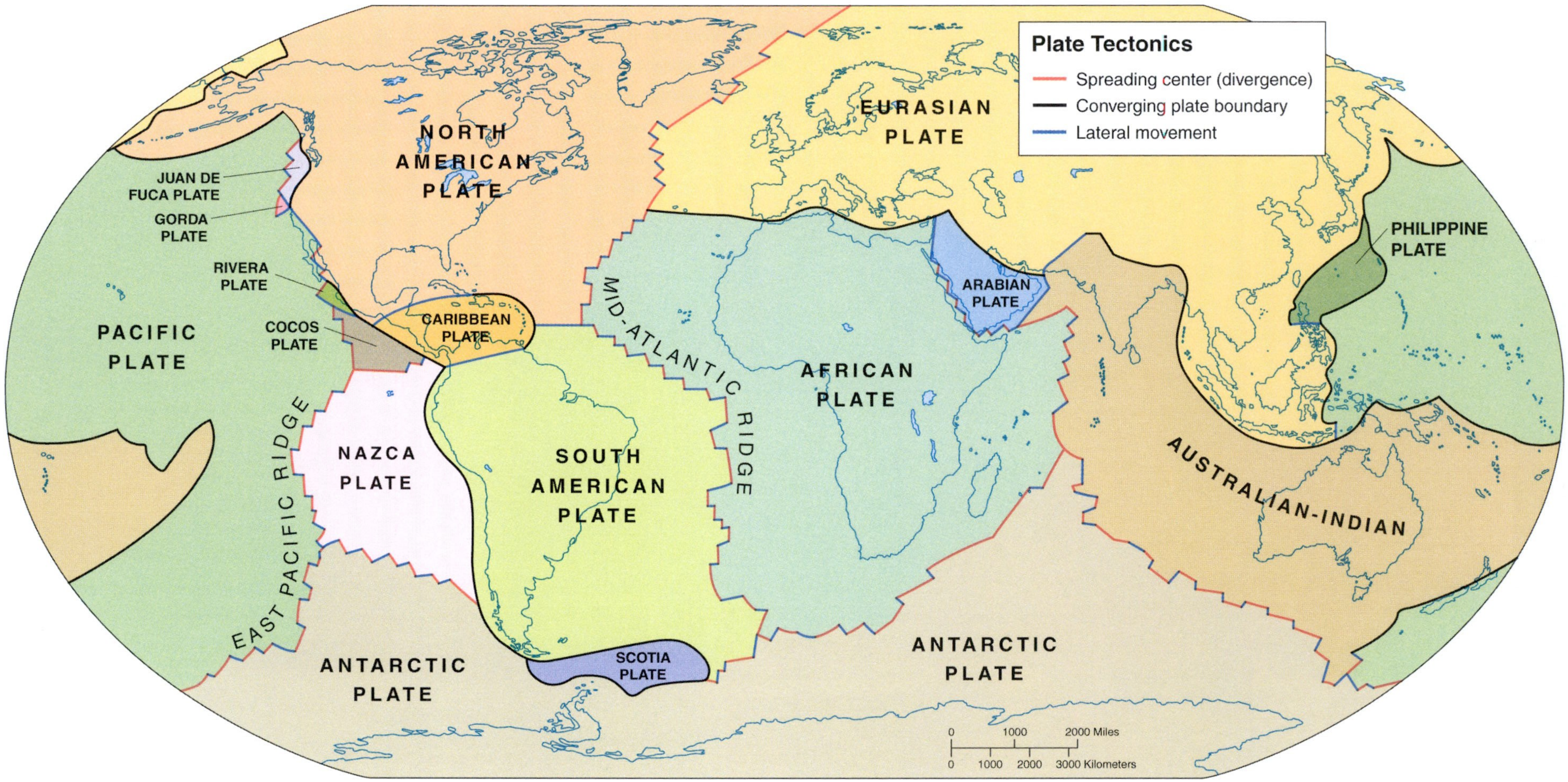

An understanding of the forces that shape the primary features of the earth's surface—the continents and ocean basins—requires a view of the earth's crust as fragments or "lithospheric plates" that shift position relative to one another. There are three dominant types of plate movement: *convergence,* in which plates move together, compressing former ocean floor or continental rocks together to produce mountain ranges, or producing mountain ranges through volcanic activity if one plate slides beneath another; *divergence,* in which the plates move away from one another, producing rifts in the earth's crust through which molten material wells up to produce new sea floors and mid-oceanic ridges; and *lateral shift,* in which plates move horizontally relative to one another, causing significant earthquake activity. All the major forms of these types of shifts are extremely slow and take place over long periods of geologic time. The movement of crustal plates, or what is known as "plate tectonics," is responsible for the present shape and location of the continents but is also the driving force behind some much shorter-term earth phenomena like earthquakes and volcanoes. A comparison of the map of plates with maps of hazards and terrain will reveal some interesting relationships.

Map 11 Topography

Topography or terrain, also called "landforms," is second only to climate as a conditioner of human activity, particularly agriculture but also the location of cities and industry. A comparison of this map of mountains, valleys, plains, plateaus, and other features of the earth's surface with a map of land use (Map 16) shows that most of the world's productive agricultural zones are located in lowland and relatively level regions. Where large regions of agricultural productivity are found, we also tend to find urban concentrations and, with cities, we find industry. There is also a good spatial correlation between the map of topography and the map showing the distribution and density of the human population (Map 15). Normally the world's major landforms are the result of extremely gradual primary geologic activity such as the long-term movement of crustal plates. This activity occurs over hundreds of millions of years. Also important is the more rapid (but still slow by human standards) geomorphological or erosional activity of water, wind, glacial ice, and waves, tides, and currents. Some landforms may be produced by abrupt or "cataclysmic" events such as a major volcanic eruption or a meteor strike, but such events are relatively rare and their effects are usually too minor to show up on a map of this scale. The study of the processes that shape topography is known as "geomorphology" and is an important branch of physical geography.

Map 12a Resources: Mineral Fuels

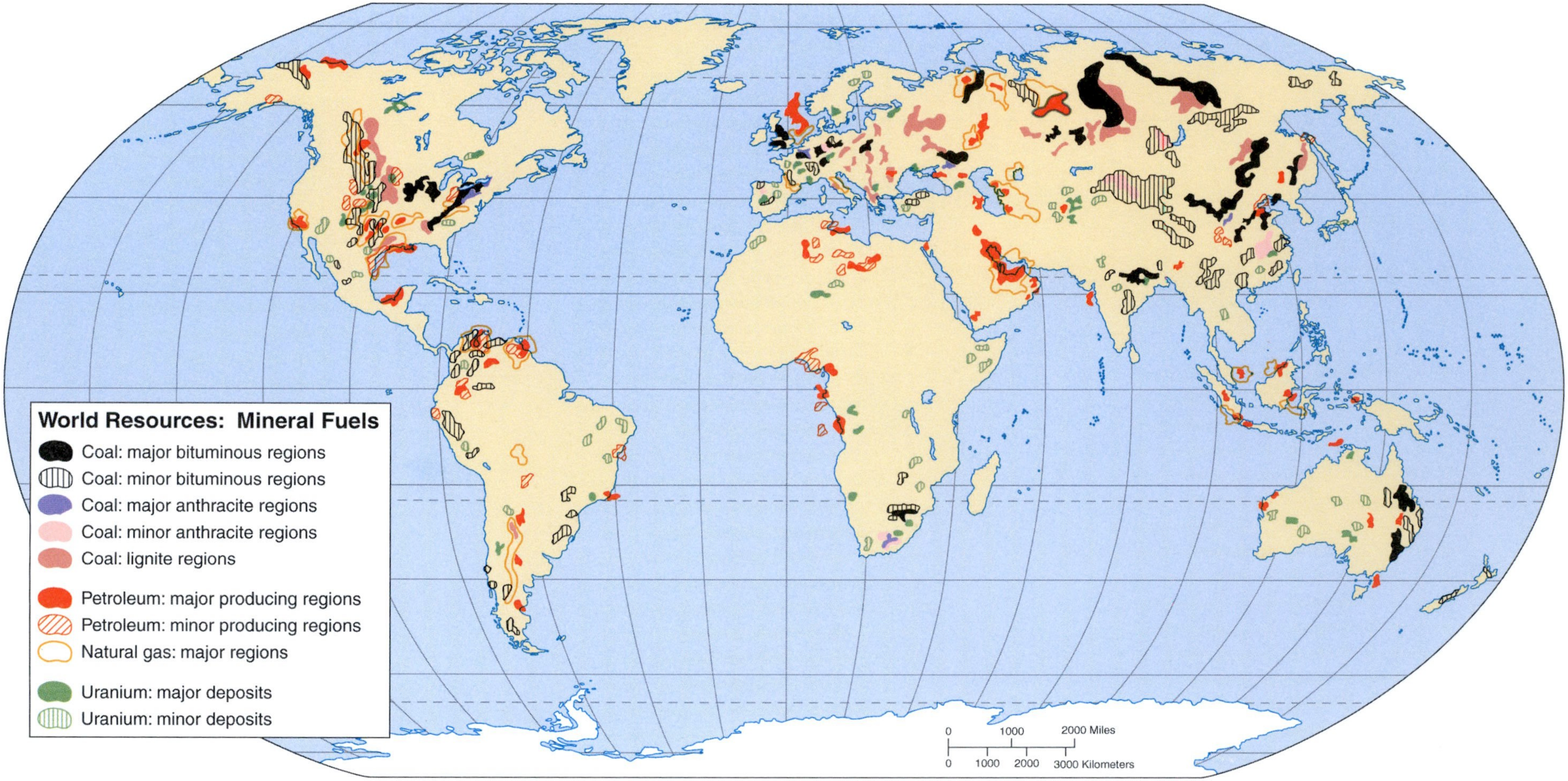

The extraction and transportation of mineral fuels rank with agriculture and forestry as "primary" human activities that impact the environment on a global scale. Nearly all of the most highly publicized environmental disasters of recent decades—the Prince William Sound oil spill or the Chernobyl nuclear accident, for example—have involved mineral fuels that were being stored, transported, or used. And the continuing extraction of mineral fuels like oil, natural gas, coal, and uranium produces high levels of atmospheric, soil, and water pollution. The location of mineral fuels tells us a great deal about where environmental degradation is likely to be occurring or to occur in the future. One need only look at the levels of atmospheric pollution and vegetative disruption in central and eastern Europe to recognize the damaging consequences of heavy reliance on coal as a domestic and industrial fuel. The location of mineral fuels also tells us something about existing or potential levels of economic development with those countries possessing abundant reserves of mineral fuels having more of a chance to maintain or attain higher levels of prosperity.

Map 12b Resources: Critical Metals

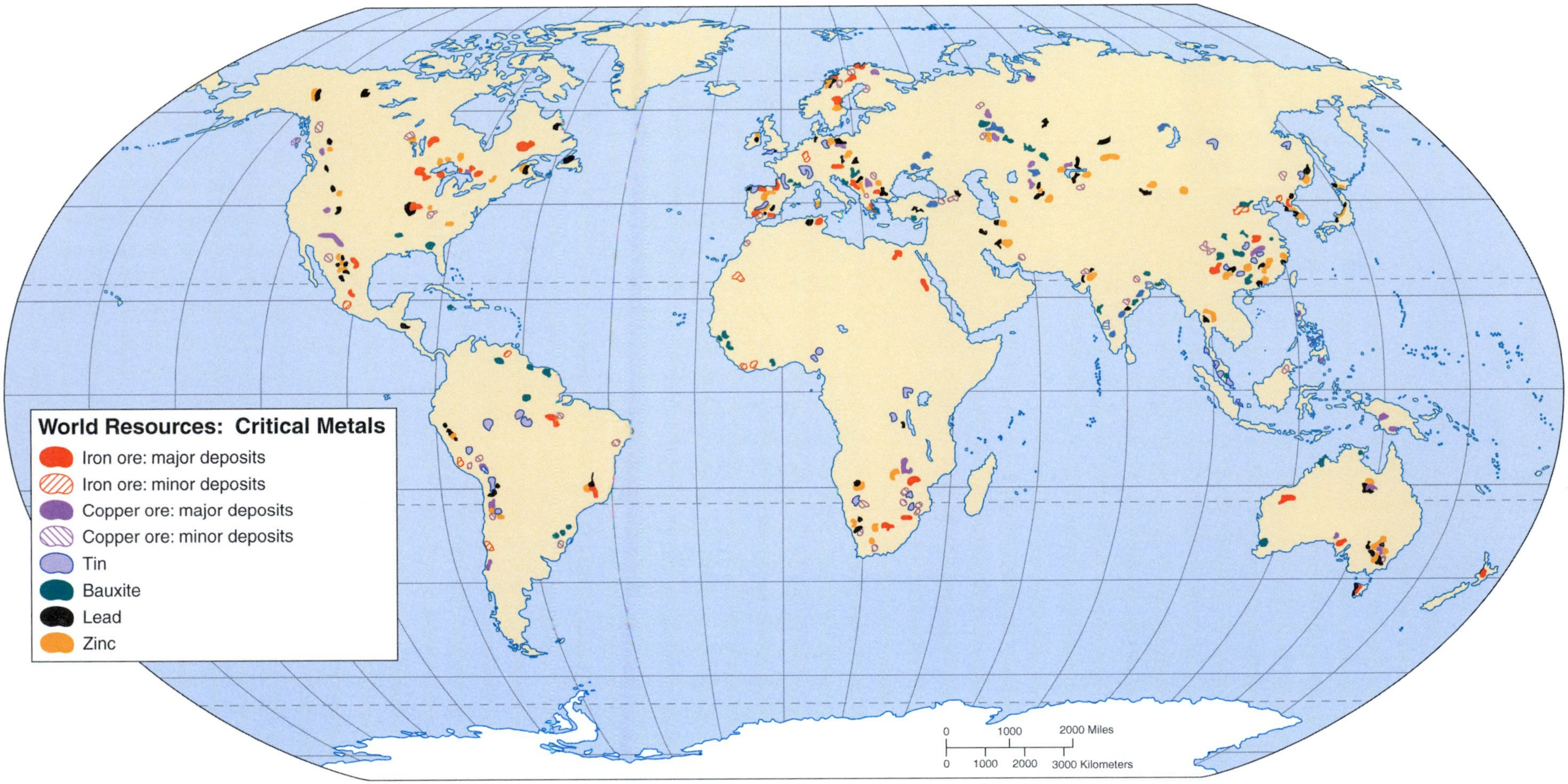

The location of deposits of critical metals such as iron, copper, tin, and others is an important determinant of the location of mining activities, like mineral fuel extraction, a "primary" economic activity. Also like mineral fuel extraction, mining for critical metallic ores makes significant environmental impact, particularly on vegetation, soils, and water resources. Some of the world's most dramatic examples of human modification of environments are located in areas of metallic ore extraction: the open pit copper mining areas of Arizona and Utah, for example. Environmental impact aside, those countries with significant critical metal deposits tend to stand a better chance of reaching higher levels of economic development, as long as they can extract and market the ores themselves rather than having the extraction process controlled by outside concerns. The average Bolivian, for example, does not benefit greatly from the fact that his/her country is an important producer of tin and other metals. Bolivia is a "colonial dependency" country and the wealth generated by metallic ore production there tends to flow out of the country to Europe and North America. On the other hand, another South American country, Brazil, is paying for much of its own current economic development by utilizing its reserves of iron and other metals and more of the wealth from the extraction of those resources stays within the country.

Map 13 Natural Hazards

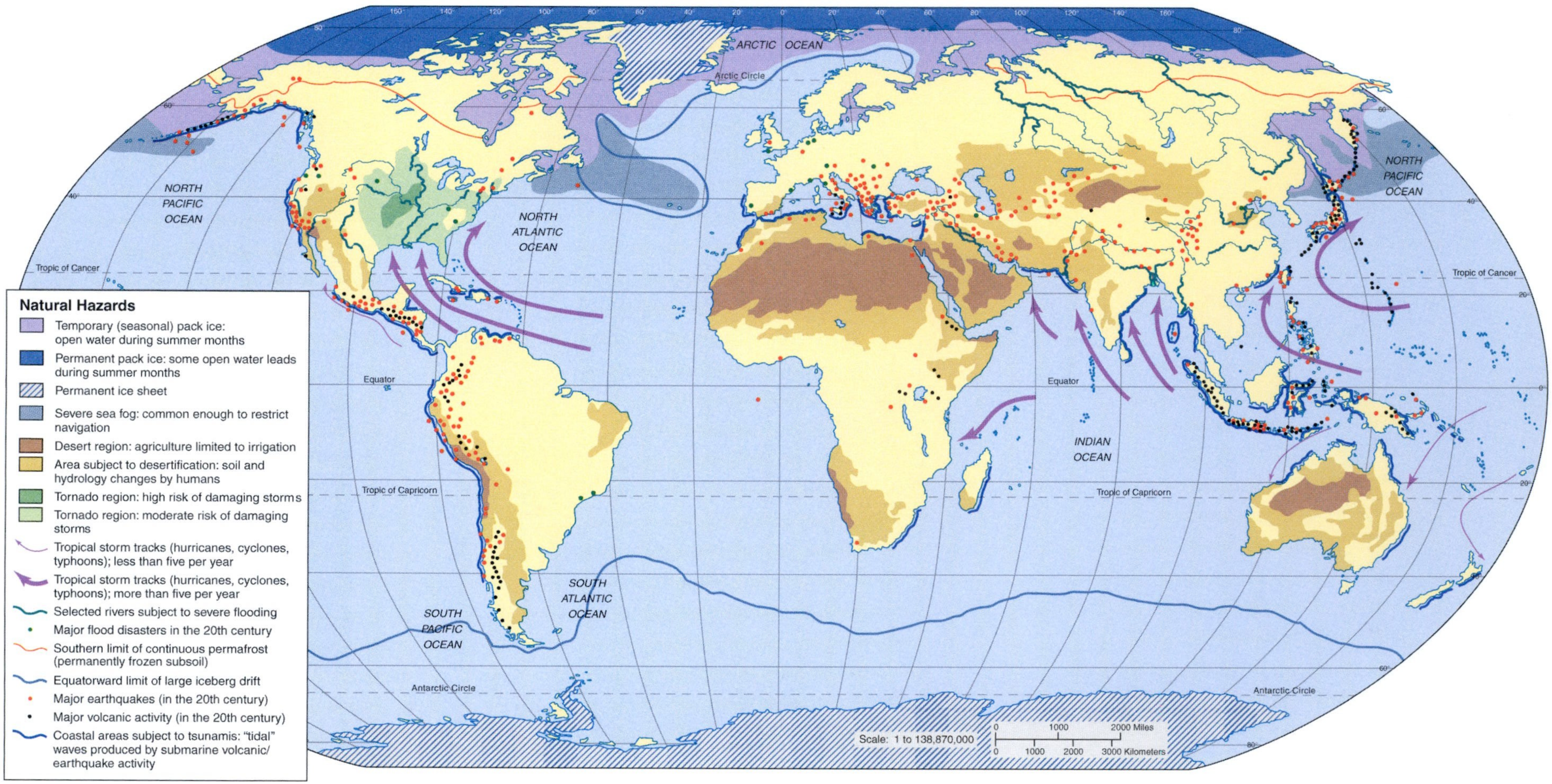

Unlike other elements of physical geography, most natural hazards are unpredictable. However, there are certain regions where the probability of the occurrence of a particular natural hazard is high. This map shows regions affected by major natural hazards at rates that are higher than the global norm. The presence of persistent natural hazards may influence the types of modifications that people make in the environment and certainly influence the styles of housing and other elements of cultural geography. Natural hazards may also undermine the utility of an area for economic purposes and some scholars suggest that regions of environmental instability may be regions of political instability as well. The study of natural hazards has become an important activity for "resource geographers" whose areas of interest overlap both human and physical fields of geography.

Unit II

Global Human Patterns

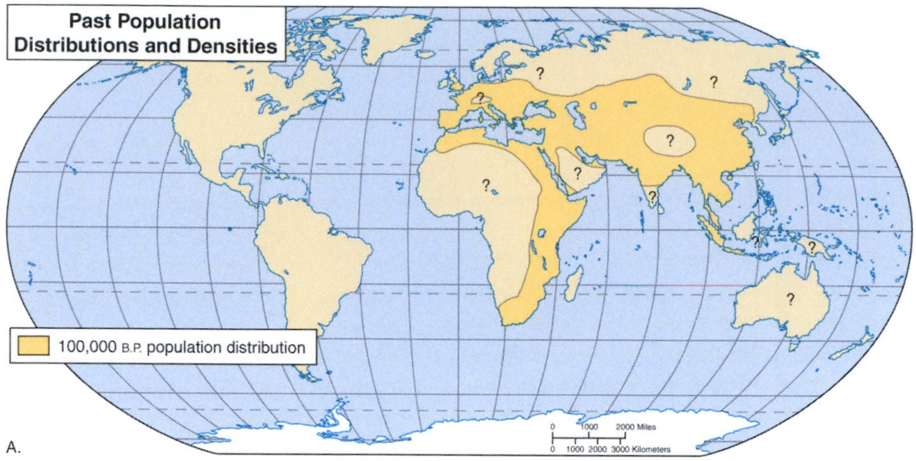

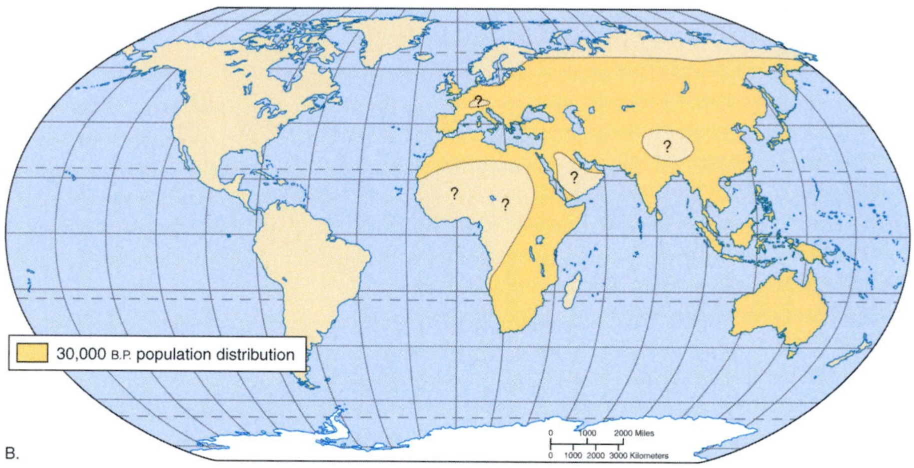

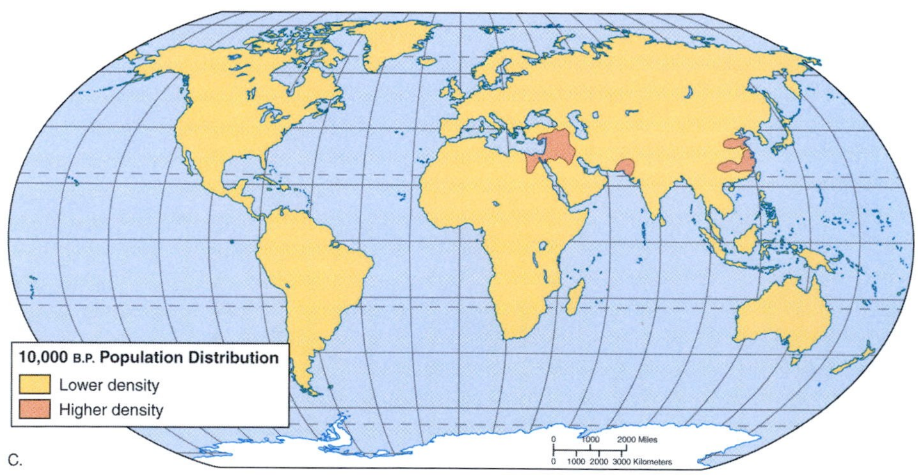

The map of the world at 100,000 B.P. (Before Present) shows the distributions of hominids who at that time had spread from their probable origin in Africa into parts of the Old World. At 30,000 B.P. few places in the Old World remained uninhabited. All the people then were hunters and gatherers who subsisted on wild foods. The environment probably was not at its carrying capacity for human populations until about 15,000 B.P.

By 10,000 B.P. hunting and gathering people had spread throughout the world. Plant and animal domestication (farming and pastoralism) had begun in some parts of the Old World perhaps as a response to behavioral changes necessitated by the population, which now exceeded the

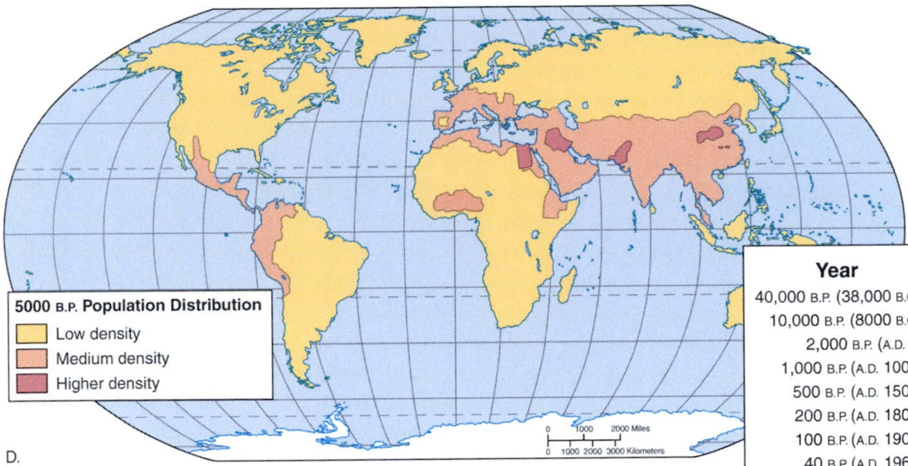

D.

Year	Population
40,000 B.P. (38,000 B.C.)	1,500,000
10,000 B.P. (8000 B.C.)	10,000,000
2,000 B.P. (A.D. 1)	200,000,000
1,000 B.P. (A.D. 1000)	275,000,000
500 B.P. (A.D. 1500)	450,000,000
200 B.P. (A.D. 1800)	910,000,000
100 B.P. (A.D. 1900)	1,600,000,000
40 B.P. (A.D. 1960)	3,000,000,000
30 B.P. (A.D. 1970)	3,610,000,000
20 B.P. (A.D. 1980)	4,400,000,000
10 B.P. (A.D. 1990)	5,300,000,000
Present	6,500,000,000 (estimated)

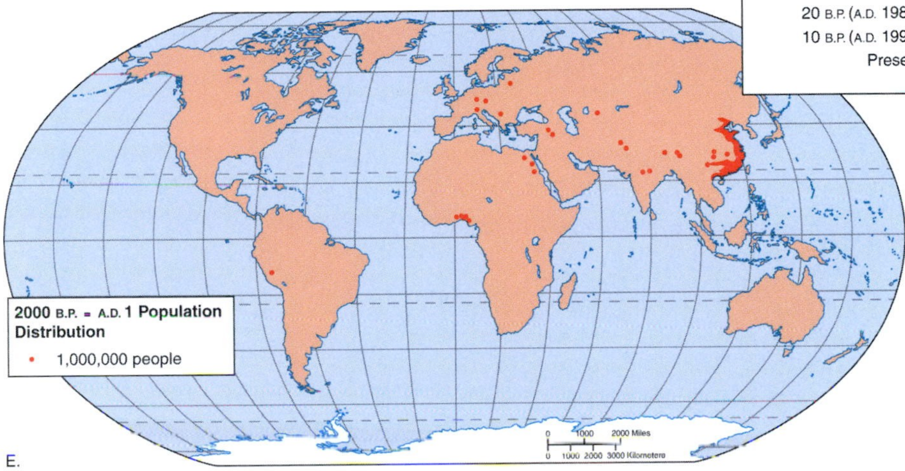

E.

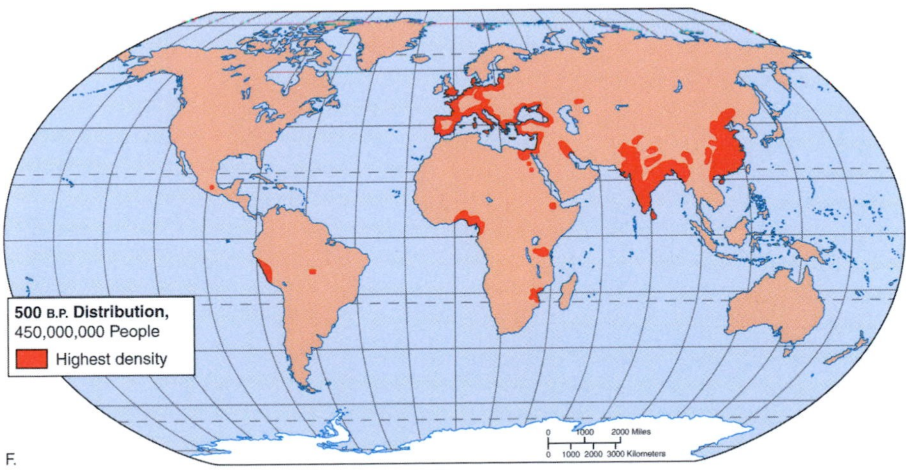

F.

environmental carrying capacity. Farming supports higher population densities than hunting and gathering. Urban civilization with cities dependent on their hinterlands had developed by 5000 B.P. The maps of A.D. 1 and A.D. 1500 approximate actual population density on a scale of one dot to every million people. A chart provides total world population figures for different time periods. Compare these maps to the contemporary population density map on the following page.

Map 15 Population Density

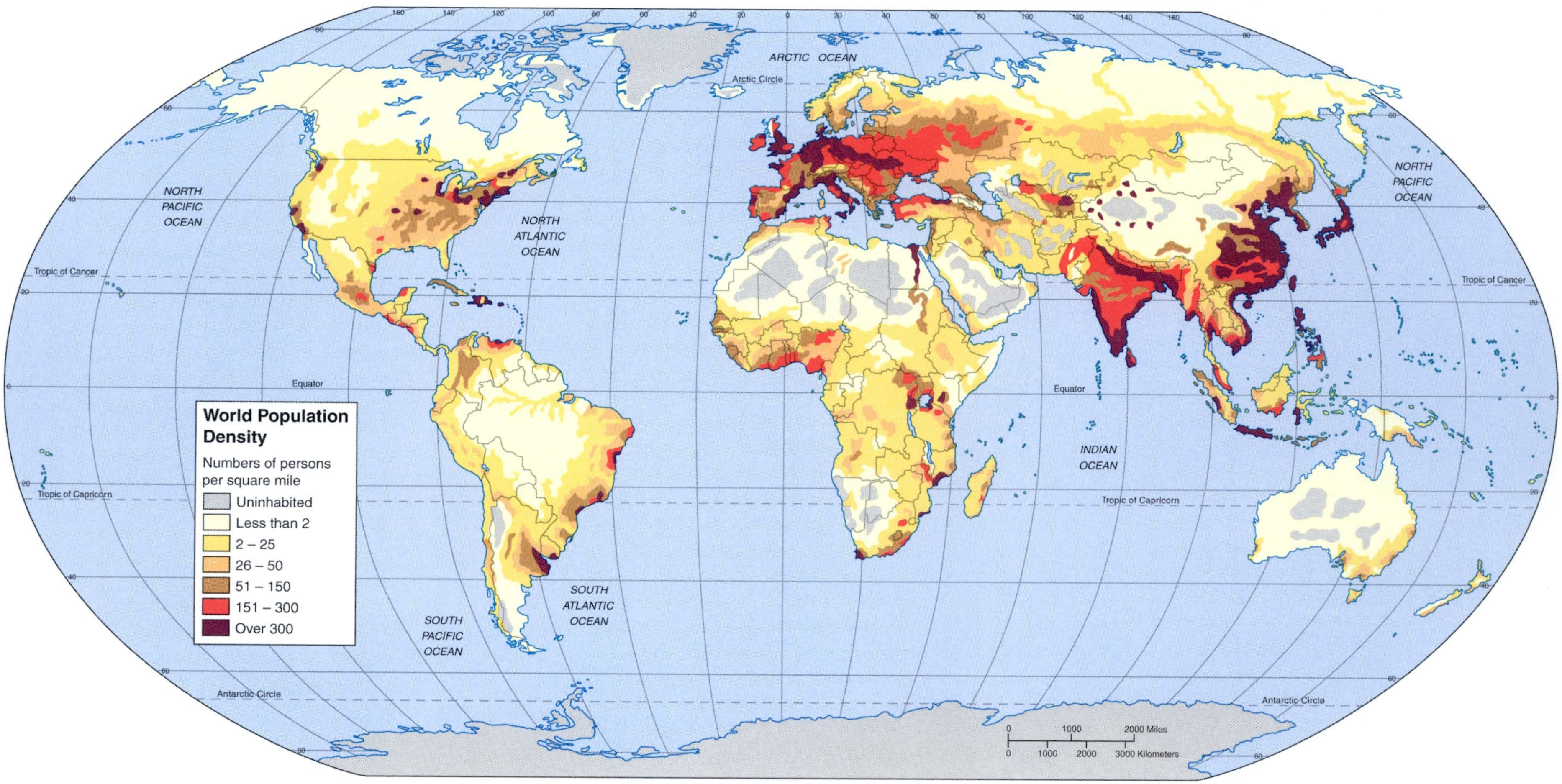

No feature of human activity is more reflective of geographic relationships than where people live. In the areas of densest populations, a mixture of natural and human factors has combined to allow maximum food production, maximum urbanization, and maximum centralization of economic activities. Three great concentrations of human population appear on the map—East Asia, South Asia, and Europe—with a fourth, lesser concentration in eastern North America. While population growth is relatively slow in three of these population clusters, in the fourth—South Asia—growth is still rapid and South Asia is expected to become even more densely populated in the early years of the twenty-first century, while density of the other regions is expected to remain about as it now appears. In Europe and North America, the relatively stable population growth rates are the result of economic development that has caused population growth to level off within the last century. In East Asia, the growth rates have also begun to decline. In the case of Japan, Taiwan, the Koreas, and other more highly developed nations of the Pacific Rim, the reduced growth is the result of economic development. In China, at least until recently, lowered population growth rates have resulted from strict family planning. The areas of future high density of population, in addition to those already existing, are likely to be in Middle and South America and in Central Africa, where population growth rates are well above the world average.

Map 16 Land Use, A.D. 1500

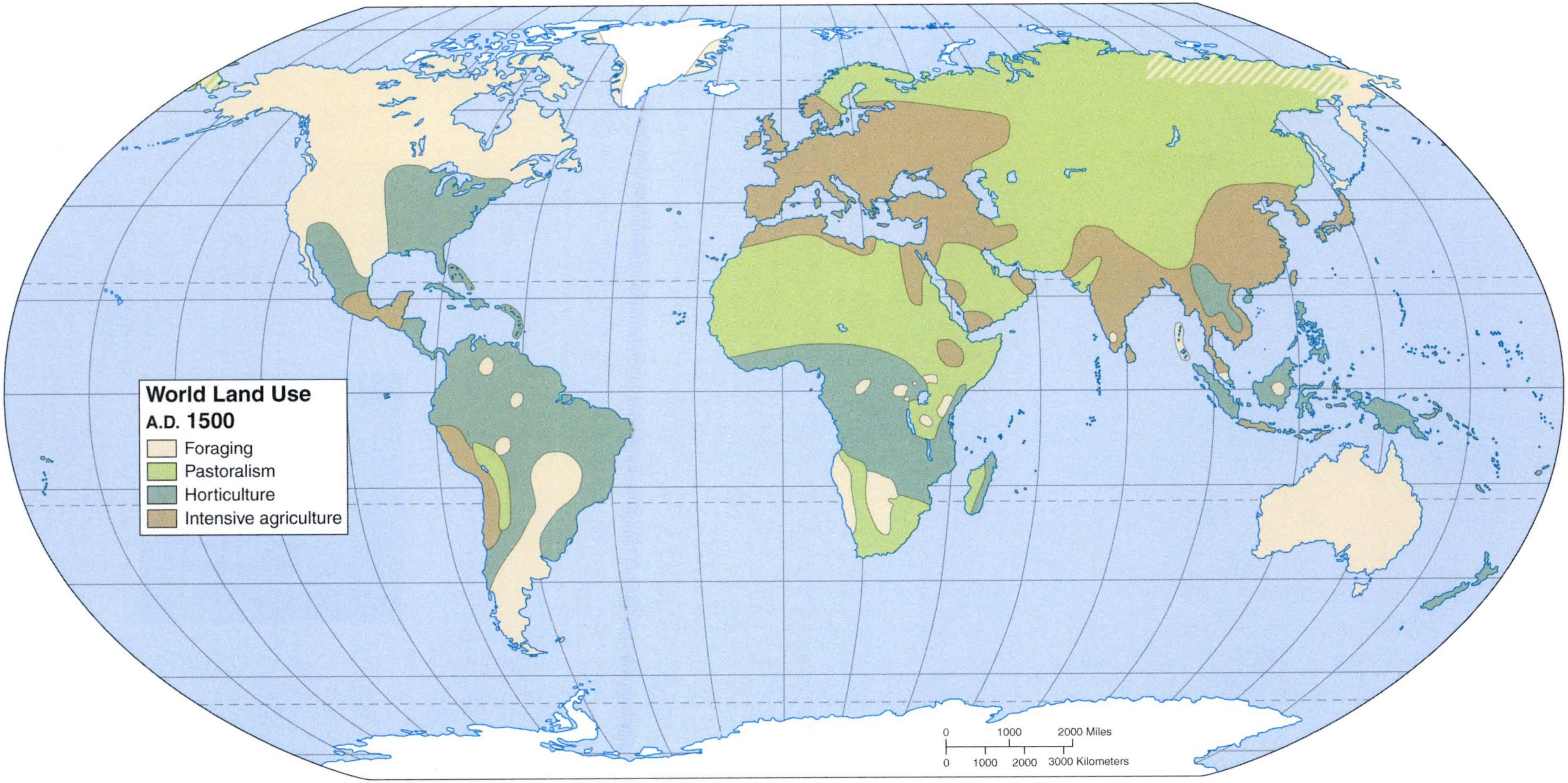

Europeans began to explore the world in the late 1400s. They encountered many independent people with self-sustaining economies at that time. Foraging people practiced hunting and gathering, utilizing the wild forms of plants and animals in their environments. Horticultural people practiced a simple form of agriculture using hoes or digging sticks as their basic tools. They sometimes cleared their land by burning and then planted crops. Pastoralists herded animals as their basic subsistence pattern. Complex state-level societies, such as the Mongols, had pastoralism as their base. Intensive agriculturalists based their societies on complicated irrigation systems and/or the plow and draft animals. Wheat and rice were two kinds of crops that supported large populations. In many of the areas of intensive agriculture—particularly in MesoAmerica, Europe, Southwest Asia, South Asia, and East Asia—complex patterns of market economies had begun to develop well before the fifteenth century and the beginnings of European expansion.

Map 17 Economic Activities

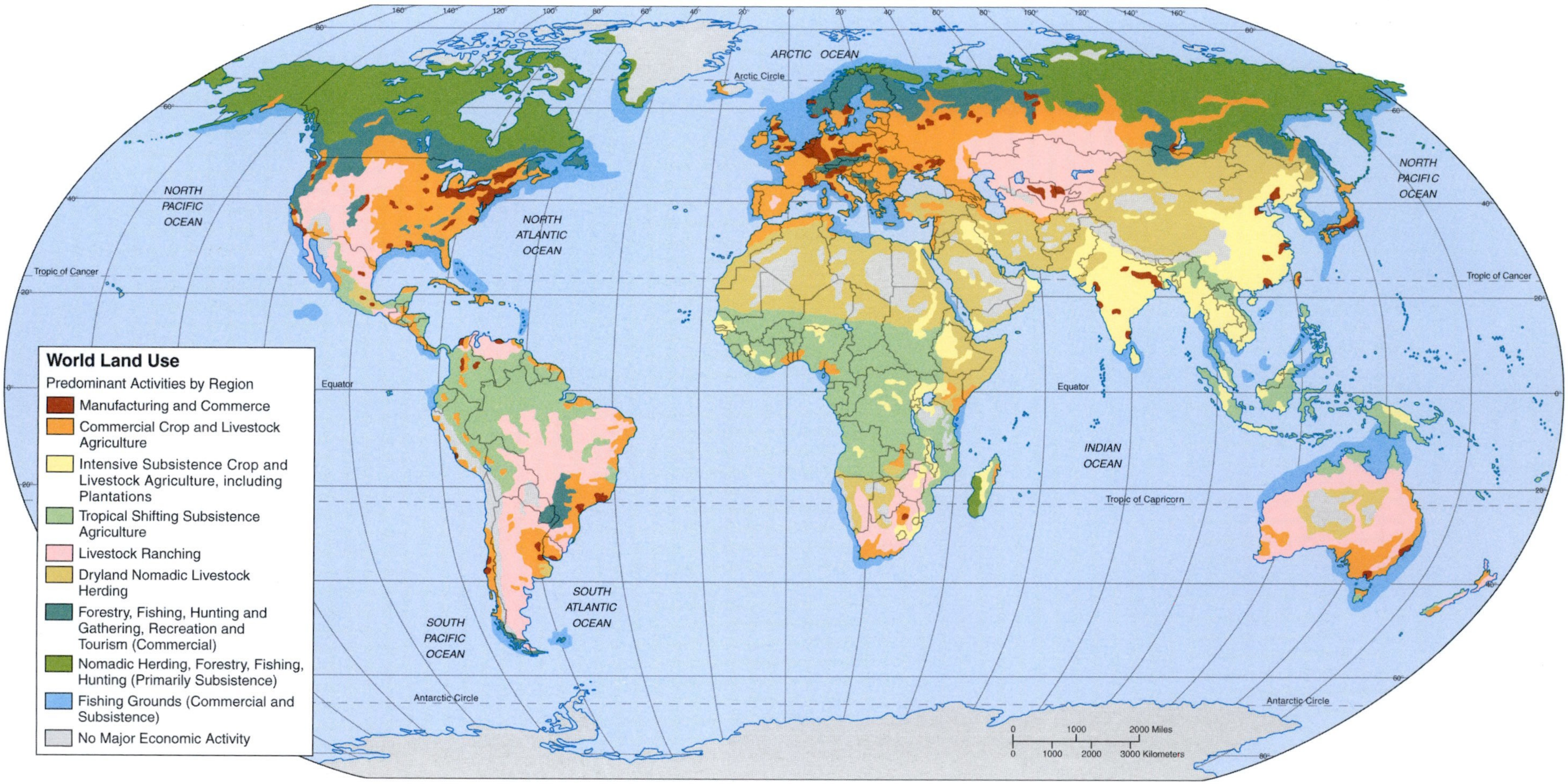

Land uses can be categorized as lying somewhere on a scale between extensive uses, in which human activities are dispersed over relatively large areas, and intensive uses, in which human activities are concentrated in relatively small areas. Many of the most important land use patterns of the world (such as urbanization, industry, mining, or transportation) are intensive and therefore relatively small in area and not easily seen on maps of this scale. Hence even in the areas identified as "Manufacturing and Commerce" on the map there are many land uses that are not strictly industrial or commercial in nature, and, in fact, more extensive land uses (farming, residential, open space) may actually cover more ground than the intensive industrial or commercial activities. On the other hand, the more extensive land uses like agriculture and forestry, tend to dominate the areas in which they are found. Thus, primary economic activities such as agriculture and forestry tend to dominate the world map of land use because of their extensive character. Much of this map is, therefore, a map that shows the global variations in agricultural patterns. Note, among other things, the differences between land use patterns in the more developed countries of the temperate zones and the less developed countries of the tropics.

Map 18 Urbanization

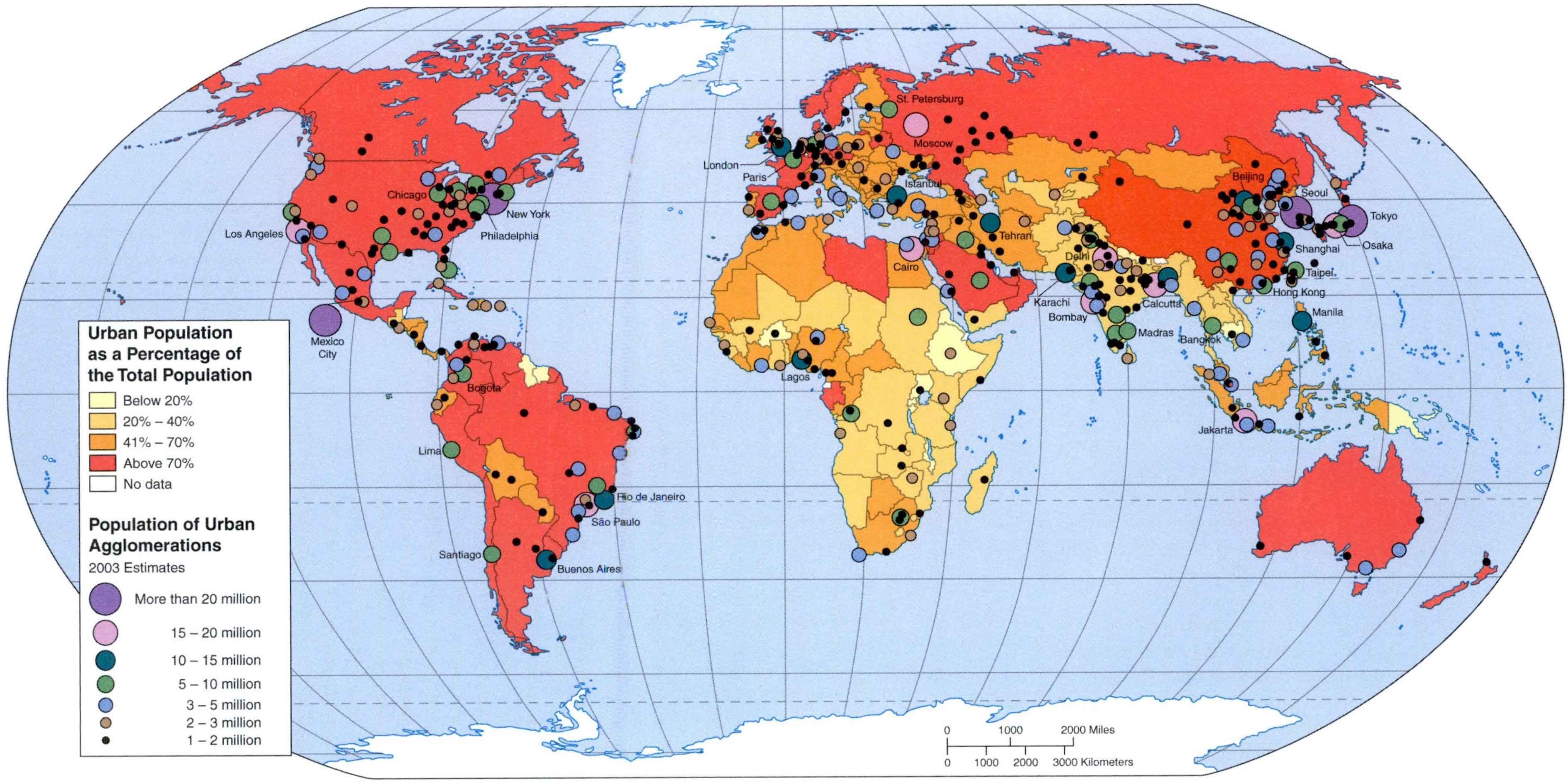

The degree to which a region's population is concentrated in urban areas is a major indicator of a number of things: the potential for environmental impact, the level of economic development, and the problems associated with human concentrations. Urban dwellers are rapidly becoming the norm among the world's people and rates of urbanization are increasing worldwide, with the greatest increases in urbanization taking place in developing regions. Whether in developed or developing countries, those who live in cities exert an influence on the environment, politics, economics, and social systems that goes far beyond the confines of the city itself. Acting as the focal points for the flow of goods and ideas, cities draw resources and people not just from their immediate hinterland but from the entire world. This process creates far-reaching impacts as resources are extracted, converted through industrial processes, and transported over great distances to metropolitan regions, and as ideas spread or *diffuse* along with the movements of people to cities and the flow of communication from them. The significance of urbanization can be most clearly seen, perhaps, in North America where, in spite of vast areas of relatively unpopulated land, well over 90 percent of the population lives in urban areas.

Map 19 Transportation Patterns

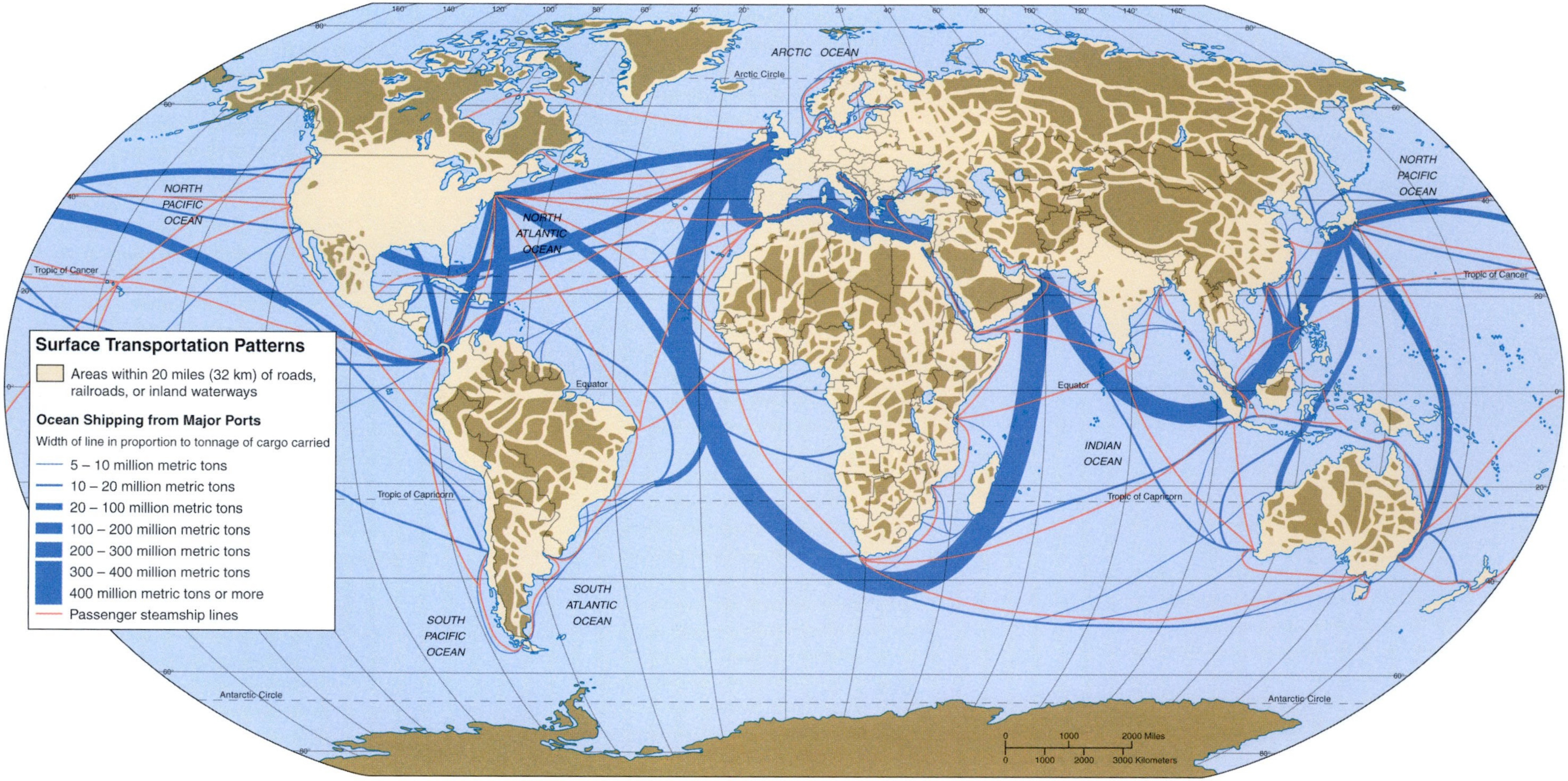

As a form of land use, transportation is second only to agriculture in its coverage of the earth's surface and is one of the clearest examples in the human world of a *network,* a linked system of lines allowing flows from one place to another. The global transportation network and its related communication web is responsible for most of the *spatial interaction,* or movement of goods, people, and ideas between places. As the chief mechanism of spatial interaction, transportation is linked firmly with the concept of a shrinking world and the development of a global community and economy.

Because transportation systems require significant modification of the earth's surface, transportation is also responsible for massive alterations in the quantity and quality of water, for major soil degradations and erosion, and (indirectly) for the air pollution that emanates from vehicles utilizing the transportation system. In addition, as improved transportation technology draws together places on the earth that were formerly remote, it allows people to impact environments a great distance away from where they live.

Map 20 Religions

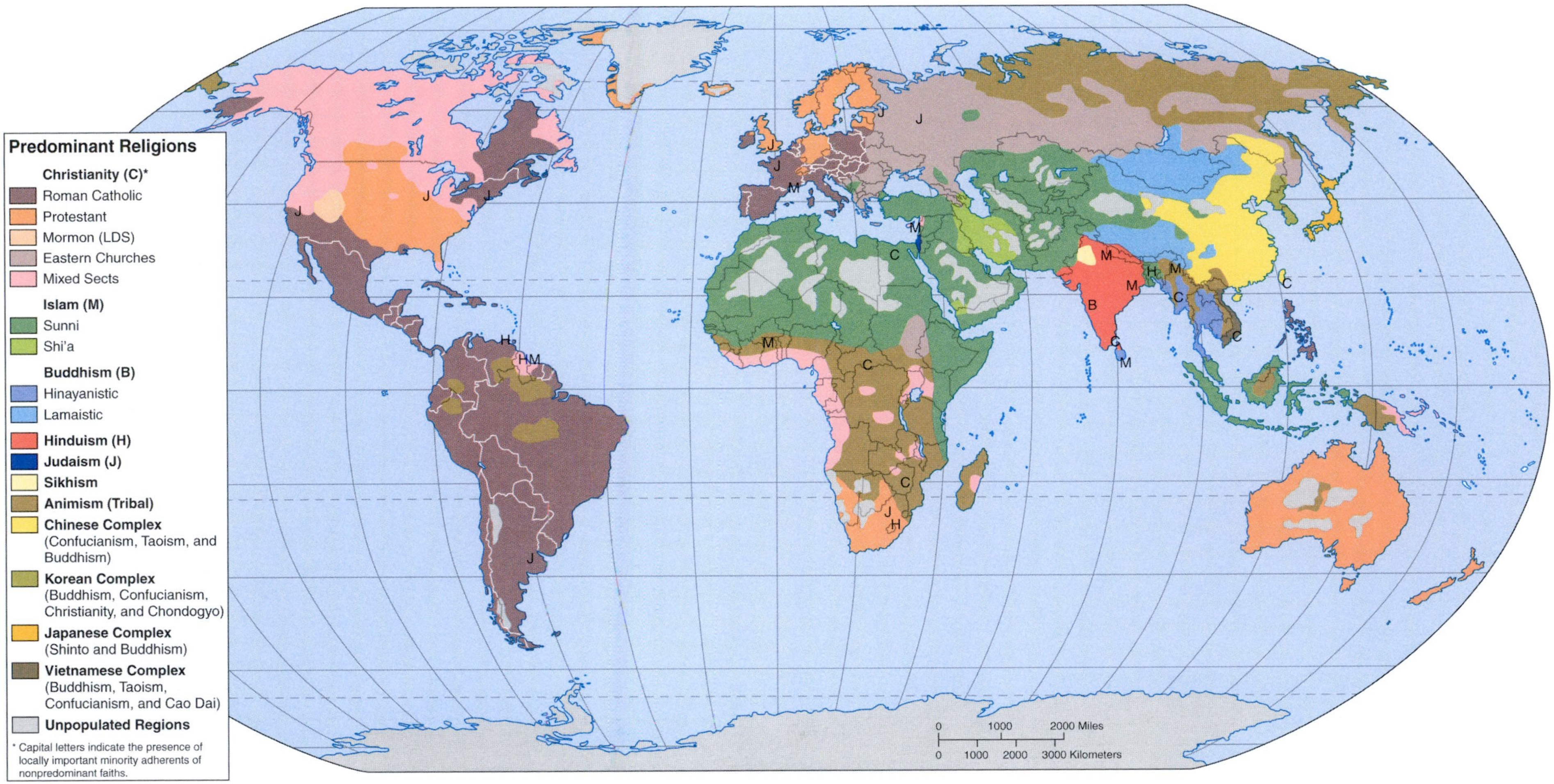

Religious adherence is one of the fundamental defining characteristics of human culture, the style of life adopted by a people and passed from one generation to the next. Because of the importance of religion for culture, a depiction of the spatial distribution of religions is as close as we can come to a map of cultural patterns. More than just a set of behavioral patterns having to do with worship and ceremony, religion is a vital conditioner of the ways that people deal with one another, with their institutions, and with the environments they occupy. In many areas of the world, the ways in which people make a living, the patterns of occupation that they create on the land, and the impacts that they make on ecosystems are the direct consequences of their adherence to a religious faith. An examination of the map in the context of international and intranational conflict will also show that tension between countries and the internal stability of states is also a function of the spatial distribution of religion.

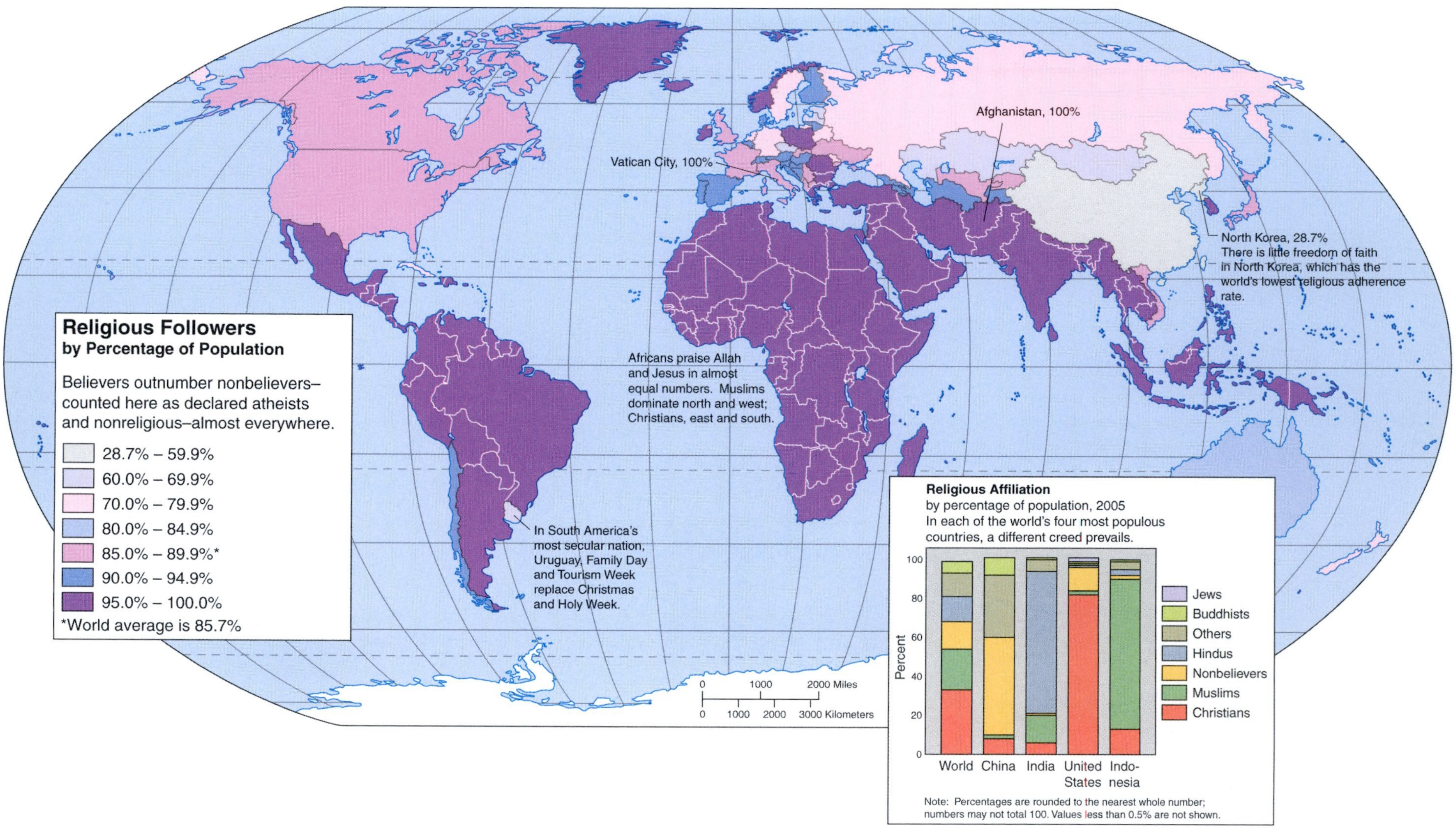

Throughout much of the past century, the numbers of proclaimed religious adherents declined as a percentage of populations as the consequence of the emergence of Communism in what became the U.S.S.R. and in China. Communism enforced atheism as a state "religion," and it is impossible to know how many Orthodox Christians actually existed in Russia or how many Christians, Muslims, Buddhists, Taoists, Confucianists, etc., existed in China. With the downfall of the old Soviet Union and the reemergence of an Orthodox Christian Russia, percentages of religious adherents in that country have increased tremendously. And as the European Union has opened its doors to "guest workers" and others from Southwest Asia and Africa, the number of Muslims has grown considerably in most of western Europe while Islam has reestablished itself as a religion among many inhabitants of the Balkan peninsula. It is worth noting that while past shifts in regional percentages of religious adherents had often been the result of missionary activity (the growing number of Christians in Africa during the twentieth century, for example), the recent changes have come about through political and economic change.

Map 22 Languages

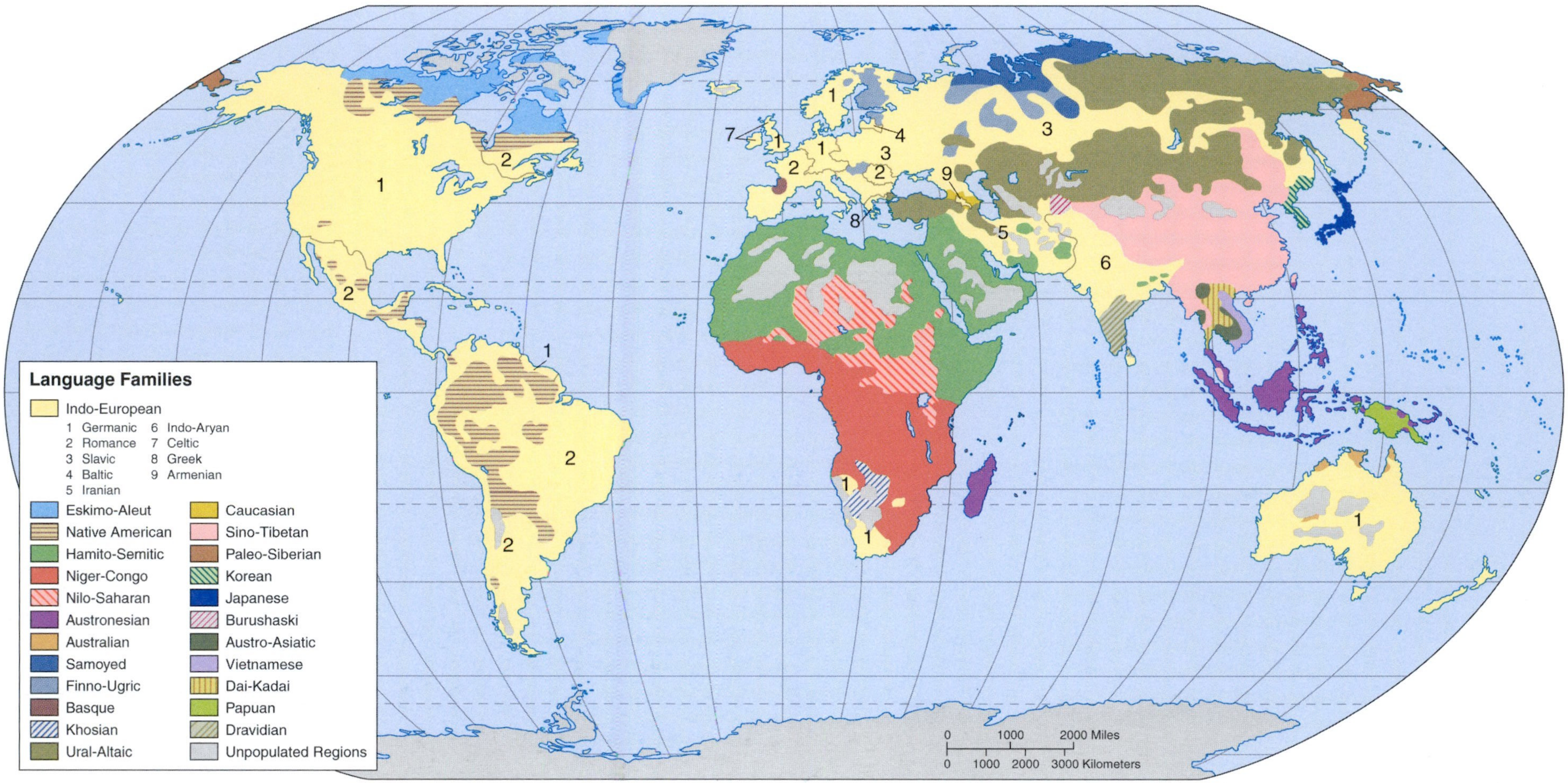

Language, like religion, is an important identifying characteristic of culture. Indeed, it is perhaps the most durable of all those identifying characteristics or *cultural traits:* language, religion, institutions, material technologies, and ways of making a living. After centuries of exposure to other languages or even conquest by speakers of other languages, the speakers of a specific tongue will often retain their own linguistic identity. Language helps us to locate areas of potential conflict, particularly in regions where two or more languages overlap, Many, if not most, of the world's conflict zones are also areas of linguistic diversity. Knowing the distribution of languages helps us to understand some of the reasons behind important current events: for example, linguistic identity differences played an important part in the disintegration of the Soviet Union in the early 1990s; and in areas emerging from recent colonial rule, such as Africa, the participants in conflicts over territory and power are often defined in terms of linguistic groups. Language distributions also help us to comprehend the nature of the human past by providing clues that enable us to chart the course of human migrations, as shown in the distribution of Indo-European, Austronesian, or Hamito-Semitic languages. Finally, because languages have a great deal to do with the way people perceive and understand the world around them, linguistic patterns help to explain the global variations in the ways that people interact.

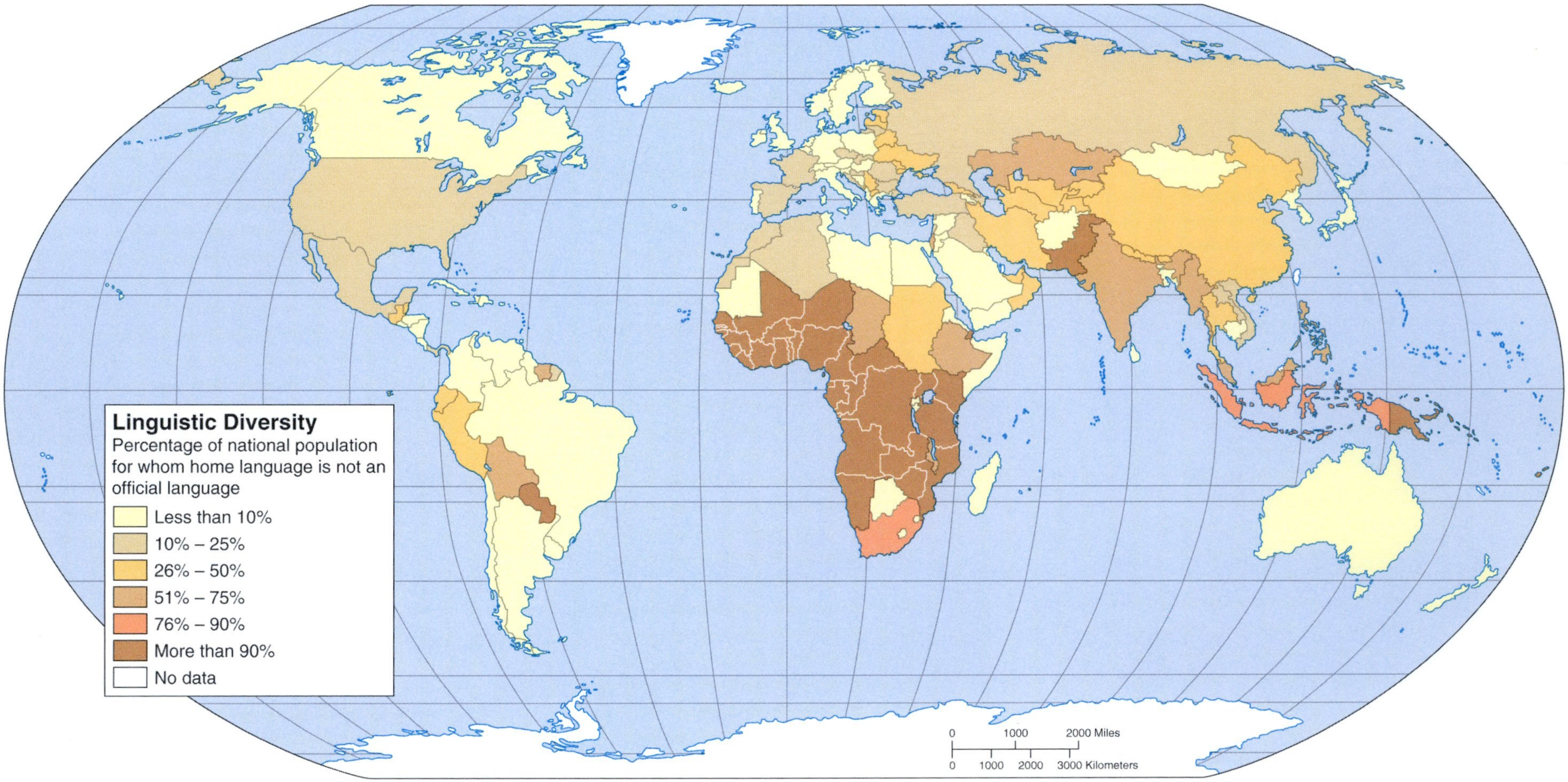

Of the world's approximately 6,000 languages, fewer than 100 are official languages, those designated by a country as the language of government, commerce, education, and information. This means that for much of the world's population, the language that is spoken in the home is different from the official language of the country of residence. The world's former colonial areas in Middle and South America, Africa, and South and Southeast Asia stand out on the map as regions in which there is significant disparity between home languages and official languages. To complicate matters further, for most of the world's population, the primary international languages of trade and tourism (French and English) are neither home nor official languages. China is a special case as the official language is the written form of Chinese while several spoken Chinese dialects such as Mandarin and Cantonese, most of them mutually unintelligible, are recognized as official languages. The formal language of government and business is Mandarin.

Map 24 Dying Languages: A Loss of Cultural Wealth

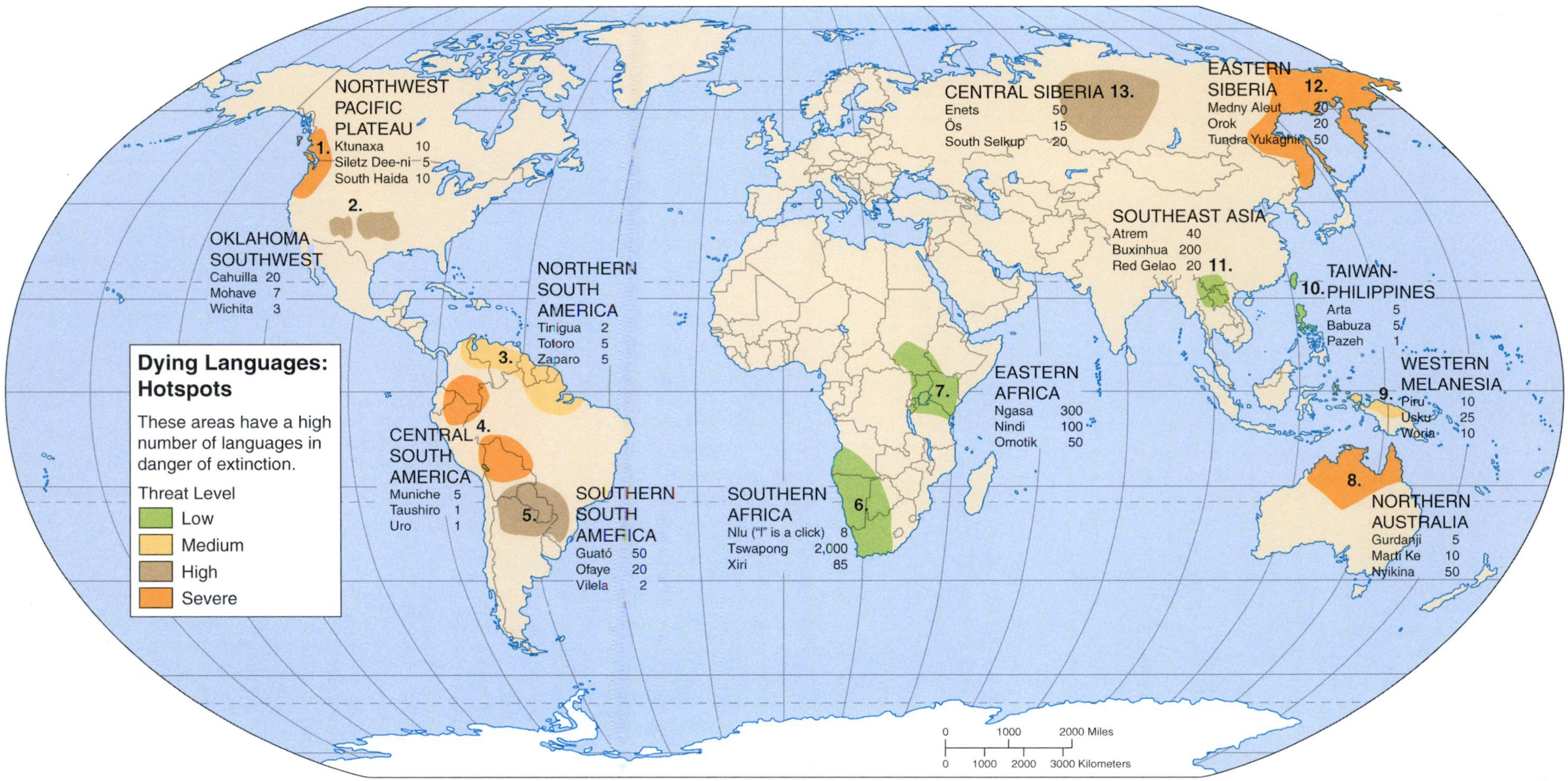

The number of languages spoken in the world of the sixteenth century was probably in excess of 10,000, possibly half of those in Africa. After four centuries of colonialism, migration, and economic and political change, less than half of the languages spoken in 1500 still exist—and many of these are in jeopardy of becoming extinct. Language is an essential marker of culture and when languages disappear, the cultures that used them tend to disappear as well, becoming assimilated—at least linguistically—into the dominant cultures of their regions. Thus, in North America, languages such as Mojave or Wichita are currently used by fewer than 10 native speakers. If those speakers die before passing their languages on, the languages become "dead" and so do many significant elements of the cultures associated with them. On many American Indian reservations there are currently important efforts to regain or restore native tongues. But these movements are active among the larger American Indian populations; while Northern Arapahoe may well be preserved, the South Haida of the Pacific Northwest may not.

Map 25 External Migrations in Modern Times

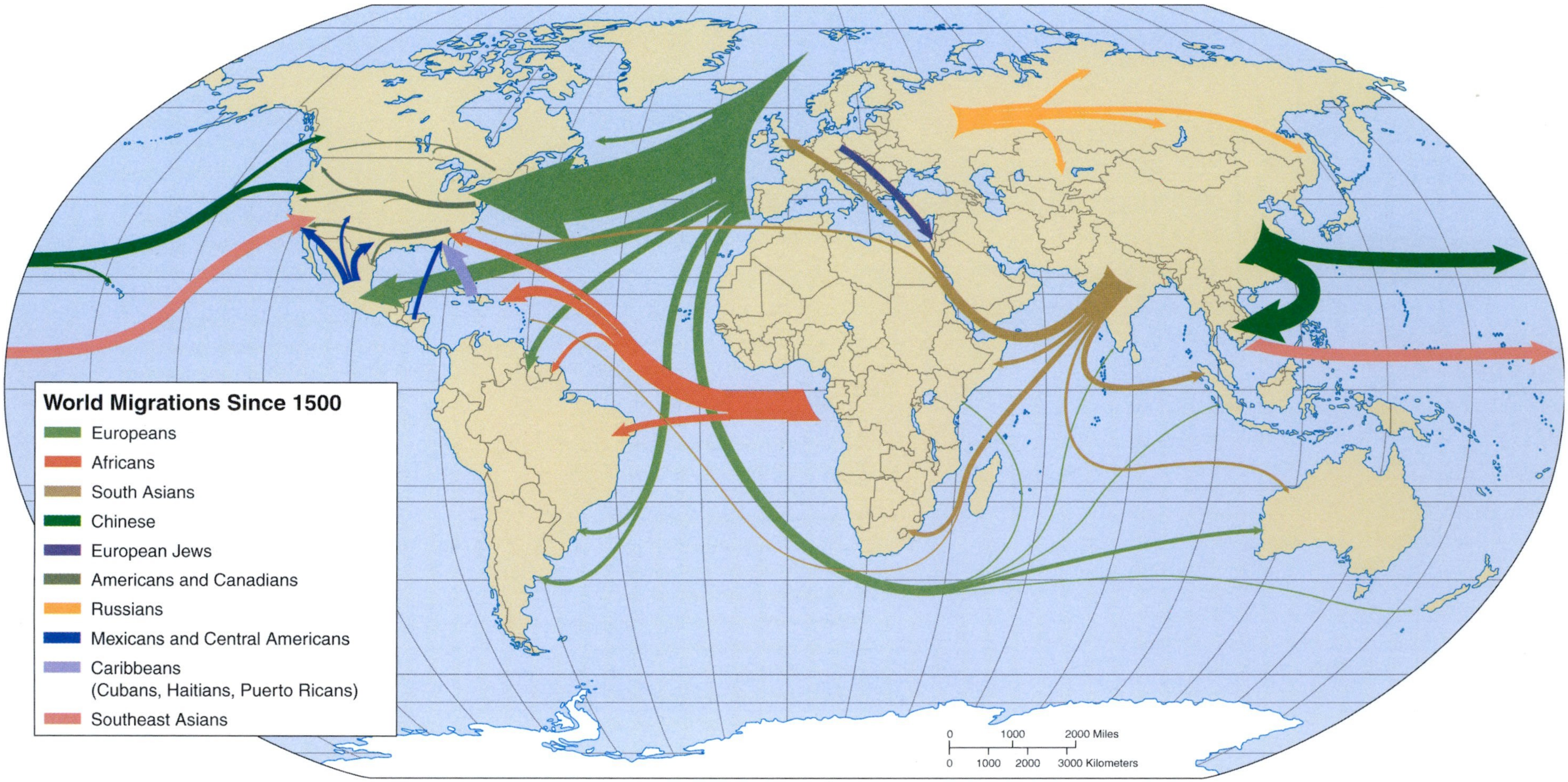

Migration has had a significant effect on world geography, contributing to cultural change and development, to the diffusion of ideas and innovations, and to the complex mixture of people and cultures found in the world today. *Internal migration* occurs within the boundaries of a country; *external migration* is movement from one country or region to another. Over the last 50 years, the most important migrations in the world have been internal, largely the rural-to-urban migration that has been responsible for the recent rise of global urbanization. Prior to the mid-twentieth century, three types of external migrations were most important: *voluntary,* most often in search of better economic conditions and opportunities; *involuntary* or *forced,* involving people who have been driven from their homelands by war, political unrest, or environmental disasters, or who have been transported as slaves or prisoners; and *imposed,* not entirely forced but which conditions make highly advisable. Human migrations in recorded history have been responsible for major changes in the patterns of languages, religions, ethnic composition, and economies. Particularly during the last 500 years, migrations of both the voluntary and involuntary or forced type have literally reshaped the human face of the earth.

Unit III

Global Demographic Patterns

Map 26 Population Growth Rates

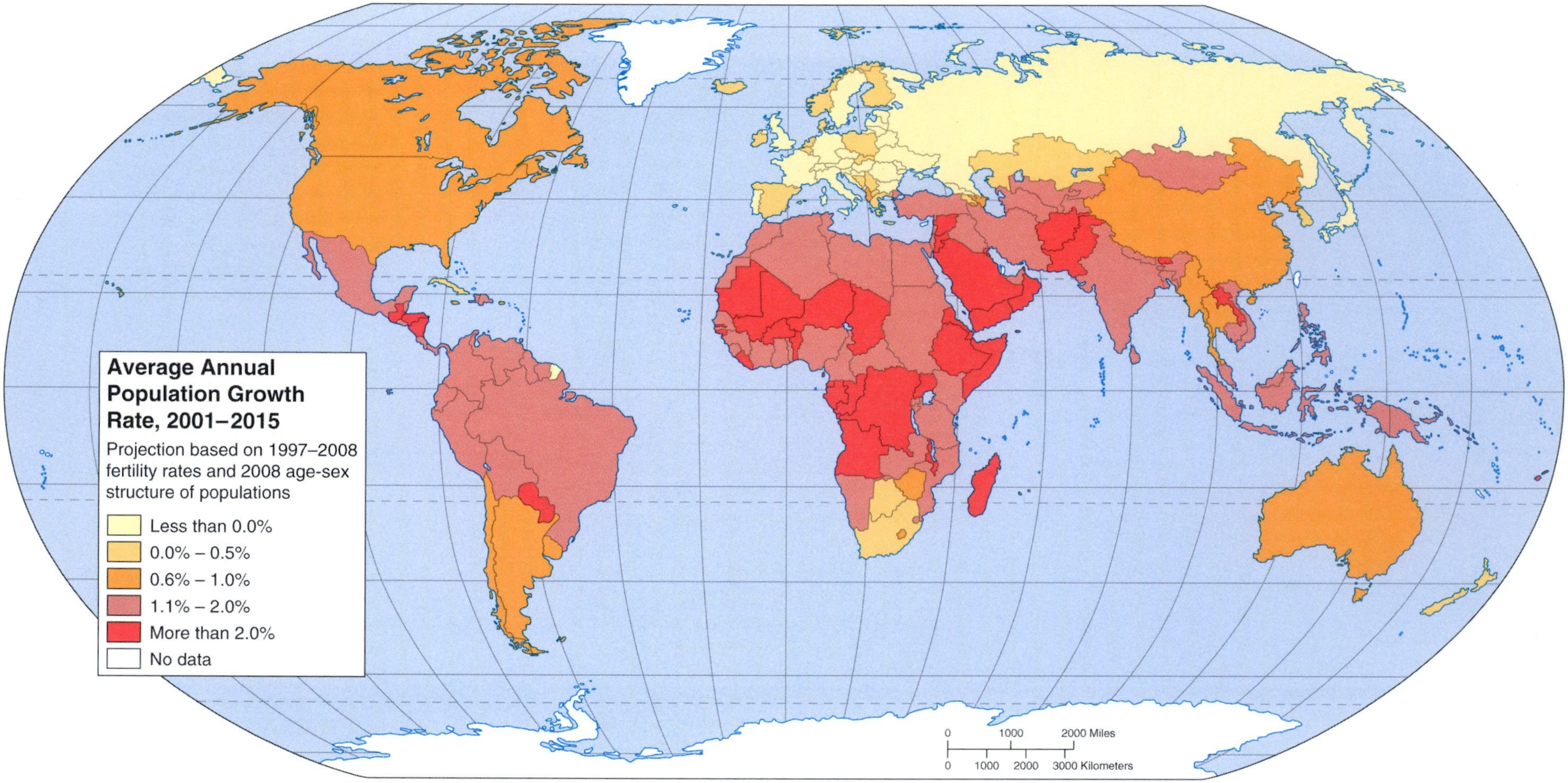

Of all the statistical measurements of human population, that of the rate of population growth is the most important. The growth rate of a population is a combination of natural change (births and deaths), in-migration, and out-migration; it is obtained by adding the number of births to the number of immigrants during a year and subtracting from that total the sum of deaths and emigrants for the same year. For a specific country, this figure will determine many things about the country's future ability to feed, house, educate, and provide medical services to its citizens. Some of the countries with the largest populations (such as India) also have high growth rates. Since these countries tend to be in developing regions, the combination of high population and high growth rates poses special problems for political stability and continuing economic development; the combination also carries heightened risks for environmental degradation. Many people believe that the rapidly expanding world population is a potential crisis that may cause environmental and human disaster by the middle of the twenty-first century.

Europe's Population Bust: A Fizzle in the Population Bomb

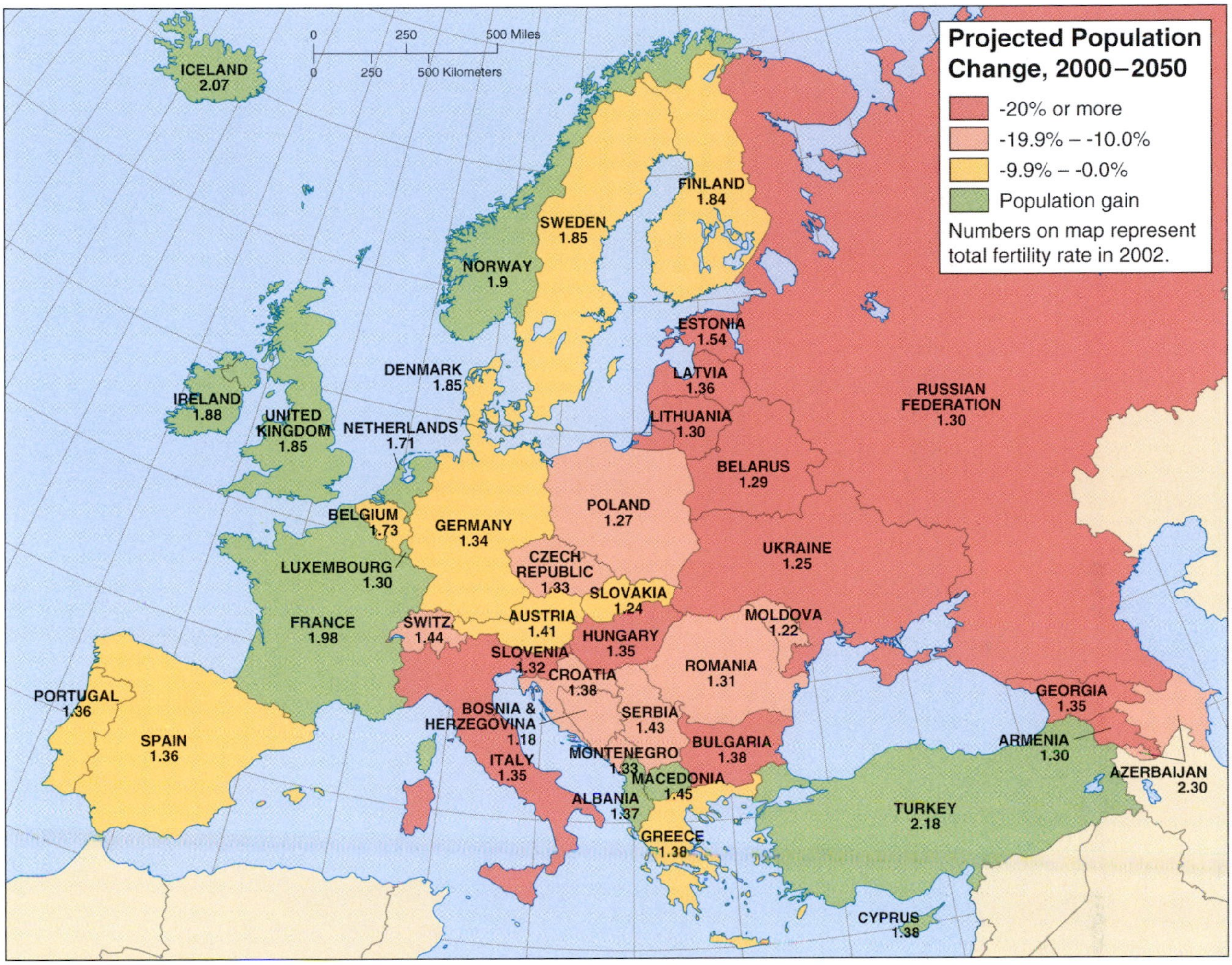

By the mid-1990s, most countries in Europe had begun to experience a phenomenon that was the logical consequence of the final stages of the "demographic transition" in which small families and increased longevity are the result of economic development, the recognition that fewer numbers of children cost less money, and the emergence of pension plans and other social mechanisms that make larger families unnecessary to protect parents in their old age. The chief manifestation of this last stage of the demographic transition was decreased fertility; indeed, in some countries (such as France) fertility rates or the numbers of children borne by the average woman were so low as to be below "replacement" levels—in other words, men and women were not producing enough offspring to replace themselves. The normal replacement fertility rate level for a wife and husband is 2.1. Throughout Europe, fertility rates are appreciably below that. This does not necessarily mean that all of the countries of Europe will experience shrinking populations. But those that do not will probably experience stable or even growing populations as the result of in-migration rather than natural increase. Even more critically, many of the in-migrants to countries like the United Kingdom are Muslims whose fertility rates are traditionally higher than their Christian neighbors. While stable or even shrinking populations may be beneficial in terms of environmental stress, etc., there are severe market and other pressures that result from aging populations: in a shrinking labor force, for example, will there be a sufficient tax base to support the national pension plans that are the norm in European countries?

Map 28a Total Fertility Rates

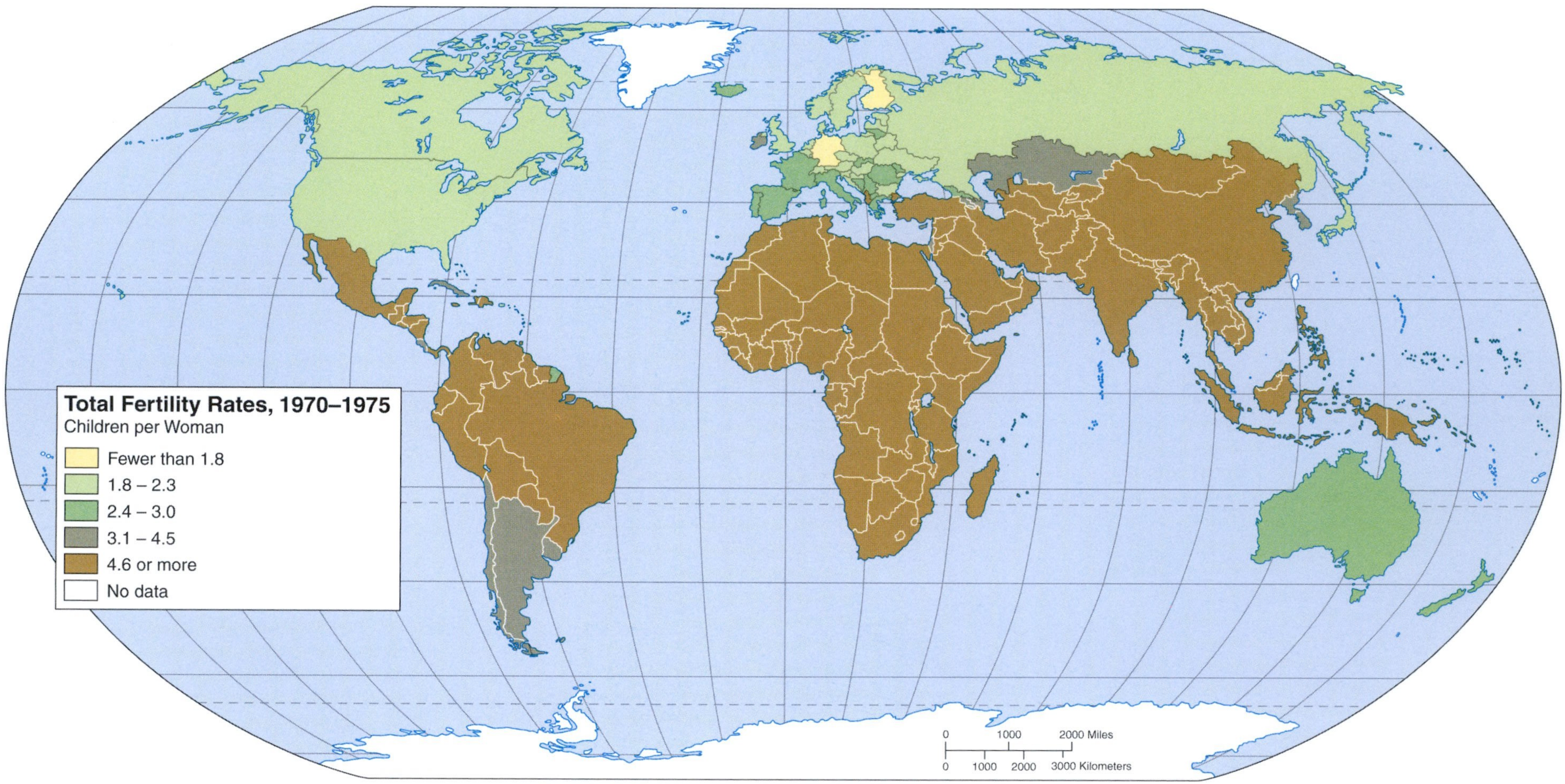

Fertility rates measure the number of children that might be expected to be given birth by a single woman during her child-bearing years. It is often a better predictor of potential population growth than such measures as birth rate. More important, it is a major indicator of the "demographic transition" or shift from a population characterized by large families and short lives to one characterized by small families and longer lives. For the last century, the first type of population profile (large families and short lives) has been evident in the lesser developed parts of the world while the developed nations had populations with the smaller family and longer life span profile. What these comparative maps show us is that over the last quarter century, the same kind of demographic transition begun in Europe and America in the nineteenth century is now beginning in Middle and South America, in Africa, and in Asia. There are still areas of concern—African fertility rates are, for example, still far too high. But particularly in South America, Southeast Asia, and East Asia, fertility rates have dropped dramatically, indicating the potential for much slower population growth in the next generation.

Map 28b Total Fertility Rates

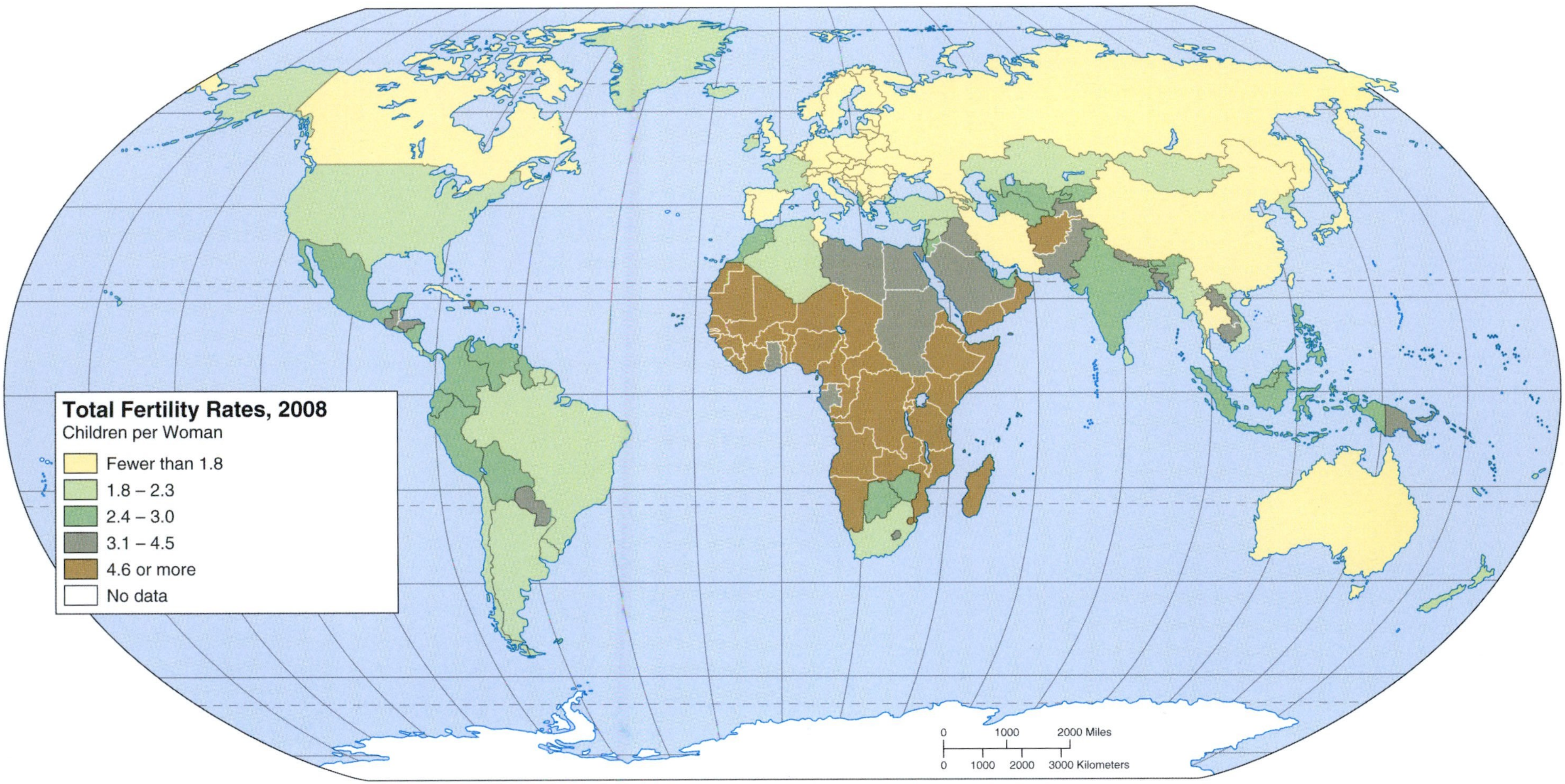

Map 29 Infant Mortality Rate

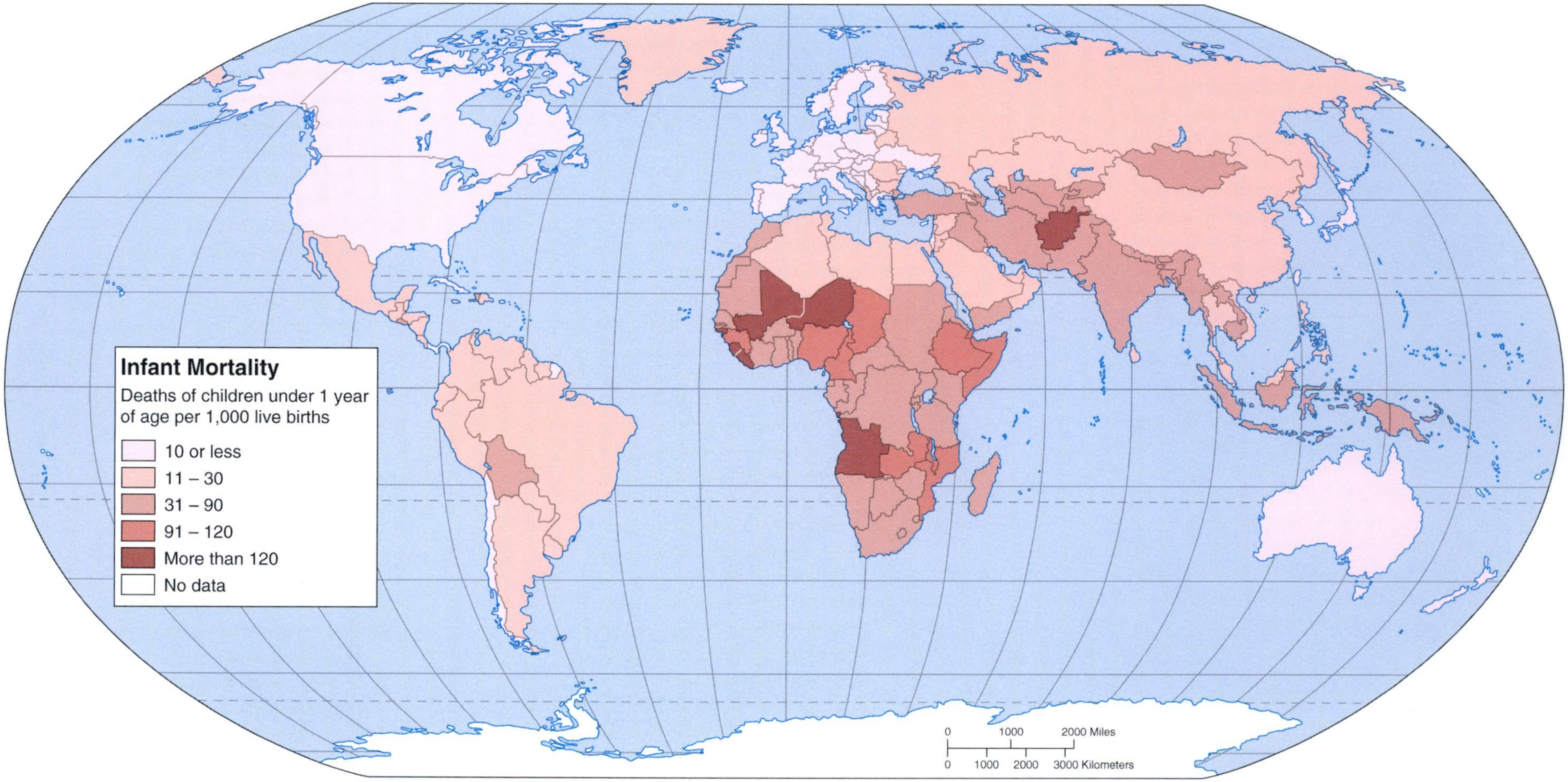

Infant mortality rates are calculated by dividing the number of children born in a given year who die before their first birthday by the total number of children born that year and then multiplying by 1,000; this shows how many infants have died for every 1,000 births. Infant mortality rates are prime indicators of economic development. In highly developed economies, with advanced medical technologies, sufficient diets, and adequate public sanitation, infant mortality rates tend to be quite low. By contrast, in less developed countries, with the disadvantages of poor diet, limited access to medical technology, and the other problems of poverty, infant mortality rates tend to be high.

Although worldwide infant mortality has decreased significantly during the last two decades, many regions of the world still experience infant mortality above the 10 percent level (100 deaths per 1,000 live births). Such infant mortality rates not only represent human tragedy at its most basic level, but also are powerful inhibiting factors for the future of human development. Comparing infant mortality rates in the midlatitudes and the tropics shows that children in most African countries are more than 10 times as likely to die within a year of birth as children in European countries.

Map 30 Child Mortality Rate

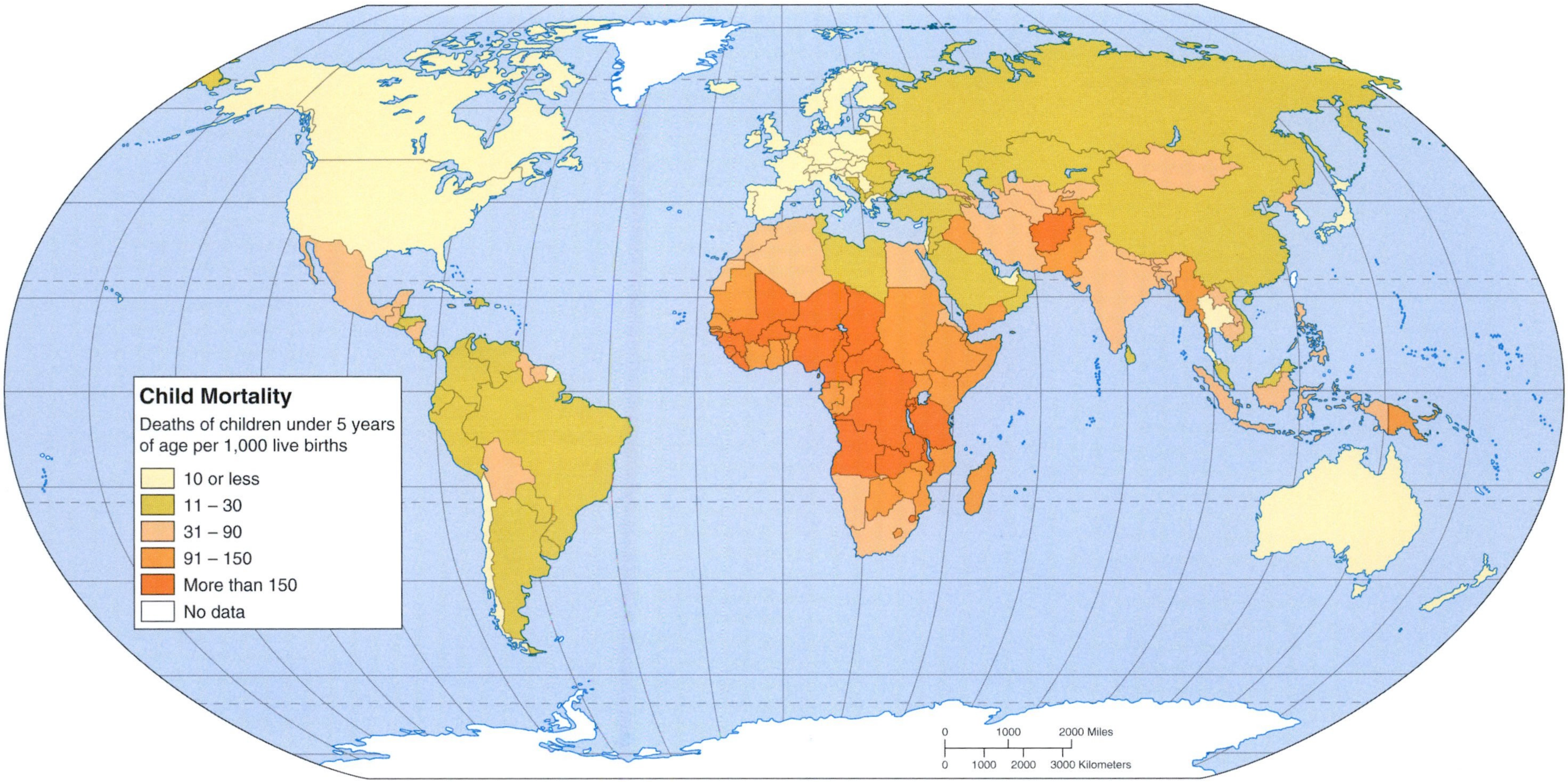

Child mortality rates are calculated by determining the probability that a child born in a specified year will die before reaching age 5, using current age-specific mortality rates for a population. The major sources of mortality rates are vital registration systems and estimates made from surveys and/or census reports. Along with infant mortality and average life-expectancy rates, child mortality rates, according to the World Bank, "are probably the best general indicators of a community's current health status and are often cited as overall measures of a population's welfare or quality of life." Where infant mortality often reflects health care conditions, child mortality is usually a reflection of the inadequacy of nutrition, leading to early deaths from nutritionally related diseases. In some less developed countries in Africa and Asia, child mortality is also an indicator of the widespread presence of infectious diseases such as malaria, tuberculosis, and HIV/AIDS.

Map 31 Child Malnutrition

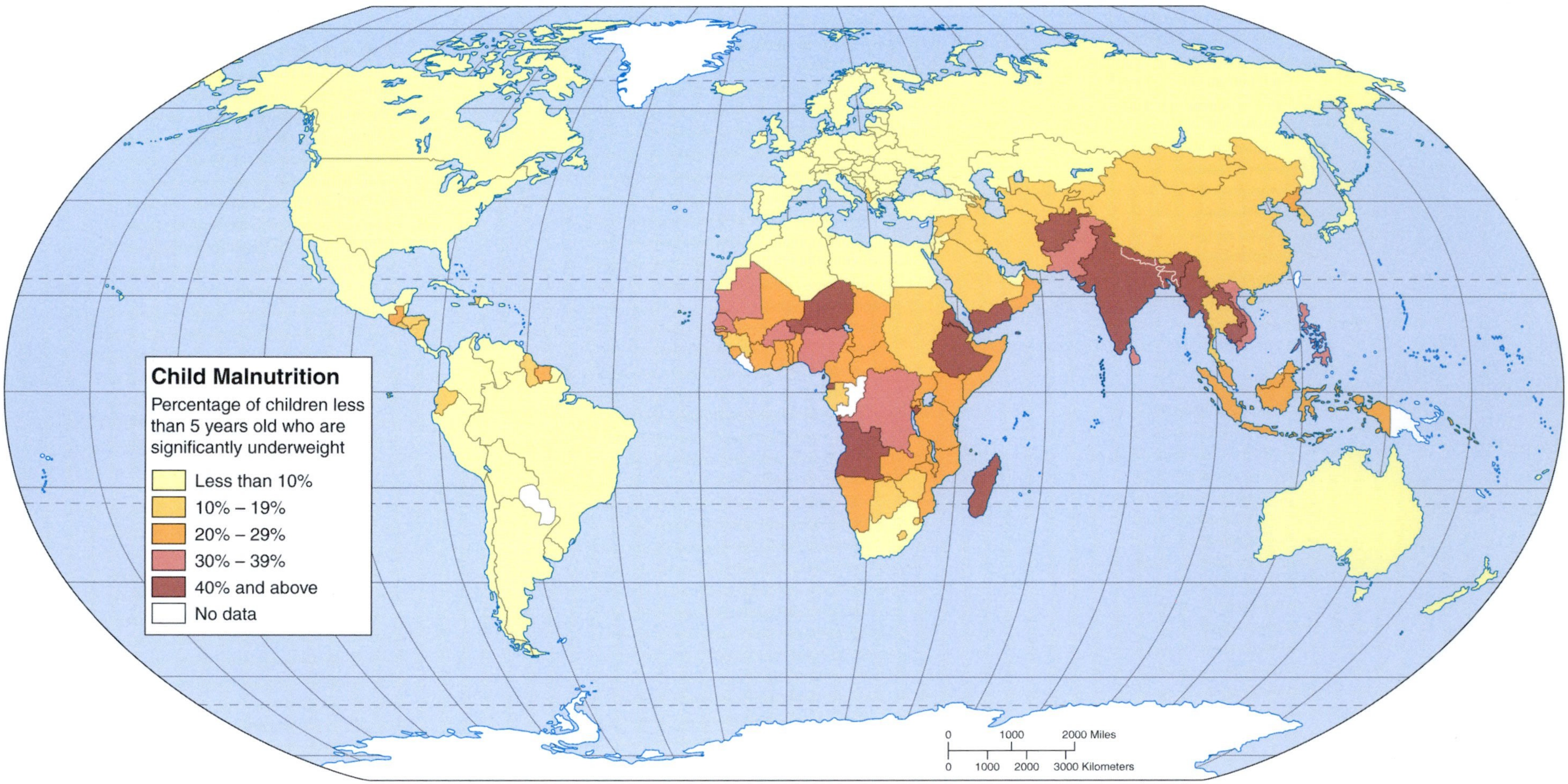

The weight of poverty is not evenly spread among the members of a population, falling disproportionately upon the weakest and most disadvantaged members of society. In most societies, these individuals are children, particularly female children. Children simply do not compete as successfully as adults for their (meager) share of the daily food supply. Where food shortages prevail, children tend to have the quality of their future lives severely compromised by poor nutrition, which, in a downward spiral, robs them of the energy necessary to compete more effectively for food. Children who are inadequately fed are less likely to do well in school, are more prone to debilitating disease, and will more often become a drain on scarce societal resources than well-fed children. Recently, health care officials in the more developed world have become concerned over the trend to "overnutrition," leading to obesity and related health problems in the world's economically developed countries. Nevertheless, child malnourishment remains one of the primary distinguishing factors between the "haves" and "have-nots."

Map 32 Primary School Enrollment

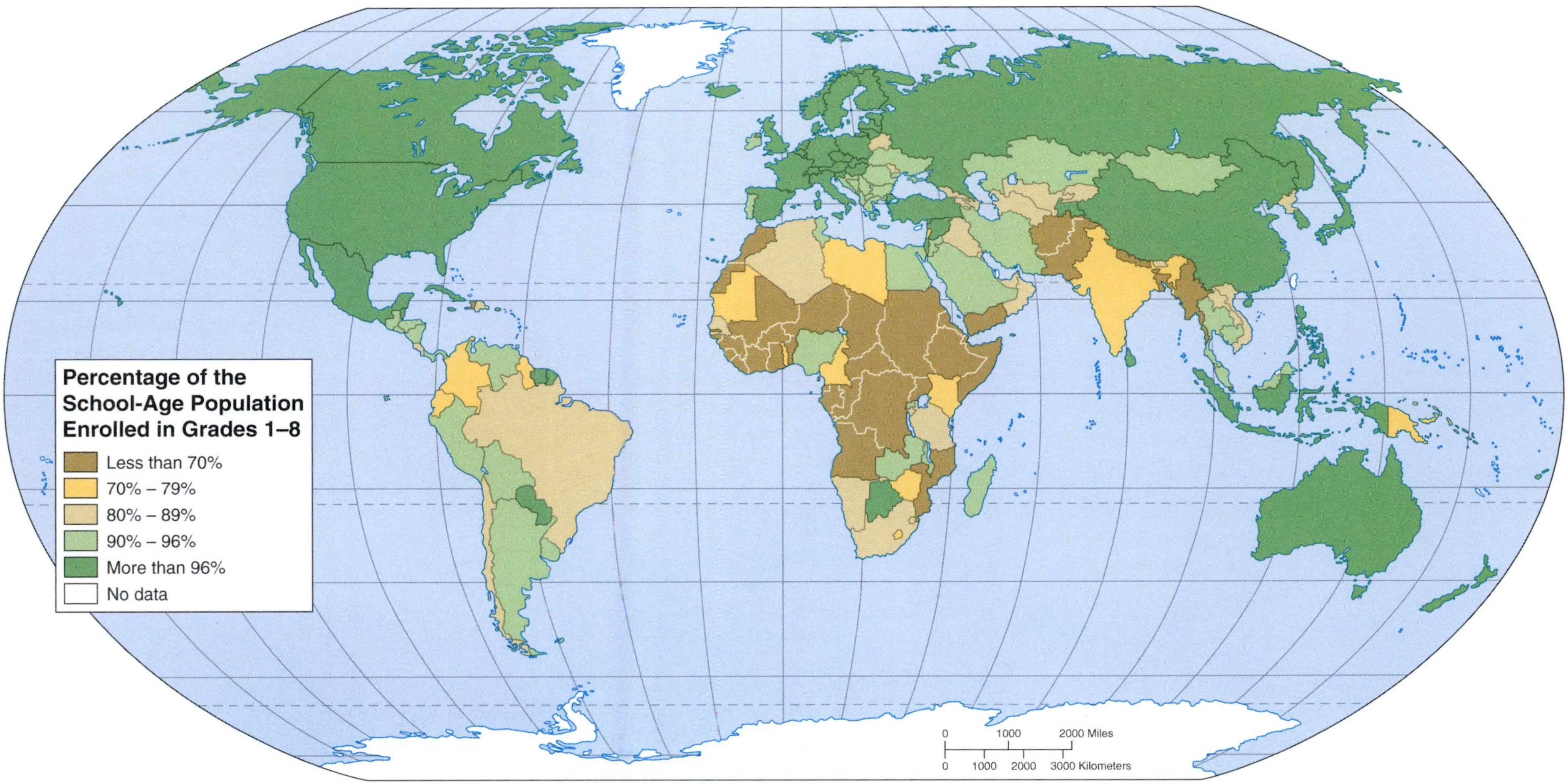

Like many of the other measures illustrated in this atlas, primary school enrollment is a clear reflection of the division of the world into "have" and "have-not" countries. It is also a measure that has changed more rapidly over the last decade than demographic and other indicators of development, as countries of even very modest means have made concerted attempts to attain relatively high percentages of primary school enrollment. That they have been able to do so is good evidence of the fact that reasonably respectable levels of human development are feasible at even modest income levels. High primary school enrollment is also a reflection of the worldwide opinion that a major element in economic development is a well-educated, literate population. The links between human progress, as typified by higher levels of education, and economic growth are not automatic, however, and those countries without programs for maintaining the headway gained by improved education may be on the road to failure in terms of economic development.

Map 33 Average Life Expectancy at Birth

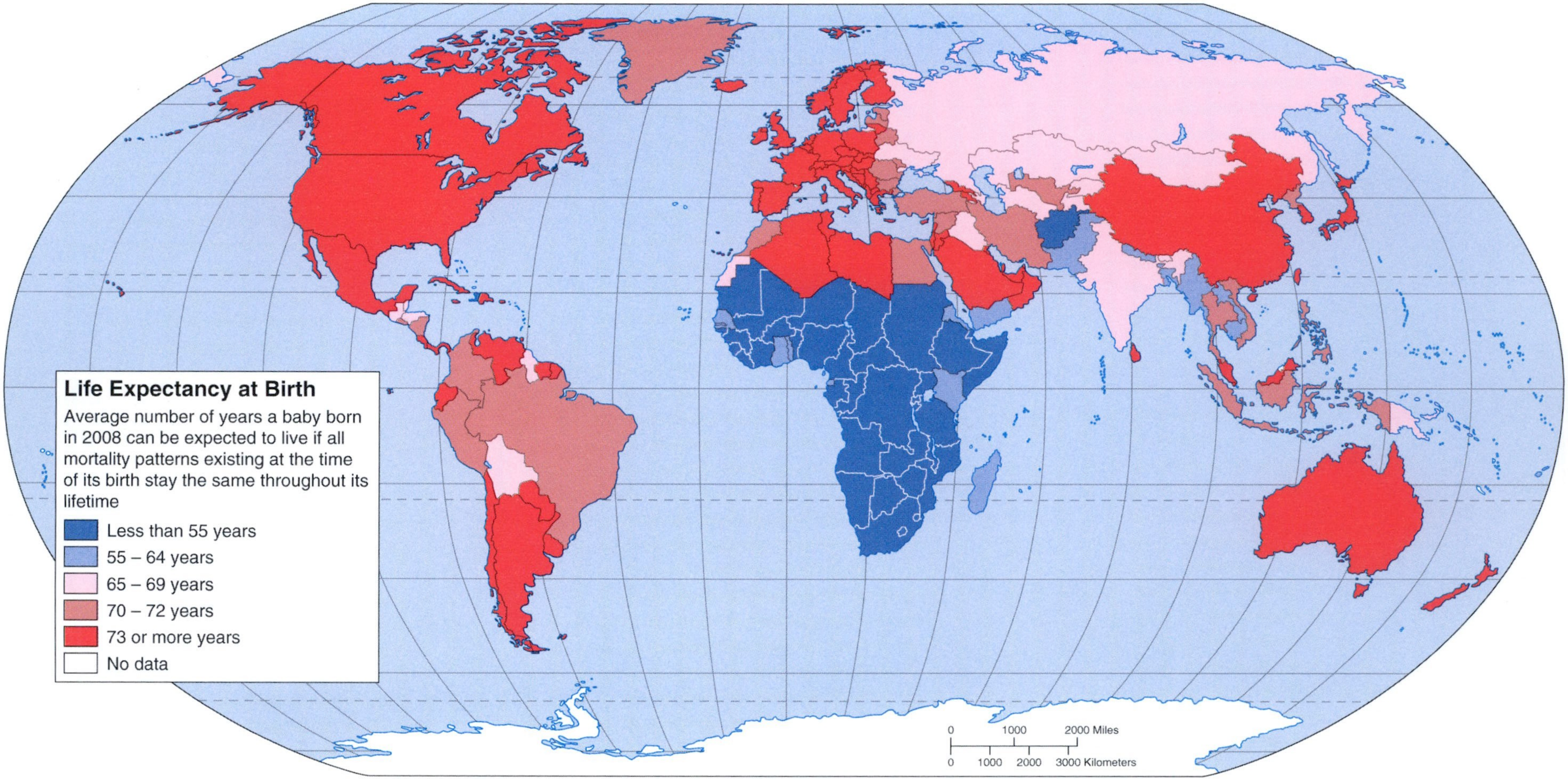

Average life expectancy at birth is a measure of the average longevity of the population of a country. Like all average measures, it is distorted by extremes. For example, a country with a high mortality rate among children will have a low average life expectancy. Thus, an average life expectancy of 45 years does not mean that everyone can be expected to die at the age of 45. More normally, what the figure means is that a substantial number of children die between birth and 5 years of age, thus reducing the average life expectancy for the entire population. In spite of the dangers inherent in misinterpreting the data, average life expectancy (along with infant mortality and several other measures) is a valid way of judging the relative health of a population. It reflects the nature of the health care system, public sanitation and disease control, nutrition, and a number of other key human need indicators. As such, it is a measure of well-being that is significant in indicating economic development and predicting political stability.

Map 34 Population by Age Group

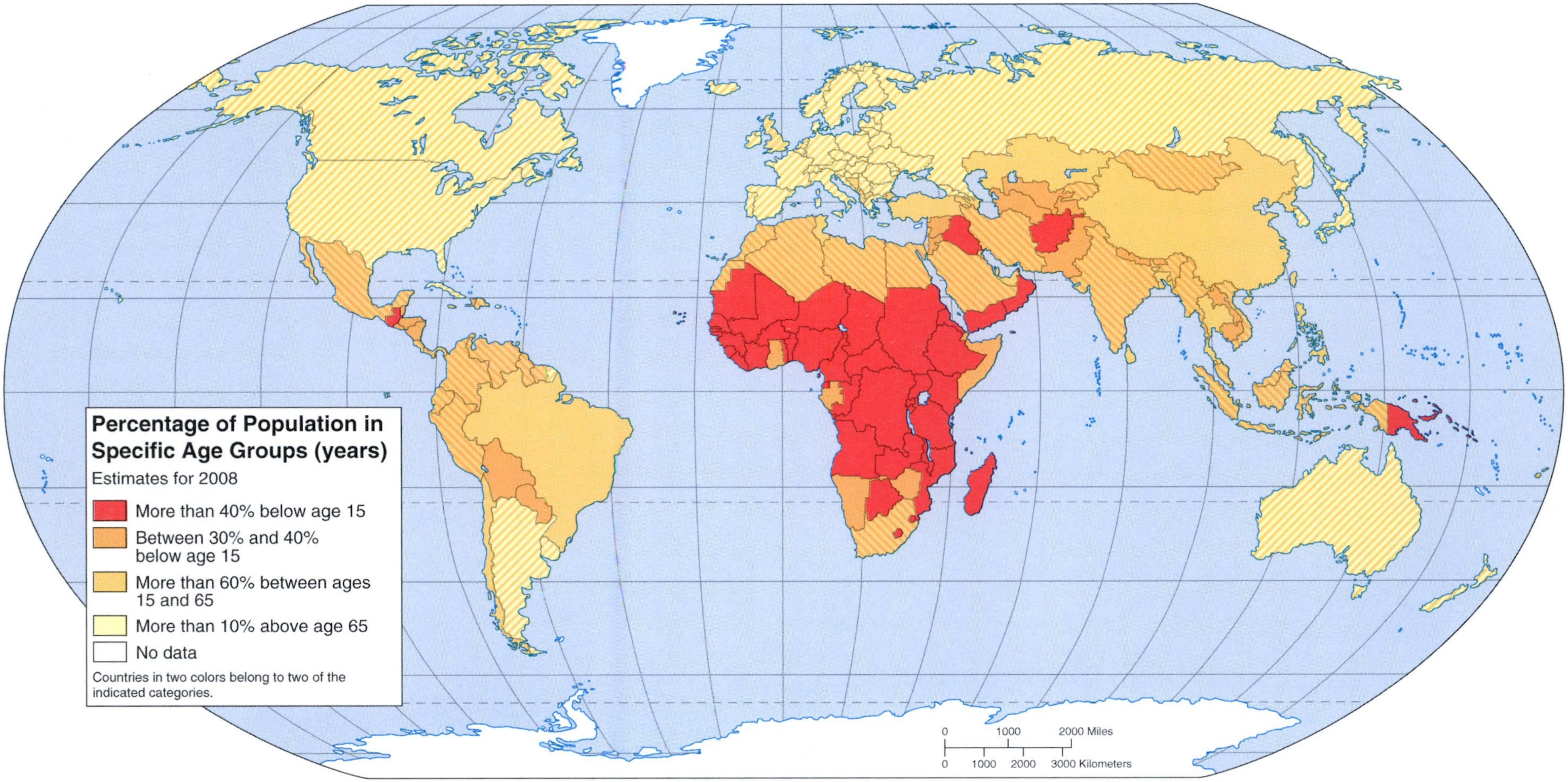

Of all the measurements that illustrate the dynamics of a population, age distribution may be the most significant, particularly when viewed in combination with average growth rates. The particular relevance of age distribution is that it tells us what to expect from a population in terms of growth over the next generation. If, for example, approximately 40–50 percent of a population is below the age of 15, that suggests that in the next generation about one-quarter of the total population will be women of childbearing age. When age distribution is combined with fertility rates (the average number of children born per woman in a population), an especially valid measurement of future growth potential may be derived. A simple example: Nigeria, with a 2002 population of 130 million, has 43.6 percent of its population below the age of 15 and a fertility rate of 5.5; the United States, with a 2002 population of 280 million, has 21 percent of its population below the age of 15 and a fertility rate of 2.07. During the period in which those women presently under the age of 15 are in their childbearing years, Nigeria can be expected to add a total of approximately 155 million persons to its total population. Over the same period, the United States can be expected to add only 61 million.

Map 35 International Migrant Populations 2005

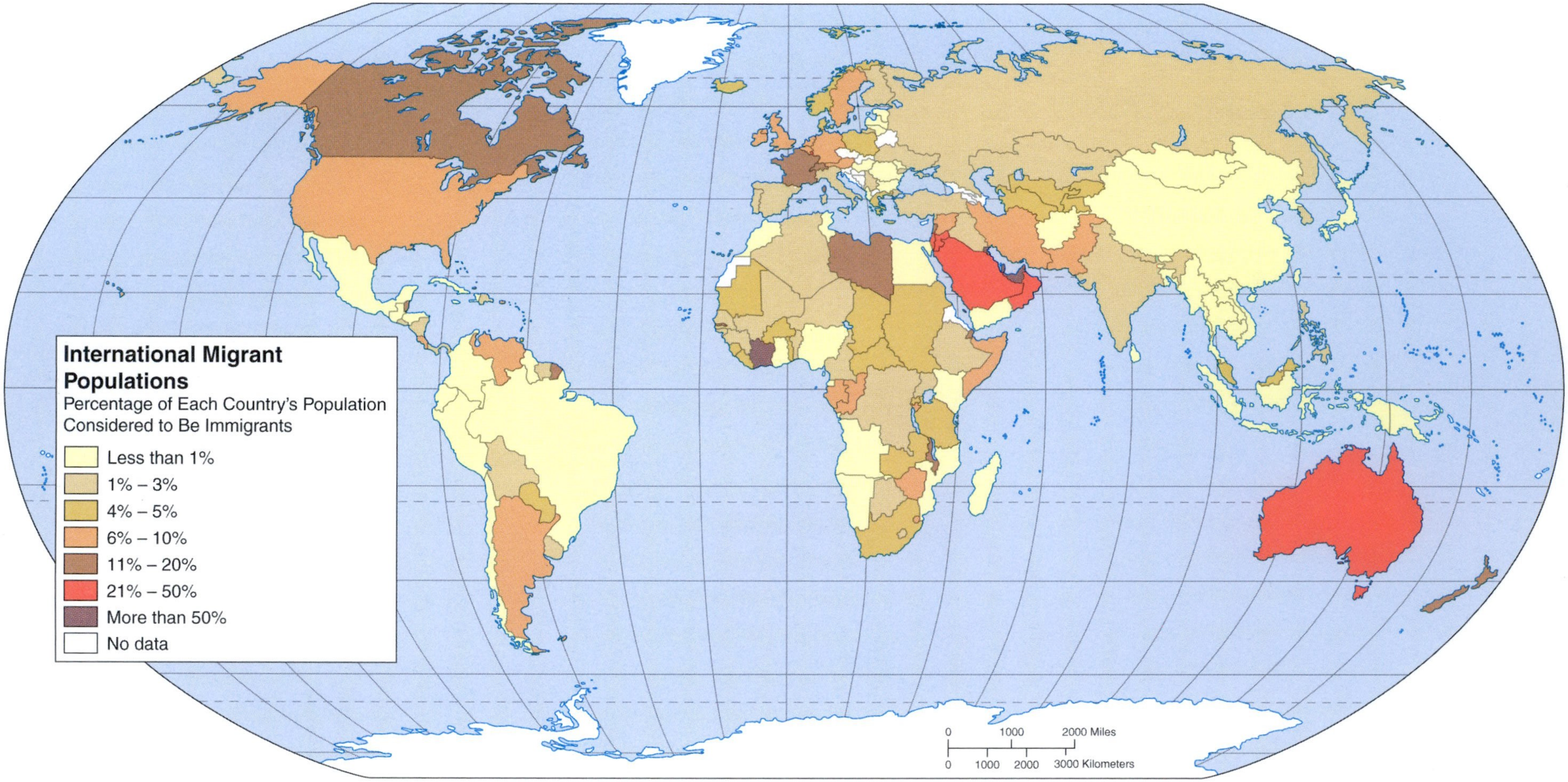

Migration—the movement from one place to another—takes a number of different forms. It may be a move within a country from an old job to a new one. It may also mean a migration, either forced or voluntary, from one country to another. The map here depicts international migration: migration between countries. Migration is distinguished from a refugee movement in that migrants are not defined as refugees granted a humanitarian and temporary protection status under international law. The nearly 2 million people who have fled Iraq for Iran are refugees who expect and hope to return to Iraq after the present civil war is ended. The Middle Americans who leave the Central American countries, Mexico, or the Caribbean for the United States plan to live in the United States permanently, although still retaining cultural and family ties to their native country. This map clearly shows that those countries viewed as having the most favorable opportunities for improvement in personal living conditions are those with the highest numbers of in-migrants; those countries that are overcrowded, with little economically upward mobility, or international conflict tend to be those with the greatest number of out-migrants.

Map 36 World Daily Per Capita Food Supply (Kilocalories)

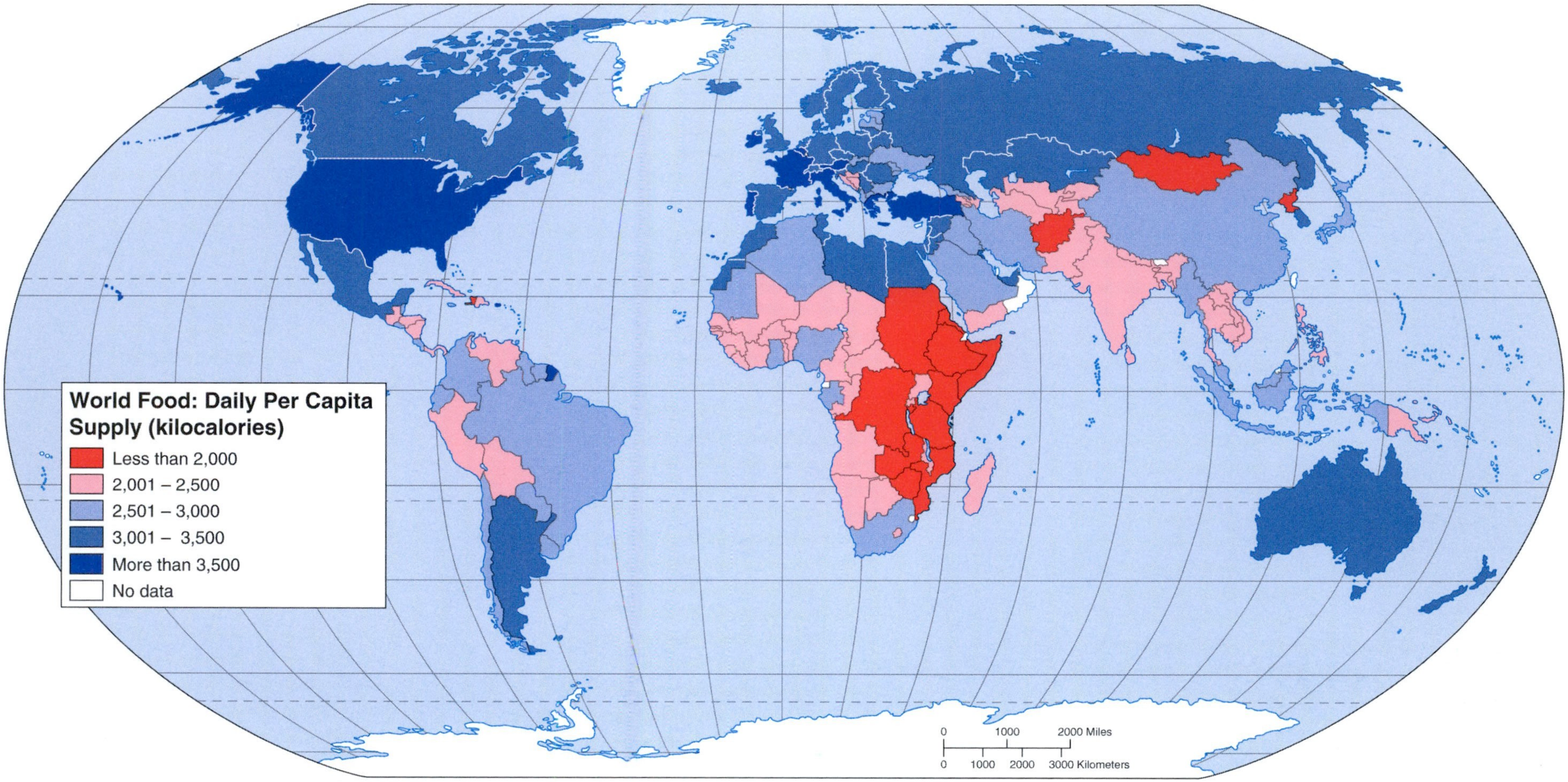

The data shown on this map, which indicate the presence or absence of critical food shortages, do not necessarily indicate the presence of starvation or famine. But they certainly do indicate potential problem areas for the next decade. The measurements are in calories from *all* food sources: domestic production, international trade, draw-down on stocks or food reserves, and direct foreign contributions or aid. The quantity of calories available is that amount, estimated by the UN's Food and Agriculture Organization (FAO), that reaches consumers. The calories actually consumed may be lower than the figures shown, depending on how much is lost in a variety of ways: in home storage (to pests such as rats and mice), in preparation and cooking, through consumption by pets and domestic animals, and as discarded foods, for example. The estimate of need is not a global uniform value but is calculated for each country on the basis of the age and sex distribution of the population and the estimated level of activity of the population. Compare this map with Map 57 for a good measure of potential problem areas for food shortages within the next decade.

Map 37 The Earth's Hungry Millions

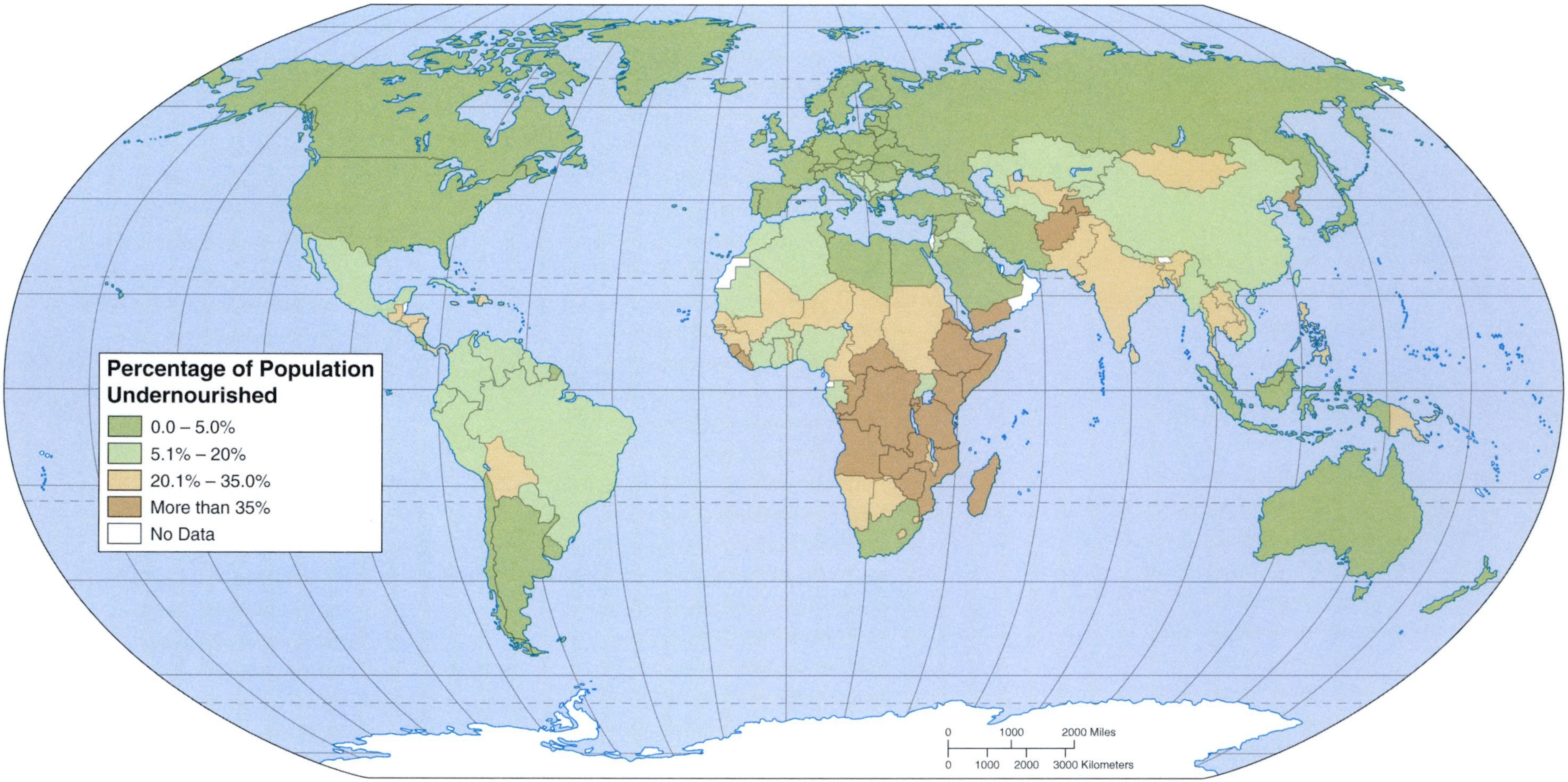

In order to maintain health and have sufficiently developed immune systems to fight off diseases, humans need to maintain a daily caloric intake of approximately 2000 calories. Current estimates are that nearly 15% of the world's population have diets that do not constitute the minimum 2000 calories for basic health maintenance. These countries are particularly susceptible to climate-induced famines which may lower already minimal food intake, or to food emergencies produced by warfare, civil unrest, economic instability, and poor governance. It is little surprise, when looking at the map, that the countries at greatest risk are in Africa and Asia where swelling populations have already approached (or, in some instances, exceeded) carrying capacity or the ability of the environment to sustain a population at a given level. The causes of national malnutrition are as numerous as the countries that experience it but, again, the map is instructive. In Africa, where HIV/AIDS is prevalent among large segments of the working-age population, food production is limited simply by poor health (which, of course, begets low harvests which contributes to poorer health). In South Asia, the problem tends to be magnified by huge populations that have outstripped the ability of the environment to support them. Bangladesh, for example, with an area about the size of West Virginia, is one of the world's ten most populous countries.

Map 38 Proportion of Daily Food Intake from Animal Products

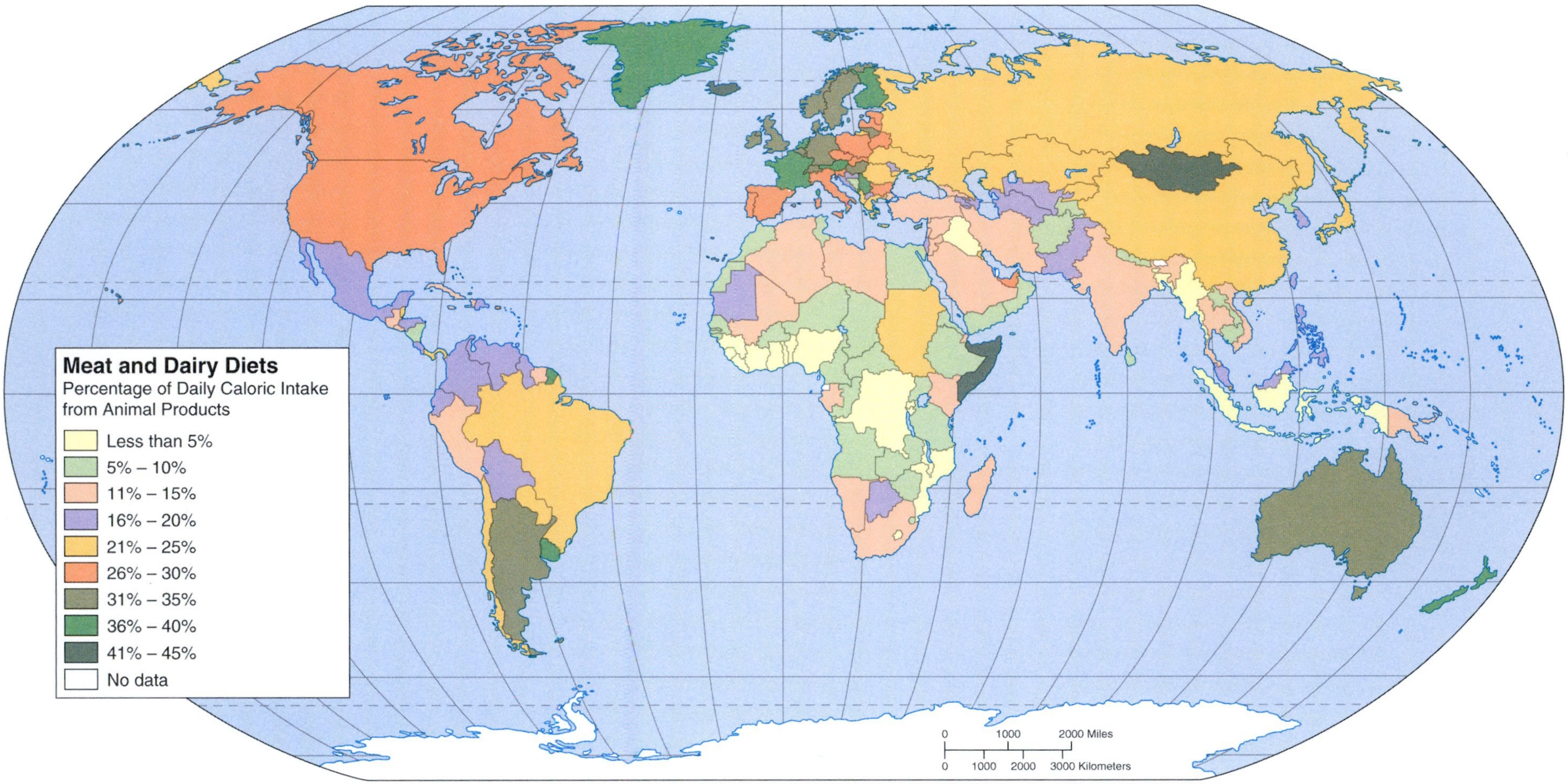

As the world grows proportionally wealthier, the percentage of the diet from meat and fish grows accordingly. While there is precious little scientific evidence to indicate that a vegetarian diet is, by definition, healthier than one richer in animal products, it is indisputable that more reliance on animal products produces a greater strain on environmental systems. Growing plants to feed animals to feed humans simply consumes a great deal more energy (in all forms) than growing plants to feed a human population directly. The long-term implications of altering land-use patterns to fit with greater wealth and, therefore, more meat consumption, are not generally favorable. And we are already beginning to notice the impact on fish population in the world's oceans as the result of an increasing demand for this food resource.

Map 39 Illiteracy Rates

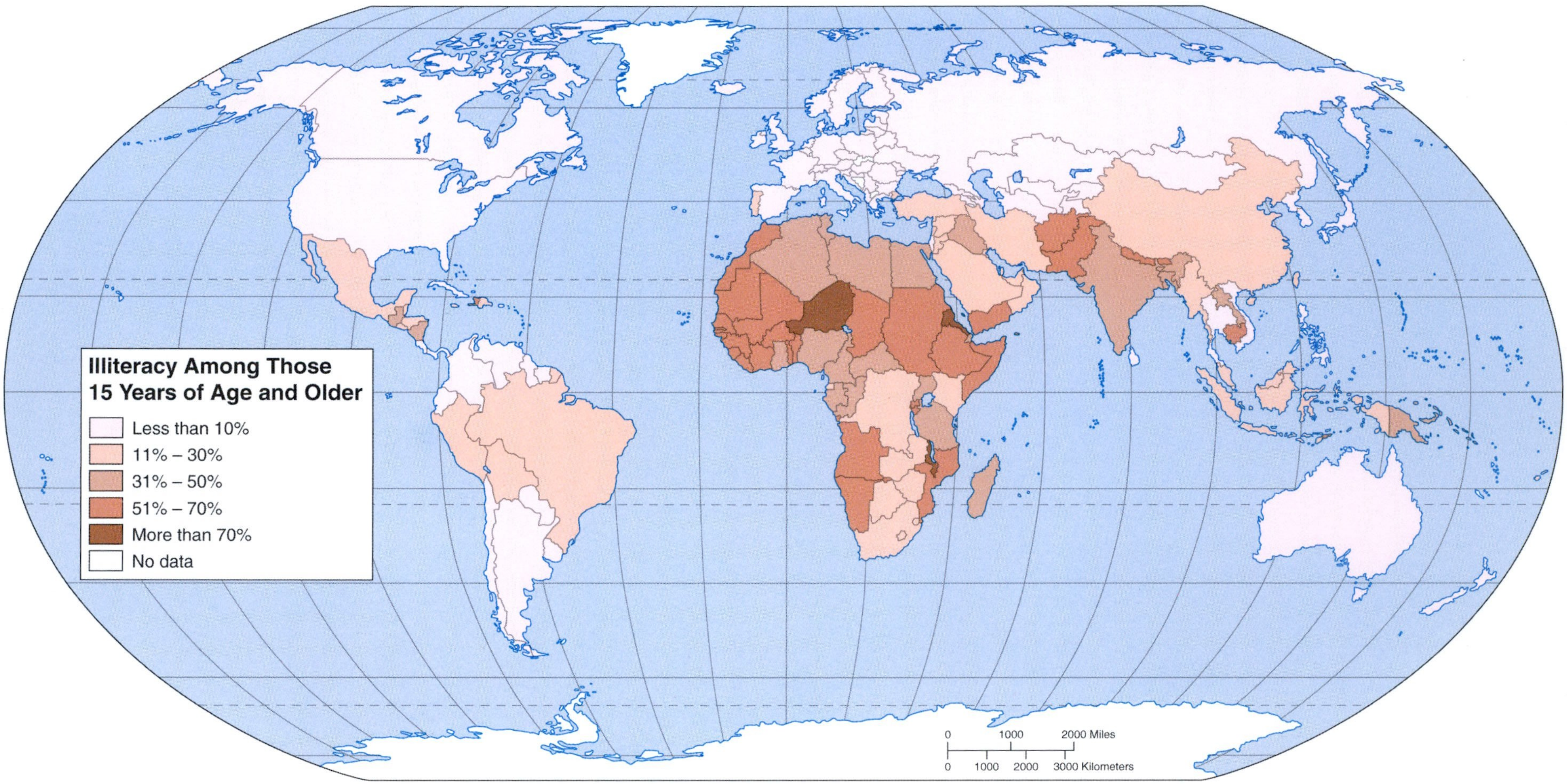

Illiteracy rates are based on the percentages of people age 15 or above (classed as adults in most countries) who are not able to write and read, with understanding, a brief, simple statement about everyday life written in their home or official language. As might be expected, illiteracy rates tend to be higher in the less developed states, where educational systems are a low government priority. Rates of literacy or illiteracy also tend to be gender-differentiated, with women in many countries experiencing educational neglect or discrimination that makes it more likely they will be illiterate. In many developing countries, between five and ten times as many women will be illiterate as men, and the illiteracy rate for women may even exceed 90 percent. Both male and female illiteracy severely compromise economic development.

Map 40 Female/Male Inequality in Education and Employment

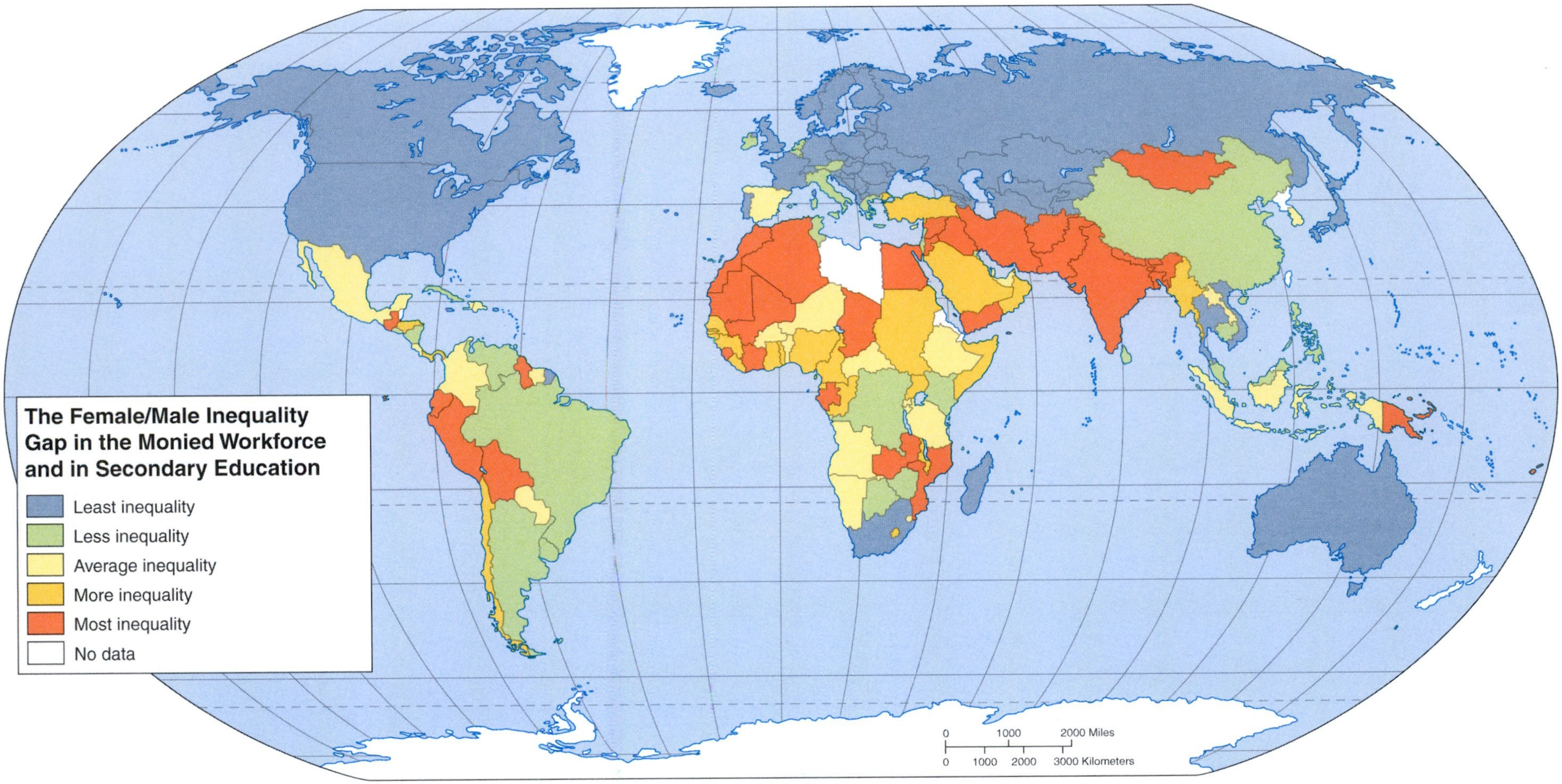

While women in developed countries, particularly in North America and Europe, have made significant advances in socioeconomic status in recent years, in most of the world they suffer from significant inequality when compared with their male counterparts. Although women have received the right to vote in most of the world's countries, in over 90 percent of these countries that right has only been granted in the last 50 years. In most regions, literacy rates for women still fall far short of those for men; in Africa and Asia, for example, only about half as many women are literate as are men. Women marry considerably younger than men and attend school for shorter periods of time.

Inequalities in education and employment are perhaps the most telling indicators of the unequal status of women in most of the world. Lack of secondary education in comparison with men prevents women from entering the workforce with equally high-paying jobs. Even where women are employed in positions similar to those held by men, they still tend to receive less compensation. The gap between rich and poor involves not only a clear geographic differentiation, but a clear gender differentiation as well.

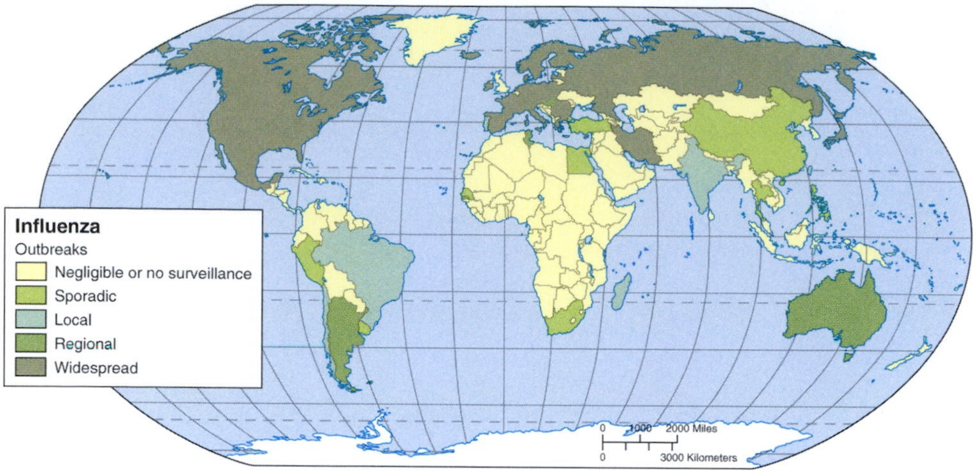

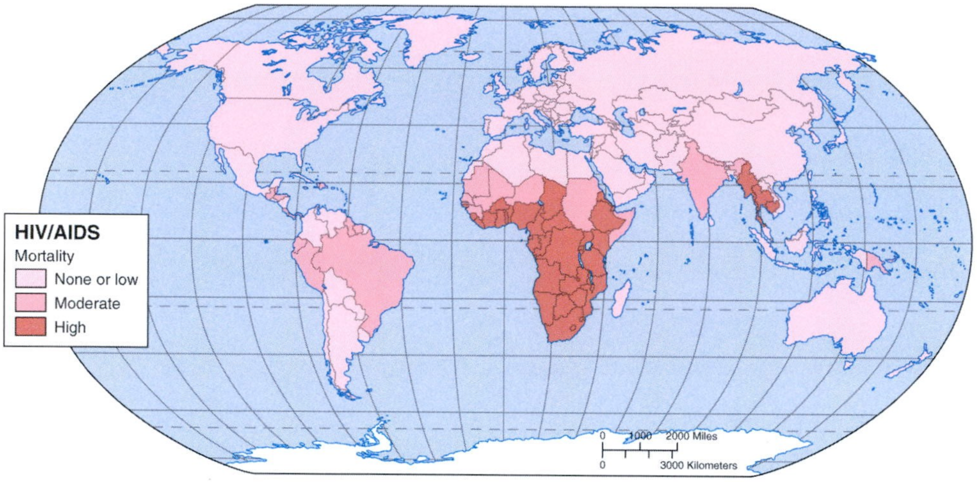

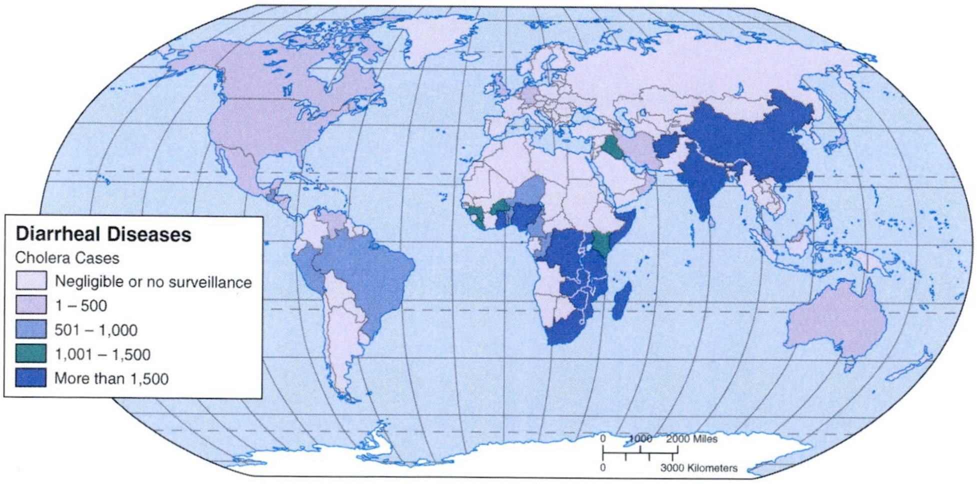

The infectious diseases shown in these maps account for more than 9 out of every 10 deaths worldwide from infectious disease. They are spread through various mechanisms, some—such as HIV/AIDS—requiring direct contact, while others such as tuberculosis or influenza may be contracted through airborne transmissions from an infected host. Still others, such as malaria and diarrheal diseases like cholera, require transmission by a vector or carrier. In malaria, the vector is the mosquito (most commonly *Aenopheles*) while the diarrheal diseases are transmitted by microscopic vectors such as bacteria or viruses that commonly live in the water that people drink. Infectious diseases such as these six do tend to have specific geographic distributions, sometimes as the result of climatic factors (malaria depends upon warm and moist conditions) and sometimes the result

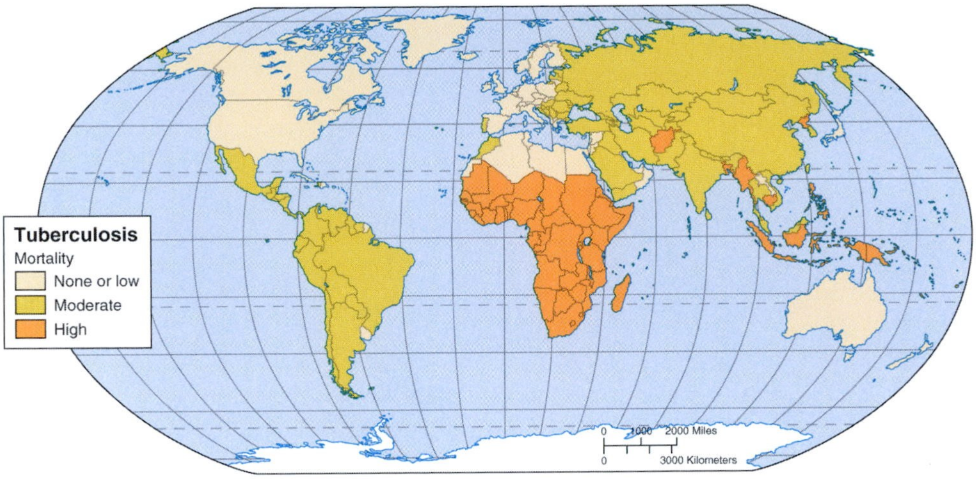

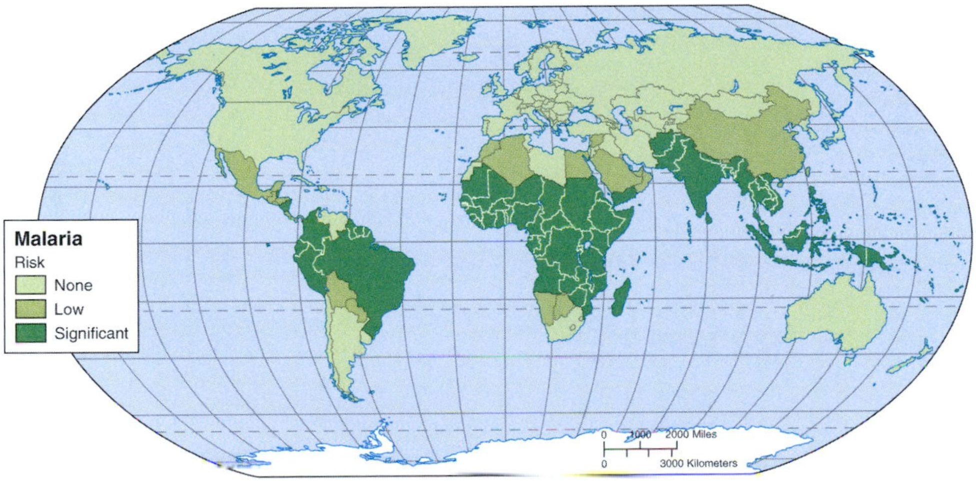

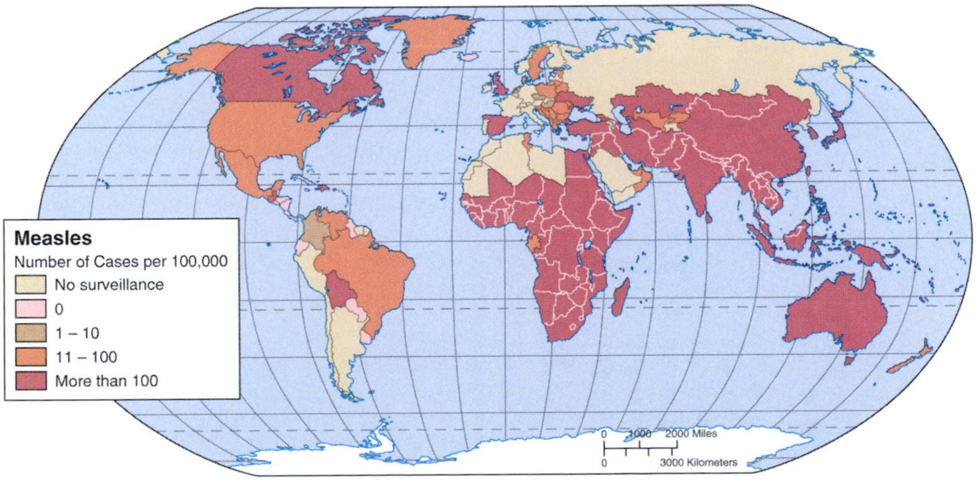

of cultural factors such as rapid and widespread travel. It is the latter that helps to explain why influenza tends to have a greater impact in the developed world, even though the influenza viruses themselves often have their origins in less developed regions of Asia. One good way to avoid influenza is to never travel on a common carrier such as an airplane, train, or bus! None of these diseases has yet been controlled (to the extent that, say, polio myelitis or smallpox have been controlled). But neither have they wrought the havoc caused by the bubonic plague in late medieval Europe. The sites where future concerns about the transmission of infectious diseases ought to be focused are the increasingly large urban concentrations in South America, Africa, and Asia.

Map 42 Global Scourges: Major Infectious Diseases

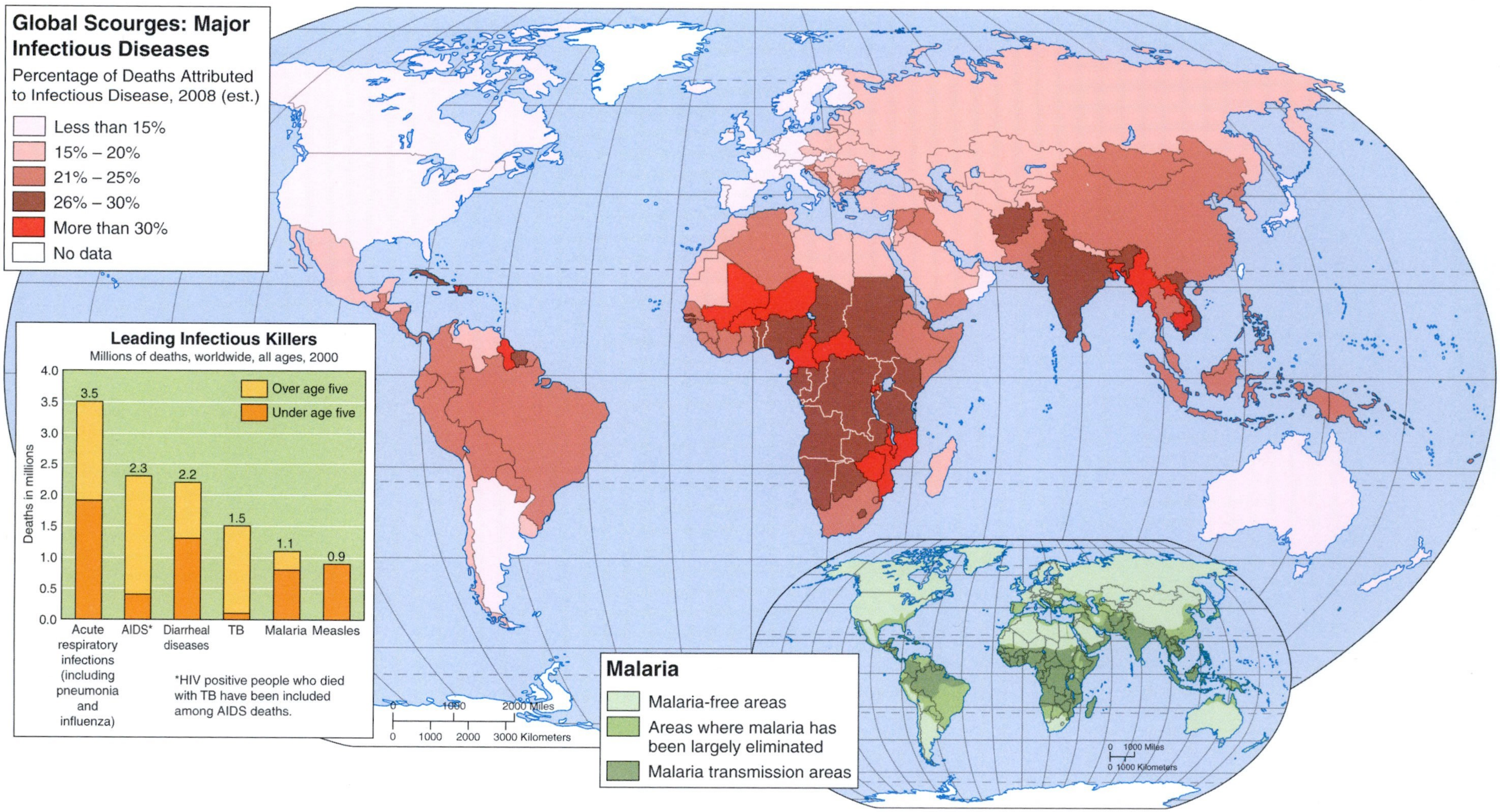

Infectious diseases are the world's leading cause of premature death and at least half of the world's population is, at any time, at risk of contracting an infectious disease. Although we often think of infectious diseases as being restricted to the tropical world (malaria, dengue fever), many if not most of them have attained global proportions. A major case in point is HIV/AIDS, which quite probably originated in Africa but has, over the last two decades, spread throughout the entire world. Major diseases of the nineteenth century, such as cholera and tuberculosis, are making a major comeback in many parts of the world, in spite of being preventable or treatable. Part of the problem with infectious diseases is that they tend to be associated with poverty (poor nutrition, poor sanitation, substandard housing, and so on) and, therefore, are seen as a problem of undeveloped countries, with the consequent lack of funding for prevention and treatment. Infectious diseases are also tending to increase because lifesaving drugs, such as antibiotics and others used in the fight against diseases, are losing their effectiveness as bacteria develop genetic resistance to them. The problem of global warming is also associated with a spread of infectious diseases as many disease vectors (certain species of mosquito, for example) are spreading into higher latitudes with increasingly warm temperatures and are spreading disease into areas where populations have no resistance to them. Infectious diseases have become something greater than simply a health issue of poor countries. They are now major social problems with potentially enormous consequences for the entire world.

Map 43 Adult HIV Prevalence 2005

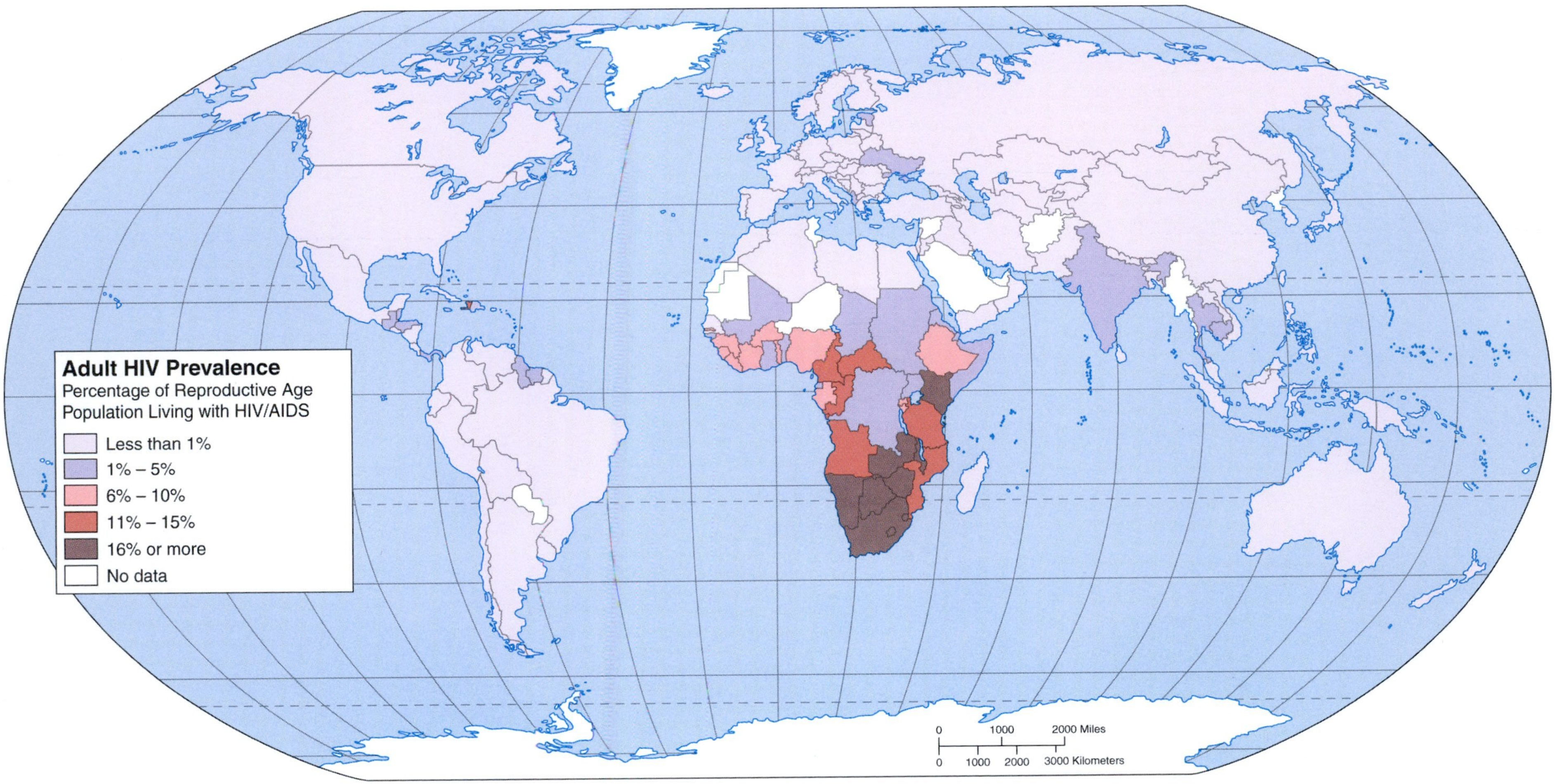

Of all the maps of the incidences of infectious disease, the global map of HIV/AIDS is the most dramatic as it impacts one continent so much more that it does other parts of the world. While there is no question that many African countries need to lower their growth rates, to do so by increasing death rates of children and young adults is dangerous. If the population profiles of countries like South Africa are plotted, they look like a bottle with a narrow bottom and a wide middle. In other words, the numbers of people who are going to be alive to be economically productive in the next generation are smaller than those in the older generations who may be forced to work well into what would be considered "old age" in order to support their sons and daughters. The stress placed by HIV/AIDS in some African countries in which more than 15% of the population are HIV positive is nearly incalculable. Try to imagine a United States with 20 million or more people between the ages of 15 and 45 dying of AIDS. And what becomes of the children orphaned by parents who have died of AIDS before seeing their offspring able to take care of themselves? In fact, many of these children are themselves born HIV positive. This is one of the world's most urgent health problems.

Map 44 The Index of Human Development

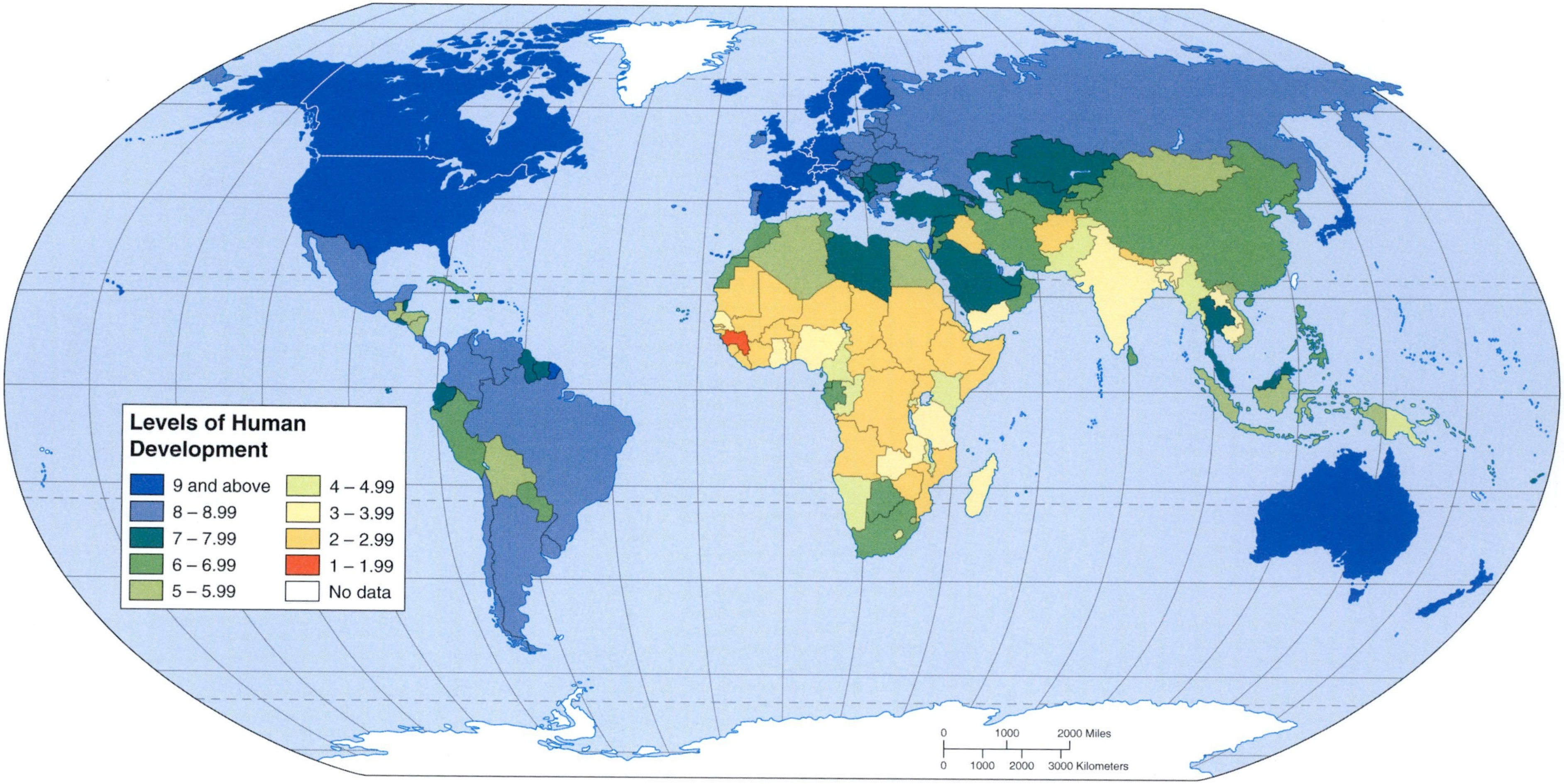

The development index upon which this map is based takes into account a wide variety of demographic, health, and educational data, including population growth, per capita gross domestic income, longevity, literacy, and years of schooling. The map reveals significant improvement in the quality of life in Middle and South America, although it is questionable whether the gains made in those regions can be maintained in the face of the dramatic population increases expected over the next 30 years. More clearly than anything else, the map illustrates the near-desperate situation in Africa and South Asia. In those regions, the unparalleled growth in population threatens to overwhelm all efforts to improve the quality of life. In Africa, for example, the population is increasing by 20 million persons per year. With nearly 45 percent of the continent's population aged 15 years or younger, this growth rate will accelerate as the women reach child-bearing age. Africa, along with South Asia, faces the very difficult challenge of providing basic access to health care, education, and jobs for a rapidly increasing population. The map also illustrates the striking difference in quality of life between those who inhabit the world's equatorial and tropical regions and those fortunate enough to live in the temperate zones, where the quality of life is significantly higher.

Map 45 Demographic Stress: The Youth Bulge

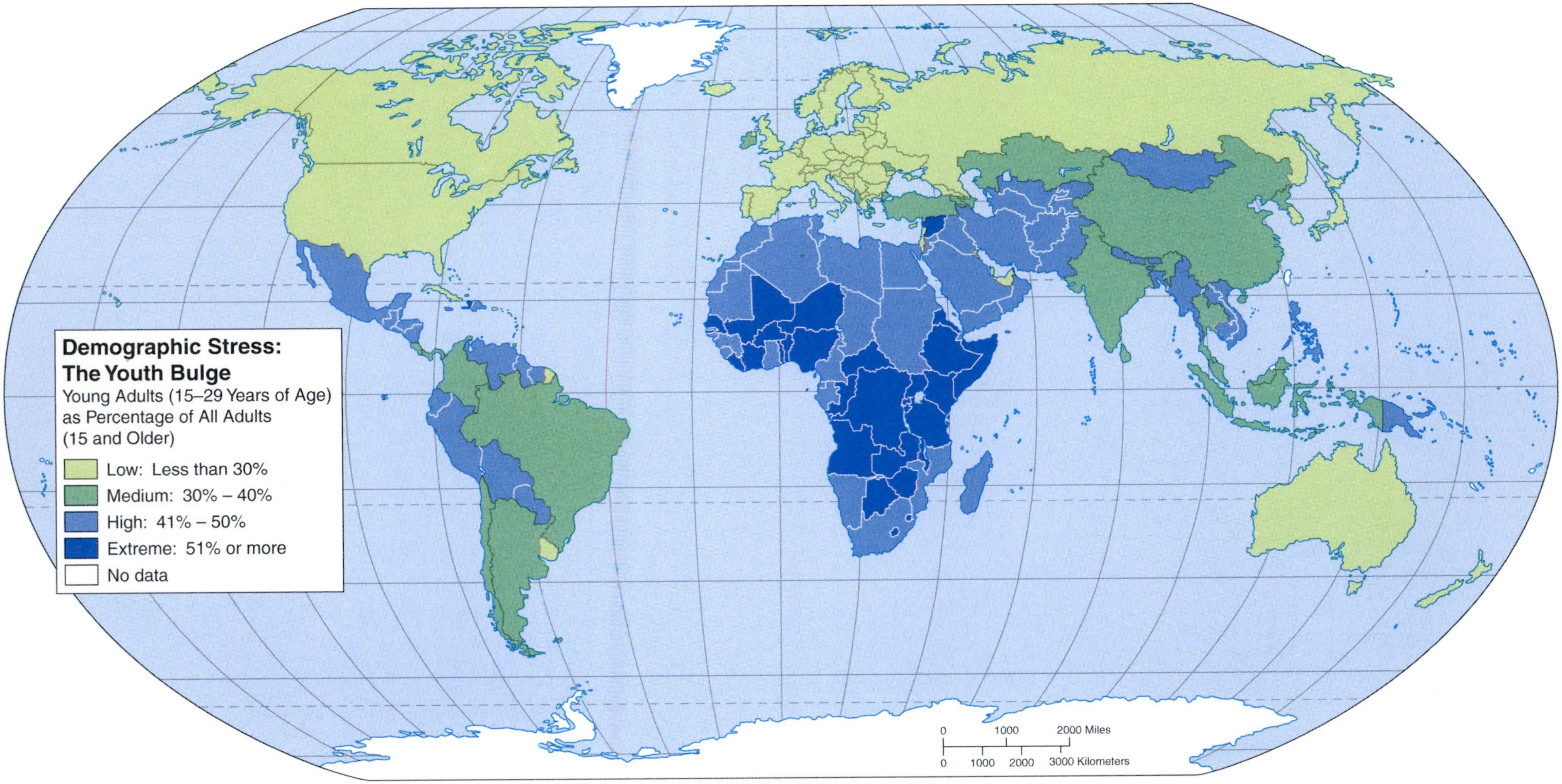

One of the greatest stresses of the demographic transition is when the death rate drops as the result of better public health and sanitation and the birth rate remains high for the same reasons it has always been high in traditional, agricultural societies: (1) the need for enough children to supply labor, which is one of the few ways to increase agricultural production in a non-mechanized agricultural system, and (2) the need for enough children to offset high infant and child mortality rates so that some children will survive to take care of parents in their old age. With declining death rates and steady birth rates, population growth skyrockets—particularly among the youngest cohorts of a population (between the ages of birth and 15). While this is, on the one hand, a demographic benefit since it increases the size of the labor force in the next generation and thereby helps to accelerate economic growth, it also means more people of child-bearing age in the next generation and, hence, greater numbers of births, which continue to swell the population in the youngest, most vulnerable, and most dependent portion of the population. The literature on population and conflict suggests that the larger the percentage of a population below the age of 25, the greater the chance for political violence and warfare. The suicide bombers of the modern Middle East are not oldsters.

Map 46 Demographic Stress: Rapid Urban Growth

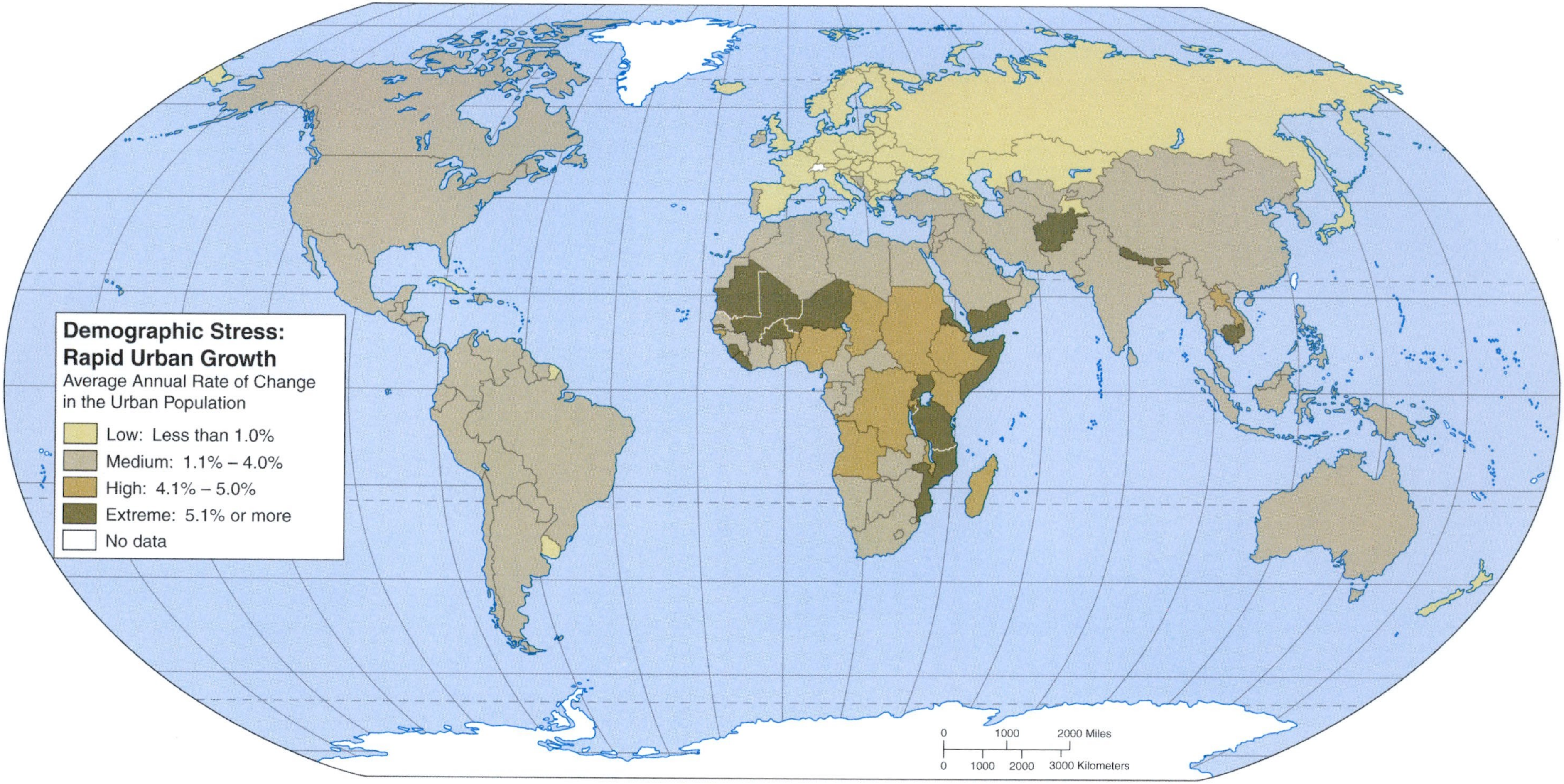

The trend toward urbanization—an increasing percentage of a country's population living in a city—is normally a healthy, modern trend. But in many of the world's developing countries, increasing urbanization is the consequence of high physiologic population densities (too many people for the available agricultural land) and the flight of poorly educated, untrained rural poor to the cities where they hope (often in vain) to find employment. High urbanization exists in many of the world's poorest countries where the urban poor live in conditions that make the worst living conditions in the inner cities of developed countries look positively luxurious. In the urban slums of South America, Africa, and South Asia, millions of people live in temporary housing of cardboard and flattened aluminum cans, with no public services such as water, sewage, or electricity. In Africa, where only 40% of the population is urbanized (in comparison with more than 90% in North America and Europe), the population of urban poor is greater than the total urban population of the United States and the European Union. This represents an increasingly destabilizing element of modern urban societies.

Map 47 Demographic Stress: Competition for Cropland

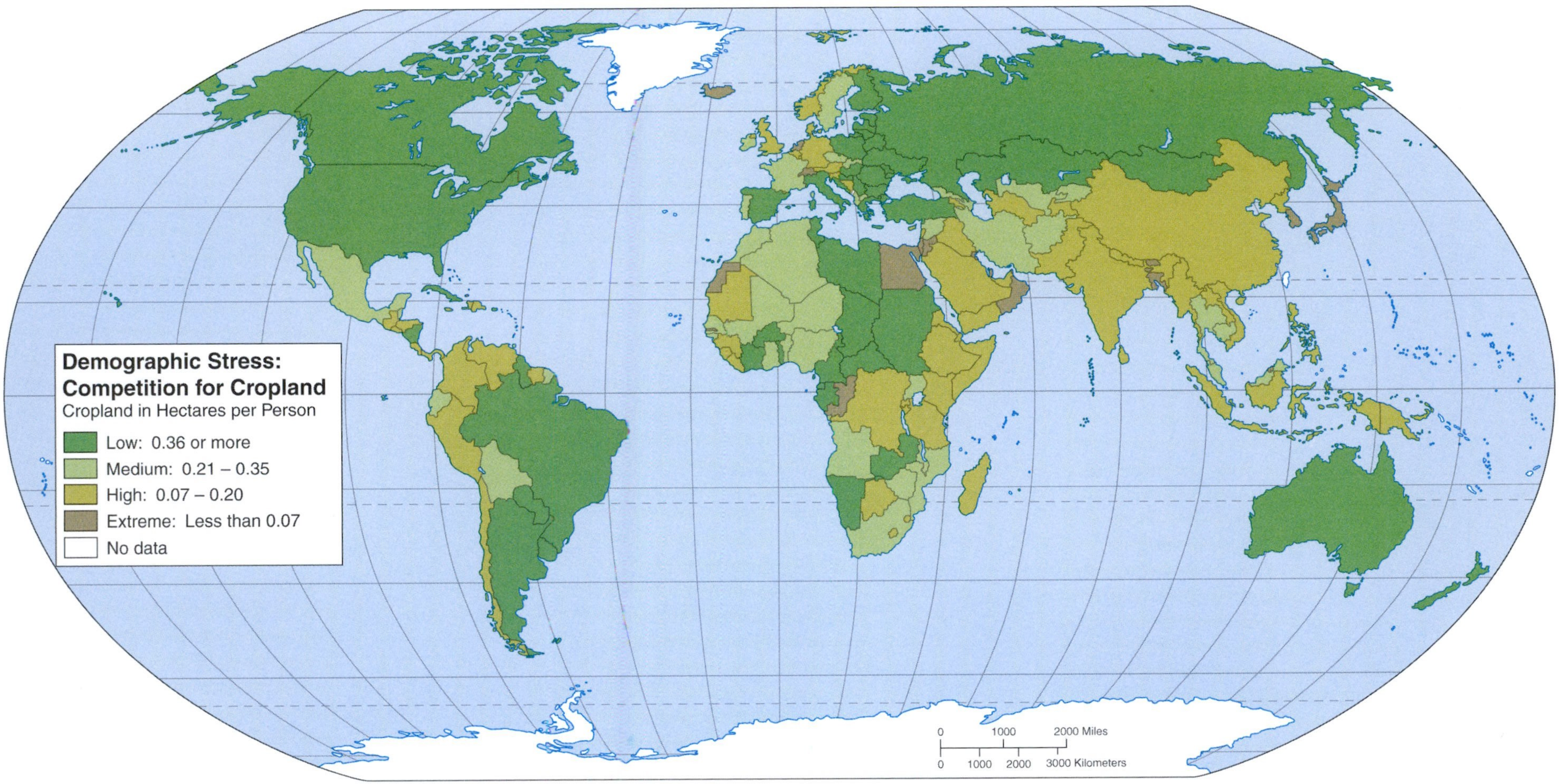

Many of the world's countries have reached the point where available acres or hectares in cropland are less than the numbers of people wishing to occupy them. For developed countries, who pay for agricultural imports with industrial exports, this is not an alarming trend and therefore countries like Germany or Italy (where the ratio between cropland and farmers is very low) have little to worry about. But in countries in South America, Africa, and South and East Asia, the trend toward more farmers and less available land *is* an alarming trend. In these developing regions, farmers depend on their crops for a relatively meager subsistence diet and—if they are lucky—a few bushels of rice or corn to take to the local market to sell or exchange for the small surpluses of other farmers. Despite the broad global trends toward urbanization and more productive agriculture (and this generally means mechanized agriculture), farm occupation and subsistence cropping remain mainstays of the economy in Africa south of the Sahara, in much of western South America, and in much of South and East Asia. Here, the increasingly small margin between the numbers of farmers and the amount of available farmland can lead to conflict among tribal communities or even among members of a single family.

Map 48 Demographic Stress: Competition for Fresh Water

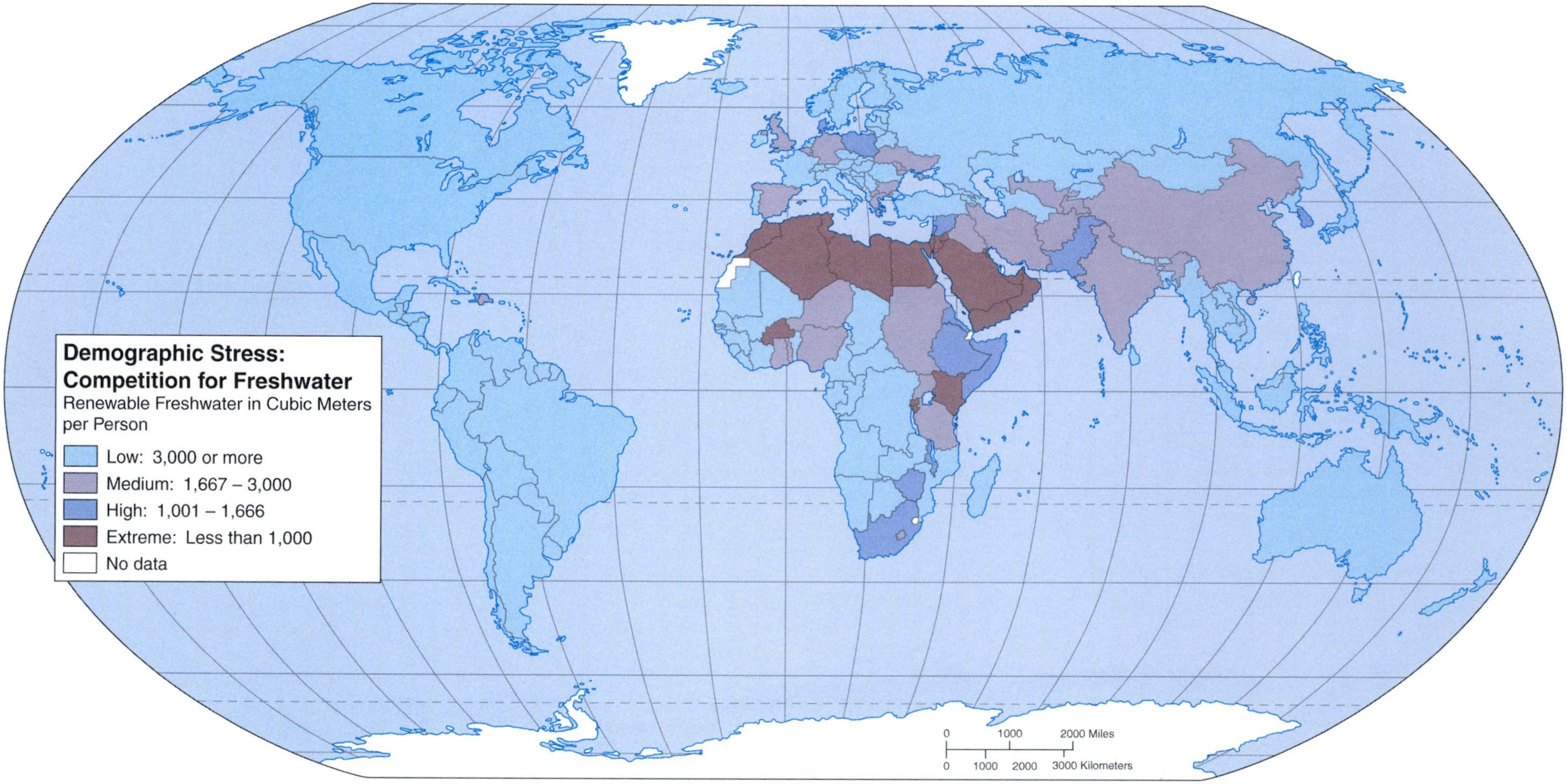

As rates of urbanization have increased, the competition between city dwellers and farmers for water has also increased. And as the population of farmers relative to the available surface or ground water supply has accelerated, so rural competition for access to fresh water for irrigation has increased. Obviously, this trend is most obvious in the world's drier regions. On this map, the greatest stress factors relating to water availability are in North and East Africa, the Middle East, and Central and East Asia. The current conflict between Israelis and Palestinians in the West Bank goes far beyond religion or politics: Some is the result of the more affluent Israeli farmers being able to drill wells to tap ground water, which lowers the water table and causes previously accessible surface wells in Palestinian villages to go dry. And part of the horrific Hutu-Tutsi civil war in Rwanda and the taking of European farmlands by the current government of Zimbabwe have more to do with the need for access to fresh water than to racial or ethnic differences. Given that fresh water is, for all practical purposes, a nonrenewable resource, more conflicts over access to water are going to erupt in the future.

Map 49 Demographic Stress: Death in the Prime of Life

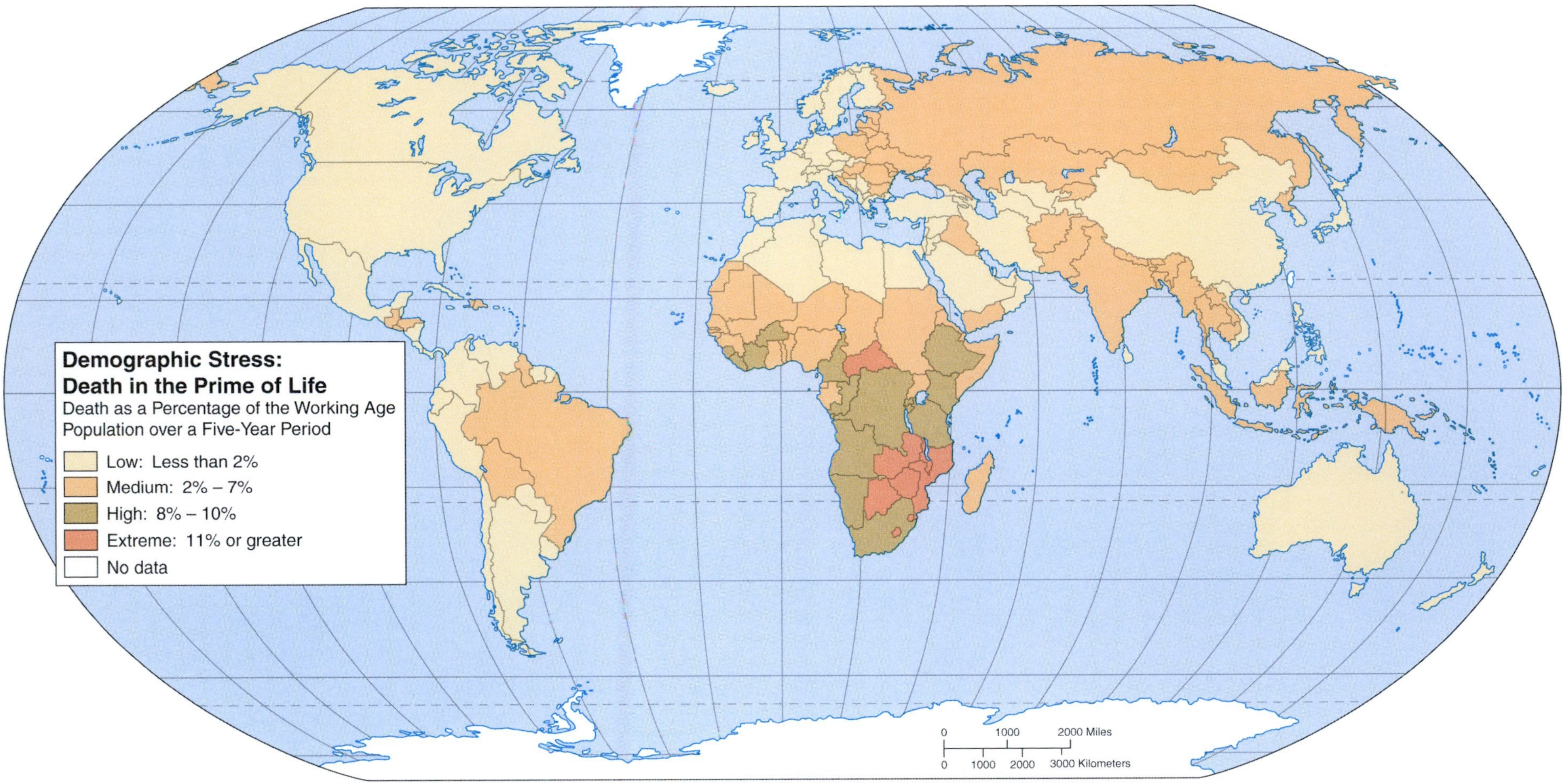

Most infectious diseases lay waste to the oldest and the youngest sectors of a population, leaving the healthier working-age population able to reproduce itself and to continue to grow an economy. Modern infectious diseases—most particularly HIV/AIDS—impacts not the youngest or oldest but those in the prime of life, in their most productive and reproductive years. Indeed, medical data from the United Nations show that 90% of HIV/AIDS-related deaths occur in people of working age. Some of the most severely impacted countries in southern Africa lose between 10% and 20% of their working age population every 5 years. Rates in the developed world are from 0.10% to 0.02% of this level. A prominent fact of life in much of the developing world—particularly in sub-Saharan Africa—is the long-distance labor migration. In this pattern, a man leaves his wife and family for extended periods of time (10 or more months of the year) to work in mines, on docks, as sailors. These men are highly vulnerable to the acquisition of HIV, which is then transmitted to wives when the husband returns home on his annual visit. Thus both the male and the female portion of the working-age population are impacted by early mortality from AIDS. None of this bodes well for economic development in those affected areas of the world.

Map 50 Demographic Stress: Interactions of Demographic Stress Factors

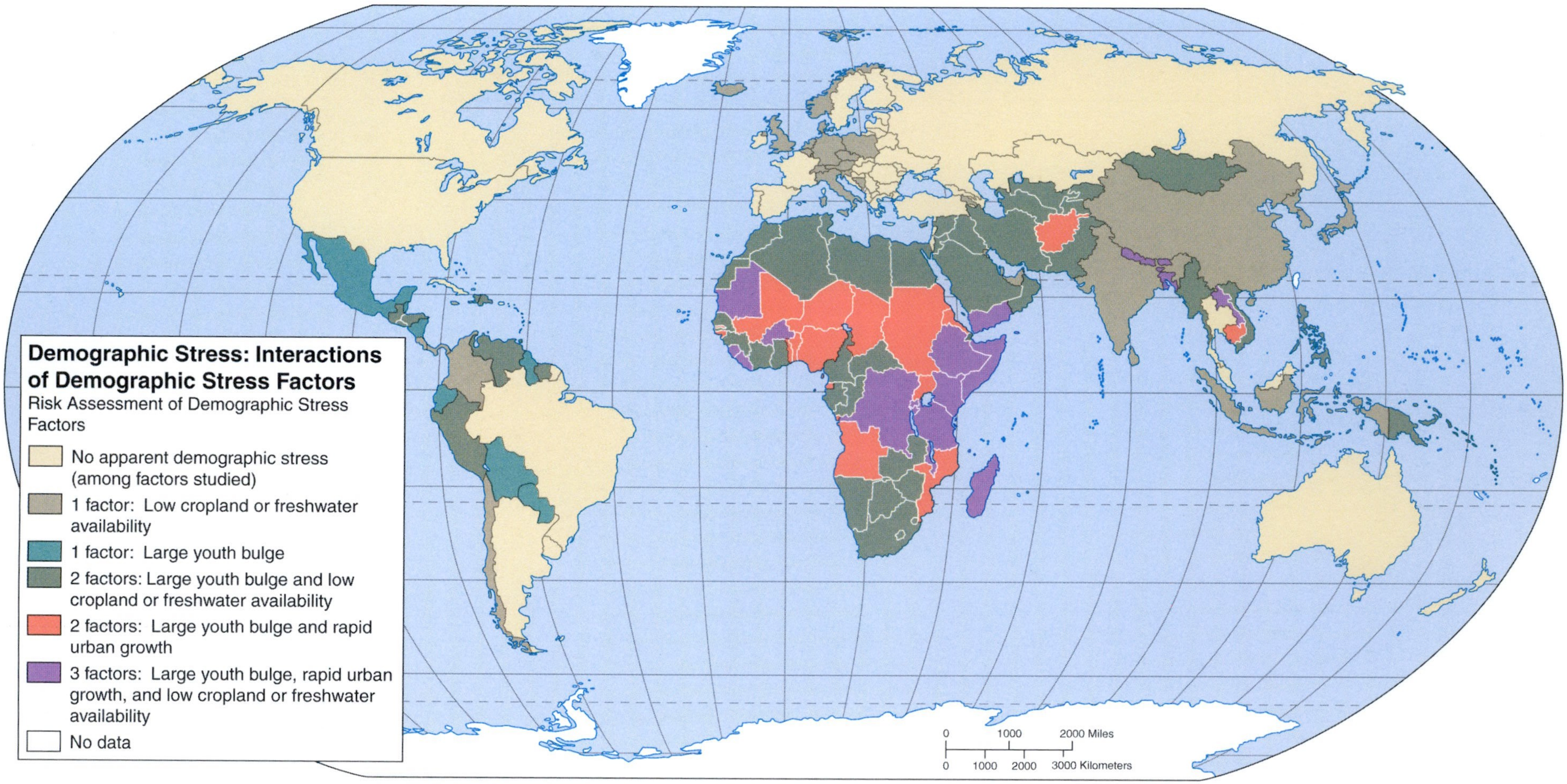

We have persisted in viewing internal and international conflicts as conflicts produced by religious differences, the acceptance or rejection of "freedom" or "democracy," or even, as argued by a prominent historian, a "Clash of Civilizations." While religious, political, or historical differences cannot be ignored in interpreting the causes of conflict, neither can simple demographic stresses produced by too many new urban dwellers without jobs or hopes of jobs, by too many farmers and too little farmland, by too many deaths among people who should be in the most productive portions of their lives. A massive study by Population Action International, from which these last half-dozen or so maps have been drawn, shows quite clearly that countries suffering from multiple demographic risk factors are, were, and will be much more prone to civil conflict than countries with only one or two demographic risk factors. A look at this map provides a predictor of future civil unrest, violence, and even civil war.

Unit IV

Global Economic Patterns

Rich and Poor Countries: Gross National Income

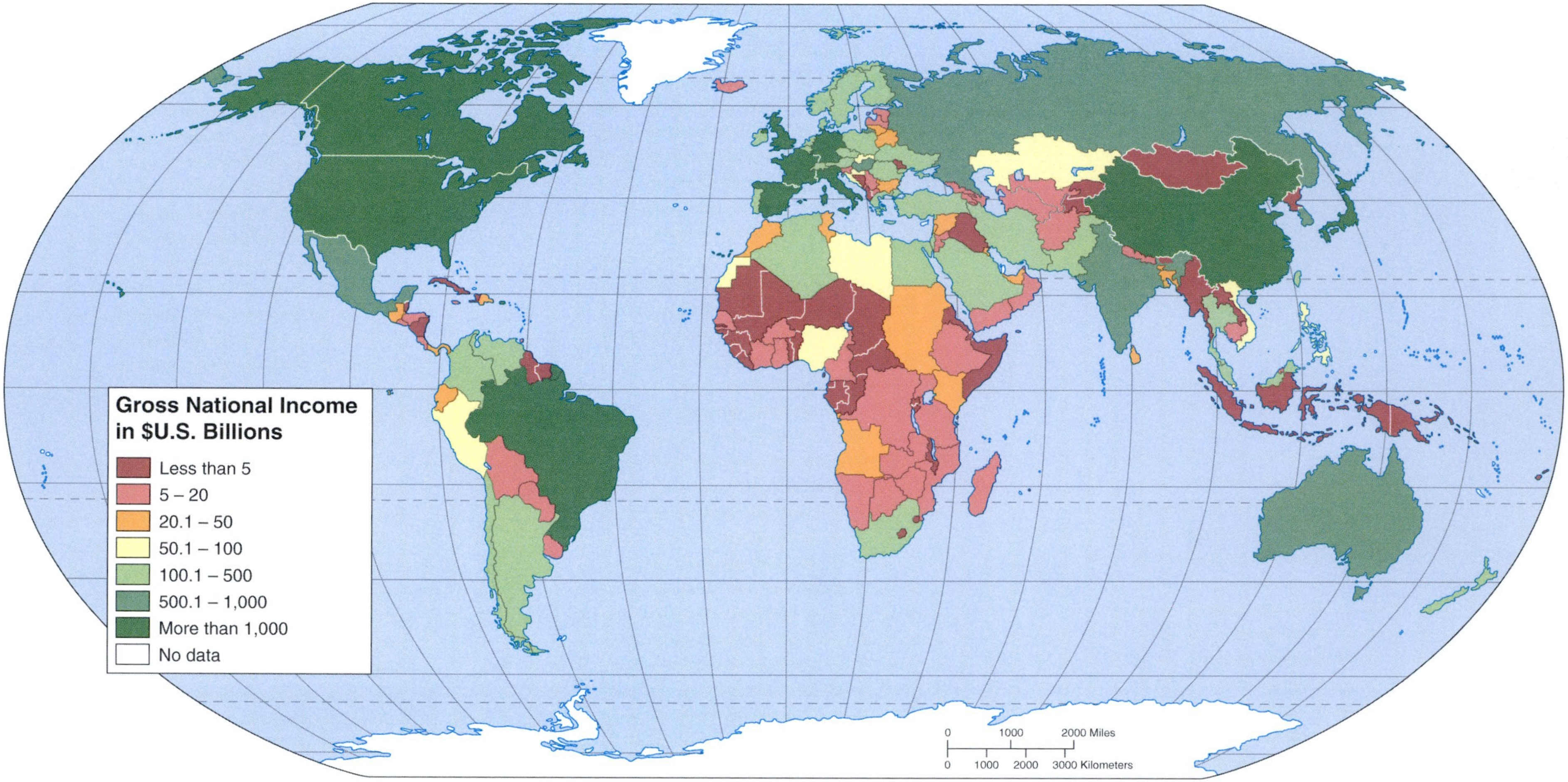

Gross National Income (GNI) is the broadest measure of national income and measures the total claims of a country's residents to all income from domestic and foreign products during a year. Although GNI is often misleading and commonly incomplete, it is often used by economists, geographers, political scientists, policy makers, development experts, and others not only as a measure of relative well-being but also as an instrument of assessing the effectiveness of economic and political policies. What is wrong with GNI? First of all, it does not take into account a number of real economic factors, such as environmental deterioration, the accumulation or degradation of human and social capital, or the value of household work. Yet in spite of these deficiencies, GNI is still a reasonable way to assess the relative wealth of nations: the vast differences in wealth that separate the poorest countries from the richest. One of the more striking features of the map is the evidence it presents that such a small number of countries possess so many of the world's riches (keeping in mind that GNI provides no measure of the distribution of wealth within a country).

Map 52 Gross National Income Per Capita

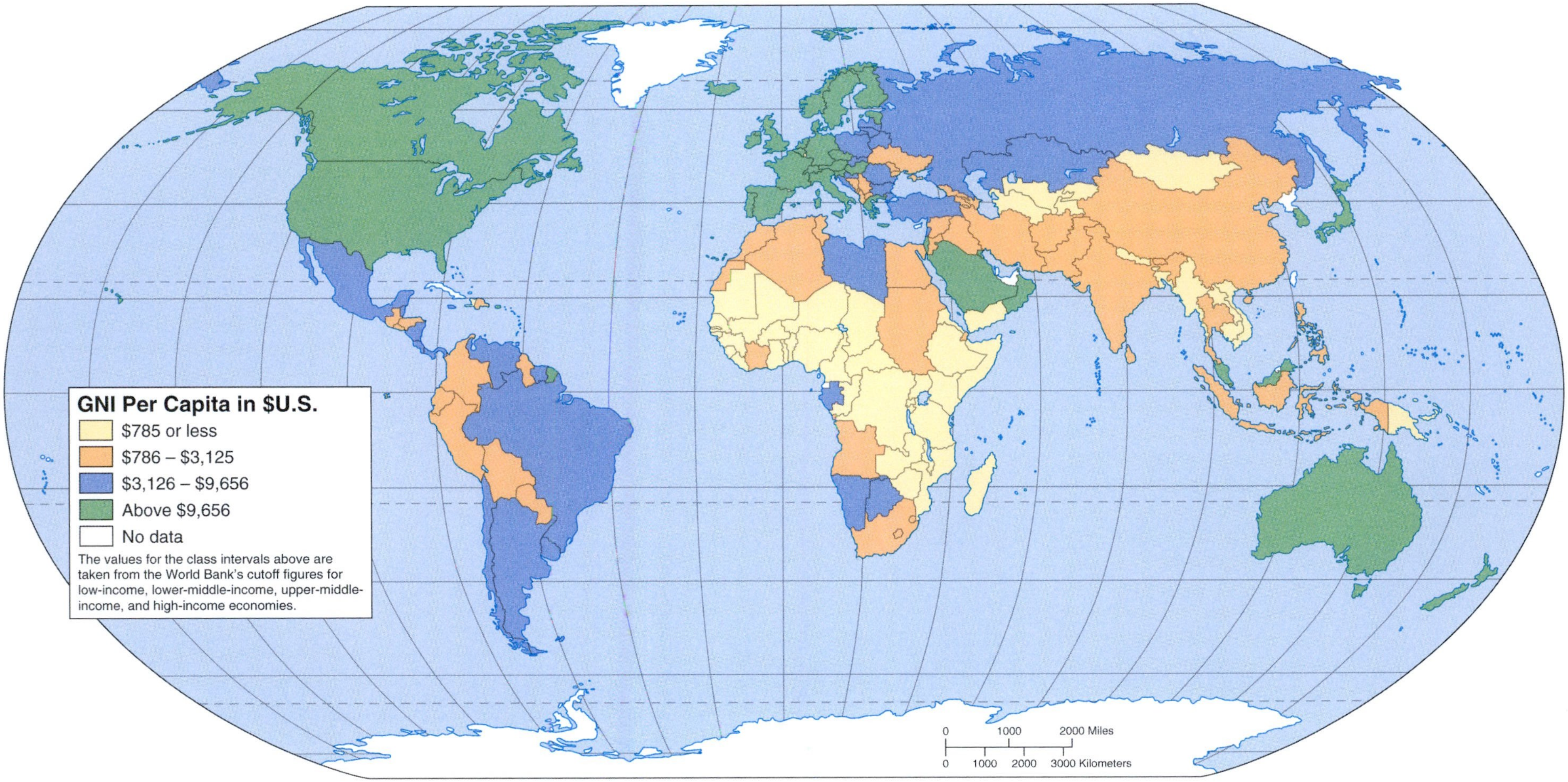

Gross National Income (GNI) in either absolute or per capita form should be used cautiously as a yardstick of economic strength because it does not measure the distribution of wealth among a population. There are countries (most notably, the oil-rich countries of the Middle East) where per capita GNI is high but where the bulk of the wealth is concentrated in the hands of a few individuals, leaving the remainder in poverty. Even within countries in which wealth is more evenly distributed (such as those in North America or Western Europe), there is a tendency for dollars or pounds sterling or euros to concentrate in the bank accounts of a relatively small percentage of the population. Yet the maldistribution of wealth tends to be greatest in the less developed countries, where the per capita GNI is far lower than in North America and Western Europe, and poverty is widespread. In fact, a map of GNI per capita offers a reasonably good picture of comparative economic well-being. It should be noted that a low per capita GNI does not automatically condemn a country to low levels of basic human needs and services. There are a few countries, such as Costa Rica and Sri Lanka, that have relatively low per capita GNI figures but rank comparatively high in other measures of human well-being, such as average life expectancy, access to medical care, and literacy.

Map 53 Economic Growth: GDP Change Per Capita

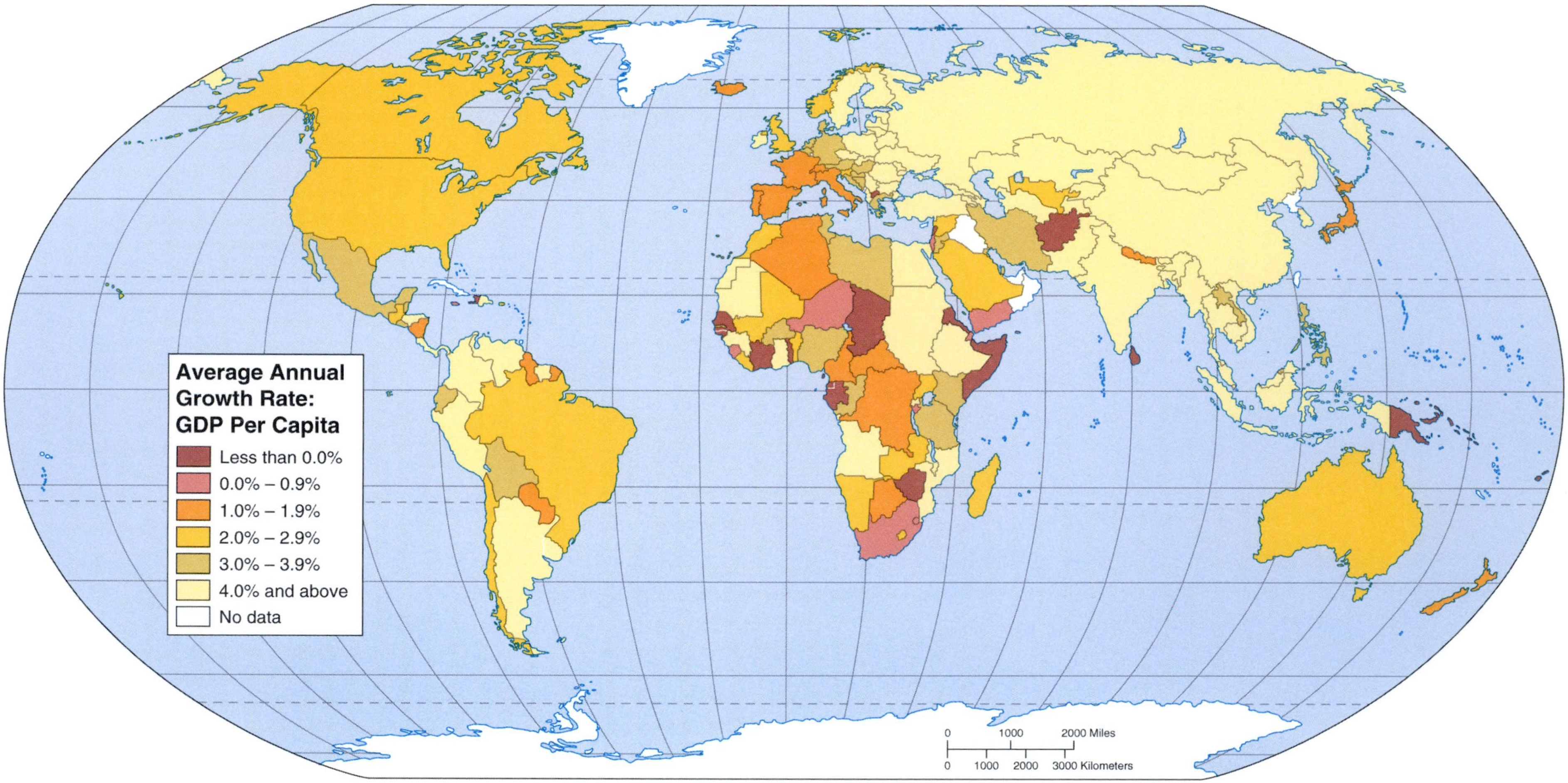

Gross Domestic Product or GDP is Gross National Income (GNI) less receipts of primary income from foreign sources. While the calculations of GDP growth per capita are complex, the growth rate is considered by the World Bank and international economists to be a particularly good measure of economic growth. One of the worldwide tendencies measured by GDP growth per capita is for continued economic development in Africa, and in South, Southeast, and East Asia where GDP grew at rates higher than the growth rates of "richer" countries in Europe and North and South America. This should not necessarily be viewed as a case of the poor catching up with the rich; in fact, it shows the huge impact that even relatively small production increases will have in countries with small GNIs and GDPs. Nevertheless, in spite of the continuing low economic growth through most of sub-Saharan Africa, the GDP growth rate of some of the world's poorer countries is an encouraging trend.

Map 54 Total Labor Force

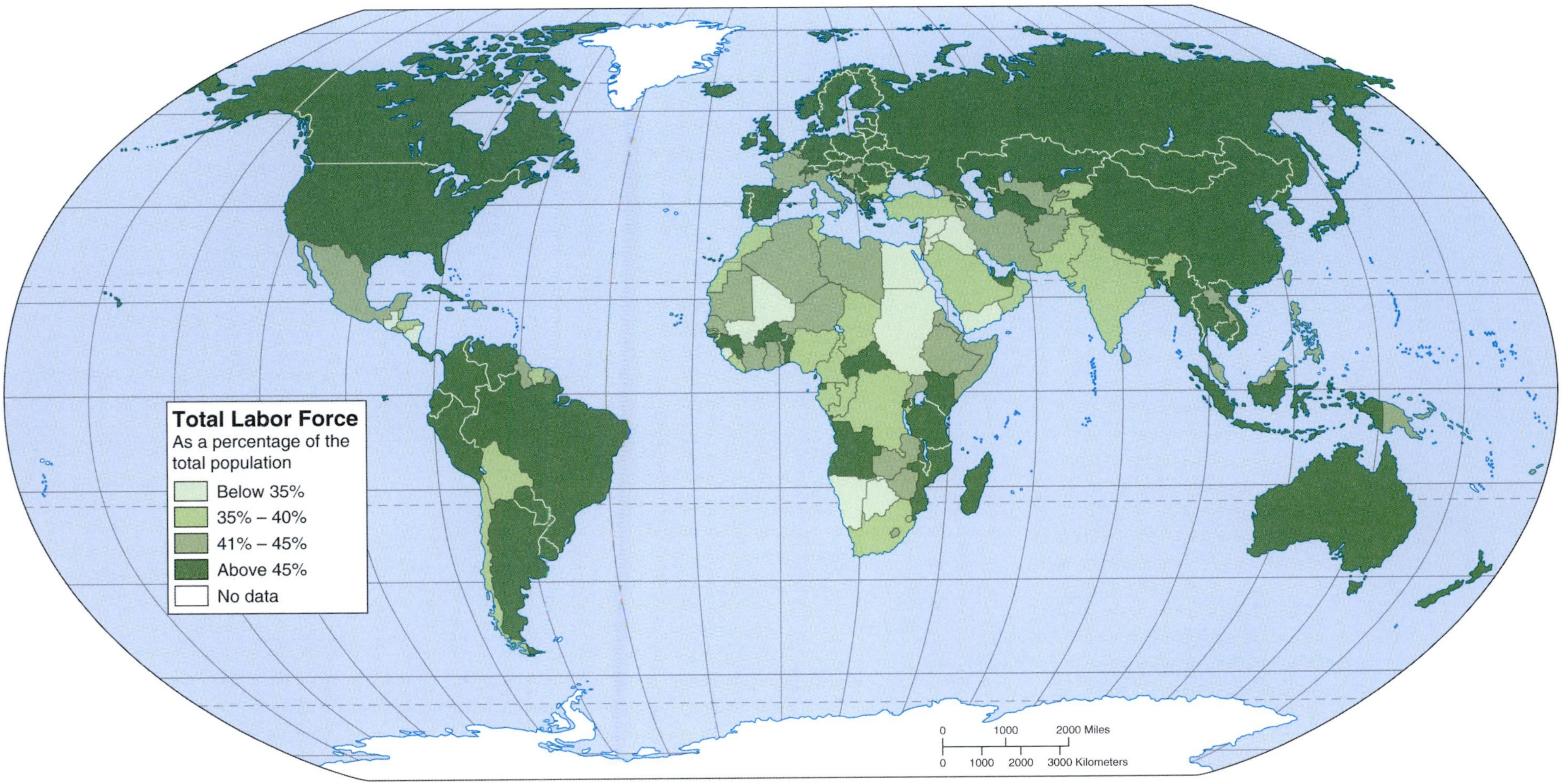

The term labor force refers to the economically active portion of a population, that is, all people who work or are without work but are available for and are seeking work to produce economic goods and services. The total labor force thus includes both the employed and the unemployed (as long as they are actively seeking employment). Labor force is considered a better indicator of economic potential than employment/unemployment figures, since unemployment figures will include experienced workers with considerable potential who are temporarily out of work. Unemployment figures will also incorporate persons seeking employment for the first time (many recent college graduates, for example). Generally, countries with higher percentages of total population within the labor force will be countries with higher levels of economic development. This is partly a function of levels of education and training and partly a function of the age distribution of populations. In developing countries, substantial percentages of the total population are too young to be part of the labor force. Also in developing countries a significant percentage of the population consists of women engaged in household activities or subsistence cultivation. These people seldom appear on lists of either employed or unemployed seeking employment and are the world's forgotten workers.

Map 55 Employment by Economic Activity

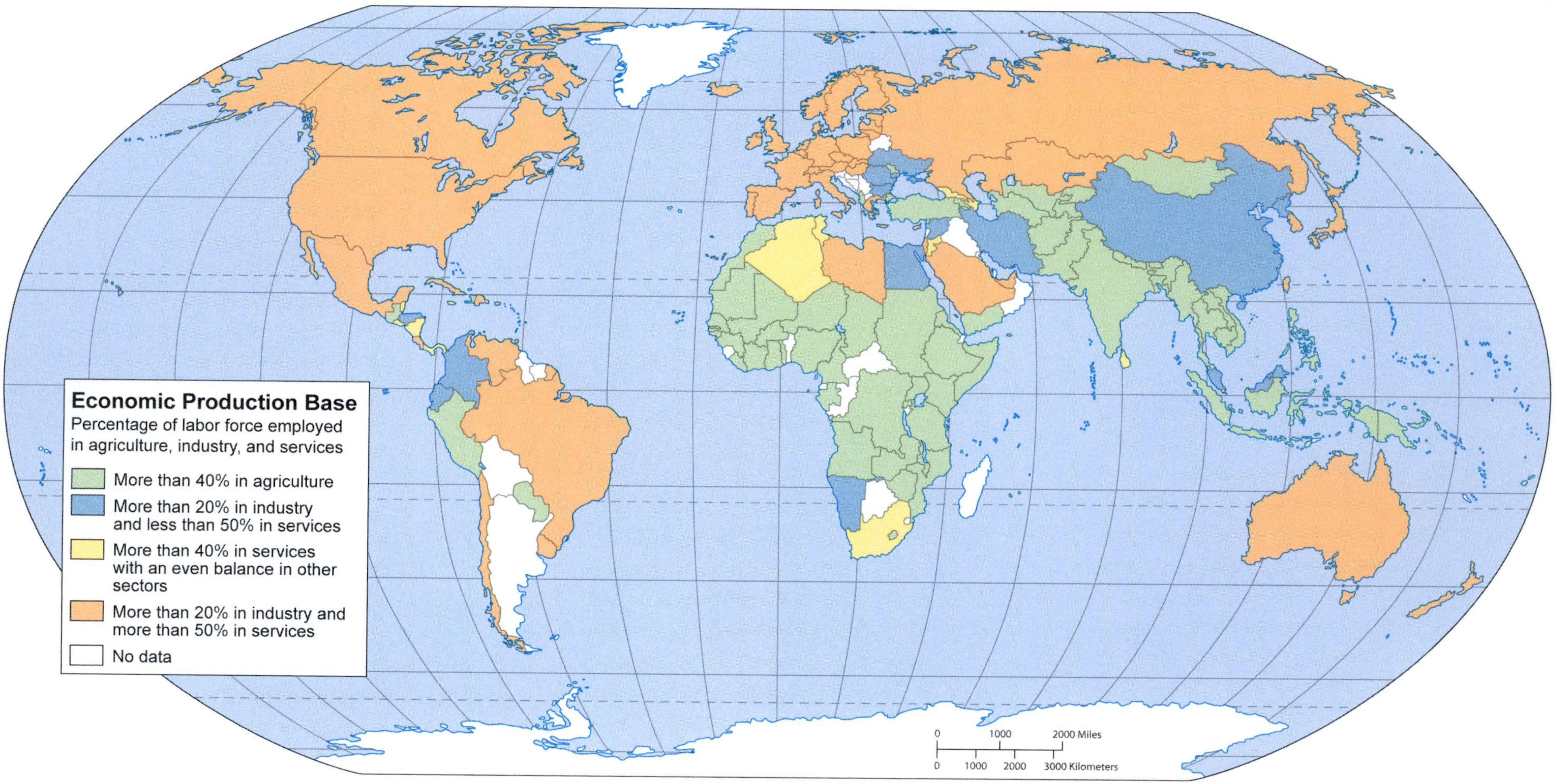

The employment structure of a country's population is one of the best indicators of the country's position on the scale of economic development. At one end of the scale are those countries with more than 40 percent of their labor force employed in agriculture. These are almost invariably the least developed, with high population growth rates, poor human services, significant environmental problems, and so on. In the middle of the scale are two types of countries: those with more than 20 percent of their labor force employed in industry and those with a fairly even balance among agricultural, industrial, and service employment but with at least 40 percent of their labor force employed in service activities. Generally, these countries have undergone the industrial revolution fairly recently and are still developing an industrial base while building up their service activities. This category also includes countries with a disproportionate share of their economies in service activities primarily related to resource extraction. On the other end of the scale from the agricultural economies are countries with more than 20 percent of their labor force employed in industry and more than 50 percent in service activities. These countries are, for the most part, those with a highly automated industrial base and a highly mechanized agricultural system (the "postindustrial," developed countries). They also include, particularly in Middle and South America and Africa, industrializing countries that are also heavily engaged in resource extraction as a service activity.

Map 56 Economic Output Per Sector

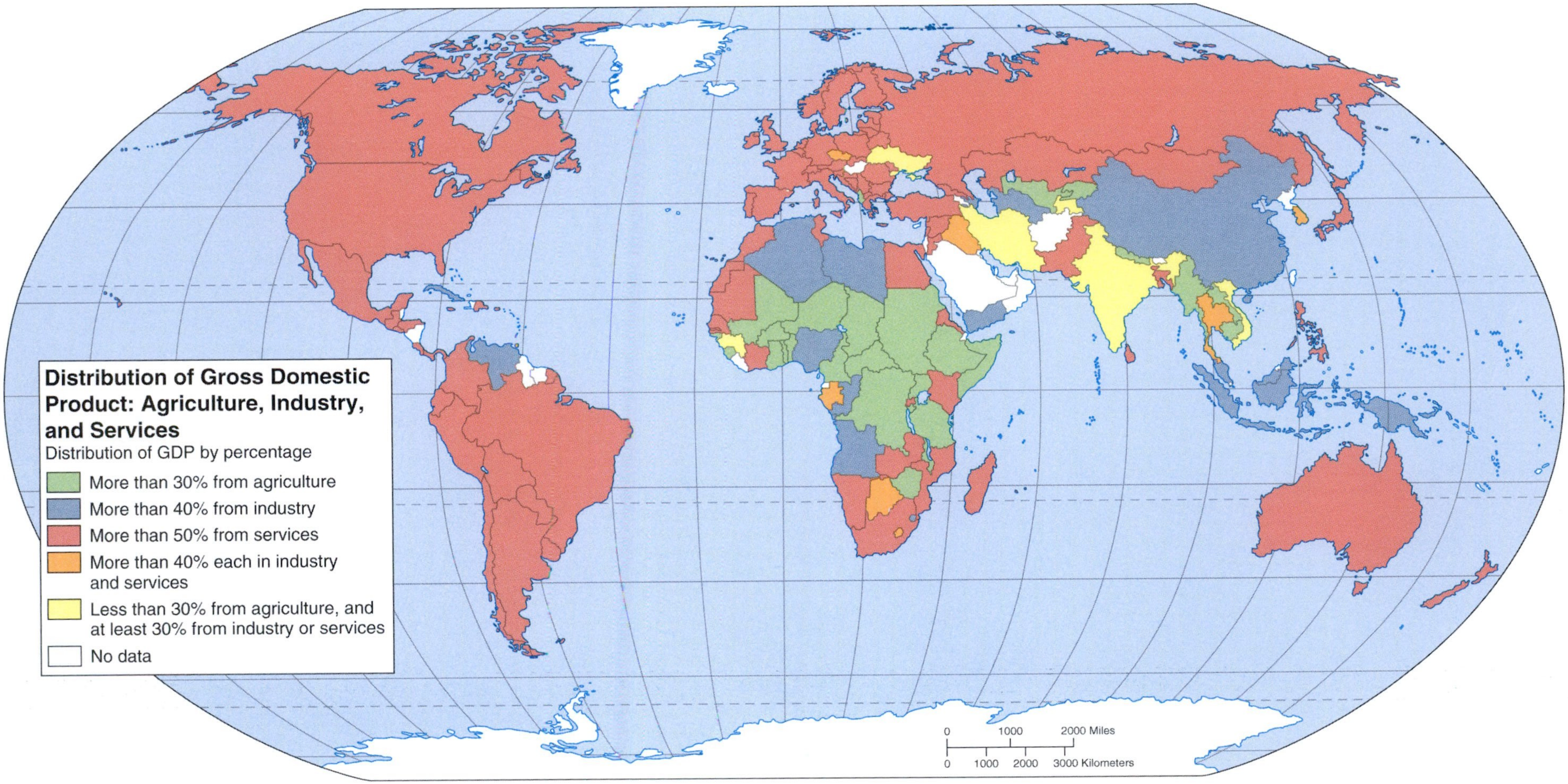

The percentage of the gross domestic product (the final output of goods and services produced by the domestic economy, including net exports of goods and nonfactor—nonlabor, noncapital—services) that is devoted to agricultural, industrial, and service activities is considered a good measure of the level of economic development. In general, countries with more than 40 percent of their GDP derived from agriculture are still in a *colonial dependency* economy—that is, raising agricultural goods primarily for the export market and dependent upon that market (usually the richer countries). Similarly, countries with more than 40 percent of GDP devoted to both agriculture and services often emphasize resource extractive (primarily mining and forestry) activities. These also tend to be *colonial dependency* countries, providing raw materials for foreign markets. Countries with more than 40 percent of their GDP obtained from industry are normally well along the path to economic development. Countries with more than half of their GDP based on service activities fall into two ends of the development spectrum. On the one hand are countries heavily dependent upon both extractive activities and tourism and other low-level service functions. On the other hand are countries that can properly be termed *postindustrial*: they have already passed through the industrial stage of their economic development and now rely less on the manufacture of products than on finance, research, communications, education, and other service-oriented activities.

Map 57 Agricultural Production Per Capita

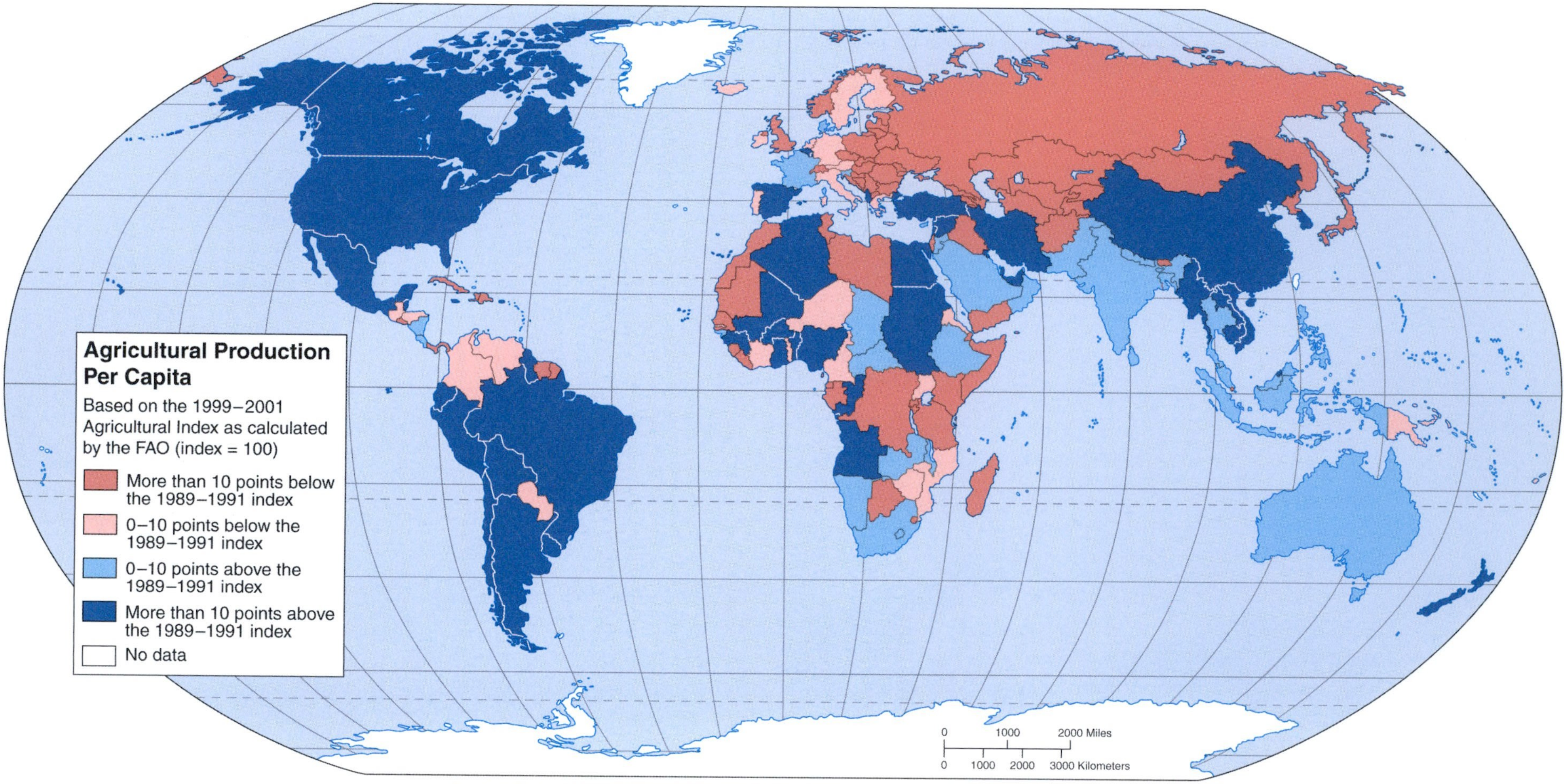

Agricultural production includes the value of all crops and livestock products originating within a country for the base year of 2002. The index value portrays the disposable output (after deductions for livestock feed and seed for planting) of a country's agriculture in comparison with the base period 1989–1991. Thus, the production values show not only the relative ability of countries to produce food but also show whether or not that ability has increased or decreased over a 10-year period. In general, global food production has kept up with or very slightly exceeded population growth. However, there are significant regional variations in the trend of food production keeping up with or surpassing population growth. For example, agricultural production in Africa and in Middle America has fallen, while production in South America, Asia, and Europe has risen. In the case of Africa, the drop in production reflects a population growing more rapidly than agricultural productivity. Where rapid increases in food production per capita exist (as in certain countries in South America, Asia, and Europe), most often the reason is the development of new agricultural technologies that have allowed food production to grow faster than population. In much of Asia, for example, the so-called Green Revolution of new, highly productive strains of wheat and rice made positive index values possible. Also in Asia, the cessation of major warfare allowed some countries (Cambodia, Laos, and Vietnam) to show substantial increases over the 1989–1991 index. In some cases, a drop in production per capita reflects government decisions to limit production in order to maintain higher prices for agricultural products. The United States and Japan fall into this category.

Map 58 Exports of Primary Products

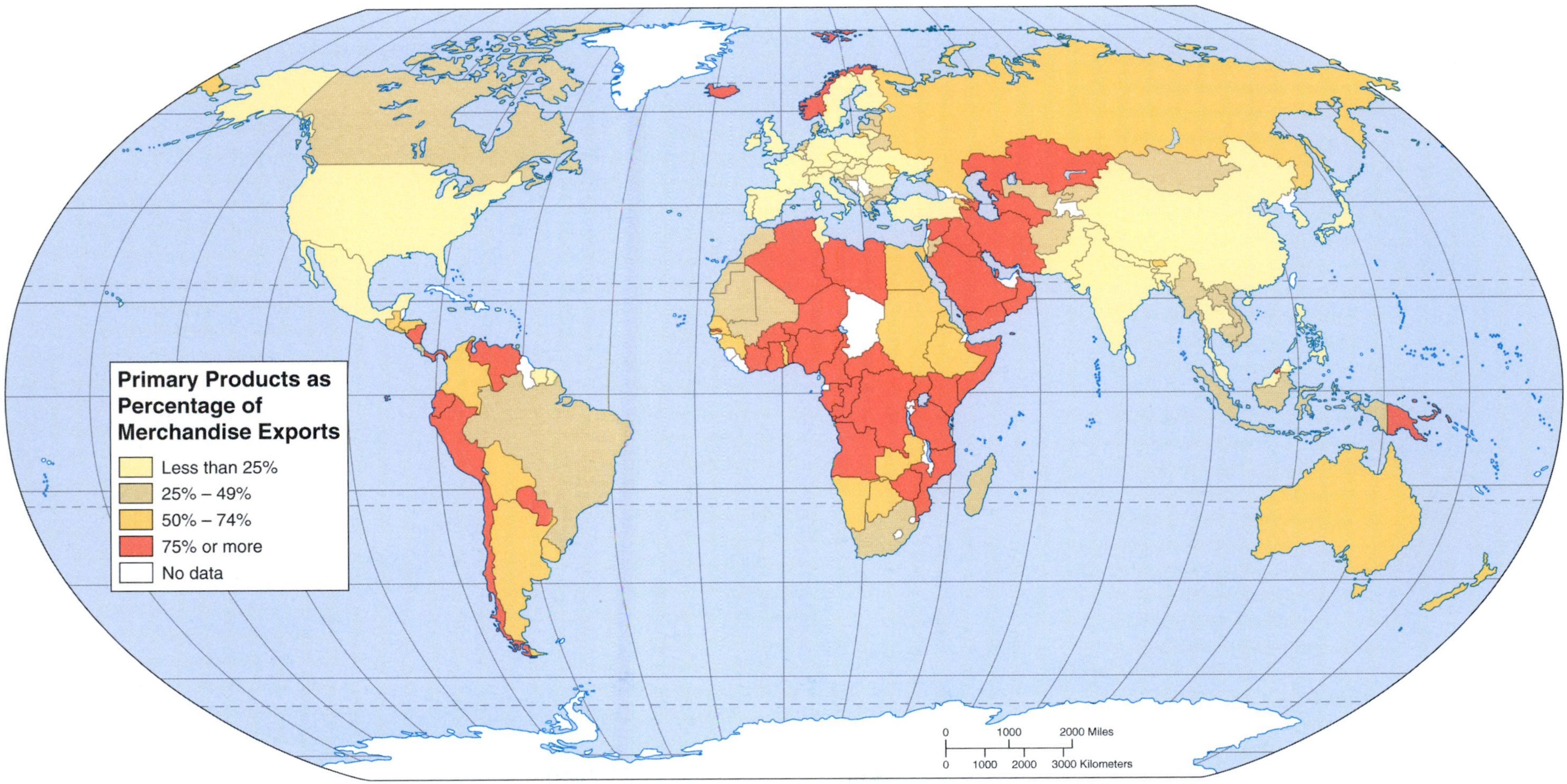

Primary products are those that require additional processing before they enter the consumer market: metallic ores that must be converted into metals and then into metal products such as automobiles or refrigerators; forest products such as timber that must be converted to lumber before they become suitable for construction purposes; and agricultural products that require further processing before being ready for human consumption. It is an axiom in international economics that the more a country relies on primary products for its export commodities, the more vulnerable its economy is to market fluctuations. Those countries with only primary products to export are hampered in their economic growth. A country dependent on only one or two products for export revenues is unprotected from economic shifts, particularly a changing market demand for its products. Imagine what would happen to the thriving economic status of the oil-exporting states of the Persian Gulf, for example, if an alternate source of cheap energy were found. A glance at this map shows that those countries with the lowest levels of economic development tend to be concentrated on primary products and, therefore, have economies that are especially vulnerable to economic instability.

Map 59 Dependence on Trade

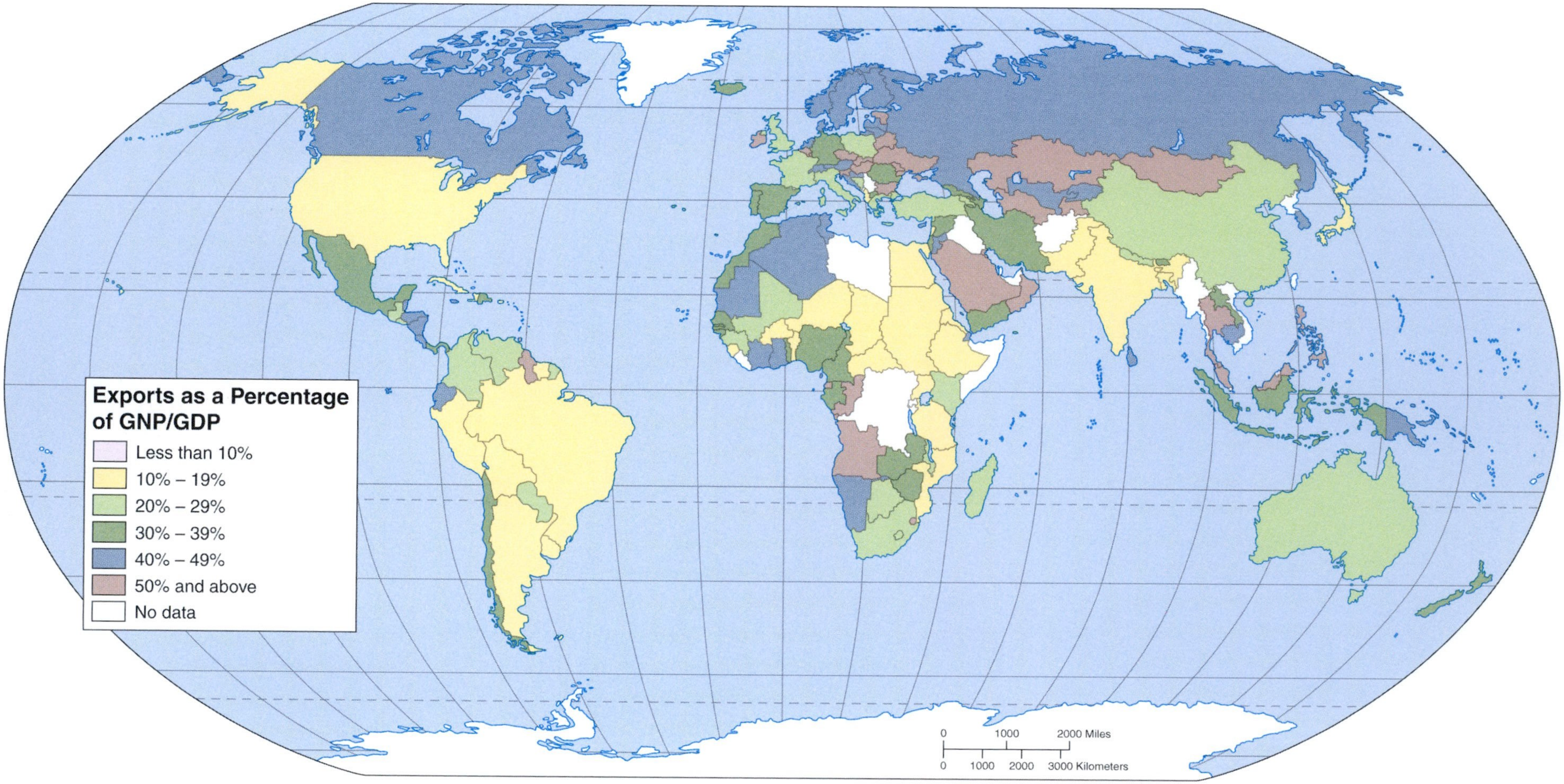

As the global economy becomes more and more a reality, the economic strength of virtually all countries is increasingly dependent upon trade. For many developing nations, with relatively abundant resources and limited industrial capacity, exports provide the primary base upon which their economies rest. Even countries like the United States, Japan, and Germany, with huge and diverse economies, depend on exports to generate a significant percentage of their employment and wealth. Without imports, many products that consumers want would be unavailable or more expensive; without exports, many jobs would be eliminated.

Map 60 The Indebtedness of States

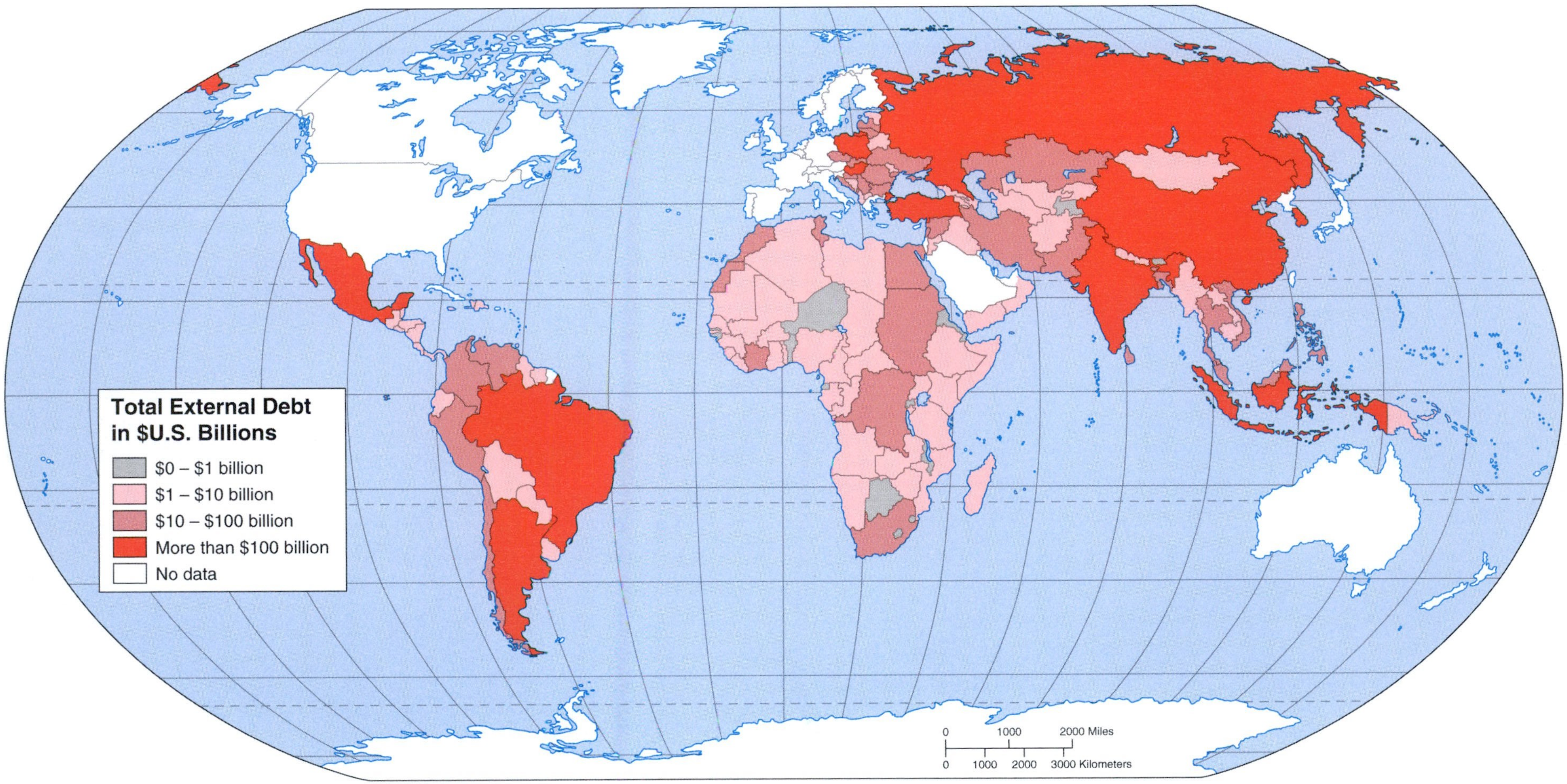

Many governments spend more on a wide variety of services and activities than they collect in taxes and other revenues. In order to finance this deficit spending, governments borrow money—often from banks or other investors outside their country. Repayment of these debts, or even meeting interest payments on them, often means expending a country's export income—in other words, exchanging a country's wealth in production or, more often, resources, for debt service. Where the debt is external, as it is in most developing countries, governments become more open to outside influence in political as well as economic terms. Even internal debt service or repayment of monies owed to investors within a country gives financial establishments a measure of influence over government decisions. The amounts of debt shown on the map indicate the total external indebtedness of states.

Map 61 Global Flows of Investment Capital

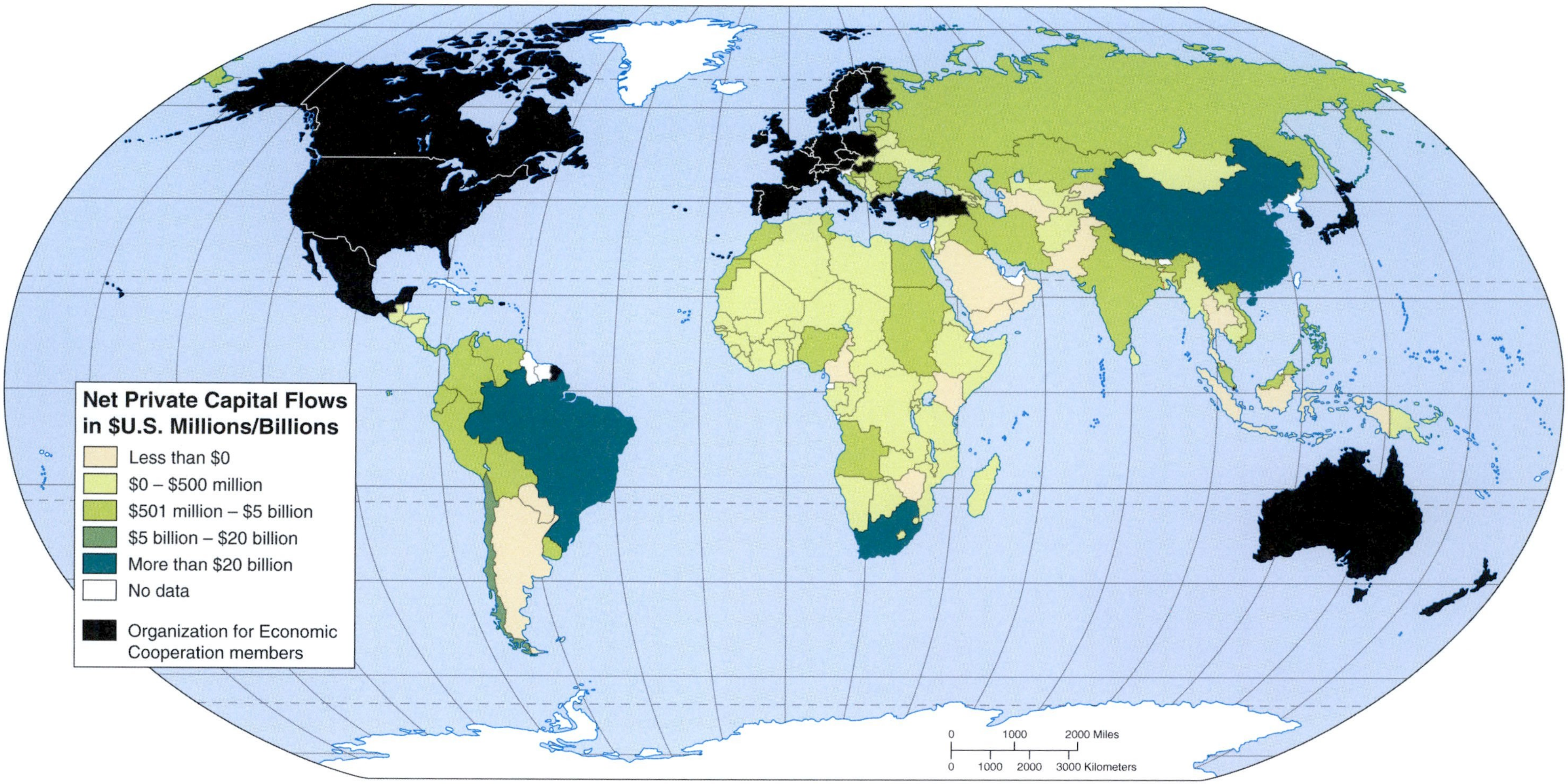

International capital flows include private debt and nondebt flows from one country to another, shown on the map as flows into a country. Nearly all of the capital comes from those countries that are members of the Organization for Economic Cooperation and Development (OECD), shown in black on the map. Capital flows include commercial bank lending, bonds, other private credits, foreign direct investment, and portfolio investment. Most of these flows are indicators of the increasing influence developed countries exert over the developing economies. Foreign direct investment or FDI, for example, is a measure of the net inflow of investment monies used to acquire long-term management interest in businesses located somewhere other than in the economy of the investor. Usually this means the acquisition of at least 10 percent of the stock of a company by a foreign investor and is, then, a measure of what might be termed "economic colonialism": control of a region's economy by foreign investors that could, in the world of the future, be as significant as colonial political control was in the past. International capital flows have increased greatly in the last decade as the result of the increasing liberalization of developing countries, the strong economic growth exhibited by many developing countries, and the falling costs and increased efficiency of communication and transportation services.

Map 62 Central Government Expenditures Per Capita

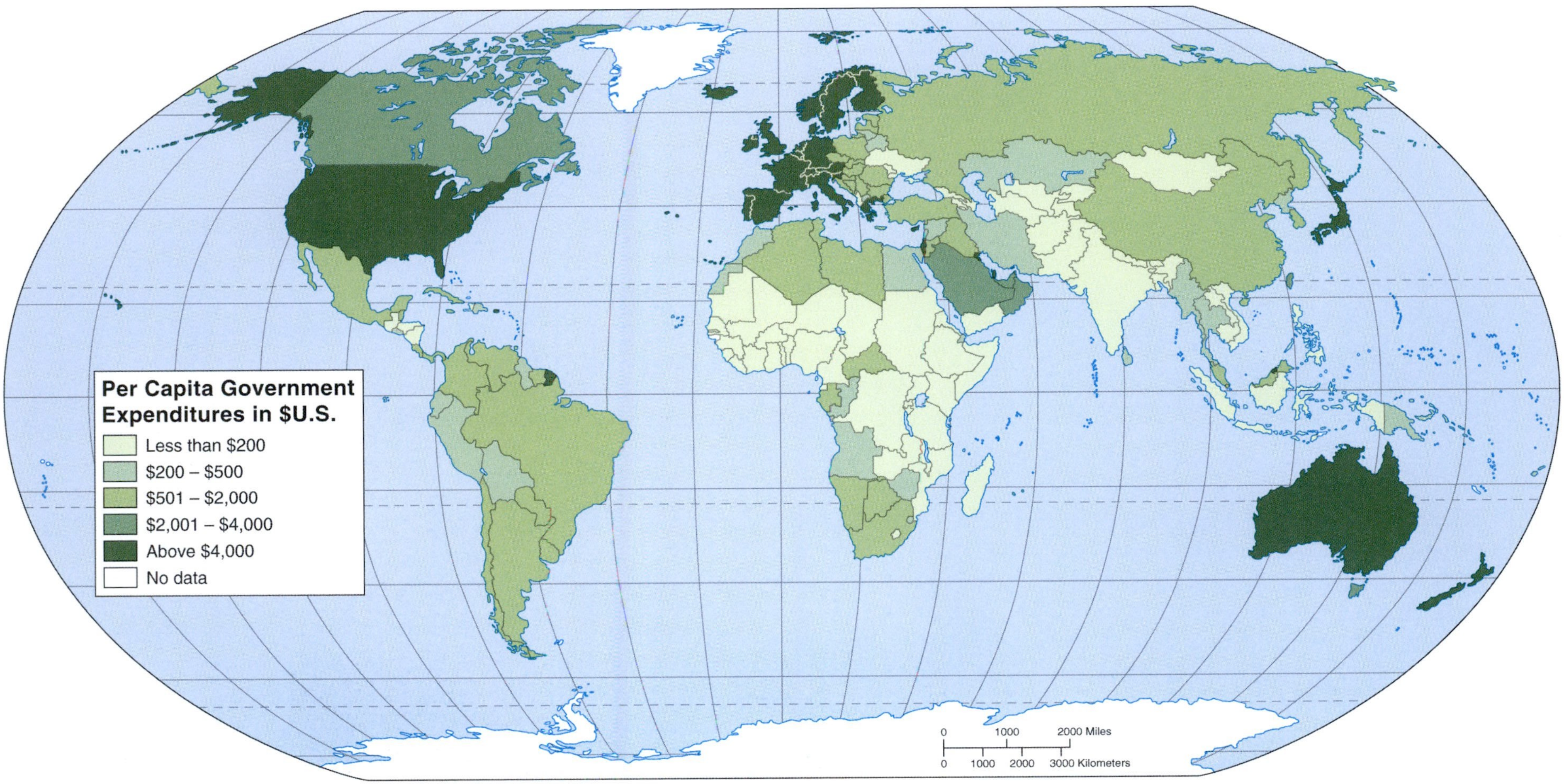

The amount of money that the central government of a country spends upon a variety of essential governmental functions is a measure of relative economic development, particularly when it is viewed on a per-person basis. These functions include such governmental responsibilities as agriculture, communications, culture, defense, education, fishing and hunting, health, housing, recreation, religion, social security, transportation, and welfare. Generally, the higher the level of economic development, the greater the per capita expenditures on these services. However, the data do mask some internal variations. For example, countries that spend 20 percent or more of their central government expenditures on defense will often show up in the more developed category when, in fact, all that the figures really show is that a disproportionate amount of the money available to the government is devoted to purchasing armaments and maintaining a large standing military force. Thus, the fact that Libya spends more than the average for Africa does not suggest that the average Libyan is much better off than the average Tanzanian. Nevertheless, this map—particularly when compared with Map 66, Energy Consumption Per Capita—does provide a reasonable approximation of economic development levels.

Map 63 Purchasing Power Parity Per Capita

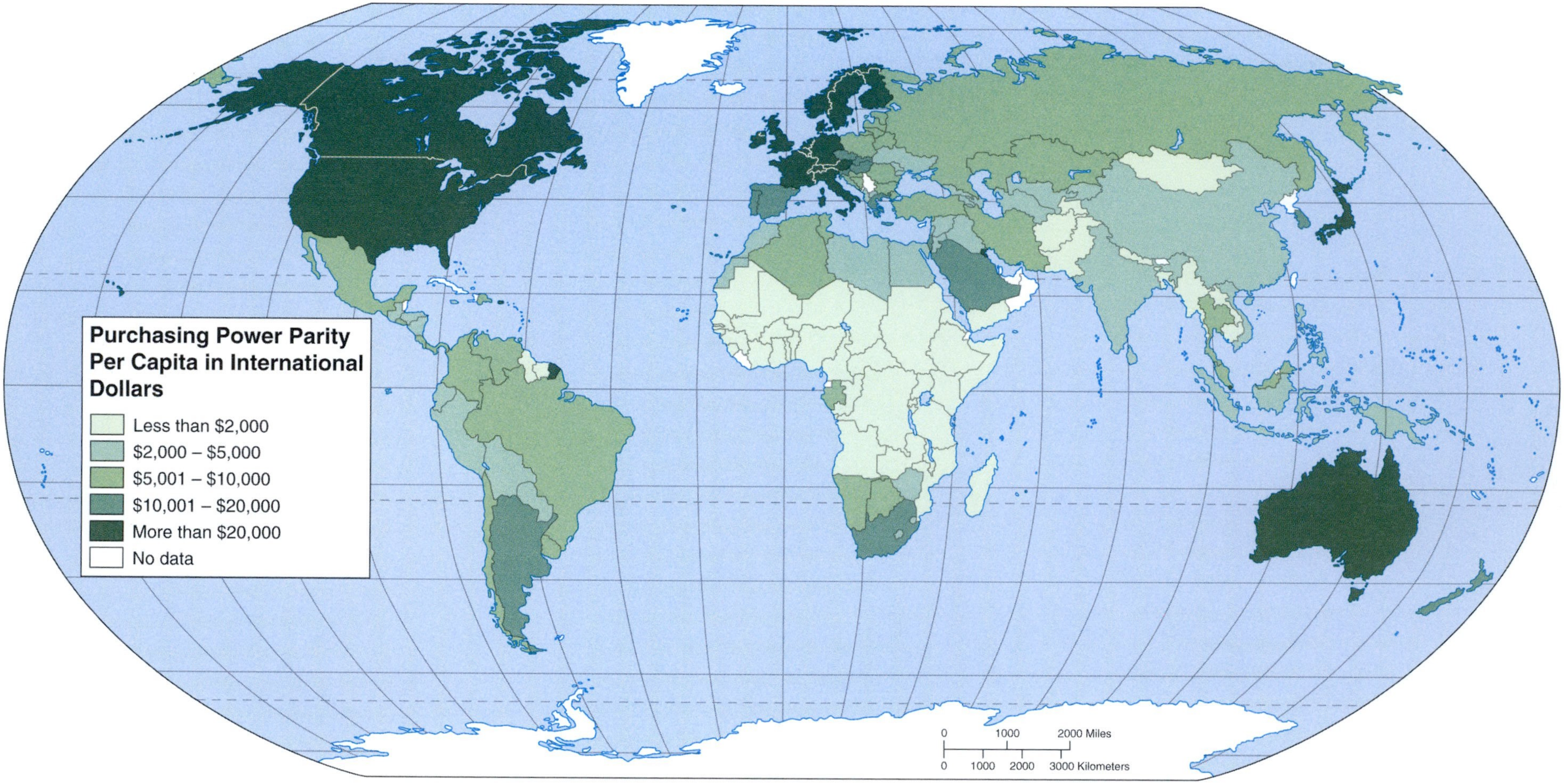

Of all the economic measures that separate the "haves" from the "have-nots," perhaps per capita Purchasing Power Parity (PPP) is the most meaningful. While per capita figures can mask significant uneven distributions within a country, they are generally useful for demonstrating important differences between countries. Per capita GNP and GDP (Gross Domestic Product) figures, and even per capita income, have the limitation of seldom reflecting the true purchasing power of a country's currency at home. In order to get around this limitation, international economists seeking to compare national currencies developed the PPP measure, which shows the level of goods and services that holders of a country's money can acquire locally. By converting all currencies to the "international dollar," the World Bank and other organizations using PPP can now show more truly comparative values, since the new currency value shows the number of units of a country's currency required to buy the same quantity of goods and services in the local market as one U.S. dollar would buy in an average country. The use of PPP currency values can alter the perceptions about a country's true comparative position in the world economy. More than per capita income figures, PPP provides a valid measurement of the ability of a country's population to provide for itself the things that people in the developed world take for granted: adequate food, shelter, clothing, education, and access to medical care. A glance at the map shows a clear-cut demarcation between temperate and tropical zones, with most of the countries with a PPP above $5,000 in the midlatitude zones and most of those with lower PPPs in the tropical and equatorial regions. Where exceptions to this pattern occur, they usually stem from a tremendous maldistribution of wealth among a country's population.

Map 64 Private Per Capita Consumption 2008

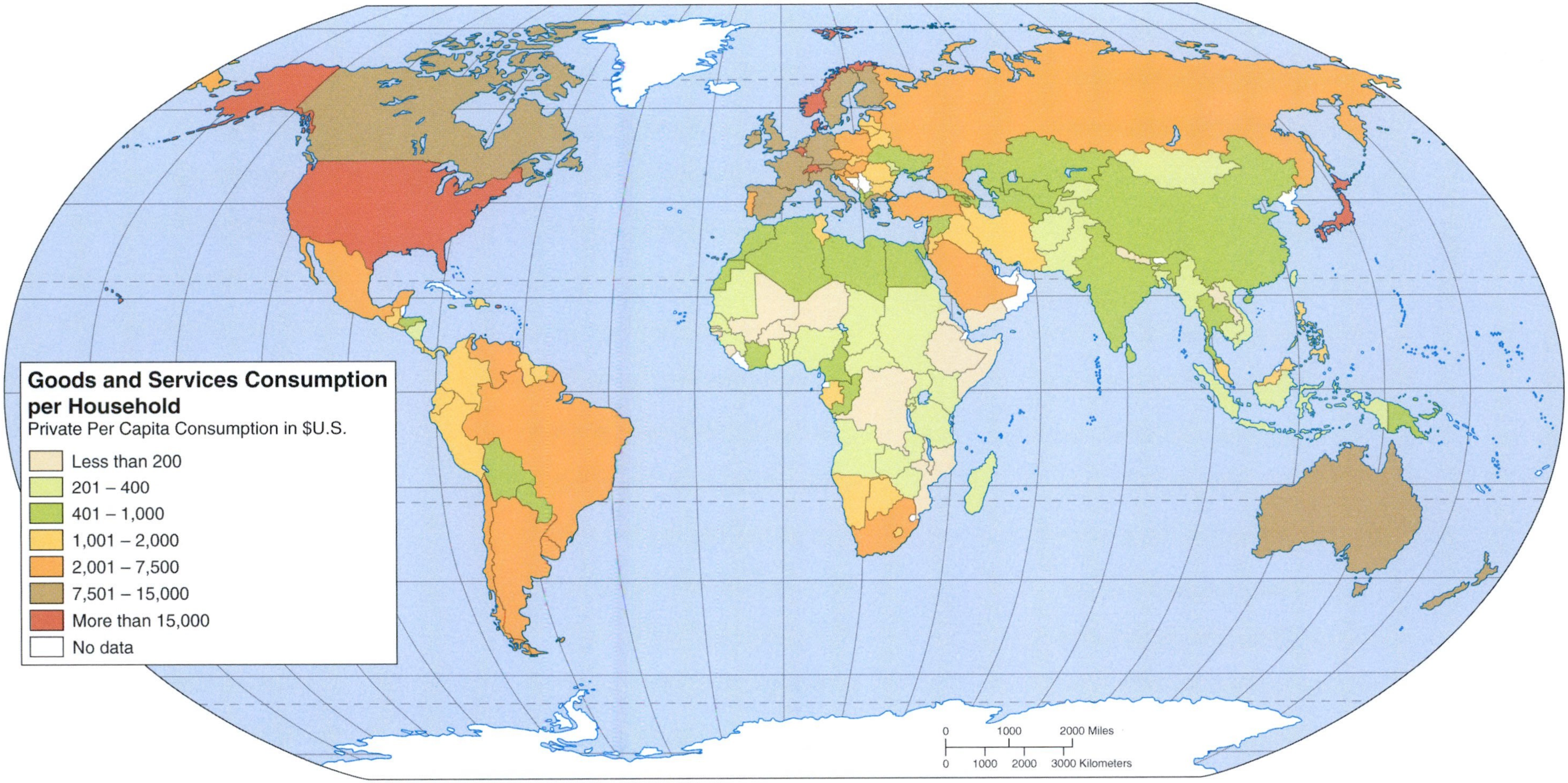

While we often examine production in analyzing global economic patterns, too seldom do we look at patterns of consumption—particularly on a per-person basis. When we do, the dramatic differences between the "rich" and "poor" (of which we are well aware) become even more dramatic. The consumption of key resources—minerals, agricultural products, harvests from the oceans, forestry products, water supplies—is among the most uneven distributions to be found among all the economic patterns of the world. Those who inhabit the wealthiest nations consume far more than their share of these key resources because they can afford to. And they produce far more than their share of the waste products, including atmospheric and oceanic pollution, of that consumption. Not only do they deplete their own resource bases but those of other countries as well. A trend that can be viewed as alarming by some and encouraging by others is that the number of "rich" or even "moderately well-off" countries is rising, and that number is rising among the world's most populous countries: India and China. What will happen to the global economy and the global environment (and we cannot separate the two) when India and China begin to consume and waste at the same levels as, say, the United States?

Map 65 Energy Production Per Capita

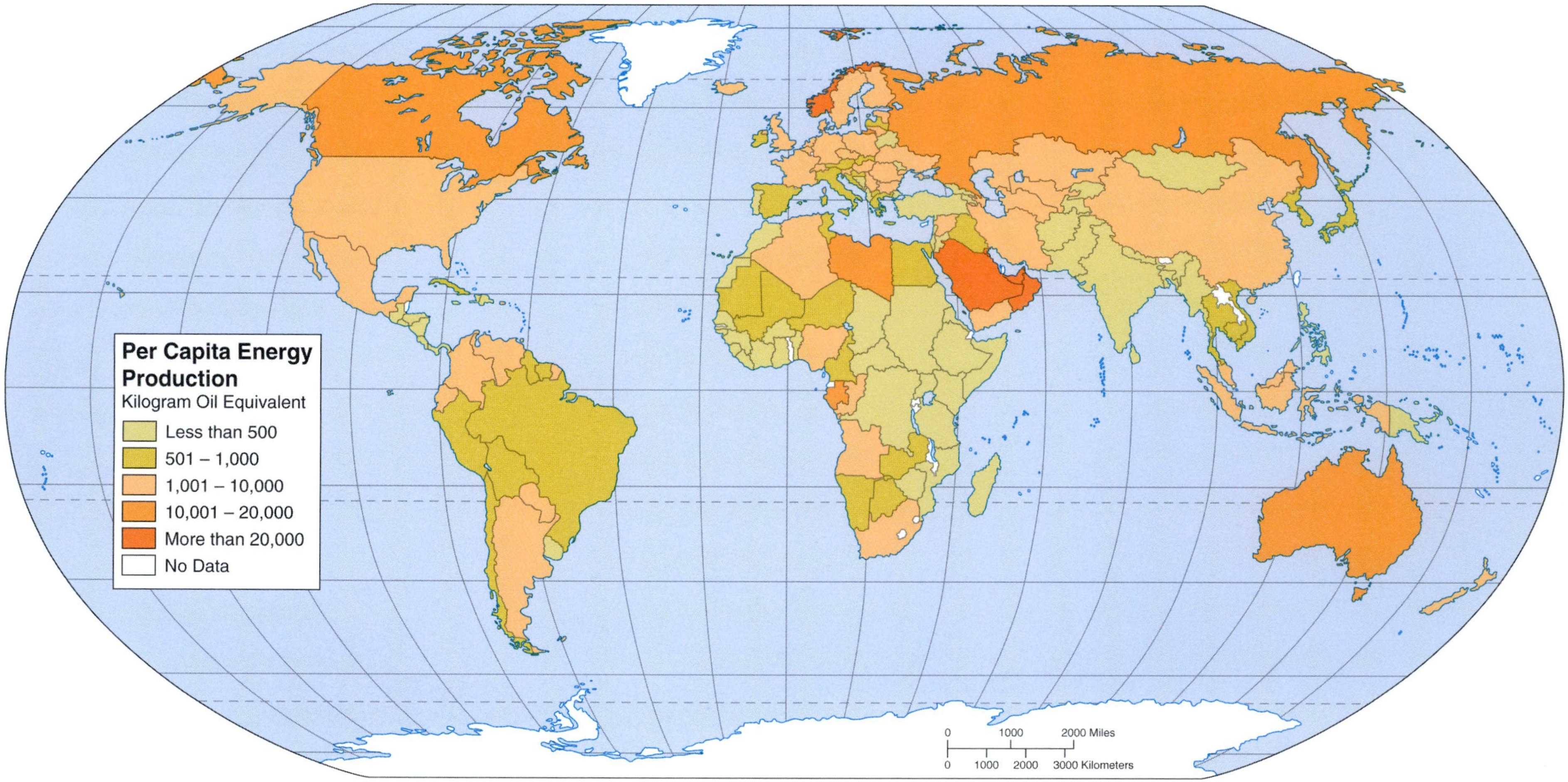

Energy production per capita is a measure of the availability of mechanical energy to assist people in their work. This map shows the amount of all kinds of energy—solid fuel (primarily coal), liquid fuel (primarily petroleum), natural gas, geothermal, wind, solar, hydroelectric, nuclear, waste recycling, and indigenous heat pumps—produced per person in each country. With some exceptions, wealthier countries produce more energy per capita than poor ones. Countries such as Japan and many European states rank among the world's wealthiest, but are energy-poor and produce relatively little of their own energy. They have the ability, however, to pay for imports. On the other hand, countries such as those of the Persian Gulf or the oil-producing states of Central and South America may rank relatively low on the scale of economic development but rank high as producers of energy. In many poor countries, especially in Central and South America, Africa, South Asia, and East Asia, large proportions of energy come from traditional fuels such as firewood and animal dung. Indeed for many in the developing world, the real energy crisis is a shortage of wood for cooking and heating.

Map 66 Energy Consumption Per Capita

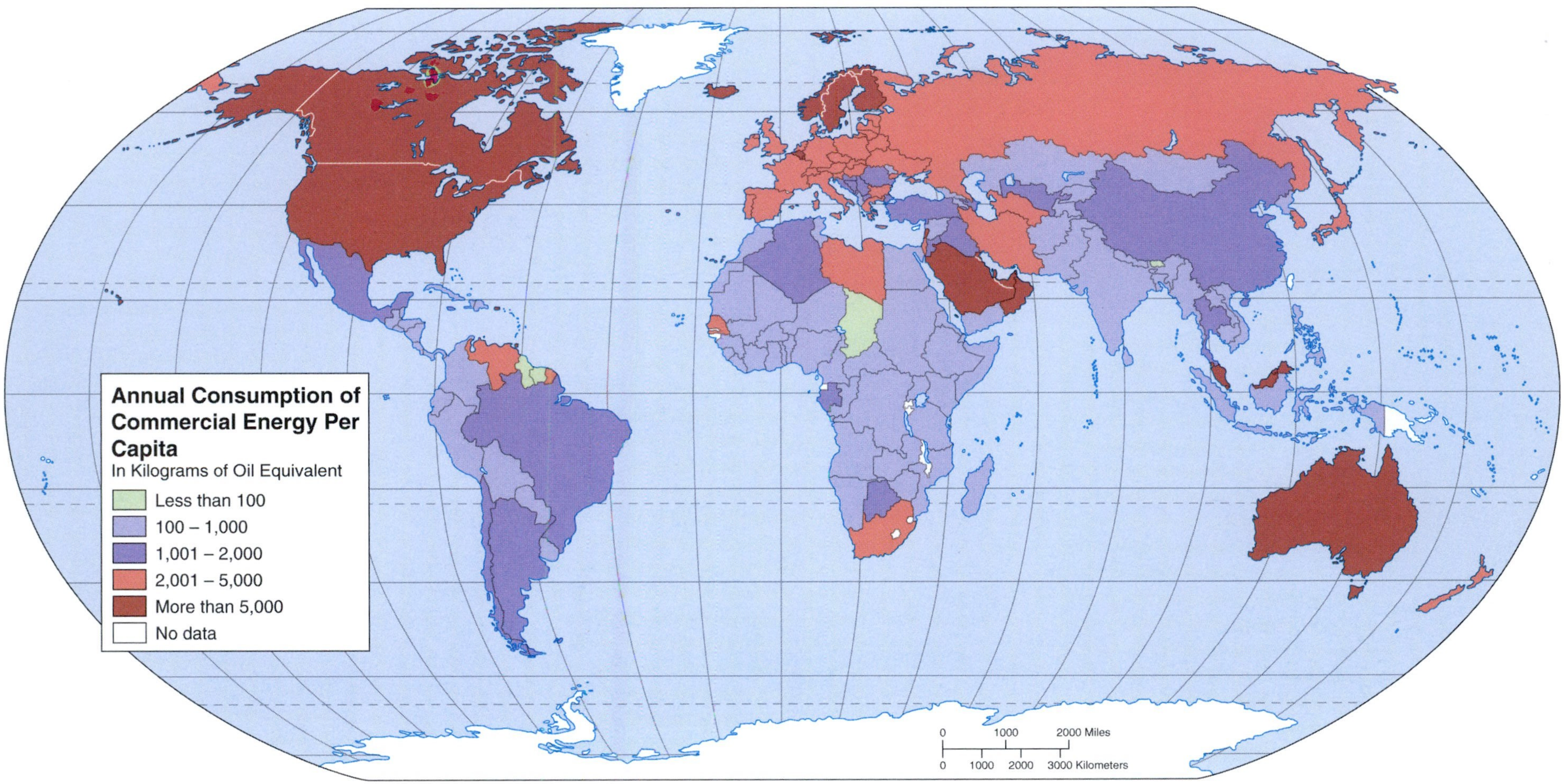

Of all the quantitative measures of economic well-being, energy consumption per capita may be the most expressive. All of the countries defined by the World Bank as having high incomes consume at least 100 gigajoules of commercial energy (the equivalent of about 3.5 metric tons of coal) per person per year, with some, such as the United States and Canada, having consumption rates in the 300 gigajoule range (the equivalent of more than 10 metric tons of coal per person per year). With the exception of the oil-rich Persian Gulf states, where consumption figures include the costly "burning off" of excess energy in the form of natural gas flares at wellheads, most of the highest-consuming countries are in the Northern Hemisphere, concentrated in North America and Western Europe. At the other end of the scale are low-income countries, whose consumption rates are often less than 1 percent of those of the United States and other high consumers. These figures do not, of course, include the consumption of noncommercial energy—the traditional fuels of firewood, animal dung, and other organic matter—widely used in the less developed parts of the world.

Map 67 Energy Dependency

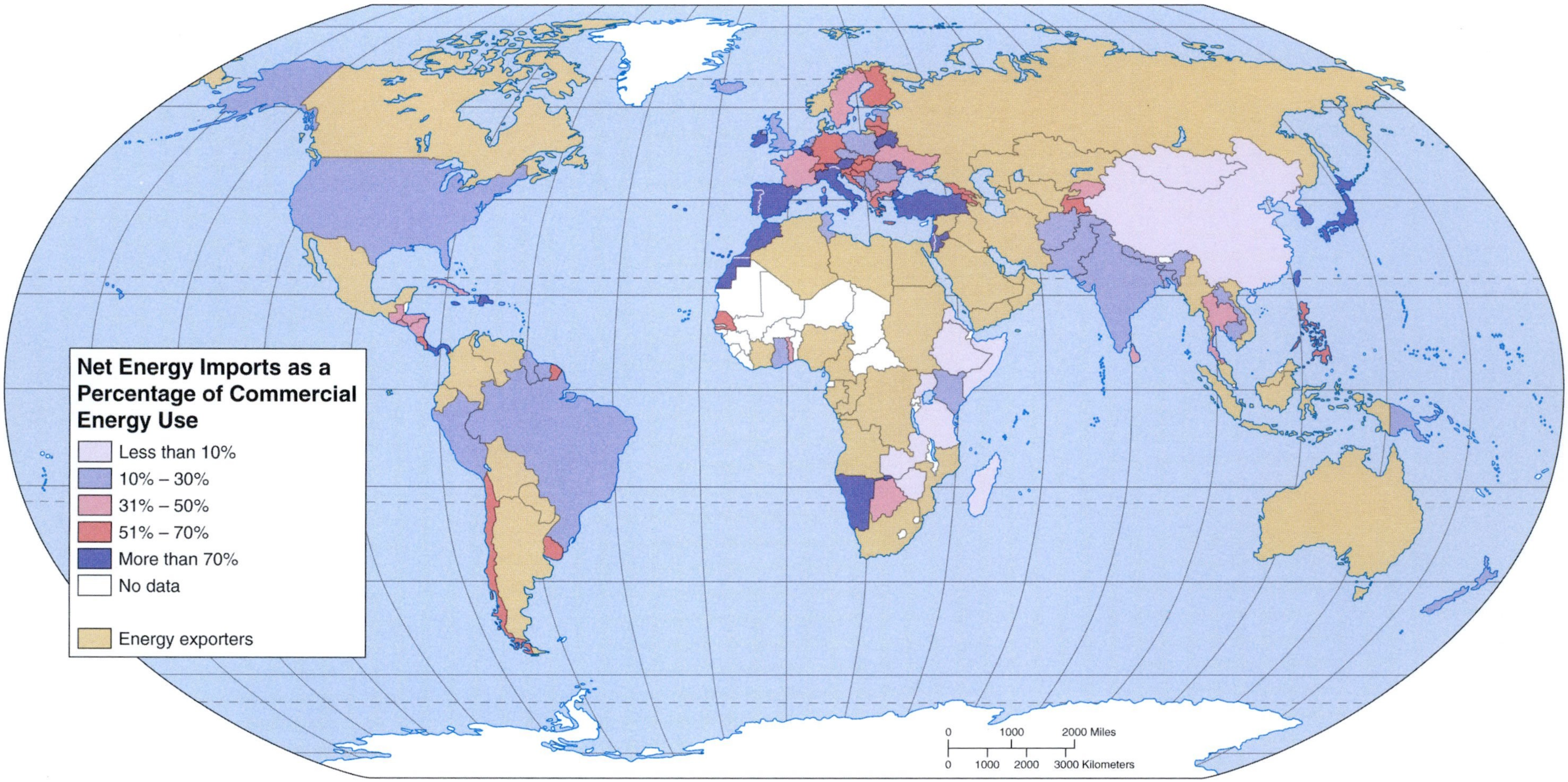

The patterns on the map show dependence on commercial energy before transformation to other end-use fuels such as electricity or refined petroleum products; energy from traditional sources such as fuelwood or dried animal dung is not included. Energy dependency is the difference between domestic consumption and domestic production of commercial energy and is most often expressed as a net energy import or export. A few of the world's countries are net exporters of energy: most are importers. The growth in global commercial energy use over the last decade indicates growth in the modern sectors of the economy—industry, transportation, and urbanization—particularly in the lesser developed countries. Still, the primary consumers of energy—and those having the greatest dependence on foreign sources of energy—are the more highly developed countries of Europe, North America, and Japan.

Map 68 Flows of Oil

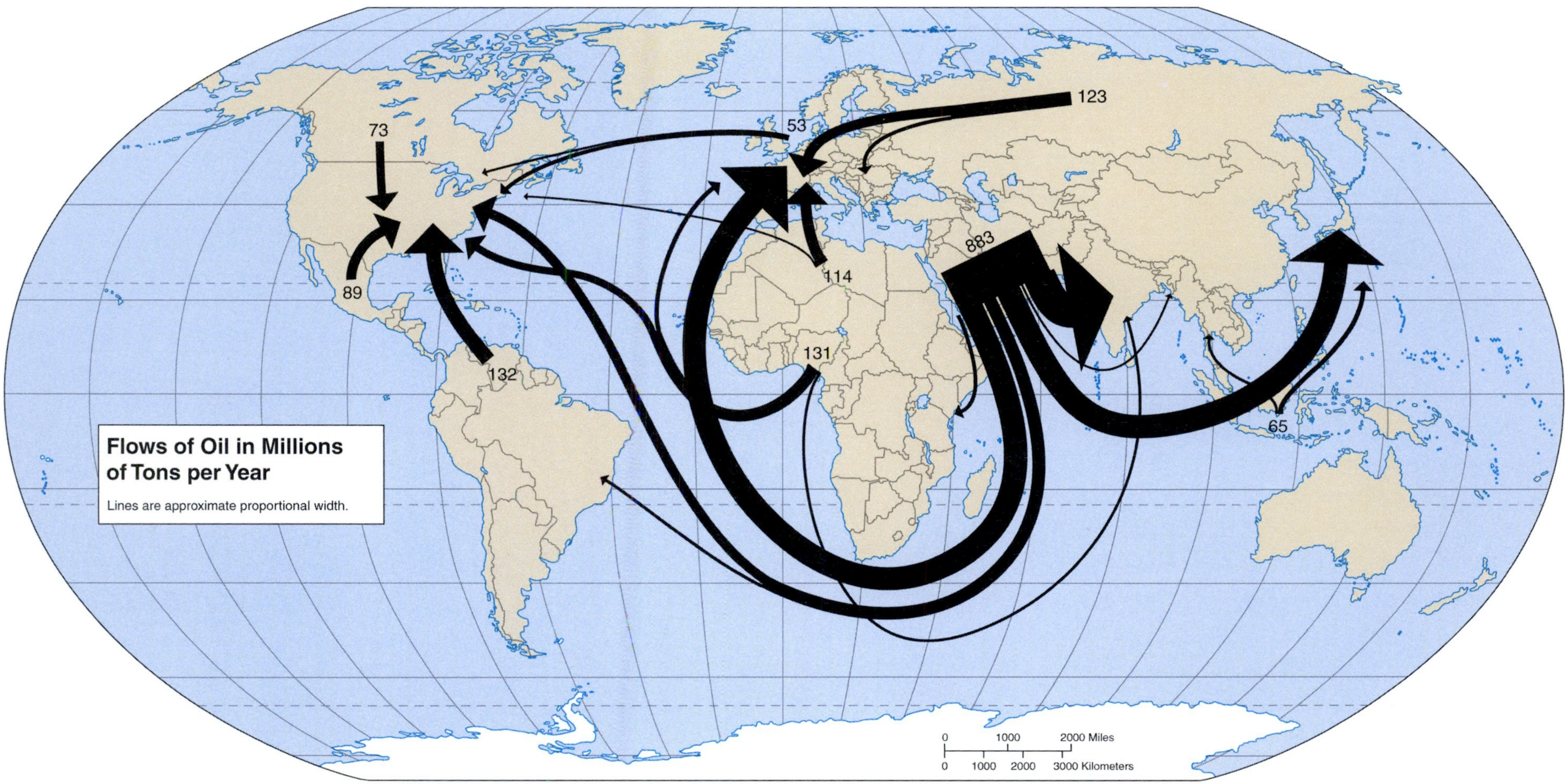

The pattern of oil movements from producing region to consuming region is one of the dominant facts of contemporary international maritime trade. Supertankers carry a million tons of crude oil and charge rates in excess of $0.10 per ton per mile, making the transportation of oil not only a necessity for the world's energy-hungry countries, but also an enormously profitable proposition. One of the major negatives of these massive oil flows is the damage done to the oceanic ecosystems—not just from the well-publicized and dramatic events like the wrecking of the *Exxon Valdez* but from the incalculable amounts of oil from leakage, scrubbings, purgings, and so on, which are a part of the oil transport technology. It is clear from the map that the primary recipients of these oil flows are the world's most highly developed economies.

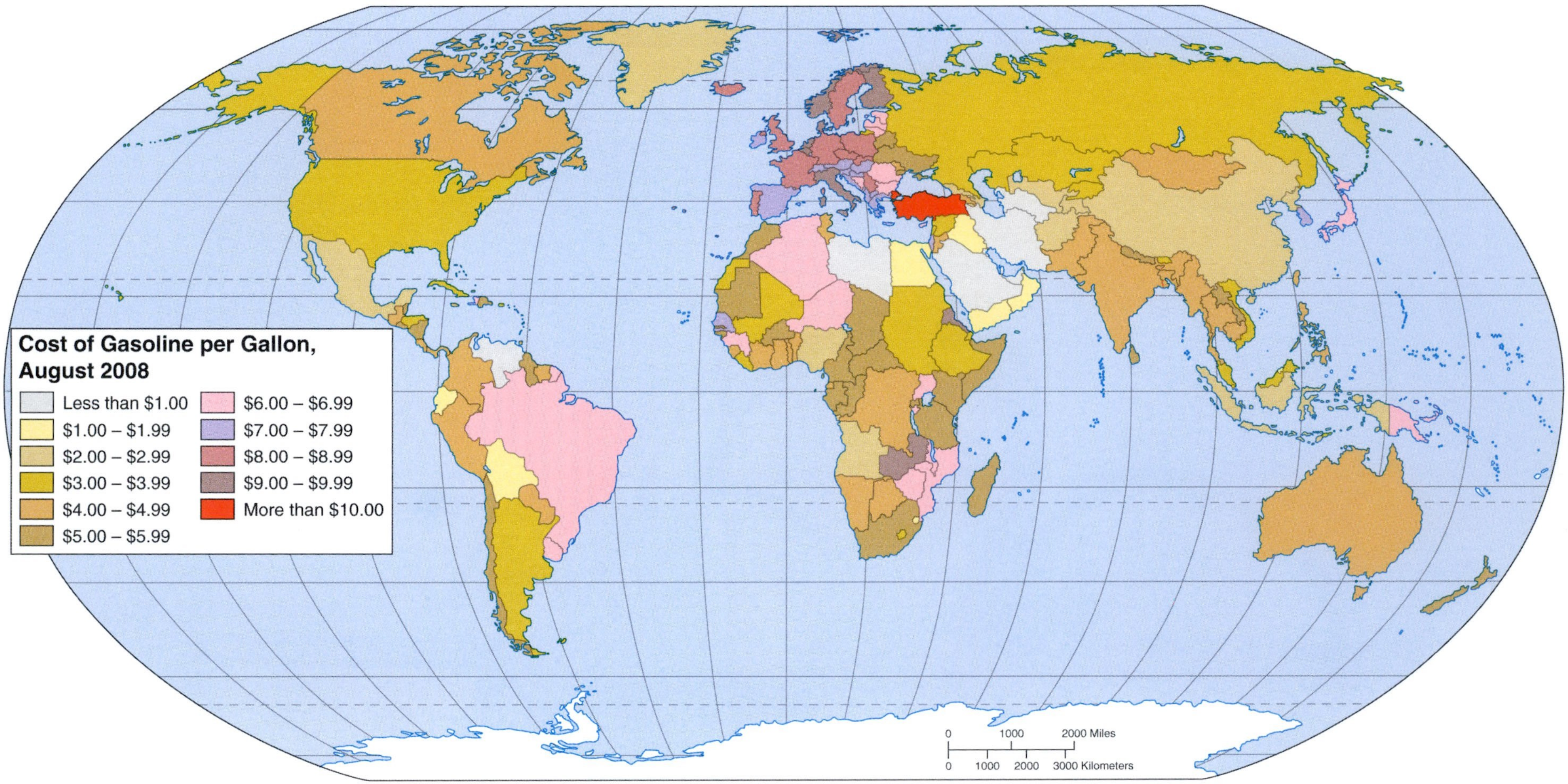

The year 2008 brought massive increases in the price of gasoline at the pump, creating considerable consternation among the American driving population, in particular. It is one thing to point out that gasoline prices are and historically have been considerably higher in Europe than in North America. But it is another to note that European spatial patterns of places of work and places of residence are significantly different than in North America. The North American metropolitan area evolved its spatial patterns in conjunction with the rise of privately owned automobiles; European cities, on the other hand, had spatial patterns well established centuries or even millennia before the automobile emerged as a common mode of transportation. As a consequence, such things as the journey to work are very different for many Europeans who can walk from where they live to where they work than it is for Americans and Canadians who often live considerable distances from their places of employment. A daily commute of 75 miles one way would not be considered unusual in America. In Europe it is virtually unheard of. There is also the component of the scale of organization of human activities: because of the very large country in which Americans live, their spatial movements are customarily more extensive than those of Europeans who live in countries the size of American states. So, yes, gasoline is much more expensive in Europe than in North America. But in North America, the increase in the cost of gasoline to and above $4.00 per gallon produces significantly more economic impact than proportionally similar increases in Europe. Similarly, the reduction in gasoline prices by the end of 2008 had a greater impact in North America—although its impact was overshadowed by larger forces.

Map 70 A Wired World: Internet Users

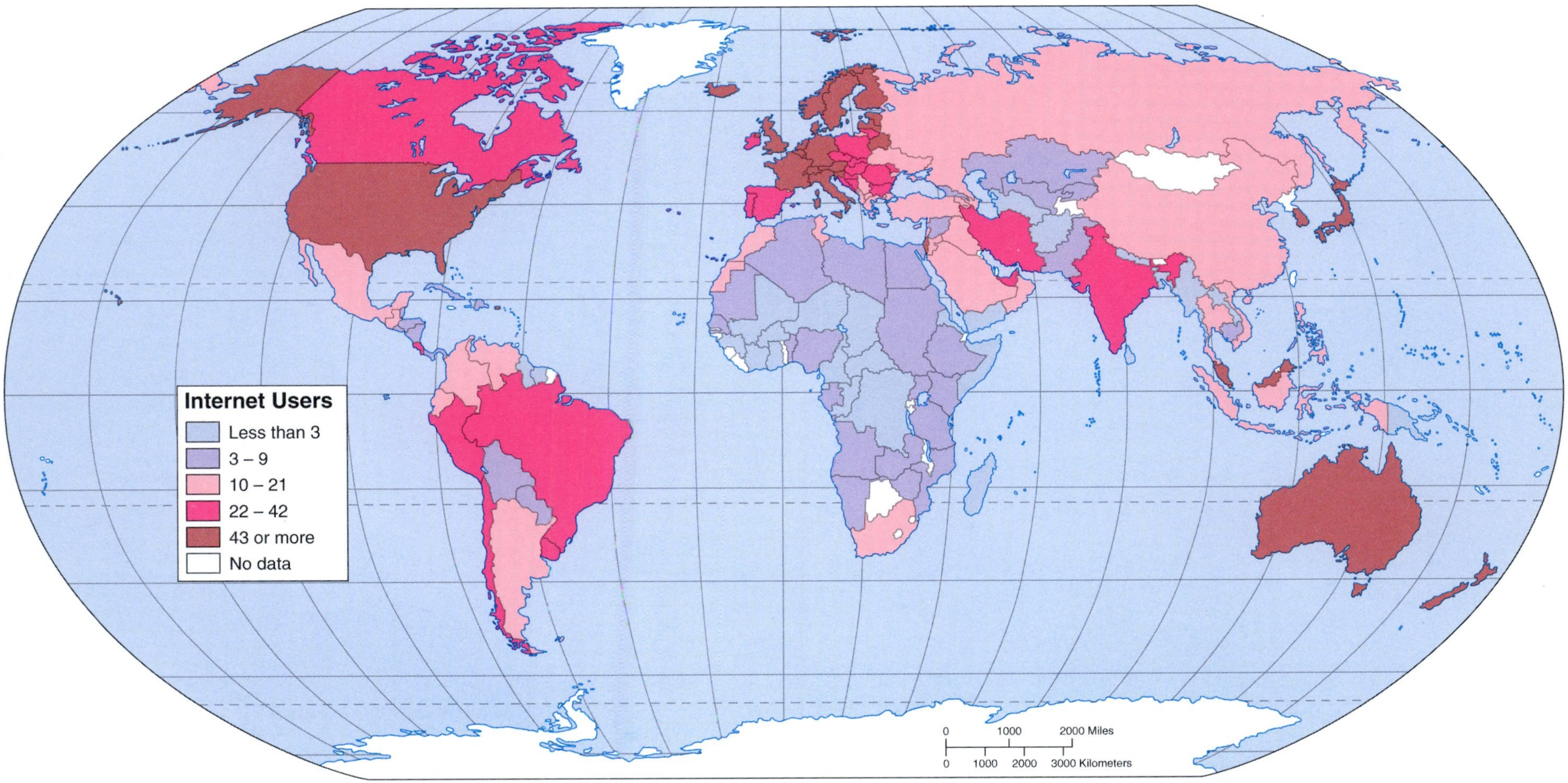

It is interesting to contemplate that a short quarter of a century ago, such a map could not have been created. The emergence of immediate, long-distance connectivity via the Internet has been one of the most important components of globalization. We now live, as author Thomas Friedman has noted, in a "flat world" where lines of connection are more important than distance and where virtually instantaneous connections have altered—perhaps forever—the way that we do business, exchange information, and transform our cultures. Some of the recent transformations we have seen in the emergence of countries like China and India as major players in the international economy are, in part, the consequence of access to the Internet. Originally conceived as a quick way for academics to exchange information, the Internet has become a cultural and social phenomenon that far exceeds its original purpose. Whether or not this will result in an eventual benefit to human well-being remains to be seen. Does the benefit of quick communication result in a cost to the complexity and richness of human cultures worldwide?

Map 71 The Rise of the Personal Computer

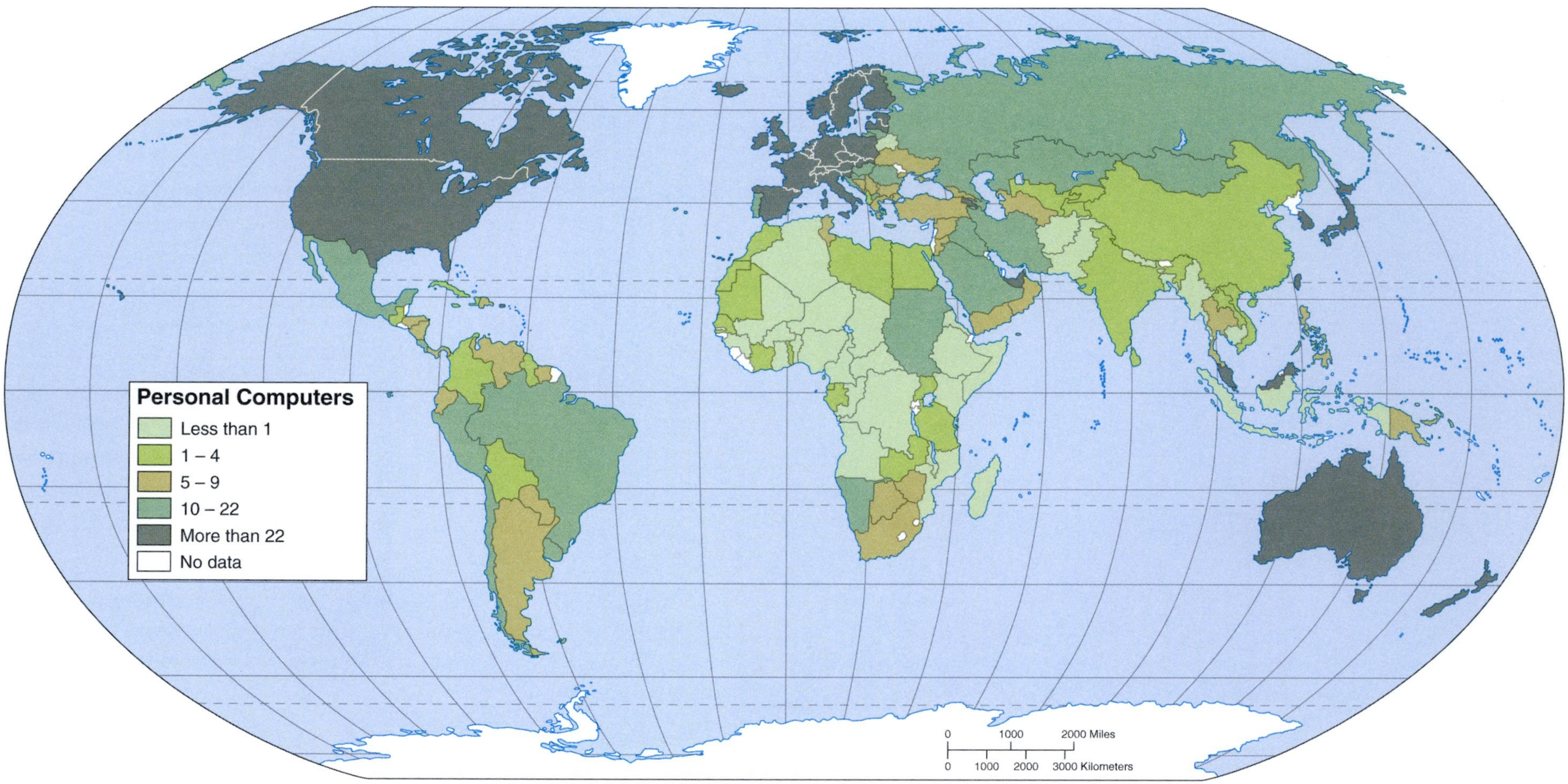

The Internet connections shown in the previous map would not be possible without personal computers—or at least not possible at their present scale. But personal computers do a great deal more than simply act as high-speed transmitters of information. They are incredibly powerful devices for the storage and analysis of data and are becoming, seemingly exponentially, even more powerful. When the use of mainframe computers became common on university campuses in the 1960s, they were used primarily for faculty and graduate student research. Now, an undergraduate can run on his or her laptop computer—in a matter of seconds—a program that would have required hours to run on an institutional computer that occupied spaces similar to those required for medium-sized classrooms. Certainly computers have changed the way that businesses are run. But they have, perhaps, changed the daily lives of personal computer users even more.

Map 72 Traditional Links: The Telephone

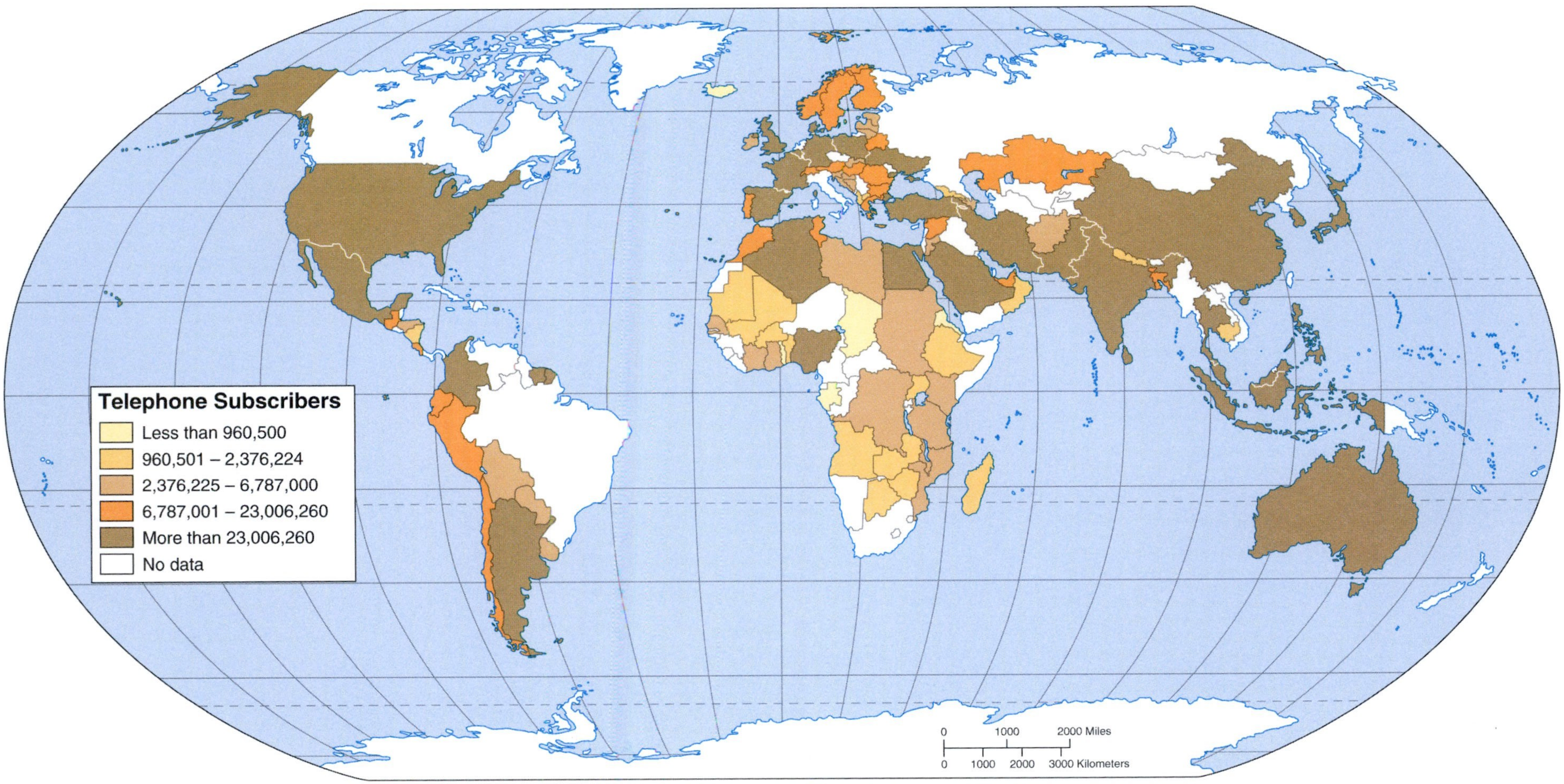

Not all of the world's communications take place via computers and the Internet. A lot of person-to-person connection is still carried out via the telephone, and access to telephone connections is perhaps as good an indication of economic development or, more important, the potential for economic development as anything else. The map clearly shows the prevalence of telephone connectivity in the developed world. But it also shows an increasingly high degree of access to telephones in major countries in the developing world, such as India and China. If these countries are to continue to develop their economies at the pace of the last decade, then their degree of communication—including access to telephones—will also have to increase. The data shown on this map include users of both land lines and cellular phones. By the end of 2008, for the first time, the number of cellular phone users worldwide exceeded the number of those using the traditional land lines. As cellular phone complexity increases to include e-mail and other computer functions, the gap between cellular users and traditional phone users can be expected to widen.

Unit V

Global Patterns of Environmental Disturbance

Map 97	Nationally Protected Areas: Parks and Preserves
Map 98	The Risks of Desertification
Map 99	Global Soil Degradation
Map 100	The Degree of Human Disturbance
Map 101	The Green and Not So Green World

Map 73 Global Air Pollution: Sources and Wind Currents

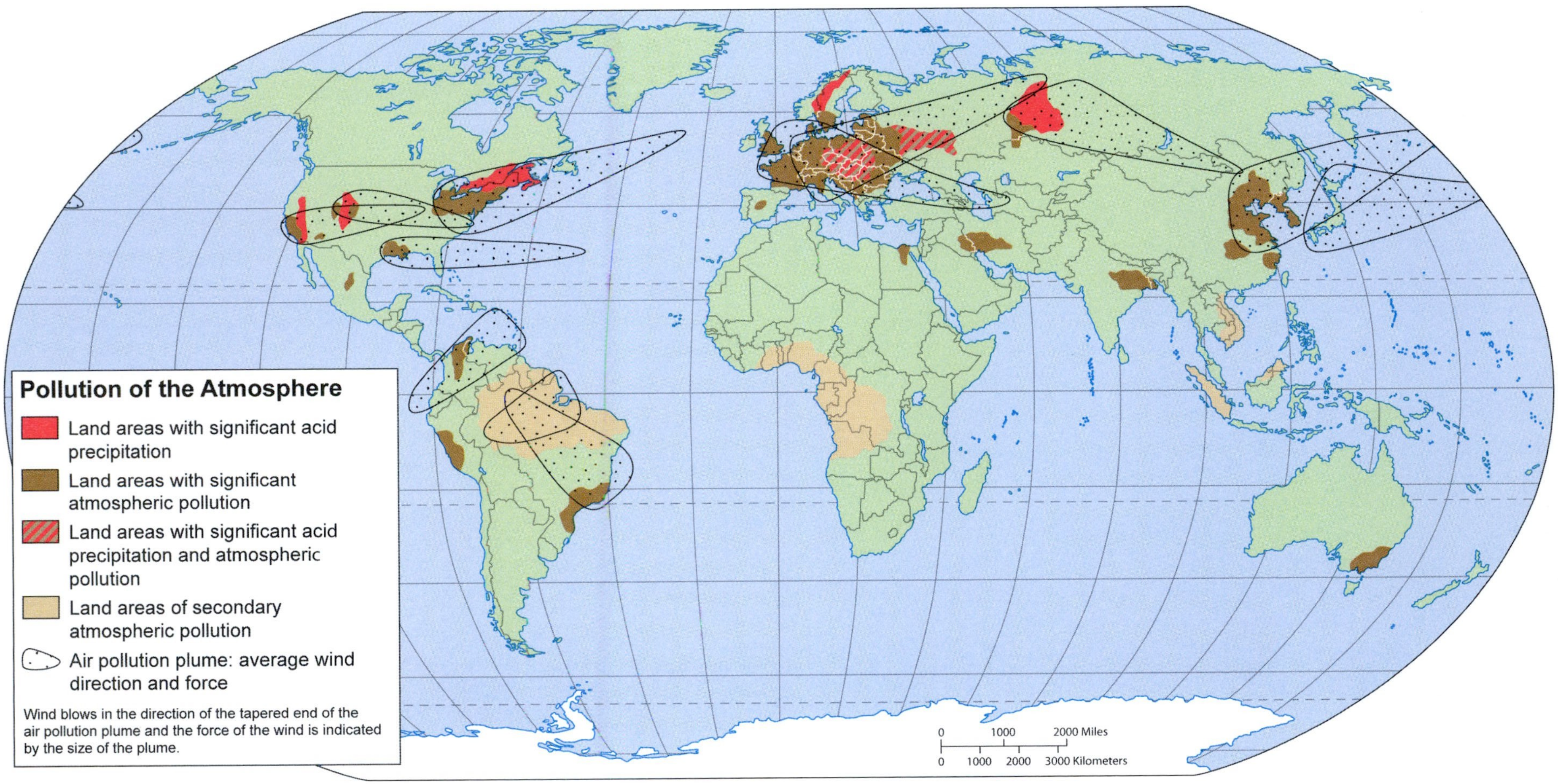

Almost all processes of physical geography begin and end with the flows of energy and matter among land, sea, and air. Because of the primacy of the atmosphere in this exchange system, air pollution is potentially one of the most dangerous human modifications in environmental systems. Pollutants such as various oxides of nitrogen or sulfur cause the development of acid precipitation, which damages soil, vegetation, and wildlife and fish. Air pollution in the form of smog is often dangerous for human health. And most atmospheric scientists believe that the efficiency of the atmosphere in retaining heat—the so-called greenhouse effect—is being enhanced by increased carbon dioxide, methane, and other gases produced by agricultural and industrial activities. The result, they fear, will be a period of global warming that will dramatically alter climates in all parts of the world.

Map 74 The Acid Deposition Problem: Air, Water, Soil

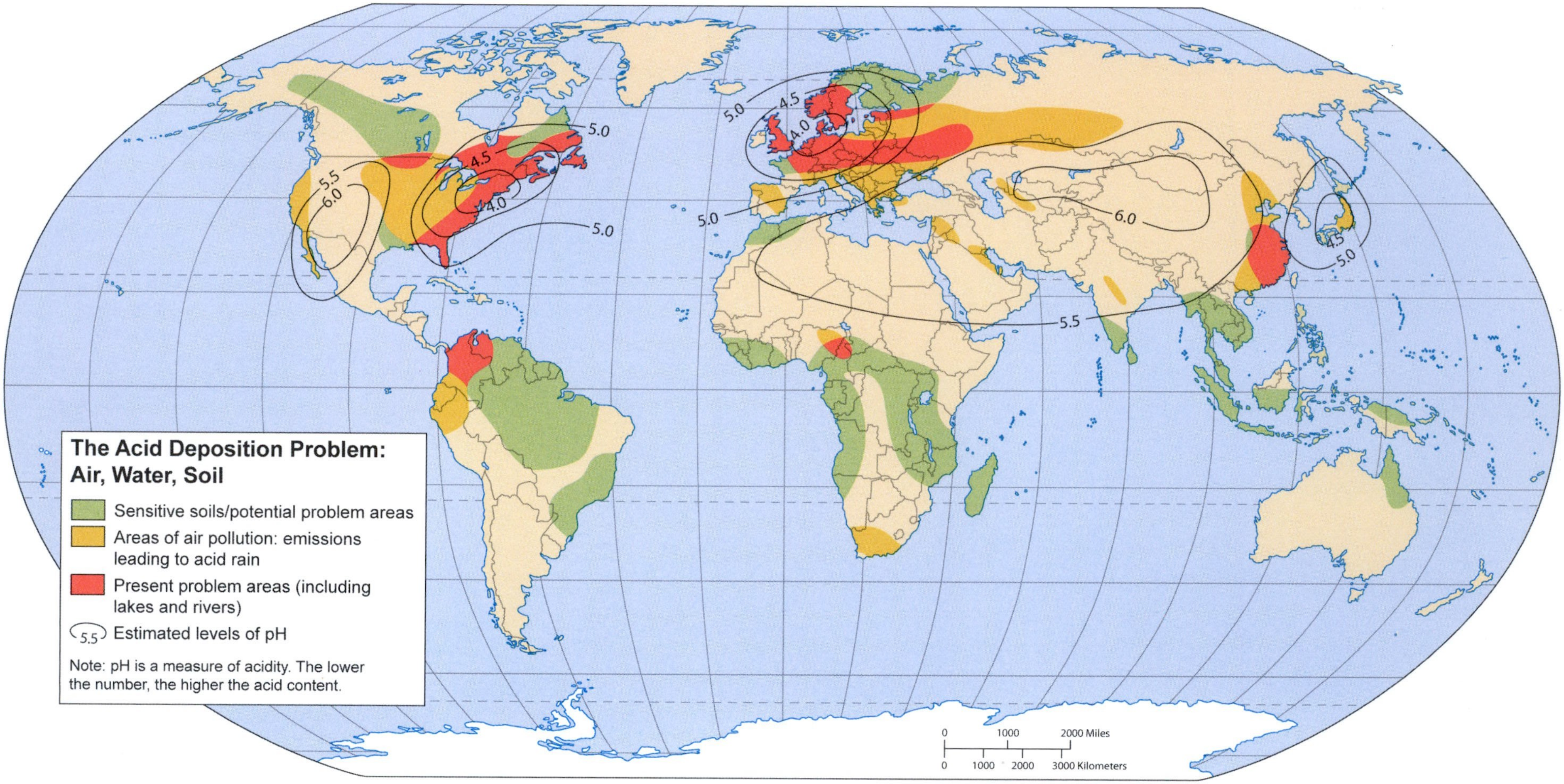

The term "acid precipitation" refers to increasing levels of acidity in snowfall and rainfall caused by atmospheric pollution. Oxides of nitrogen and sulfur resulting from incomplete combustion of fossil fuels (coal, oil, and natural gas) combine with water vapor in the atmosphere to produce weak acids that then "precipitate" or fall along with water or ice crystals. Some atmospheric acids formed by this process are known as "dry-acid" precipitates and they too will fall to earth, although not necessarily along with rain or snow. In some areas of the world, the increased acidity of streams and lakes stemming from high levels of acid precipitation or dry acid fallout has damaged or destroyed aquatic life. Acid precipitation and dry acid fallout also harm soil systems and vegetation, producing a characteristic burned appearance in forests that lends the same quality to landscapes that forest fires would. The region most dramatically impacted by acid precipitation is Central Europe where decades of destructive environmental practices, including the burning of high sulfur coal for commercial, industrial, and residential purposes, has produced the destruction of hundreds of thousands of acres of woodlands—a phenomenon described by the German foresters who began their study of the area following the lifting of the Iron Curtain as "Waldsterben": Forest Death.

Map 75 Major Polluters and Common Pollutants

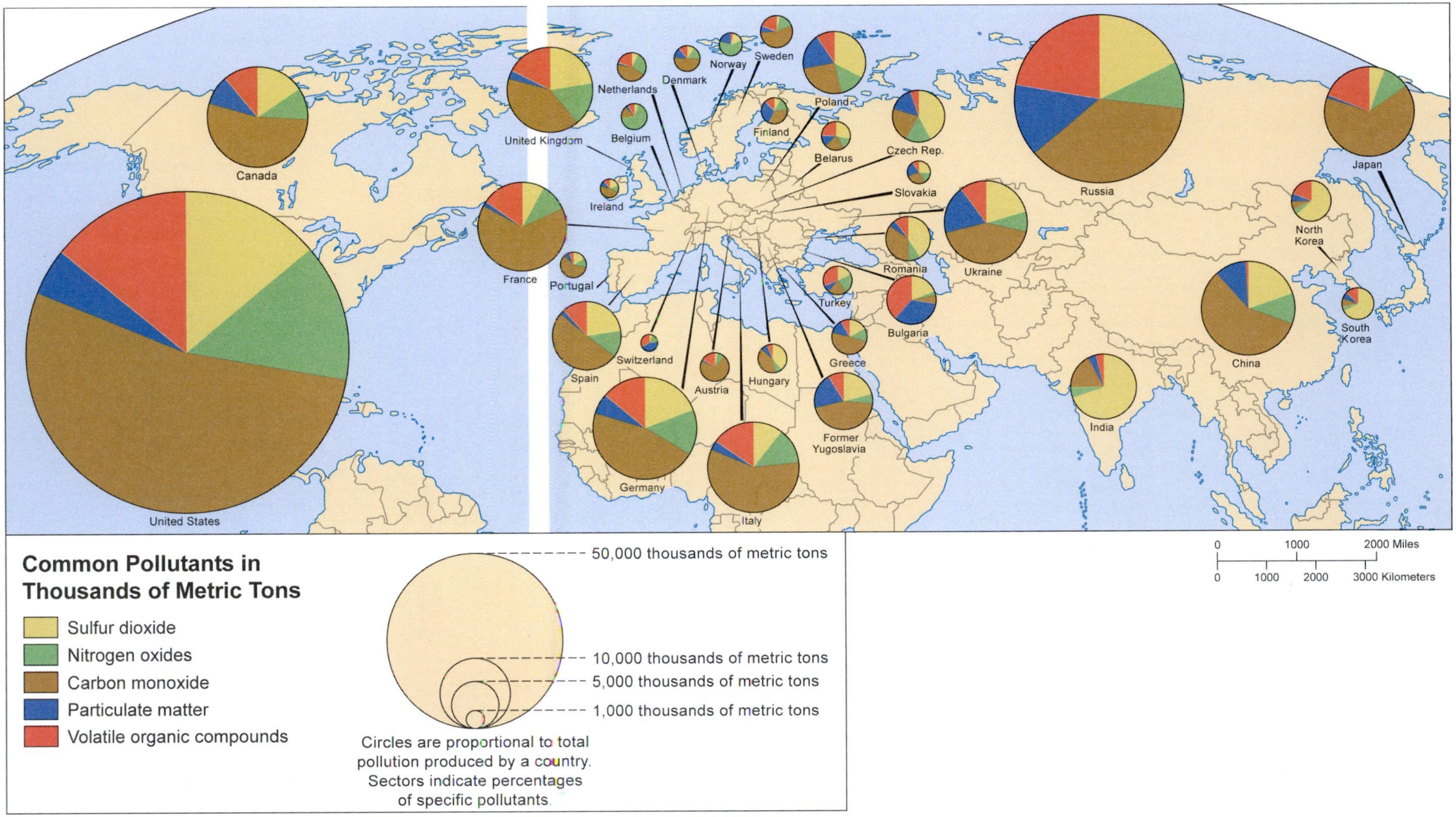

More than 90 percent of the world's total of anthropogenic (human-generated) air pollutants come from the heavily populated industrial regions of North America, Europe, South Asia (primarily in India), and East Asia (mainly in China, Japan, and the two Koreas). This map shows the origins of the five most common pollutants: sulfur dioxide, nitrogen oxide, carbon monoxide, particulate matter, and volatile organic compounds. These substances are produced both by industry and by the combustion of fossil fuels that generate electricity and power trains, planes, automobiles, buses, and trucks. In addition to combining with other components of the atmosphere and with one another to produce smog, they are the chief ingredients in acid accumulations in the atmosphere, which ultimately result in acid deposition, either as acid precipitation or dry acid fallout. Like other forms of pollutants, these air pollutants do not recognize political boundaries, and regions downwind of major polluters receive large quantities of pollutants from areas over which they often have no control.

Map 76 Global Carbon Dioxide Emissions

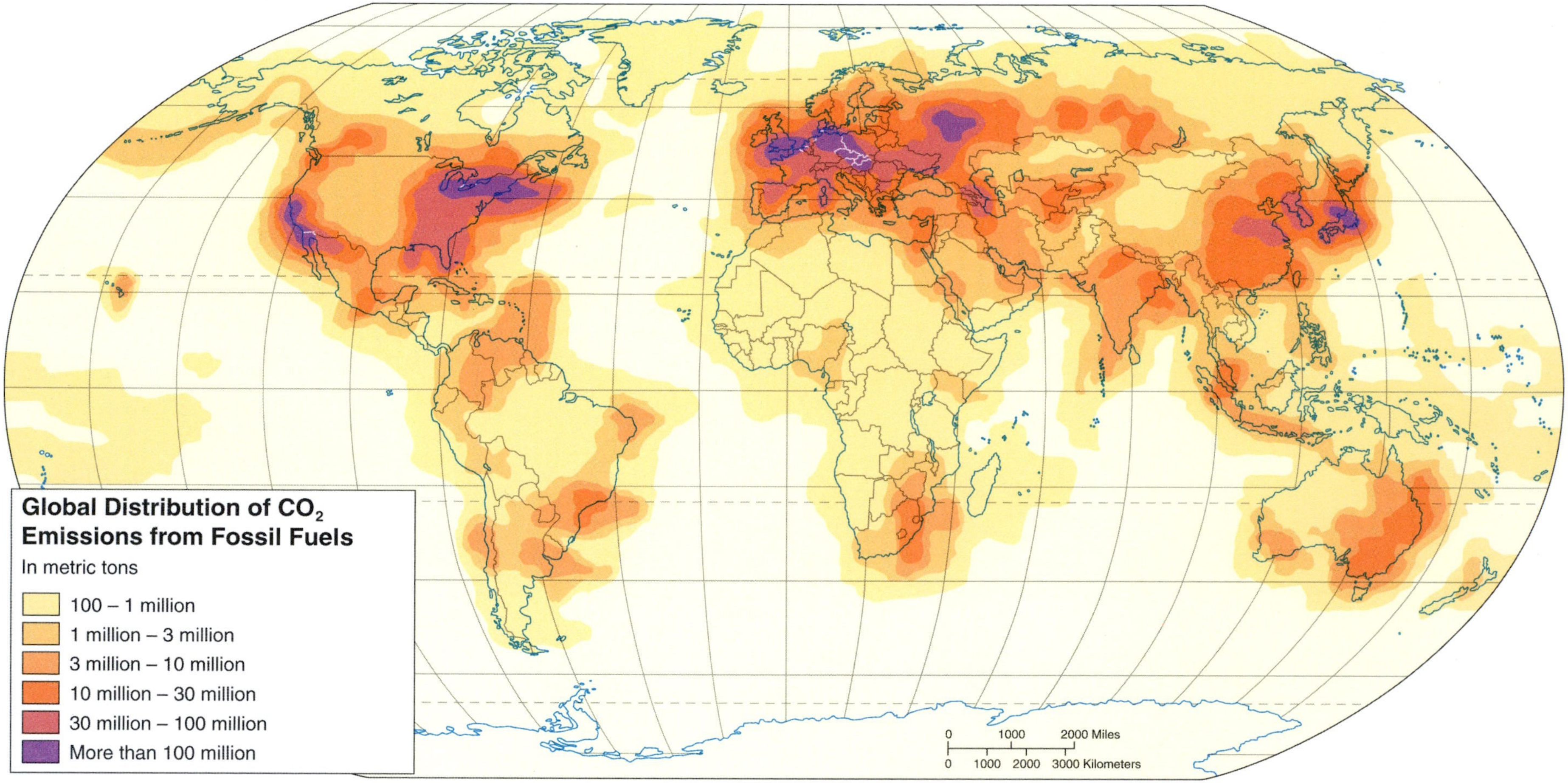

One of the most important components of the atmosphere is the gas carbon dioxide (CO_2), the byproduct of animal respiration, decomposition, and combustion. During the past 200 years, atmospheric CO_2 has risen dramatically, largely as the result of the tremendous increase in fossil fuel combustion brought on by the industrialization of the world's economy and the burning and clearing of forests by the expansion of farming. While CO_2 by itself is relatively harmless, it is an important "greenhouse gas." The gases in the atmosphere act like the panes of glass in a greenhouse roof, allowing light in but preventing heat from escaping. The greenhouse capacity of the atmosphere is crucial for organic life and is a purely natural component of the global energy cycle. But too much CO_2 and other greenhouse gases such as methane could cause the earth's atmosphere to warm up too much, producing the global warming that atmospheric scientists are concerned about. Researchers estimate that if greenhouse gases such as CO_2 continue to increase at their present rates, the earth's mean temperature could rise between 1.5 and 4.5 degrees Celsius by the middle of the present century. Such a rise in global temperatures would produce massive alterations in the world's climate patterns.

Map 77 The Earth Warms: 1976–2006

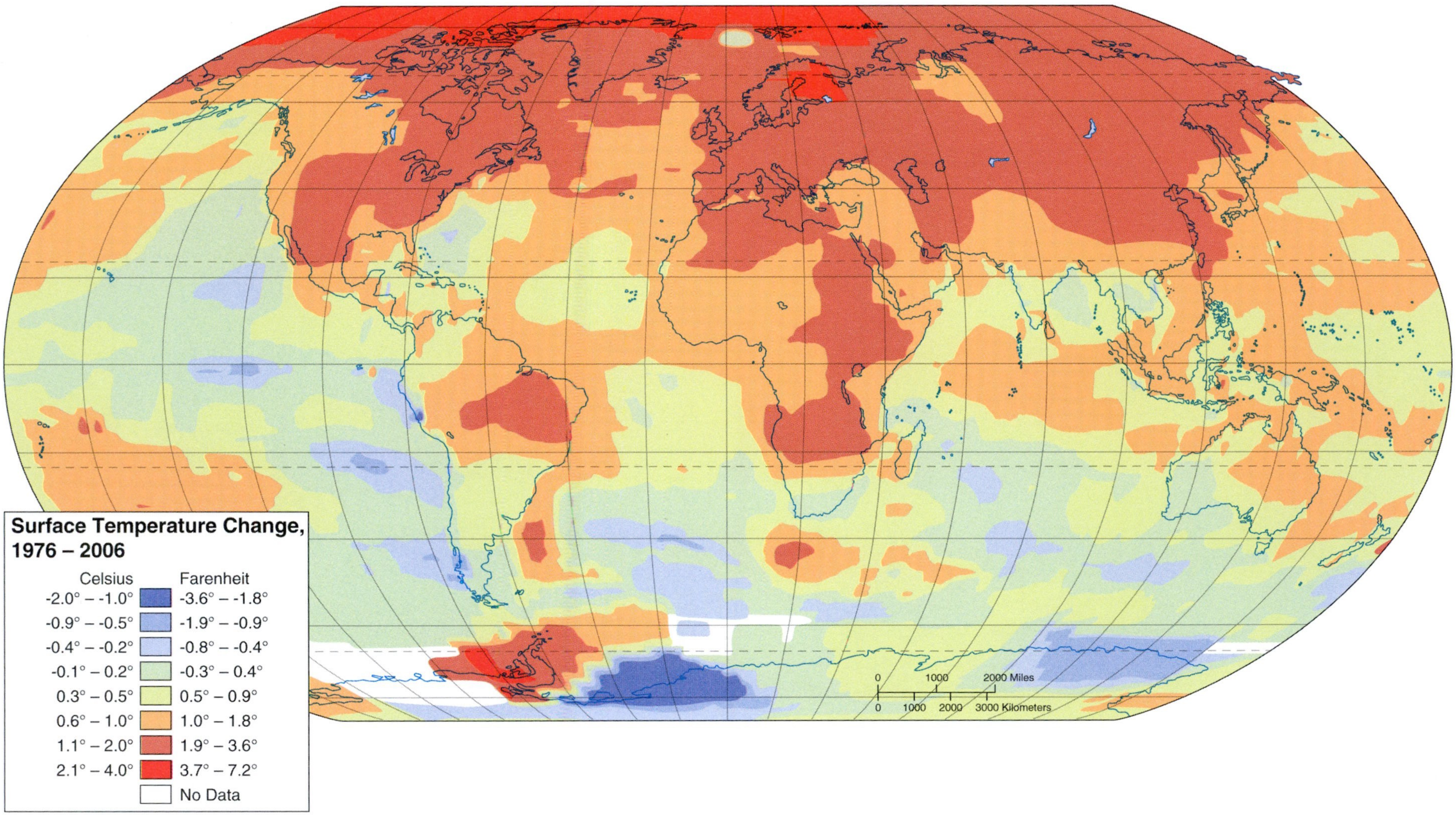

While there may still be disagreement about the nature of present climate change and its link to human activities, that disagreement is largely political and economic, rather than scientific. The evidence from a generation of scientific studies of the atmosphere is compelling: the world is warming more rapidly than at any time in the past and the bulk of this warming trend is the consequence of human alterations in the atmospheric proportion of the "greenhouse gases," such as carbon dioxide and methane, that help to trap heat radiated from the earth and release it back into space more slowly. The atmosphere is a massive heat engine: short wave (light) energy from the sun passes through the atmosphere where some of it is absorbed and converted to heat; a larger percentage of the light energy is absorbed by the body of the earth where it is converted to heat to warm the atmosphere. The "greenhouse effect," as this process is known, is a good thing. Were it not for the greenhouse effect, life on Earth as we know it would be impossible. The temperature changes shown on this map are the result of an accelerating greenhouse effect and, in turn, cause changes in precipitation patterns (see next map), storms of greater intensity and frequency, and rising sea levels. The Earth has experienced these changes before. What makes the current changes different is that they are occurring over decades rather than over millennia, making it very difficult (if not impossible) for natural and human systems to adapt to them.

Map 78 Global Precipitation Changes: 1976–2006

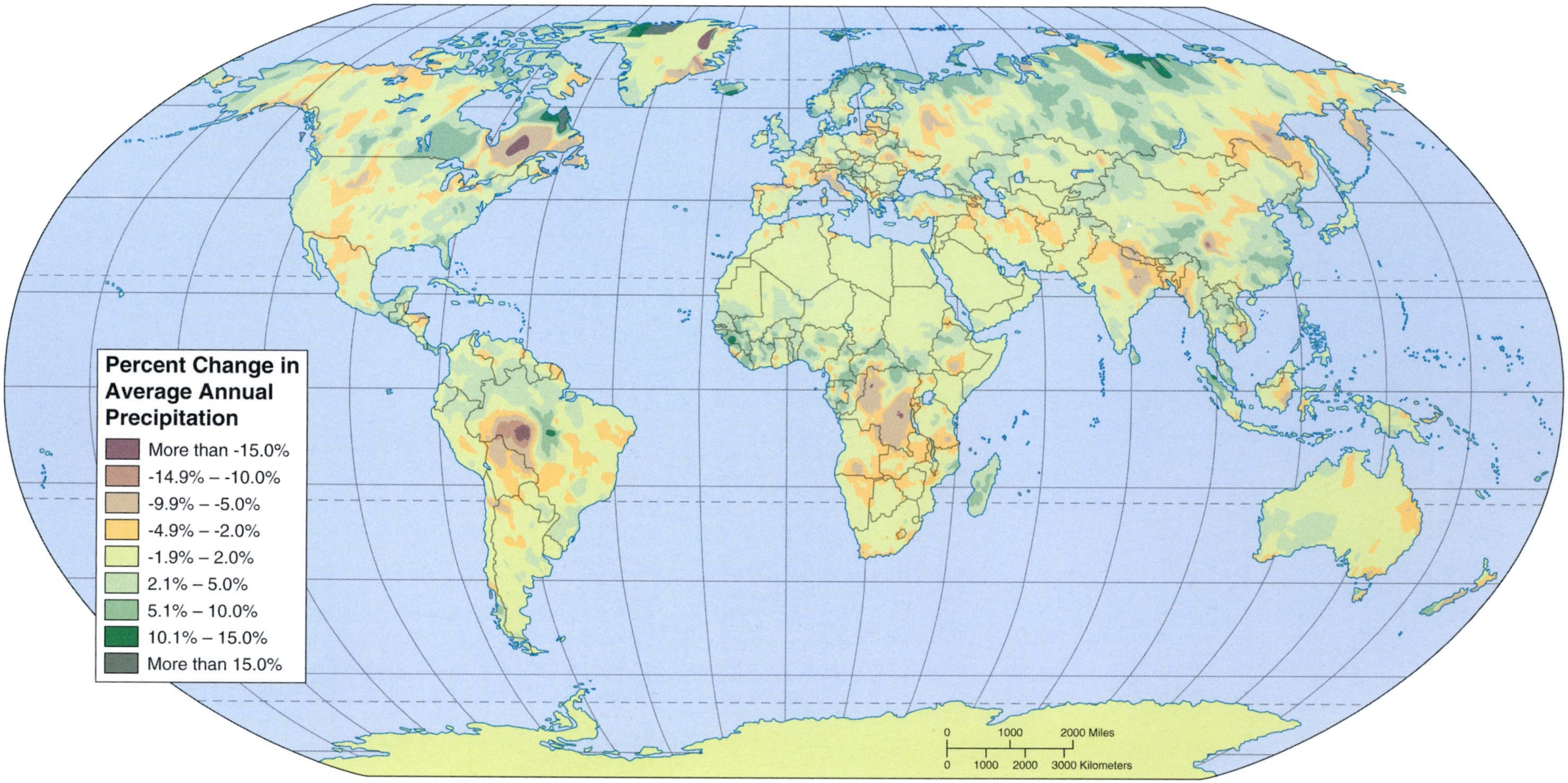

The links between global temperature change and alterations in the geographical distribution of precipitation are so complex as to be only marginally understood. What can be said with certainty is that increasing global temperatures will produce altered precipitation patterns. We have already begun to experience some of this in the form of prolonged droughts in the interior regions of North America; increased precipitation in the previously semi-arid region of the Sahel in Africa; and severe and long-term drought in the southern portions of the Amazon Basin, adjacent areas in South America, and in those sections of Africa south of the Sahel. There is an approximate tendency for dry regions to become drier and for wet regions to become wetter with increasing temperatures, although this is not universally the case. The primary problem is that persistent, long-term changes in precipitation patterns will cause significant disruptions in established agricultural patterns such as those in the United States, China, India, and Brazil—four of the world's most productive agricultural countries.

Map 79 Potential Global Temperature Change

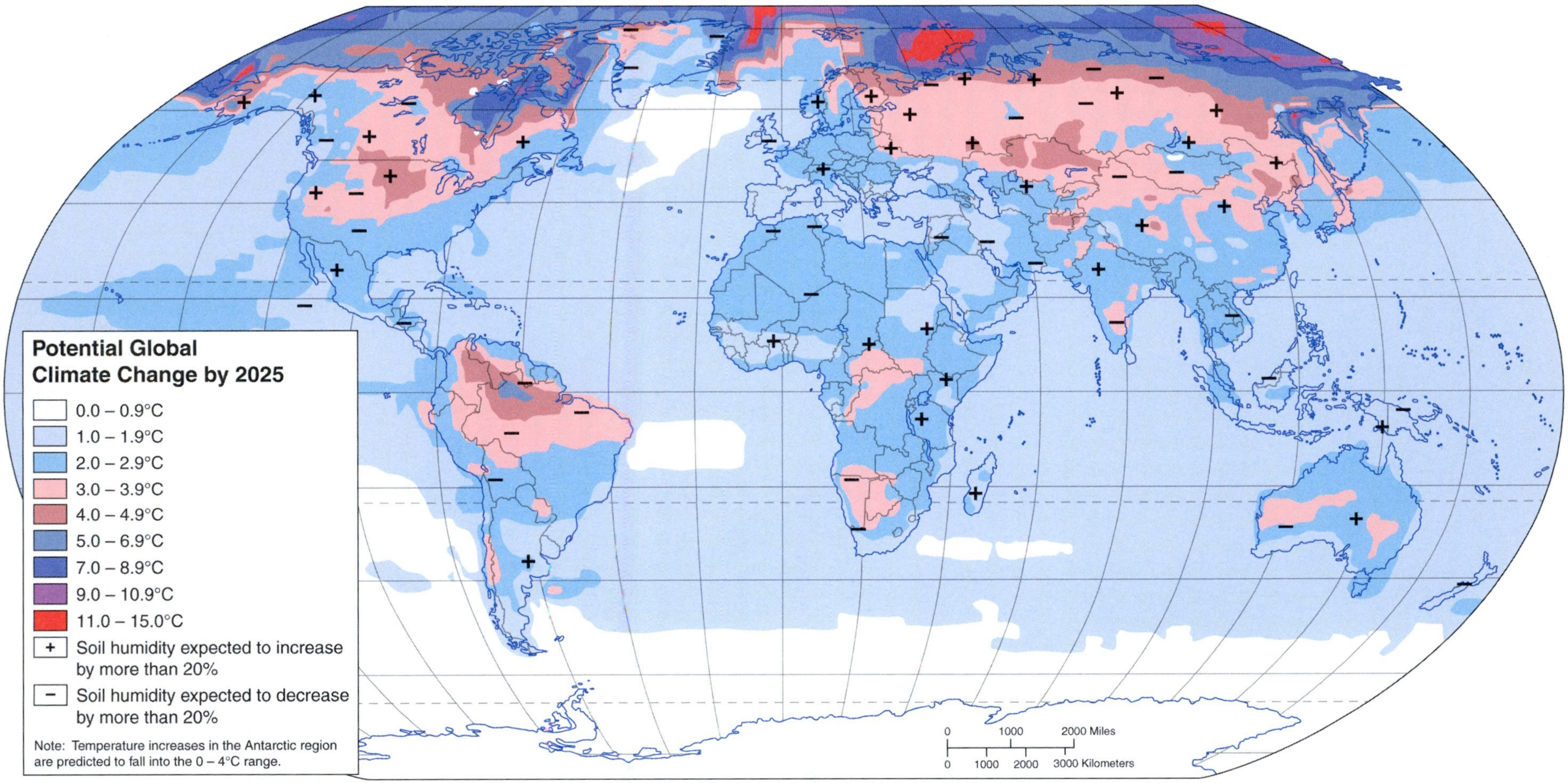

By the end of the first quarter of the twenty-first century, the world's population will probably be facing climatic conditions quite different from those of the last quarter of the twentieth century: significantly increased temperatures, particularly in the higher latitudes (closer to the poles), and significantly altered precipitation patterns with some of the world's great agricultural regions experiencing long-term and persistent drought. The increase of temperatures in the higher latitudes means greater melting of the Arctic polar ice sheet (ice floating on water), greater melting of the Greenland ice sheet (ice of incredible thickness now resting on land), and greater melting of the Antarctic ice sheet—similar in structure but larger than that of Greenland. There are some benefits of polar ice melting (as long as you are not a polar bear), most notably the opening of sea lanes across the Arctic Ocean—the final establishment of the Northwest Passage sought for by European explorers for several centuries. But there are many more hazards in the form of even small increases in sea level—a meter or two would be enough to produce a major increase in damage by even moderate tropical storms, let alone hurricanes. And many low-lying coastal and island communities run the risk of complete inundation by rising sea waters. Is all this inevitable just because a computer model such as the one that generated the above map predicts it? No. But ignoring the evidence gathered and analyzed so carefully by objective scientists would be both foolhardy and arrogant.

Map 80 Water Resources: Availability of Renewable Water Per Capita

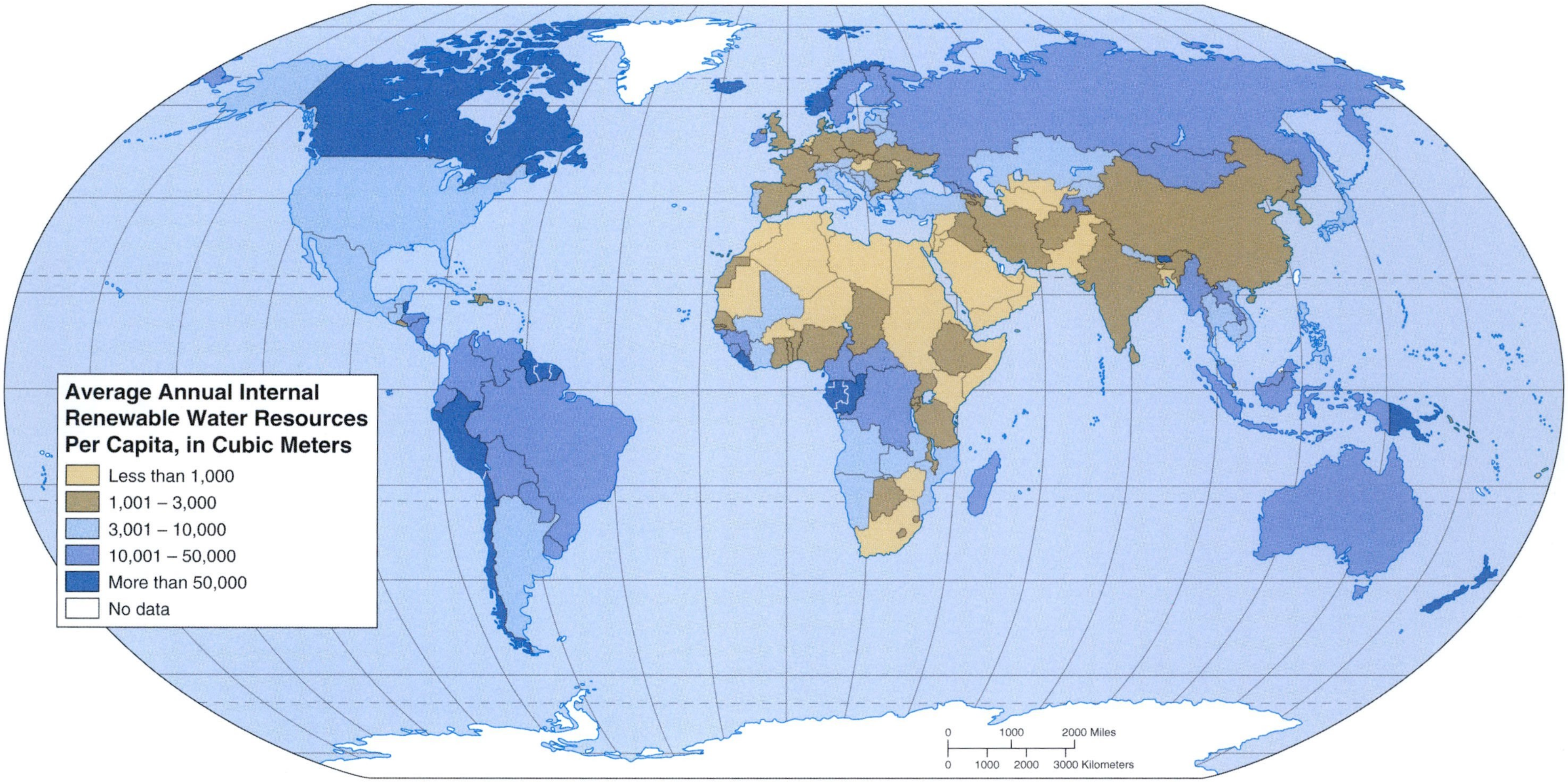

Renewable water resources are usually defined as the total water available from streams and rivers (including flows from other countries), ponds and lakes, and groundwater storage or aquifers. Not included in the total of renewable water would be water that comes from such nonrenewable sources as desalinization plants or melted icebergs. While the concept of renewable or flow resources is a traditional one in resource management, in fact, few resources, including water, are truly renewable when their use is excessive. The water resources shown here are indications of that principle. A country like the United States possesses truly enormous quantities of water. But the United States also uses enormous quantities of water. The result is that, largely because of excessive use, the availability of renewable water is much less than in many other parts of the world where the total supply of water is significantly less.

Map 81 Water Resources: Annual Withdrawal Per Capita

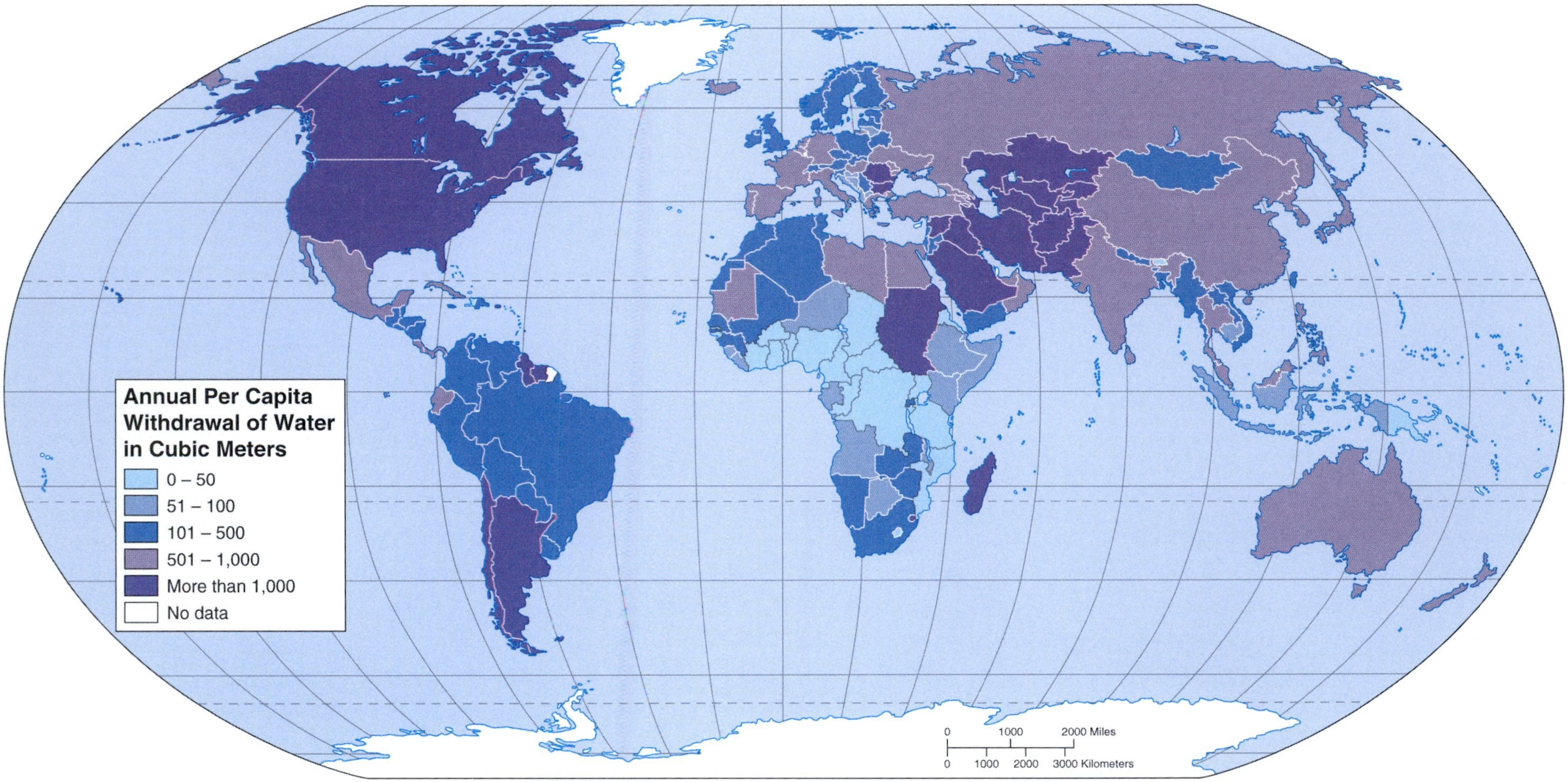

Water resources must be viewed like a bank account in which deposits and withdrawals are made. As long as the deposits are greater than the withdrawals, a positive balance remains. But when the withdrawals begin to exceed the deposits, sooner or later (depending on the relative sizes of the deposits and withdrawals) the account becomes overdrawn. For many of the world's countries, annual availability of water is insufficient to cover the demand. In these countries, reserves stored in groundwater are being tapped, resulting in depletion of the water supply (think of this as shifting money from a savings account to a checking account). The water supply can maintain its status as a renewable resource only if deposits continue to be greater than withdrawals, and that seldom happens. In general, countries with high levels of economic development and countries that rely on irrigation agriculture are the most spendthrift when it comes to their water supplies.

Map 82 Water Stress: Shortage, Abundance, and Population Density

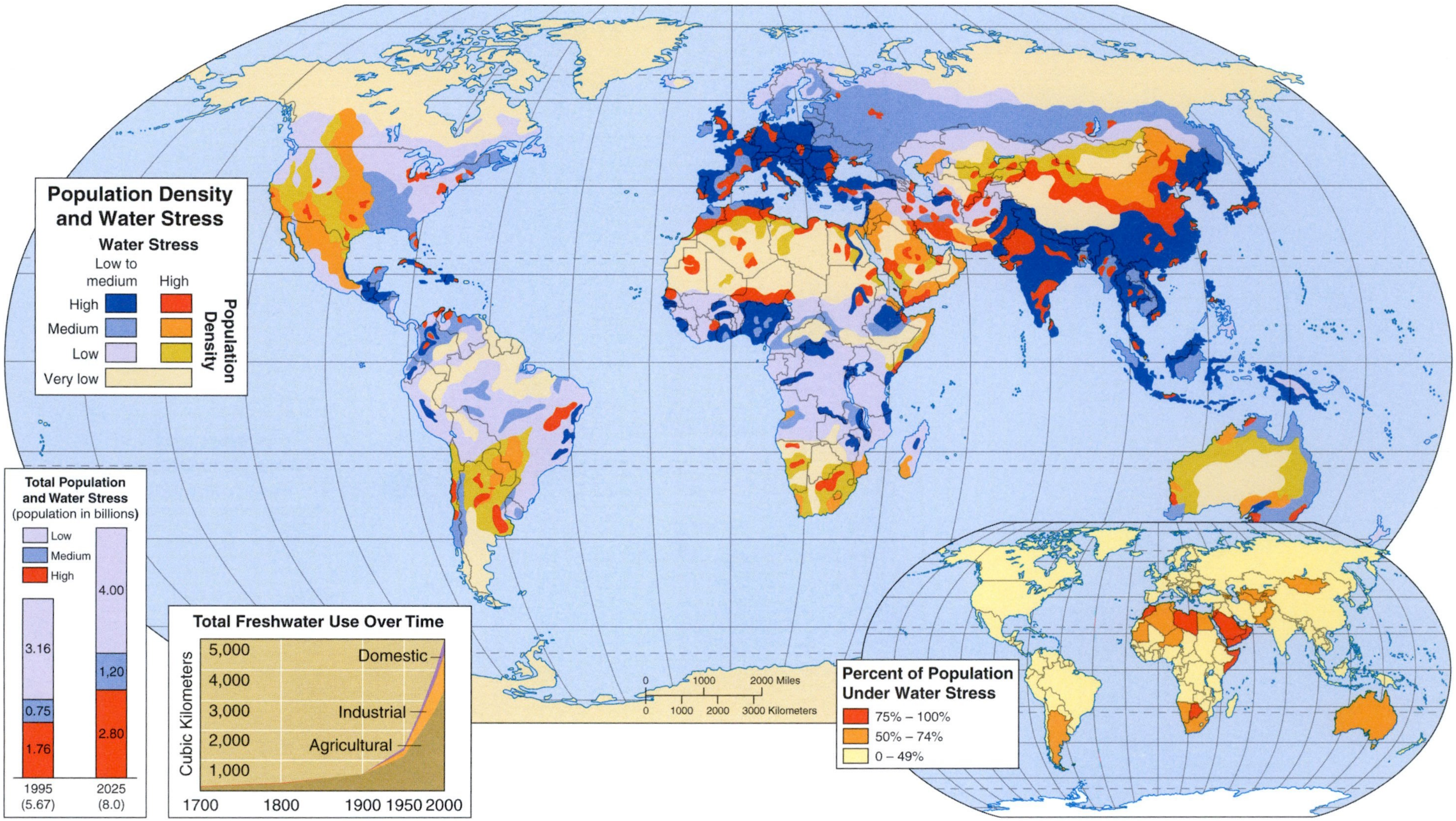

Maps such as the previous two, based on national-level data for water consumption and availability, should be used only to obtain national-level understanding. Information on water withdrawal and availability are regionally and locally based geographic phenomena and are linked not just with water supplies but with the density of human populations. Even areas (such as New England in the United States) in which water availability is high and withdrawal rates are relatively low show areas of stress in regions of high population density (cities such as Boston). This map, originally produced by scientists at the University of New Hampshire, attempts to show those areas of the world where populations will tend to be at high, medium, and low risk of stress because of water availability. It is important to note that many of the world's prime agricultural regions, such as the Great Plains of the United States or the Argentine Pampas, show the potential for high risk of water stress in the immediate future. Why is this important? Because the greatest single use of water on the planet is for irrigation (nearly 70 percent of the world's water use) and it is the continued expansion of irrigation systems that allows the increase in agricultural production that feeds the Earth's more than 6 billion persons.

Map 83 A New Environmental Stress? Consumption of Bottled Water

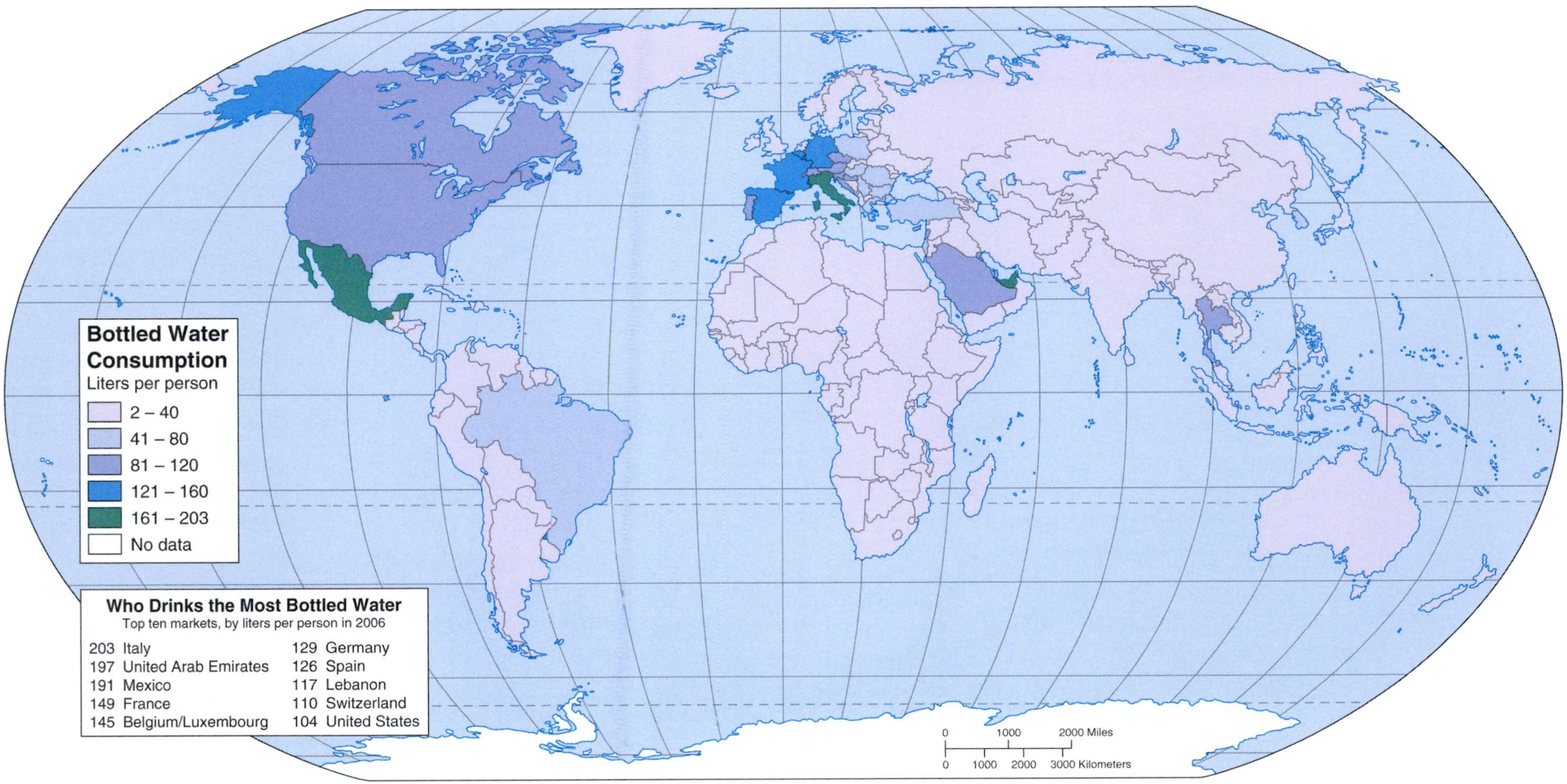

It is probably a truism that most people who commonly use bottled water seldom think of the environmental costs of doing so. Bottled water consumption tends to correlate spatially with higher levels of income (there are, of course, exceptions) and the highest rates of bottled water consumption tend to be in North America and Europe. The consumption of bottled water is one of the most costly environmental practices developed within the last quarter century. Most bottled water comes in plastic bottles that are seldom recycled (it has been estimated that in the United States alone, over 60 million plastic water bottles are simply thrown away each day). Enormous quantities of petroleum are required to produce and ship water bottles and bottled water is more costly, ounce-for-ounce, than gasoline—even when gasoline exceeds $4.00 per gallon. There are much less costly alternatives, including reusable plastic bottles with built-in filters that can be used to hold 200 gallons of water before the filters need to be replaced. In much of the developed world (where the bulk of bottled water is consumed), the easiest solution is to simply drink water from the tap—which is where, advertising notwithstanding, most bottled water comes from originally anyway.

Map 84 Pollution of the Oceans

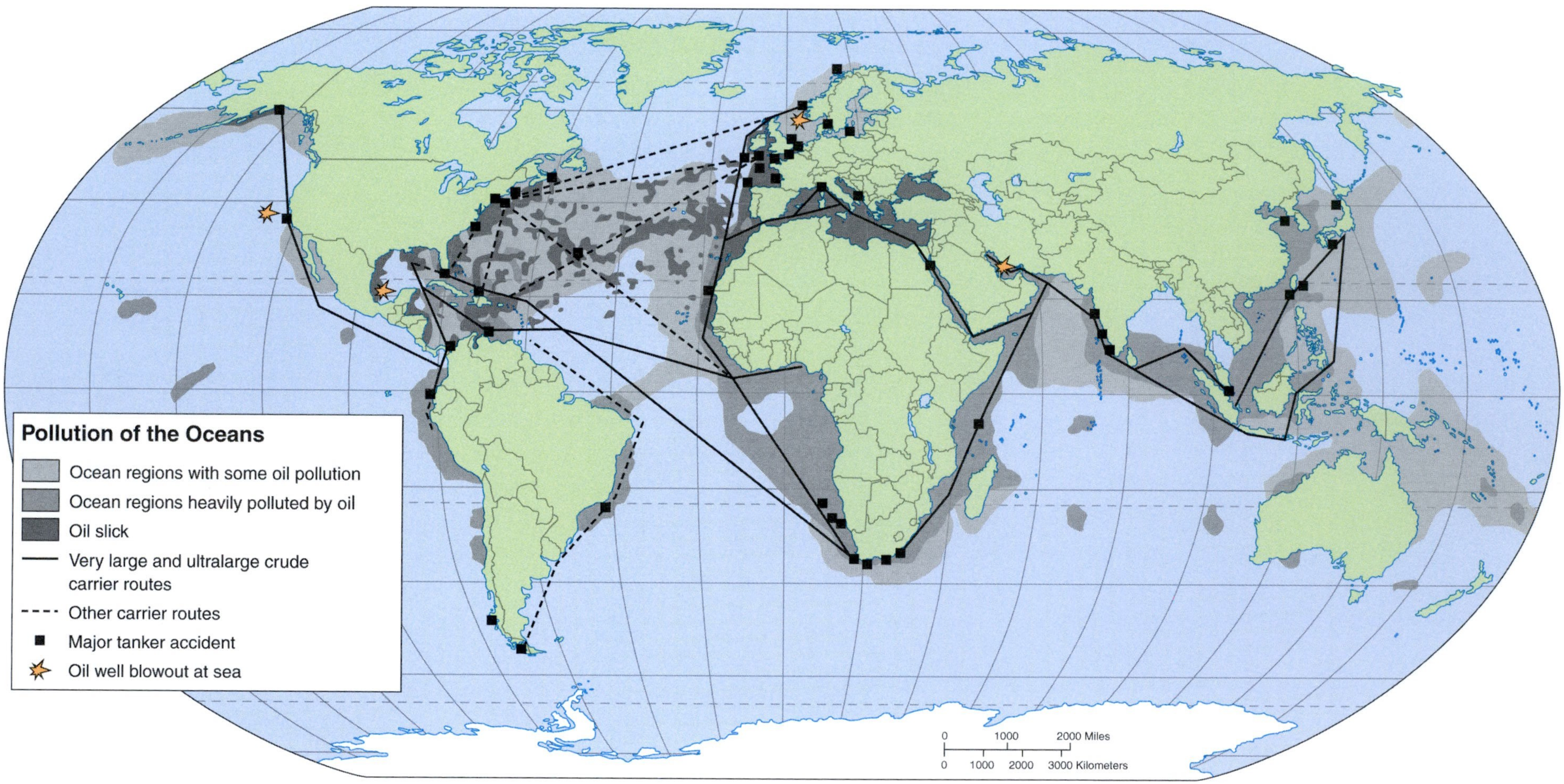

The pollution of the world's oceans has long been a matter of concern to physical geographers, oceanographers, and other environmental scientists. The great circulation systems of the ocean are one of the controlling factors of the earth's natural environment, and modifications to those systems have unknown consequences. This map is based on what we can measure: (1) areas of oceans where oil pollution has been proven to have inflicted significant damage to ocean ecosystems and life forms (including phytoplankton, the oceans' primary food producers, equivalent to land-based vegetation) and (2) areas of oceans where unusually high concentrations of hydrocarbons from oil spills may have inflicted some damage to the oceans' biota. A glance at the map shows that there are few areas of the world's oceans where some form of pollution is not a part of the environmental system. What the map does not show in detail, because of the scale, are the dramatic consequences of large individual pollution events: the wreck of the *Exxon Valdez* and the polluting of Prince William Sound, or the environmental devastation produced by the 1991 Gulf War in the Persian Gulf.

Map 85 Food Supply from Marine and Freshwater Systems

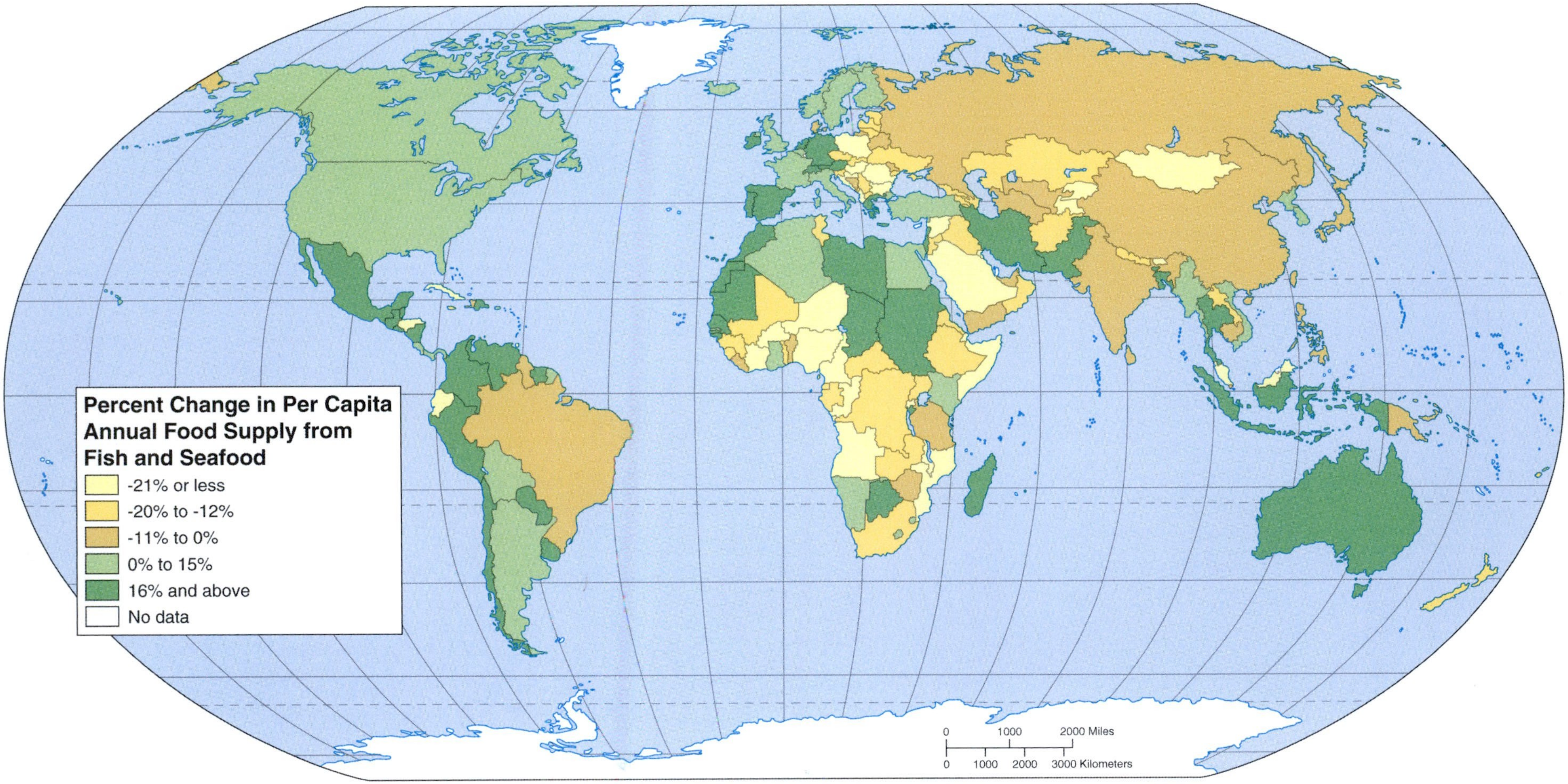

Not that many years ago, food supply experts were confidently predicting that the "starving millions" of the world of the future could be fed from the unending bounty of the world's oceans. While the annual catch from the sea helped to keep hunger at bay for a time, by the late 1980s it had become apparent that without serious human intervention in the form of aquaculture, the supply of fish would not be sufficient to offset the population/food imbalance that was beginning to affect so many of the world's regions. The development of factory-fishing with advanced equipment to locate fish and process them before they went to market increased the supply of food from the ocean, but in that increase was sown the seeds of future problems. The factory-fishing system, efficient in terms of economics, was costly in terms of fish populations. In some well-fished areas, the stock of fish that was viewed as near infinite just a few decades ago has dwindled nearly to the point of disappearance. This map shows both increases and decreases in the amount of individual countries' food supplies from the ocean. The increases are often the result of more technologically advanced fishing operations. The decreases are usually the result of the same thing: increased technology has brought increased harvests, which has reduced the supply of fish and shellfish and that, in turn, has increased prices. Most of the countries that have experienced sharp decreases in their supply of food from the world's oceans are simply no longer able to pay for an increasingly scarce commodity.

Map 86 Cropland Per Capita: Changes 1996–2007

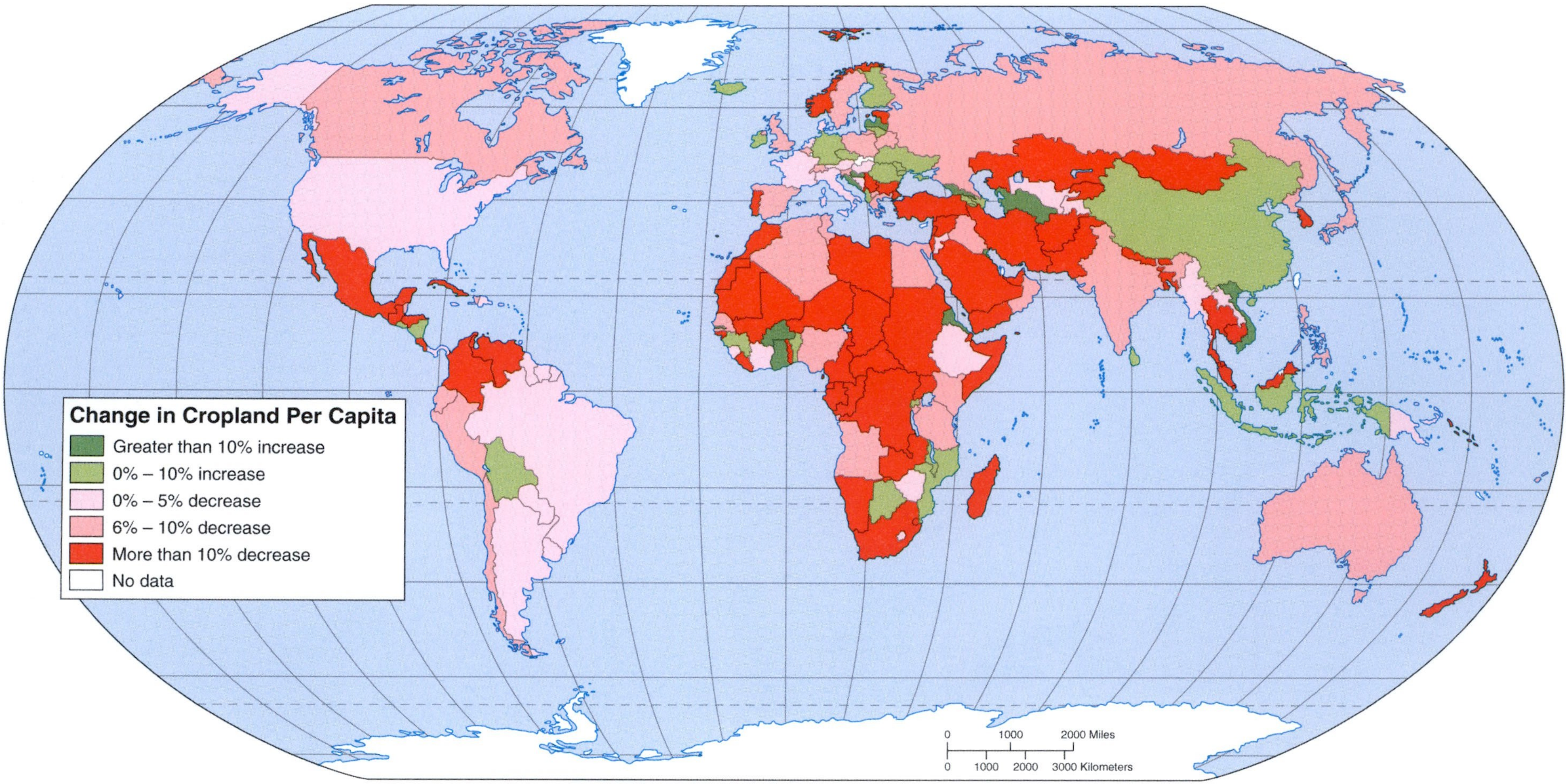

As population has increased rapidly throughout the world, the area of cultivated land has increased at the same time; in fact, the amount of farmland per person has gone up slightly. Unfortunately, the figures that show this also tell us that since most of the best (or even good) agricultural land in 1985 was already under cultivation, most of the agricultural area added since the mid 1990s involves land that would have been viewed as marginal by the fathers and grandfathers of present farmers—marginal in that it was too dry, too wet, too steep to cultivate, too far from a market, and so on. The continued expansion of agricultural area is one reason that serious famine and starvation have struck only a few regions of the globe. But land, more than any other resource we deal with, is finite, and the expansion cannot continue indefinitely. Future gains in agricultural production are most probably going to come through more intensive use of existing cropland, heavier applications of fertilizers and other agricultural chemicals, and genetically engineered crops requiring heavier applications of energy and water, than from an increase in the amount of the world's cropland.

Map 87 Food Staples Under Stress: The Fight Against Genetic Simplicity

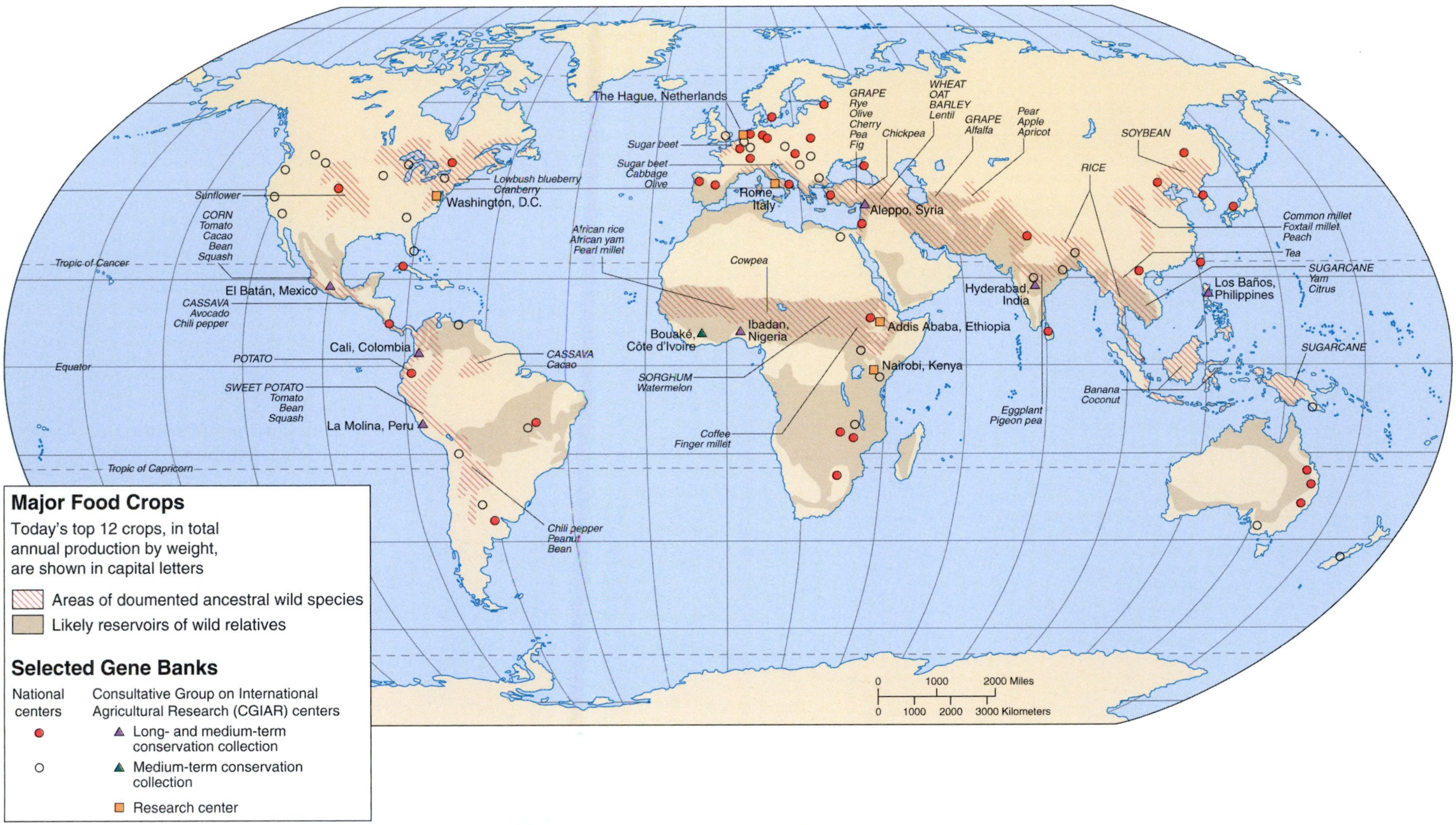

The point has been made elsewhere in this atlas that the loss of biological diversity is one of the greatest threats to ecosystem stability. That is certainly true when we speak of the threatened loss of diversity among those plants that provide the Earth's population with the bulk of its food supply. As farming becomes more specialized and certain strains of wheat, rice, corn, etc., achieve dominance because of their increased productivity levels, there is great danger in "putting all our eggs in one basket." Plant geneticists are attempting to protect the world against the loss of genetic diversity and the potential starvation that could result from crop failure of the favored strains by preserving as much of the genetic material of both domesticated relatives and wild ancestors of the most-used grains and other crops. Protecting the world's genetic reserves of plants is more than just an academic exercise. It is insurance against the biological degradation that future population growth is going to create and, therefore, provides some measure of assurance that future food production will be able to meet future demand.

Map 88 The Rise in Organic Farming

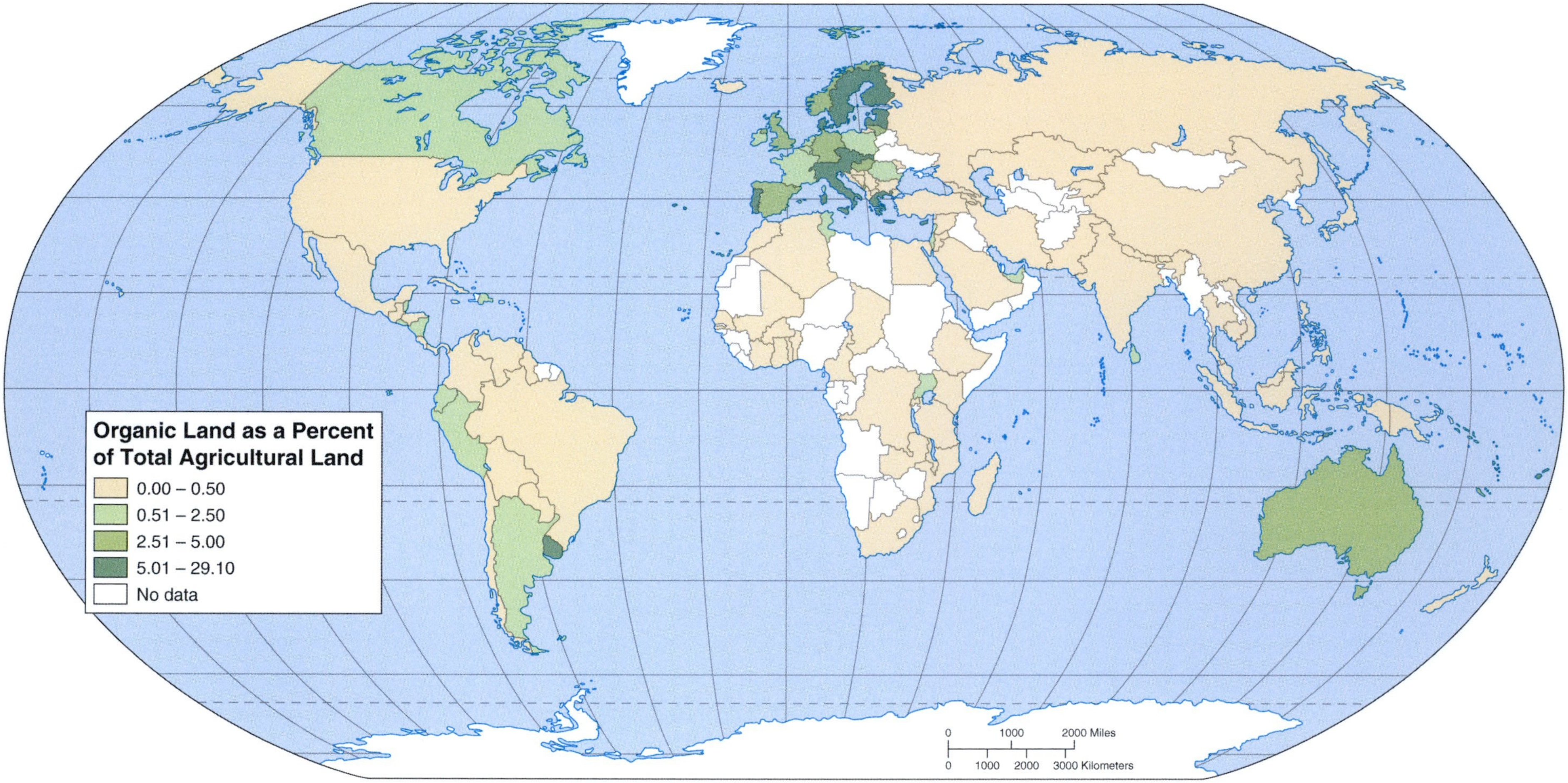

Both the demand for and the amount of land in organic crops have been increasing dramatically over the past decade. Part of both the increased demand and the increased amount of organically-farmed land is the consequence of growing affluence and greater awareness of both the benefits of organically-grown fruits and vegetables and the hazards of pesticides, herbicides, artificial fertilizers, and other agricultural chemicals that are part of "traditional" commercial agriculture in the developed world. There is also increasing awareness that the long-held assumption that organically-grown produce has to be much more expensive is, put simply, false. Organic farmers, particularly those who have become part of organic farming co-ops and have established marketing arrangements with the growing organic food outlets, can certainly be competitive with their more conventional counterparts. It is a myth that food that is actually good for you is priced out of the reach of all but the most affluent. This is simply not the case and as that recognition grows, we are likely to see greater percentages of agricultural land being devoted to organic farming.

Map 89 World Pastureland 2005

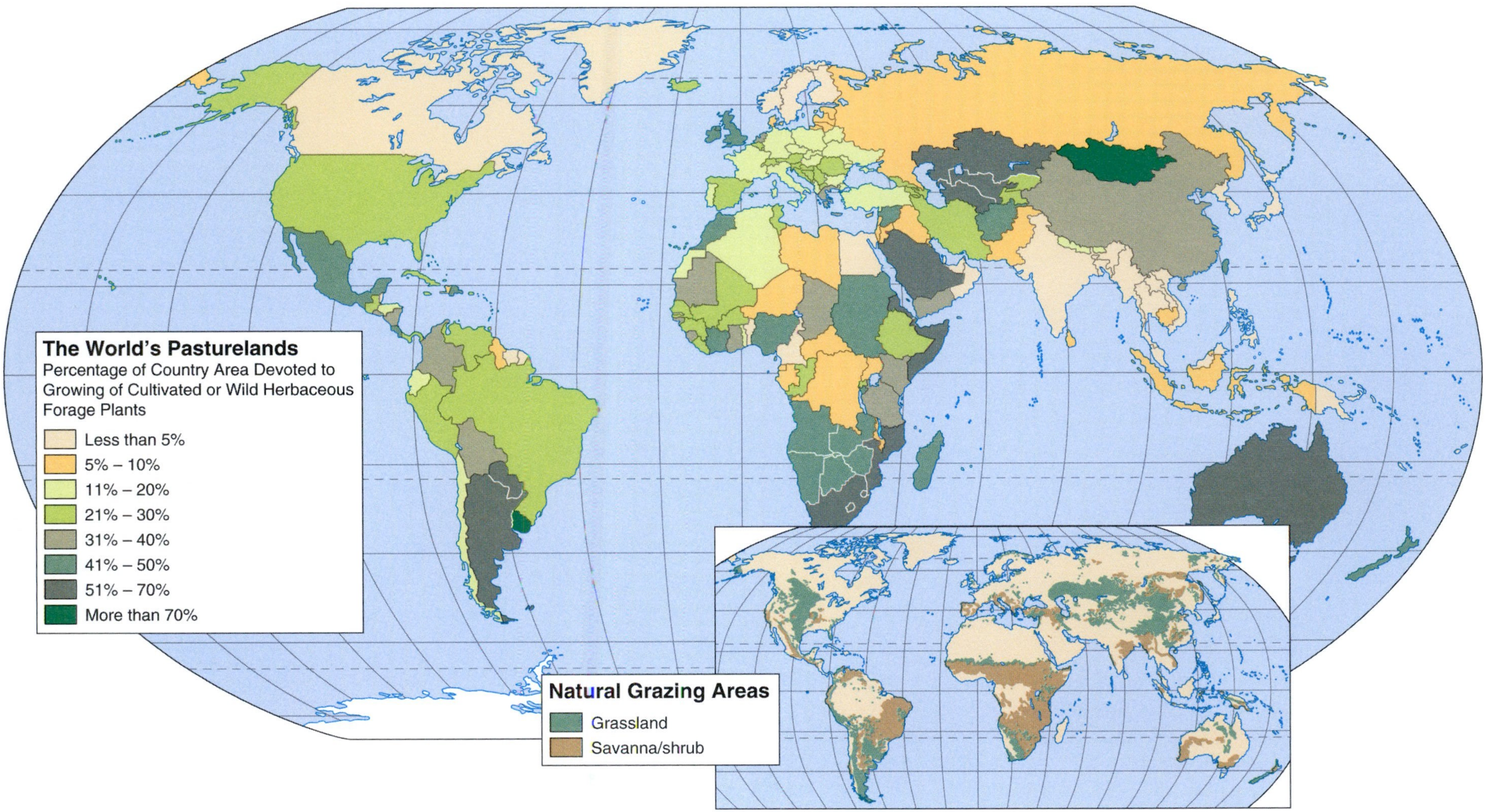

More than 25 percent of the world's surface is considered pastureland—either native wild grasses or human made pastures created by clearing forests and planting grass and other herbaceous forage plants for livestock feed. Two types of pasture predominate: the prairie and steppe grasses of the mid-latitudes and the savanna grasses of the sub-tropics and tropics. These two primary grassland biomes are shown in the inset map. The larger map depicts pastureland as a proportion of land area of individual countries. You will note that many of the countries with high levels of pastureland (Argentina and Australia, for example) are among the world's leading exporters of livestock products. Other countries with high levels of pastureland (Saudi Arabia and the Central Asian states) consume the bulk of their products domestically. The significance of the distribution and use of pastureland is that—whether it is in the African Sahel, China, Brazil, or the United States—the world's pasturelands are deteriorating rapidly under increasing demands to produce more animal products than even a wealthier world can afford to pay for (see Map 38). Grassland degradation, particularly in areas where pastoral nomads use their animals as their chief source of food and income, creates woody scrublands where the carrying capacity for grazing animals is sharply diminished. In short, most of the world's pasturelands are overgrazed, and the world's supply of animal products is in jeopardy.

Map 90 Fertilizer Use 2007

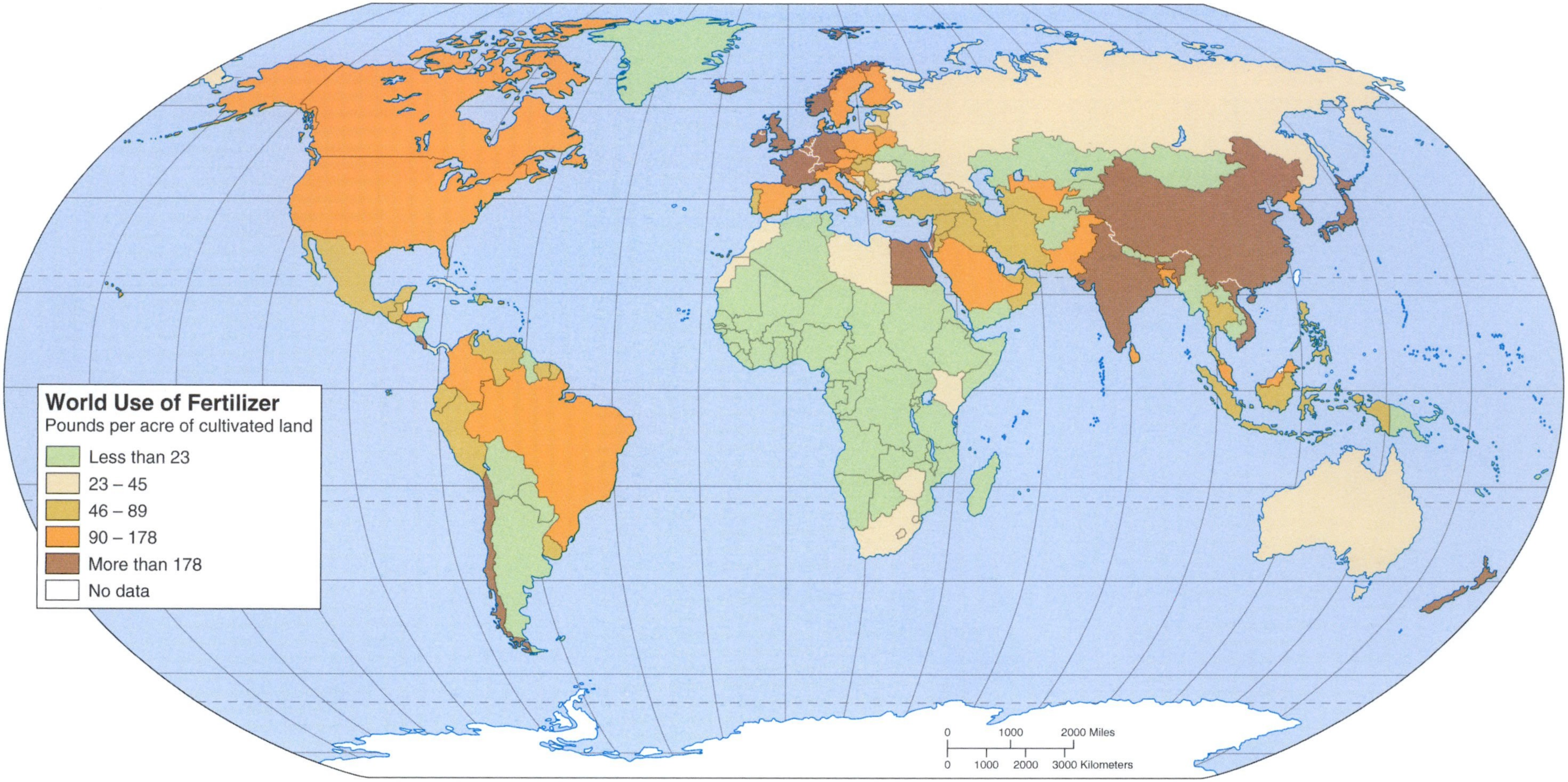

The use of fertilizer to maintain the productivity of agricultural lands is a wonderful agricultural invention—as long as the fertilizers used are natural rather than artificial. In most of the world's developed countries, such as those in Europe and North America, the use of animal manure to fertilize fields has decreased dramatically over the past century, in favor of artificial fertilizers that are cheaper and easier to use and—what is most important—increase crop yields more dramatically. The danger here is that artificial fertilizers normally have high concentrations of nitrates that tend to convert to nitrites in the soil, reducing the ability of soil bacteria to extract "free" nitrogen from the atmosphere. As more artificial fertilizer is used, natural soil fertility is decreased, creating the demand for more artificial fertilizers. In some areas, overuse of artificial fertilizers has actually created soils that are too "hot" chemically to produce crops. Countries with high fertilizer use in the developing world—northern South America, Southwest, South, and East Asia—still tend to use more natural fertilizers. But as farmers in those areas gain more ability to buy and use artificial fertilizers, their soils will also begin to suffer from overfertilization. Global agriculture needs to come to grips with the need to maintain productivity but to do so in a manner that is sustainable.

Map 91 Annual Deforestation Rates 2000–2005

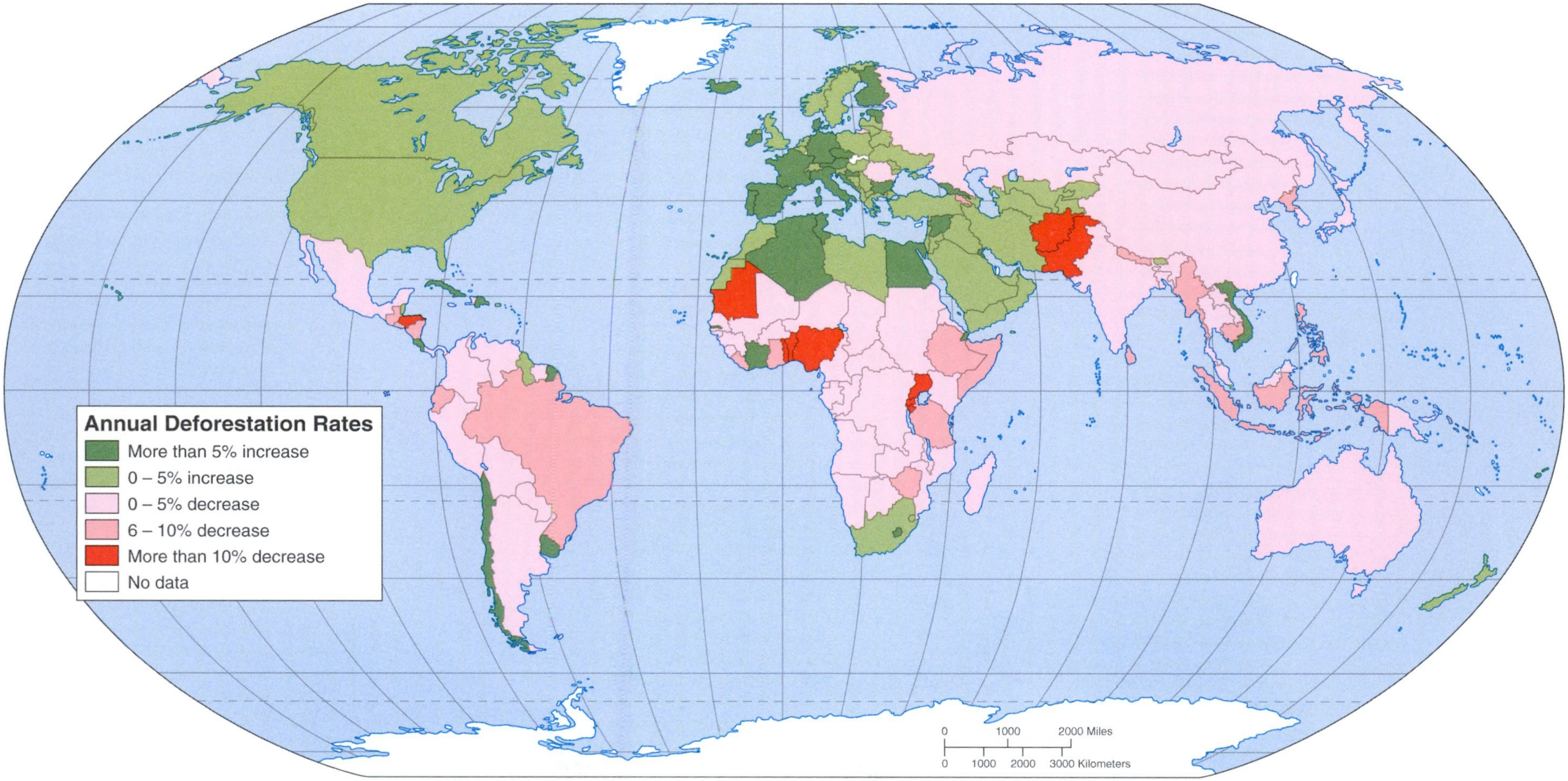

One of the most discussed environmental problems is that of deforestation. For most people, deforestation means clearing of tropical rain forests for agricultural purposes. Yet nearly as much forest land per year—much of it in North America, Europe, and Russia—is impacted by commercial lumbering as is cleared by tropical farmers and ranchers. Even in the tropics, much of the forest clearance is undertaken by large corporations producing high-value tropical hardwoods for the global market in furniture, ornaments, and other fine wood products. Still, it is the agriculturally driven clearing of the great rain forests of the Amazon Basin, west and central Africa, Middle America, and Southeast Asia that draws public attention. Although much concern over forest clearance focuses on the relationship between forest clearance and the reduction in the capacity of the world's vegetation system to absorb carbon dioxide (and thus delay global warming), of just as great concern are issues having to do with the loss of biodiversity (large numbers of plants and animals), the near-total destruction of soil systems, and disruptions in water supply that accompany clearing.

Map 92 World Timber Production 2005

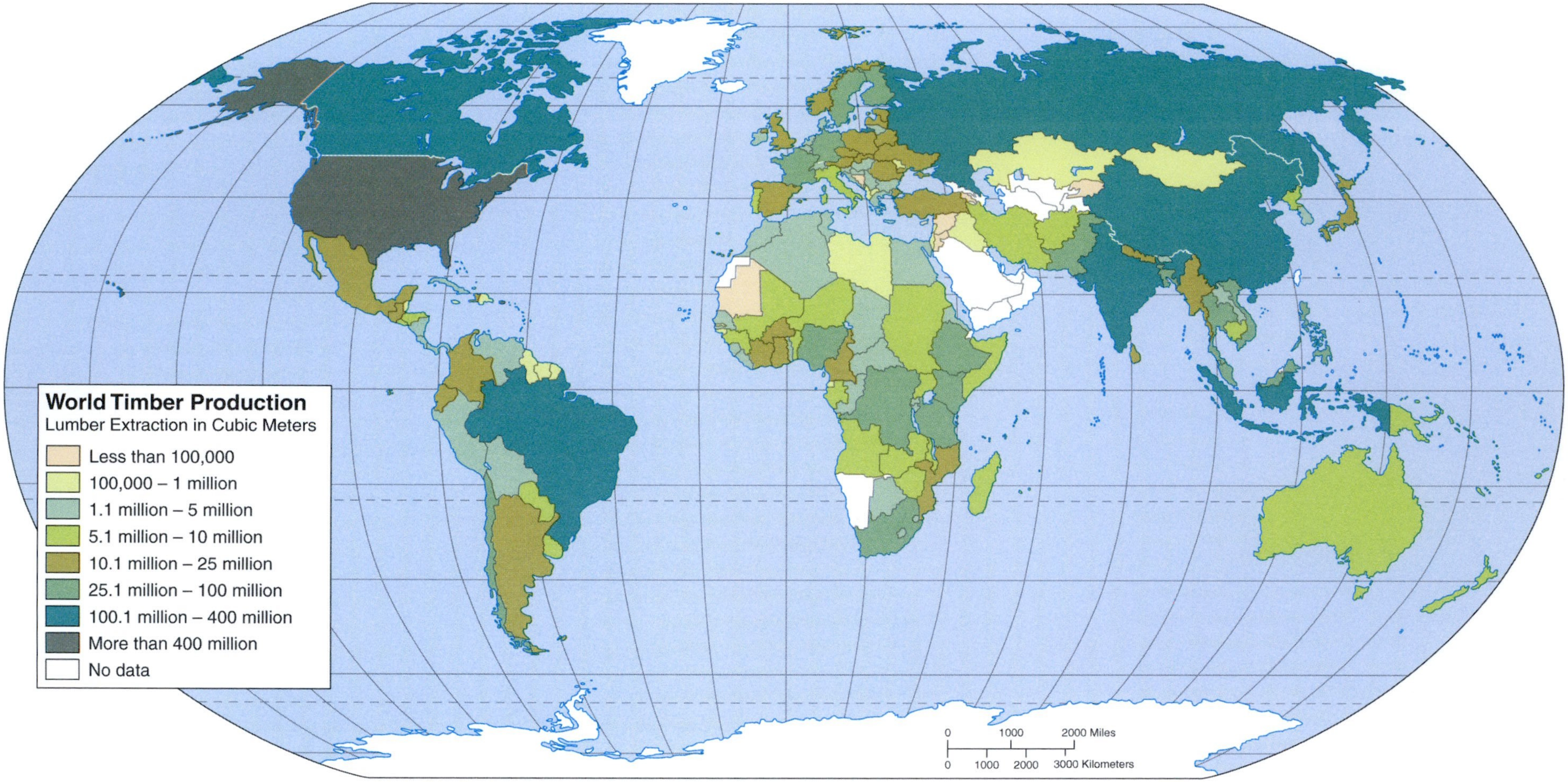

With the exception of water and arable land, forests have played a larger part in the transition from hunting-gathering to the emergence of modern industrial economies than any other natural resource. The chief uses of forests are for lumber, fuelwood, and pulp and paper products. The per capita use of wood does not tend to vary greatly between the developed and developing countries. In developed areas such as North America, Europe, and Russia, the bulk of forest products is used for lumber (much of it exported to forest-poor areas) and for pulp and paper products. In the developing countries in South America, Africa, and Asia (excluding Russia), more wood is used for fuel than for any other purpose. In those parts of the world where forests are used for lumber and pulp-paper products, more-or-less sustainable forest systems have been developed. These systems are often artificially maintained and are much simpler (and, hence, more fragile) ecosystems than the rich, complex natural ecosystems they have replaced. Where forests provide the chief source of fuel, the forest systems tend to remain healthier but are shrinking in size in proportion to the increase in human populations. Of the major forest ecosystems, those in greatest jeopardy are the tropical forests that are being cleared for agricultural land and the boreal or northern forests of Canada and Russia that are being simplified by overcutting.

Map 93 The Loss of Biodiversity: Globally Threatened Animal Species 2007

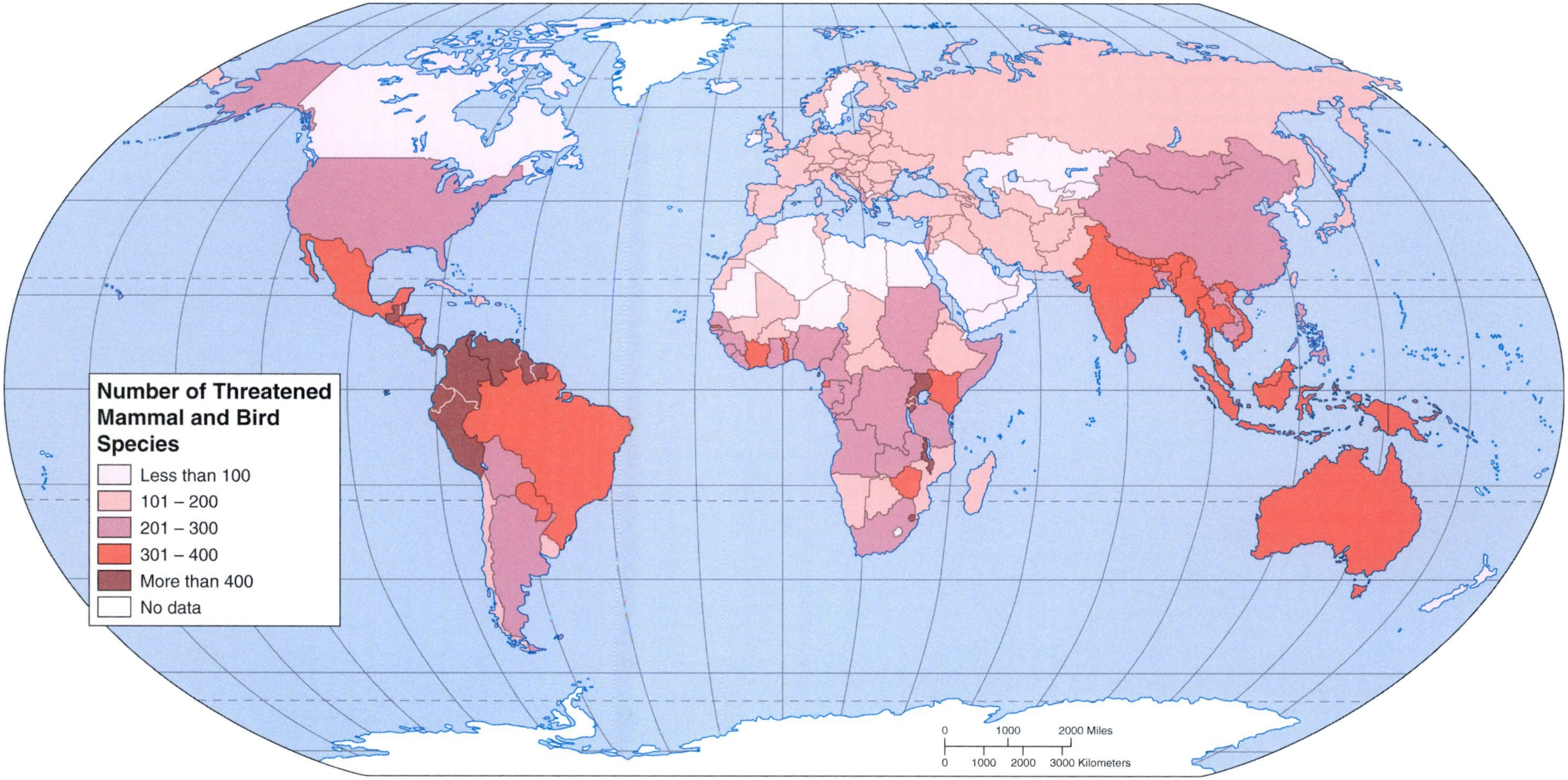

Threatened species are those in grave danger of going extinct. Their populations are becoming restricted in range, and the size of the populations required for sustained breeding is nearing a critical minimum. *Endangered species* are in immediate danger of becoming extinct. Their range is already so reduced that the animals may no longer be able to move freely within an ecozone, and their populations are at the level where the species may no longer be able to sustain breeding. Most species become threatened first and then endangered as their range and numbers continue to decrease. When people think of animal extinction, they think of large herbivorous species like the rhinoceros or fierce carnivores like lions, tigers, or grizzly bears. Certainly these animals make almost any list of endangered or threatened species. But there are literally hundreds of less conspicuous animals that are equally threatened. Extinction is normally nature's way of informing a species that it is inefficient. But conditions in the late twentieth century are controlled more by human activities than by natural evolutionary processes. Species that are endangered or threatened fall into that category because, somehow, they are competing with us or with our domesticated livestock for space and food. And in that competition the animals are always going to lose.

Map 94 Disappearing Giants: The Reduction in "Conservative" Species

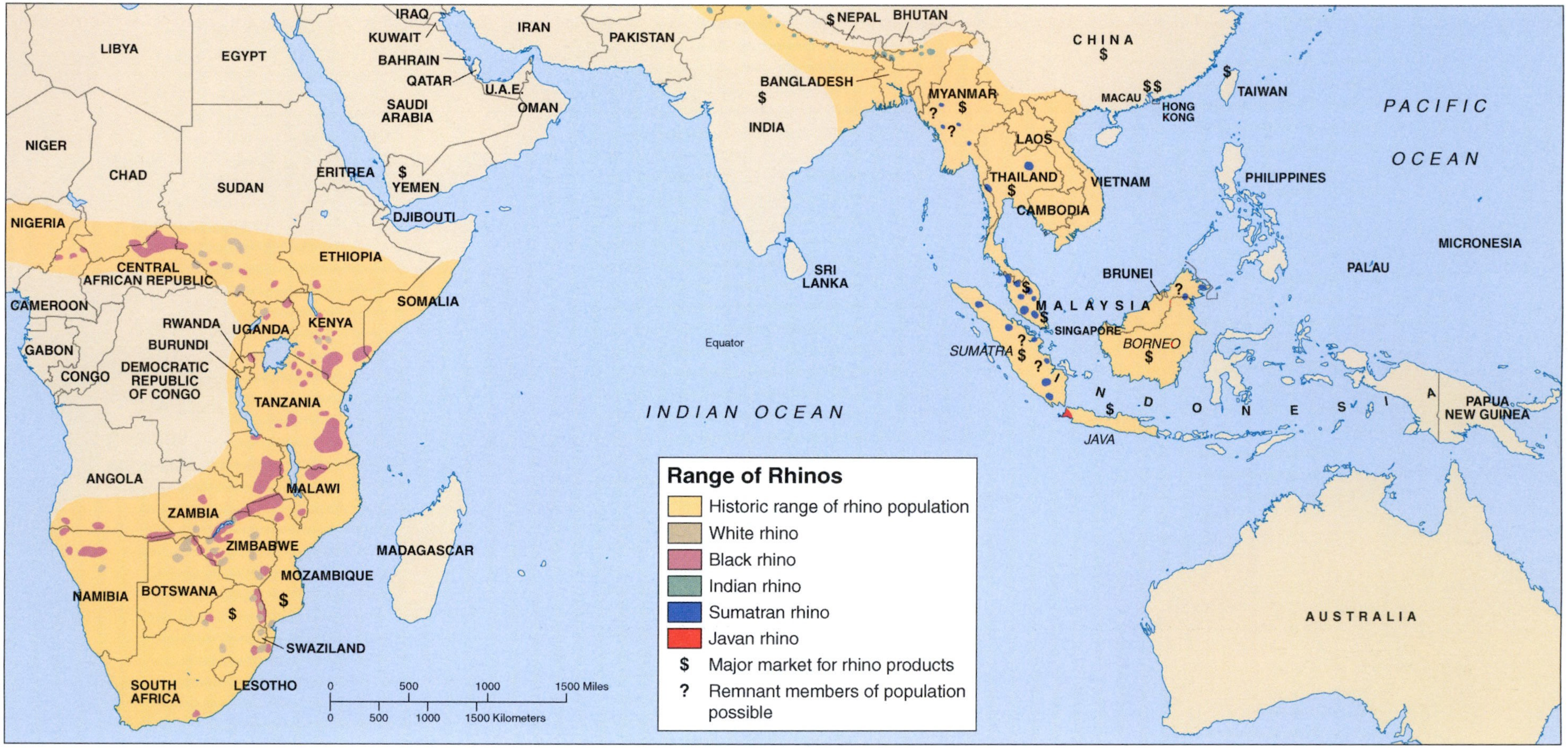

The species singled out for special attention have been given that attention not just because they represent major species in greatest danger of extinction but because they are signal species (indicating the health of ecosystems) as well as conservative or aristocratic species. These species often suffer most from human infringement on traditional land or from overhunting. Rhinos, elephants, and tigers fall into the conservative or aristocratic categories: they are large, they breed slowly, they do not produce large numbers of offspring in a single breeding season, they may have long gestation periods, etc. Conservative species are also often those that are both in great demand for their hides, horns, tusks, or fur (in other words, non-edible products that demand extremely high prices in the luxury market), and that tend to compete most directly with humans for the same resources—even if that resource is simply space. Part of the rapid loss of such aristocratic species as rhinos, elephants, and tigers is the value of ivory, tiger hides, or rhinoceros horn in the global market; but part of the rapid loss is that these species occupy space that is also needed for cropland or grazing land. In other words, these species compete with humans for the same resources and, in any such competition, it is the humans who are going to win. And a win for humans represents a huge loss for those animals that may have formerly been the ecological dominants in their home ranges.

MEDITERRANEAN SEA
MOROCCO
TUNISIA
ALGERIA
LIBYA
EGYPT
WESTERN SAHARA
MAURITANIA
MALI
NIGER
CHAD
SUDAN
ERITREA
DJIBOUTI
ETHIOPIA
SOMALIA
SENEGAL
GAMBIA
GUINEA-BISSAU
GUINEA
SIERRA LEONE
LIBERIA
CÔTE D'IVOIRE
GHANA
TOGO
BENIN
BURKINA FASO
NIGERIA
CAMEROON
CENTRAL AFRICAN REPUBLIC
EQUATORIAL GUINEA
GABON
CONGO
DEMOCRATIC REPUBLIC OF CONGO
UGANDA
KENYA
RWANDA
BURUNDI
TANZANIA
ANGOLA
ZAMBIA
MALAWI
ZIMBABWE
MOZAMBIQUE
MADAGASCAR
NAMIBIA
BOTSWANA
SOUTH AFRICA
SWAZILAND
LESOTHO
ATLANTIC OCEAN
INDIAN OCEAN
0 500 1000 Miles
0 500 1000 Kilometers
Range of Elephants
Remnant populations: may be below critical mass for reproduction
Higher density populations at or above critical mass for reproduction
1976 population = 1,500,000
2006 population = 500,000

0 500 1000 Miles
0 500 1000 Kilometers
RUSSIA
Lake Baikal
KAZAKHSTAN
Lake Balkash
ARAL SEA
UZBEKISTAN
CASPIAN SEA
CASPIAN TIGER
Panthera tigris virgata
EXTINCT
KYRGYZSTAN
TAJIKISTAN
TURKMENISTAN
AFGHANISTAN
IRAN
PAKISTAN
MONGOLIA
AMUR TIGER
Panthera tigris altaica
N. KOREA
S. KOREA
JAPAN
CHINA
CHINESE TIGER
Panthera tigris amoyensis
NEPAL
BHUTAN
INDIA
BANGLADESH
MYANMAR
INDOCHINESE TIGER
Panthera tigris corbetti
LAOS
THAILAND
VIETNAM
CAMBODIA
TAIWAN
PHILIPPINES
ATLANTIC OCEAN
PALAU
INDIAN TIGER
Panthera tigris tigris
SRI LANKA
INDIAN OCEAN
SUMATRA
SUMATRAN TIGER
Panthera tigris sumatrae
MALAYSIA
BRUNEI
INDONESIA
JAVA
BALI
JAVAN TIGER
Panthera tigris sondaica
EXTINCT
BALINESE TIGER
Panthera tigris balica
EXTINCT
Range of Tigers
100 years ago
Today
Project tiger reserve

Map 95 The Loss of Biodiversity: Globally Threatened Plant Species 2007

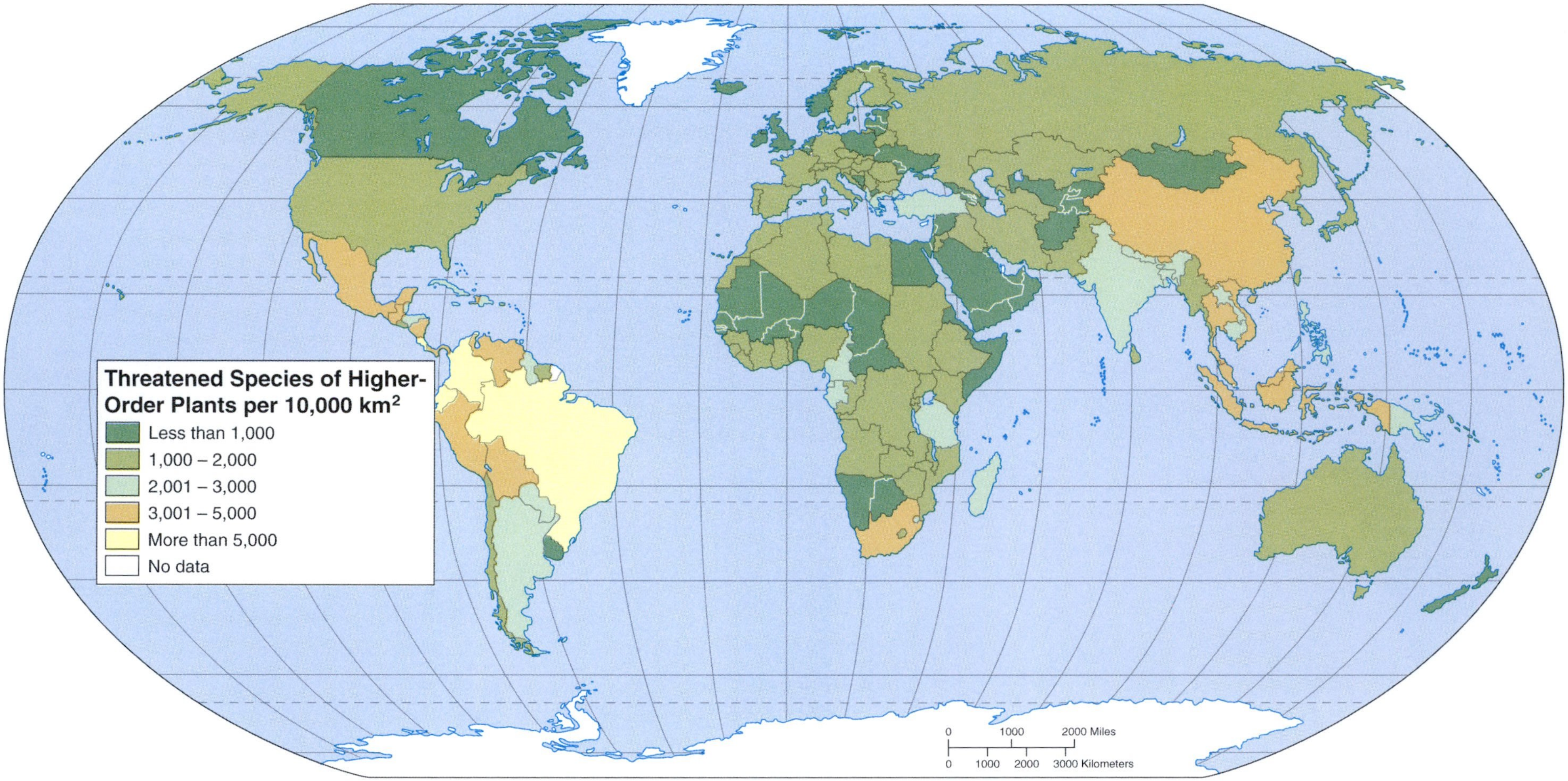

While most people tend to be more concerned about the animals on threatened and endangered species lists, the fact is that many more plants are in jeopardy, and the loss of plant life is, in all ecological regions, a more critical occurrence than the loss of animal populations. Plants are the primary producers in the ecosystem; that is, plants produce the food upon which all other species in the food web, including human beings, depend for sustenance. It is plants from which many of our critical medicines come, and it is plants that maintain the delicate balance between soil and water in most of the world's regions. When environmental scientists speak of a loss of biodiversity, what they are most often describing is a loss of the richness and complexity of plant life that lends stability to ecosystems. Systems with more plant life tend to be more stable than those with less. For these and other reasons, the scientific concern over extinction is greater when applied to plants than to animals. It is difficult for people to become as emotional over a teak tree as they would over an elephant. But as great a tragedy as the loss of the elephant would be, the loss of the teak would be greater.

Map 96 The Areas of Greatest Loss and Hotspots of Biodiversity

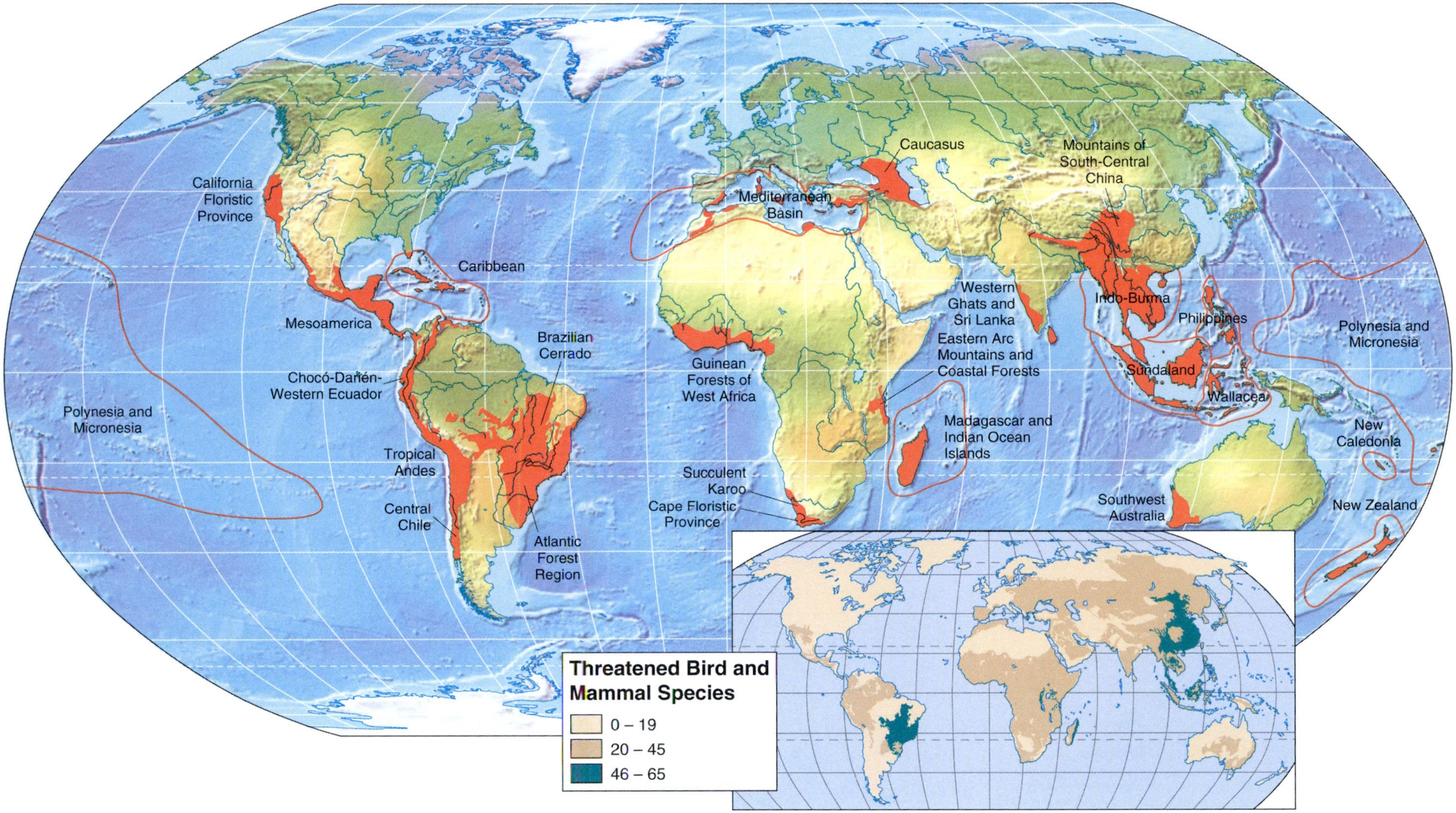

Where we have normally thought of tropical forest basins such as Amazonia as the world's most biologically diverse ecosystems, recent research has discovered the surprising fact that a number of hotspots of biological diversity exist outside the major tropical forest regions. These hotspot regions contain slightly less than 2 percent of the world's total land area but may contain up to 60 percent of the total world's terrestrial species of plants and animals. Geographically, the hotspot areas are characterized by vertical zonation (that is, they tend to be hilly to mountainous regions), long known to be a factor in biological complexity. They are also in coastal locations or near large bodies of water, locations that stimulate climatic variability and, hence, biological complexity. Although some of the hotspots are sparsely populated, others, such as Sundaland, are occupied by some of the world's densest populations. Protection of the rich biodiversity of these hotspots is, most biologists feel, of crucial importance to the preservation of the world's biological heritage. The greatest actual or potential losses of bird and mammal species tend to occur in those areas of the world where there is the most active competition for living spaces between humans and animals. Thus highly populated areas tend to suffer the greatest loss of native species. China, Brazil, and Indonesia stand out on the inset map; each of these countries ranks in the world's top 10 in total population.

Map 97 Nationally Protected Areas: Parks and Preserves

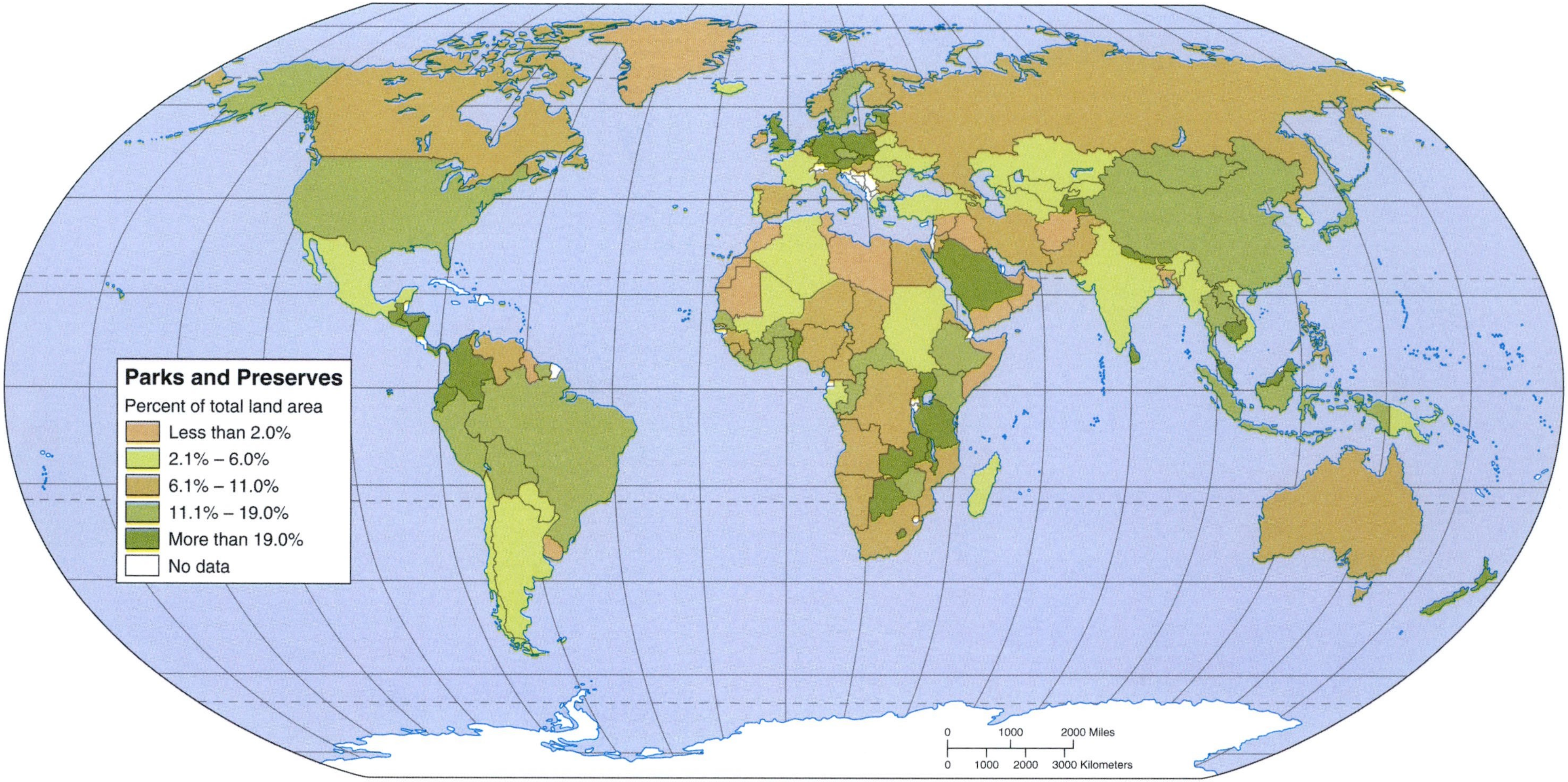

Interpreting the patterns on this map ranges from the obvious to the obscure. Some of the countries with the lowest level of nationally protected land are obvious—Iraq, Afghanistan, Somalia—where internal conflict trumps environmental protection. Other countries with low levels, such as Libya, suggest there is little perceived need for protecting lands in a desert environment (quite the opposite is true). For large countries such as Brazil, China, and the United States (three of the five largest countries in the world), to have such high percentages of nationally protected lands may indicate a high level of conservation awareness—or it may suggest that these countries have recognized the economic benefits of preserved lands in terms of increased tourism, etc. Or it may be a combination of the two. Certainly in the United States (where the first nationally-protected areas were established) the conservation ethic was the driving mechanism but tourism became increasingly important during the latter half of the twentieth century. Interestingly enough, those countries with the greatest percentages of their land areas in nationally protected spaces may not be countries that many of us would normally predict as being environmentally conservative or preservative. It is heartening that desert countries like Saudi Arabia have recognized the fragility of desert environments by protecting them. It is also heartening that those African countries that serve as the last refuges of the great animals have recognized the importance of attempting to preserve those species by setting aside important national parks and preserves. Countries such as Tanzania and Zambia deserve special credit for setting aside large tracts of land as wildlife refuges and preserves.

Map 98 The Risks of Desertification

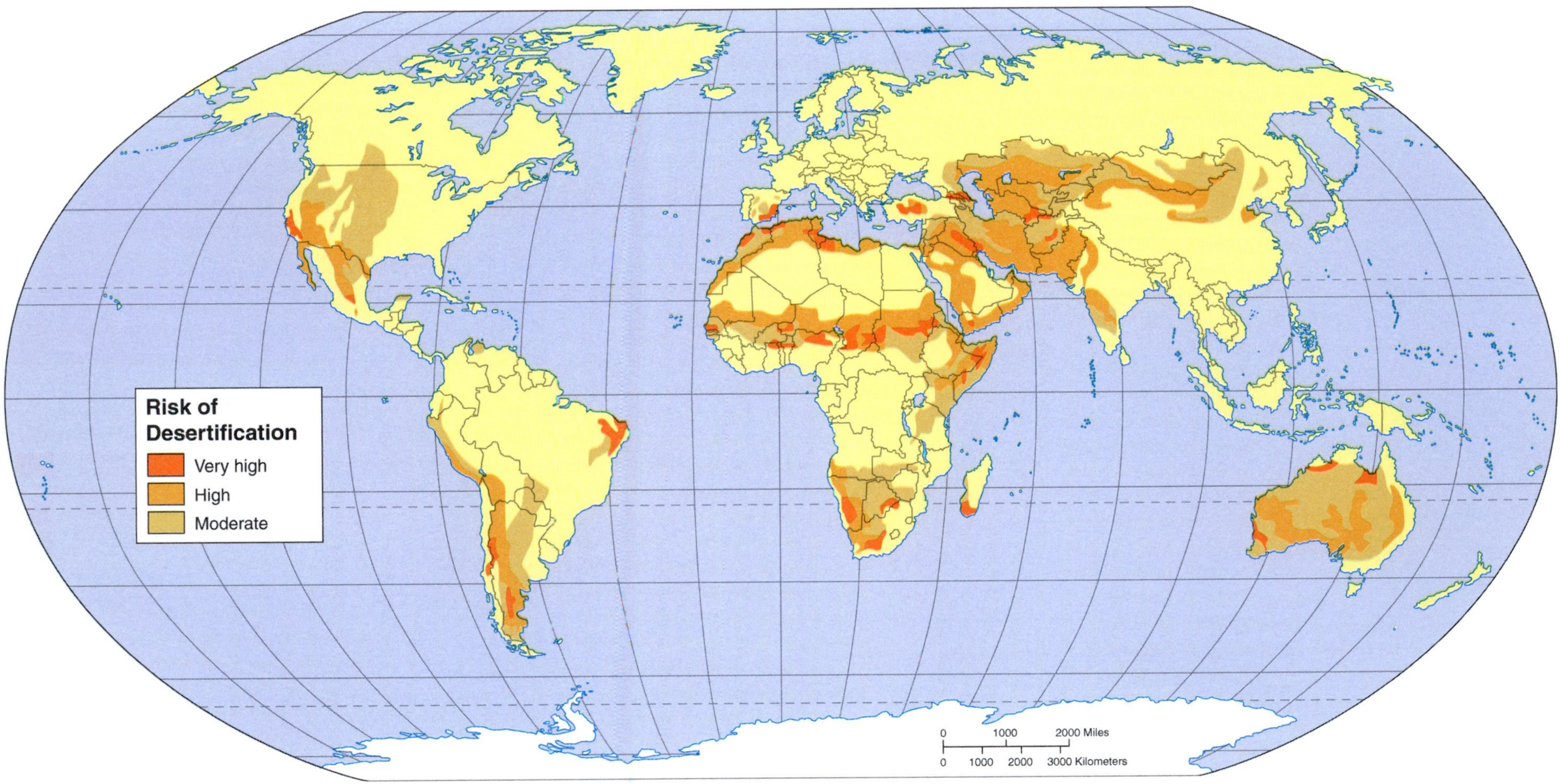

The awkward-sounding term *desertification* refers to a reduction in the food-producing capacity of drylands through vegetation, soil, and water changes that culminate in either a drier climate or in soil and plant systems that are less efficient in their use of water. Most of the world's existing drylands—the shortgrass steppes, the tropical savannas, the bunchgrass regions of the desert fringe—are fairly intensively used for agriculture and are, therefore, subject to the kinds of pressures that culminate in desertification. Most desertification is a natural process that occurs near the margins of desert regions. It is caused by dehydration of the soil's surface layers during periods of drought and by high water loss through evaporation in an environment of high temperature and high winds. This natural process is greatly enhanced by human agricultural activities that expose topsoil to wind and water erosion. Among the most important practices that cause desertification are (1) overgrazing of rangelands, resulting from too many livestock on too small an area of land; (2) improper management of soil and water resources in irrigation agriculture, leading to accelerated erosion and to salt buildup in the soil; (3) cultivation of marginal terrain with soils and slopes that are unsuitable for farming; (4) surface disturbances of vegetation (clearing of thorn scrub, mesquite, chaparral, and similar vegetation) without soil protection efforts being made or replanting being done; and (5) soil compaction by agricultural implements, domesticated livestock, and rain falling on an exposed surface.

Map 99 Global Soil Degradation

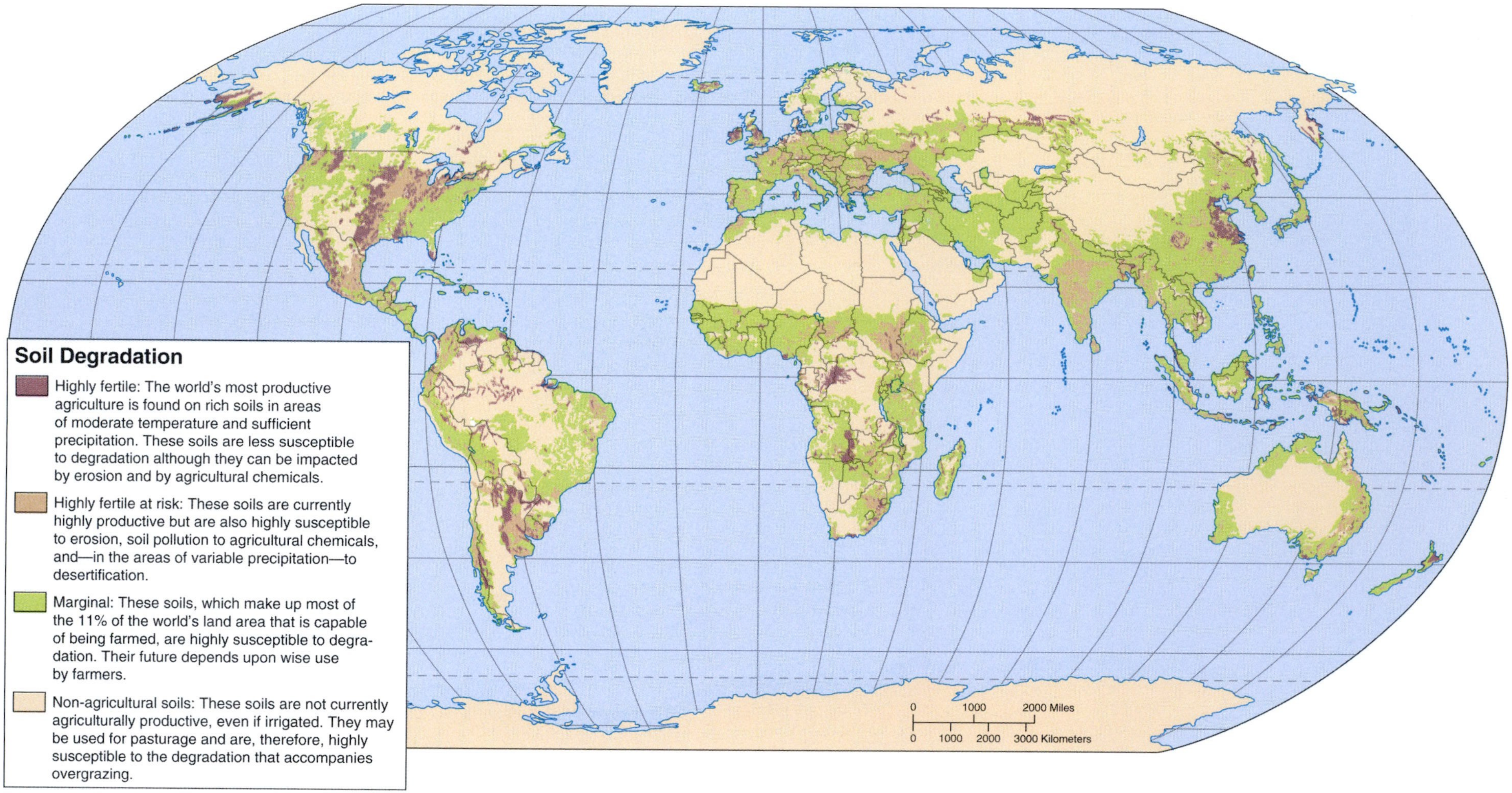

Recent research has shown that more than 3 billion acres of the world's surface suffer from serious soil degradation, with more than 22 million acres so severely eroded or poisoned with chemicals that they can no longer support productive crop agriculture. Most of this soil damage has been caused by poor farming practices, overgrazing of domestic livestock, and deforestation. These activities strip away the protective cover of natural vegetation forests and grasslands, allowing wind and water erosion to remove the topsoil that contains necessary nutrients and soil microbes for plant growth. But millions of acres of topsoil have been degraded by chemicals as well. In some instances these chemicals are the result of overapplication of fertilizers, herbicides, pesticides, and other agricultural chemicals. In other instances, chemical deposition from industrial and urban wastes and from acid precipitation has poisoned millions of acres of soil. As the map shows, soil erosion and pollution are not problems just in developing countries with high population densities and increasing use of marginal lands. They also afflict the more highly developed regions of mechanized, industrial agriculture. While many methods for preventing or reducing soil degradation exist, they are seldom used because of ignorance, cost, or perceived economic inefficiency.

Map 100 The Degree of Human Disturbance

The data on human disturbance have been gathered from a wide variety of sources, some of them conflicting and not all of them reliable. Nevertheless, at a global scale this map fairly depicts the state of the world in terms of the degree to which humans have modified its surface. The almost pristine areas, covered with natural vegetation, generally have population densities under 10 persons per square mile. These areas are, for the most part, in the most inhospitable parts of the world: too high, too dry, too cold for permanent human habitation in large numbers. The partially transformed areas are normally agricultural areas, either subsistence (such as shifting cultivation) or extensive (such as livestock grazing). They often contain areas of secondary vegetation, regrown after removal of original vegetation by humans. They are also sometimes marked by a density of livestock in excess of carrying capacity, leading to overgrazing, which further alters the condition of the vegetation. The almost fully transformed areas are those of permanent and intensive agriculture and urban settlement. The primary vegetation of these regions has been removed, with no evidence of regrowth or with current vegetation that is quite different from natural (potential) vegetation. Soils are in a state of depletion and degradation, and, in drier lands, desertification is a factor of human occupation. The disturbed areas match closely those areas of the world with the densest human populations.

Map 101 The Green and Not So Green World

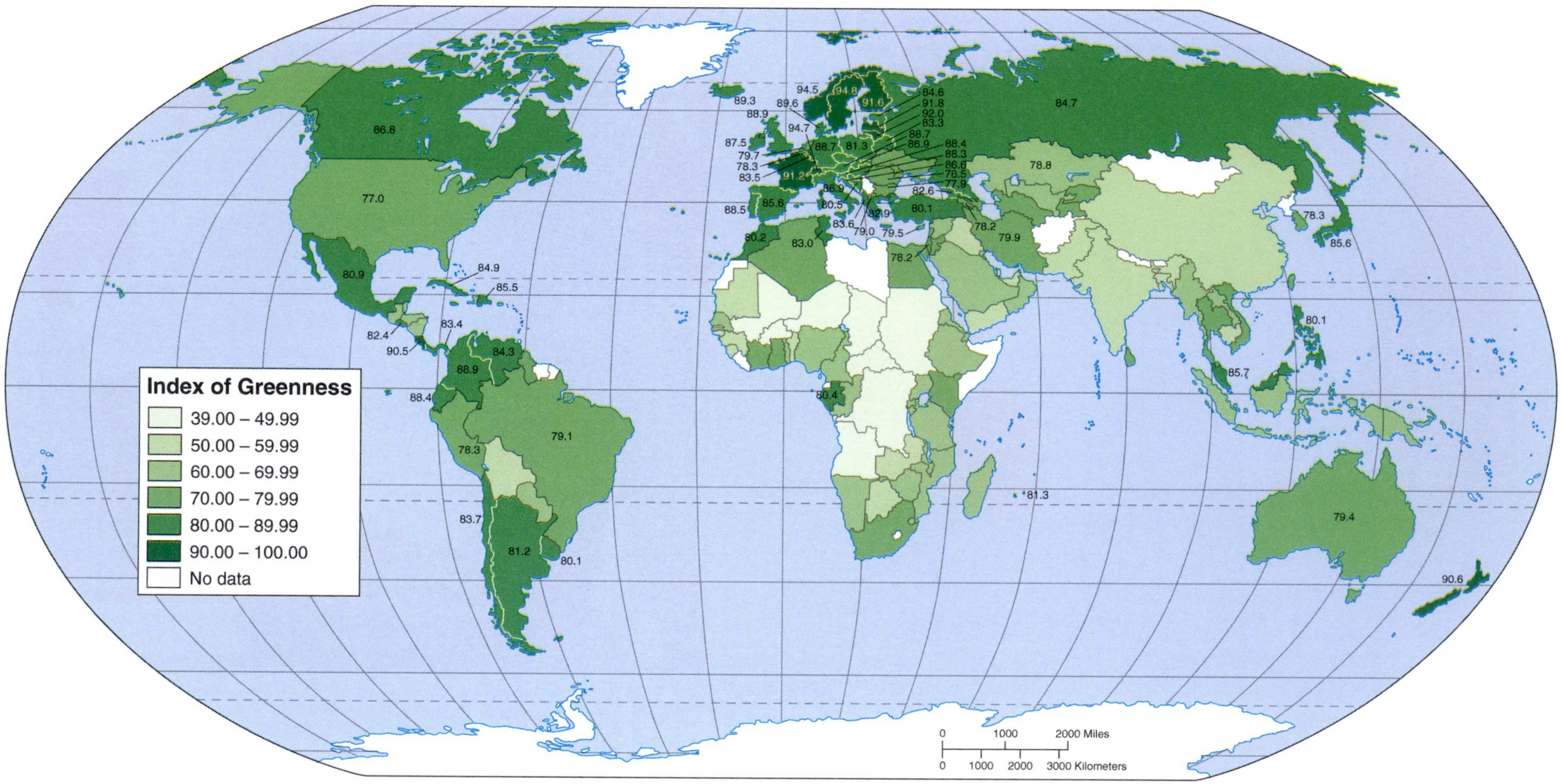

The index of "greenness" ranks countries according to their greenhouse gas emissions, the quality of their water resources, their protection of plant and animal habitats, and other factors. As might be expected, the "greenest" countries also tend to be relatively affluent and in cooler climates. There are obvious exceptions to this generalization (Mexico, for example) but, in general, we find countries like Sweden (no. 1), Switzerland, Norway, Finland, France, New Zealand, and Denmark ranking high on the index while poorer countries in warmer climates (much of South and East Asia, much of Sub-Saharan Africa) generally tend to rank lower on the index. There is a pretty clear correlation between affluence and "greenness"—that is, wealthy countries not only recognize the importance of "being green" but have the ability to pay for it. Still, some of the world's more affluent countries, like the United States and Australia, rank rather lower on the scale. The relatively low ranking (not quite in the top third of the countries in the index) attained by the United States—despite its excellent record on controlling water and air pollution—results from its persistent reliance on fossil fuels and a correspondingly high release of greenhouse gas emissions. Sweden, whose per capita income is similar to that of the United States, uses renewable resources (particularly hydroelectric power) to keep its emissions low and its ranking on the scale of "greenness" high. Similarly, France (no. 7) maintains a high rank because of a reliance on nuclear power, which curbs greenhouse gas emissions. In Sub-Saharan Africa, the level of carbon dioxide emissions is low but water quality and sanitation are poor. Rapid industrialization in countries like Brazil, India, and China have impacted emissions levels, as well as water and sanitation levels. These countries appear rather lower on the index (China, for example, ranks 134 among 186 countries). The rankings shown on the map are based on 2008 data; it will be interesting to see how the patterns have changed in 20 years.

Unit VI

Global Political Patterns

Map 102 Political Systems

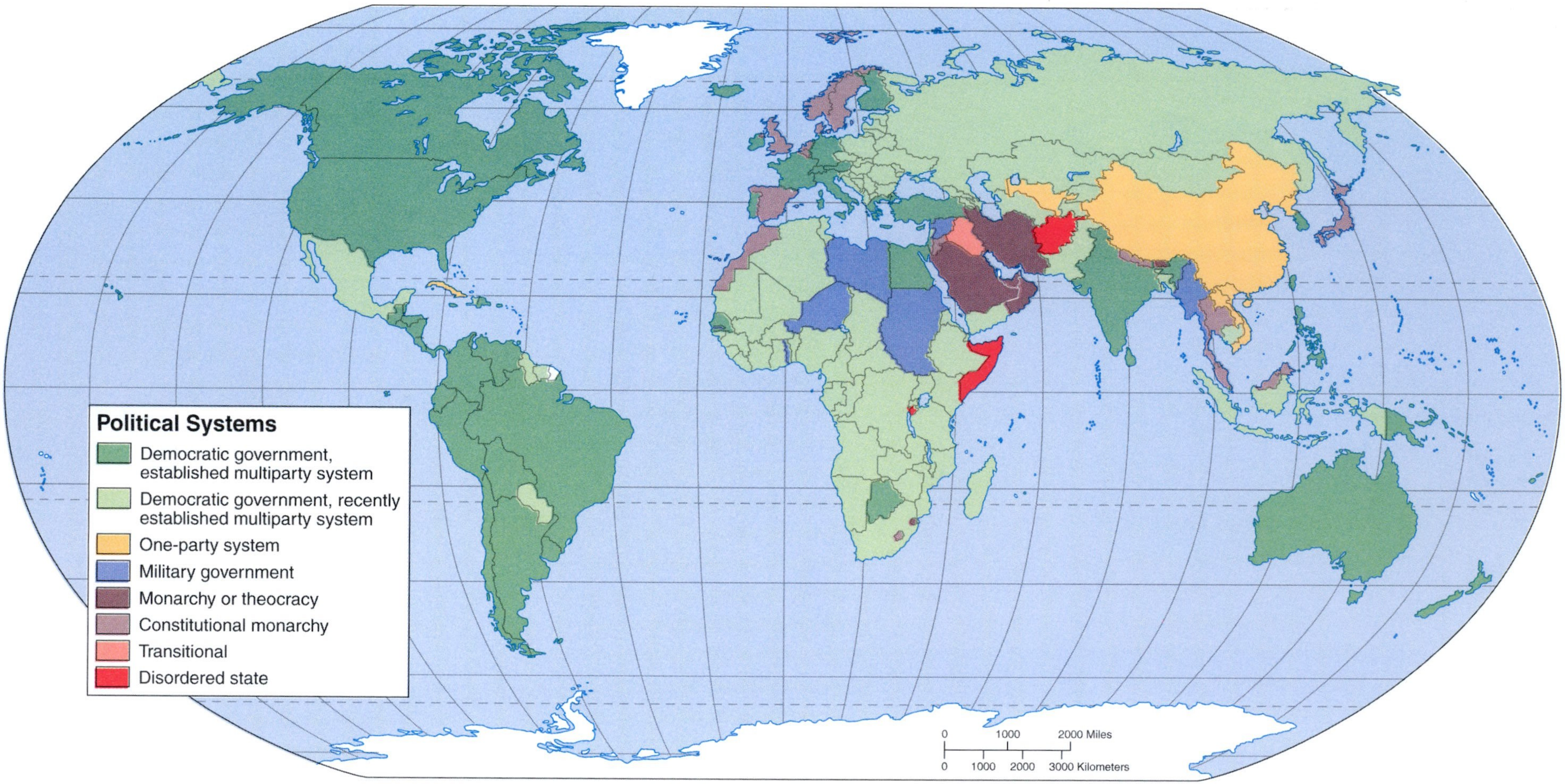

World political systems have changed dramatically during the last decade and may change even more in the future. The categories of political systems shown on the map are subject to some interpretation: established multiparty democracies are those in which elections by secret ballot with adult suffrage are and have been long-term features of the political landscape; recently established multiparty democracies are those in which the characteristic features of multiparty democracies have only recently emerged. The former Soviet satellites of eastern Europe and the republics that formerly constituted the USSR are in this category; so are states in emerging regions that are beginning to throw off the single-party rule that often followed the violent upheavals of the immediate postcolonial governmental transitions. The other categories are more or less obvious. One-party systems are states where single-party rule is constitutionally guaranteed or where a one-party regime is a fact of political life. Monarchies are countries with heads of state who are members of a royal family. In a constitutional monarchy, such as the U.K. and the Netherlands, the monarchs are titular heads of state only. Theocracies are countries in which rule is within the hands of a priestly or clerical class; today, this means primarily fundamentalist Islamic countries such as Iran. Military governments are frequently organized around a junta that has seized control of the government from civil authority; such states are often technically transitional, that is, the military claims that it will return the reins of government to civil authority when order is restored. Finally, disordered states are countries so beset by civil war or widespread ethnic conflict that no organized government can be said to exist within them.

Map 103 Sovereign States: Duration of Independence

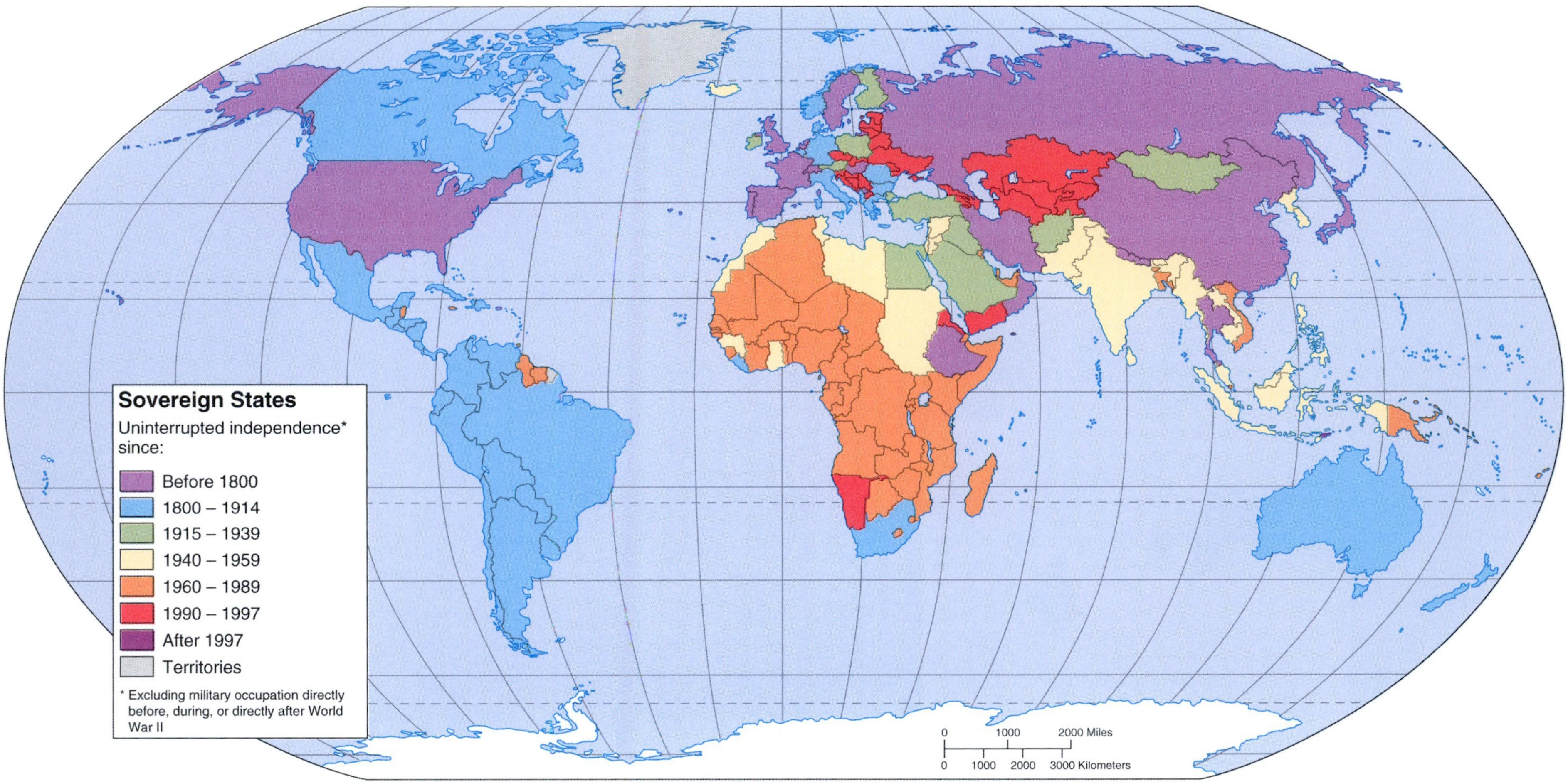

Most countries of the modern world, including such major states as Germany and Italy, became independent after the beginning of the nineteenth century. Of the world's current countries, only 27 were independent in 1800. (Ten of the 27 were in Europe; the others were Afghanistan, China, Colombia, Ethiopia, Haiti, Iran, Japan, Mexico, Nepal, Oman, Paraguay, Russia, Taiwan, Thailand, Turkey, the United States, and Venezuela). Following 1800, there have been five great periods of national independence. During the first of these (1800–1914), most of the mainland countries of the Americas achieved independence. During the second period (1915–1939), the countries of Eastern Europe emerged as independent entities. The third period (1940–1959) includes World War II and the years that followed, when independence for African and Asian nations that had been under control of colonial powers first began to occur. During the fourth period (1960–1989), independence came to the remainder of the colonial African and Asian nations, as well as to former colonies in the Caribbean and the South Pacific. More than half of the world's countries came into being as independent political entities during this period. Finally, in the last decade of the twentieth century, the breakup of the existing states of the Soviet Union, Yugoslavia, and Czechoslovakia created 22 countries where only 3 had existed before.

Agriculture is the basis of the development of the state, a form of complex political organization. Geographers and anthropologists believe that agriculture allowed for larger concentrations of population. Farmers do not need to be as mobile as hunters and gatherers to make a living, and people living sedentarily can have larger families than those constantly on the move. Ideas about access to land and ownership also change as people develop the social and political hierarchies that come with the transition from a hunting-gathering society to an agricultural one. Social stratification based on wealth and

power creates different classes or groups, some of whom no longer work the land. An agricultural surplus supports those who perform other functions for society, such as artisans and craftspeople, soldiers and police, priests and kings. Thus, over time, egalitarian hunters and gatherers shifted to state-level societies in some parts of the world. The first true states are shown in the rust-colored areas of this map.

Map 105 Organized States and Chiefdoms, A.D. 1500

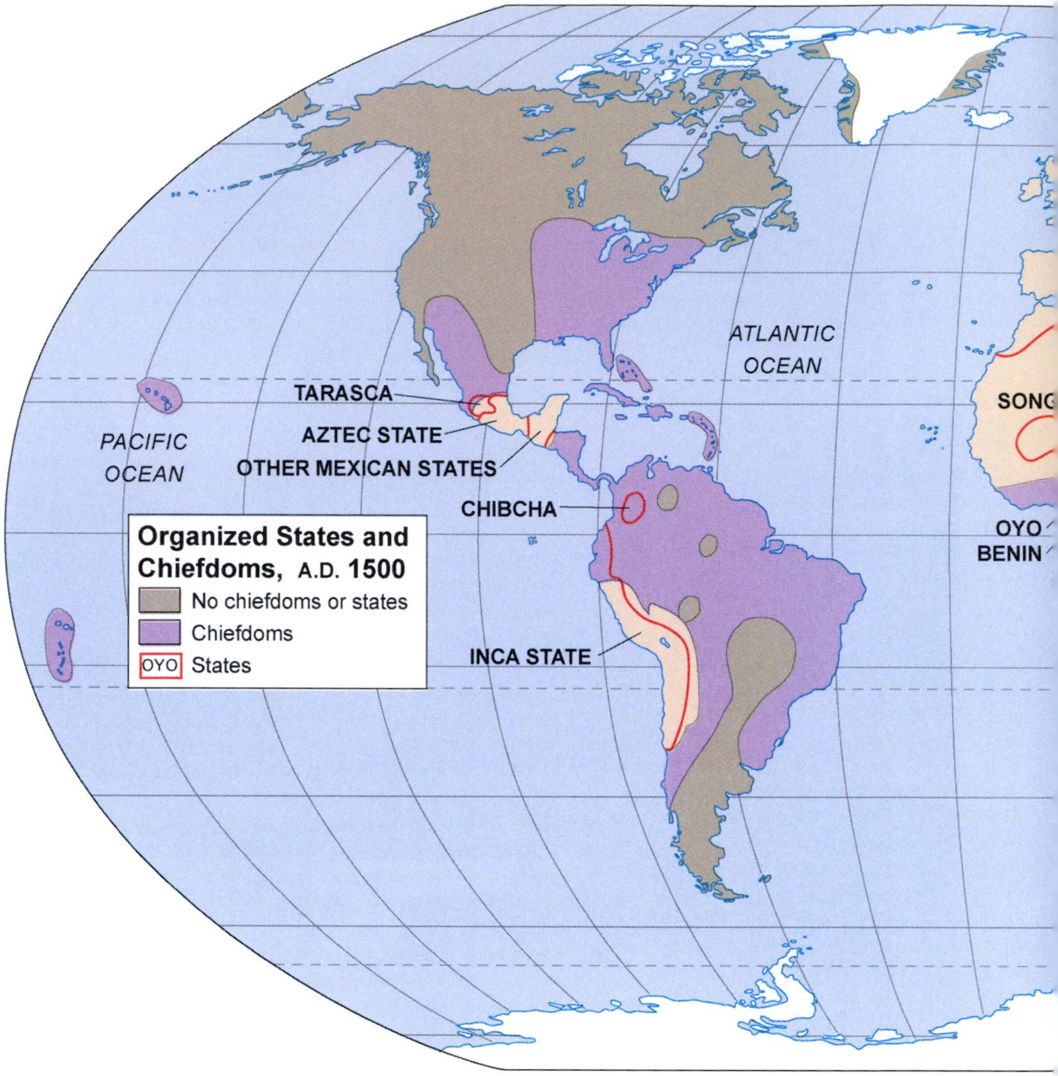

As Europeans began to expand outward through exploration and settlement from the late fifteenth to the eighteenth centuries, it was inevitable that they would find the complex political organizations of chiefdoms and states in many different parts of the world. Both chiefdoms and states are large-scale forms of political organization in which some people have privileged access to land, power, wealth, and prestige. Chiefdoms are kin-based societies in which wealth is distributed from upper to lower classes. States are organized on the basis of socioeconomic classes, headed by a

centralized government that is led by an elite. States in non-European areas included, just as they did in Europe, full-time bureaucracies and specialized subsystems for such activities as military action, taxation, operation of state religions, and social control. In many of the colonial regions of the world after the fifteenth century, Europeans actually found it easier to gain control of organized states and chiefdoms because those populations were already accustomed to some form of institutionalized central control.

Map 106 European Colonialism 1500–2000

European nations have controlled many parts of the world during the last 500 years. The period of European expansion began when European explorers sailed the oceans in search of new trading routes and ended after World War II when many colonies in Africa and Asia gained independence. The process of colonization was very complex but normally involved the acquisition, extraction, or production of raw materials (including minerals, forest products, products from the sea, agricultural products, and animal furs/pelts) from the areas being controlled by the European colonial power in exchange for items of European manufacture. The concept of colonial dependency implied an economic structure in which the European country obtained raw materials from the colonial country in exchange for those manufactured items upon which populations in the colonial areas quickly came to depend. The colors on this map represent colonial control at its maximum extent and do not take into account shifting colonial control. In North America, for example, "New France" became British territory and "Louisiana" became Spanish territory after the Seven Years' (French and Indian) War. There are also two non-European countries—the United States and Japan—shown on the map as colonial powers. Both countries followed the classical European-style colonial model, albeit over relatively small areas and for relatively short periods of time.

Map 107 Post–Cold War International Alliances

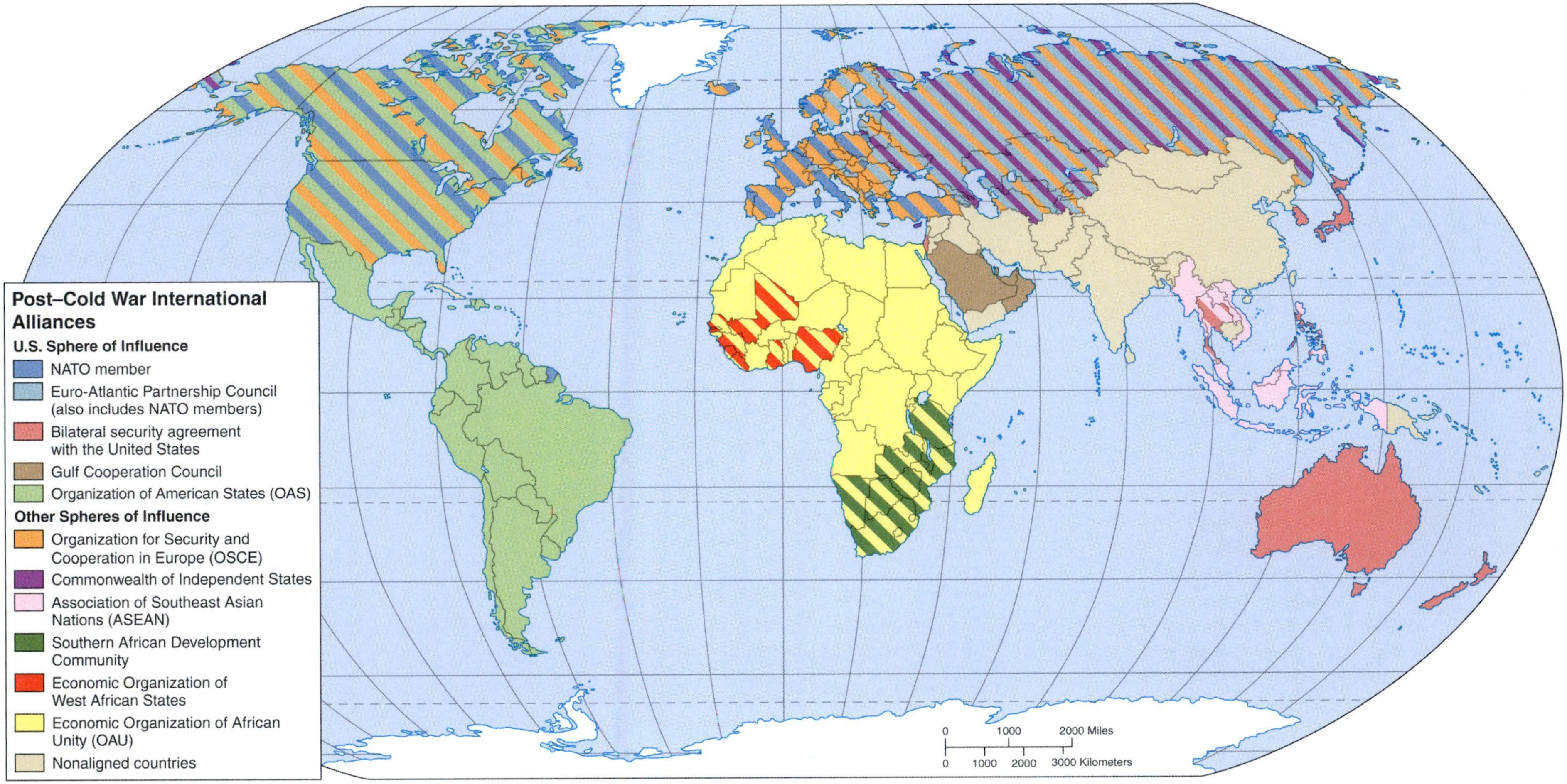

When the Warsaw Pact dissolved in 1992, the North Atlantic Treaty Organization (NATO) was left as the only major military alliance in the world. Most former Warsaw Pact members and a number of independent states that were formerly republics of the old USSR have joined NATO; others are under consideration for membership. The bipolar division of the world into two major military alliances is over, at least temporarily, leaving the United States alone as the world's dominant political and military power. Other international alliances, such as the Commonwealth of Independent States (including some of the former republics of the Soviet Union), will continue to exist but will be relatively unimportant compared with NATO. The CIS has recently seen its membership decline as Georgia has left the Commonwealth following its conflict with Russia in the summer of 2008. It may well be that during the first few decades of the twenty-first century economic alliances will begin to overshadow military ones in their relevance for the world's peoples.

Map 108 International Conflicts in the Post–World War II World

	Conflict[1]	Start Date	Major Belligerent Countries[2] (in alphabetical order)
1	Palestine	1948	Egypt, Iraq, Israel, Jordan, Lebanon, Syria
2	Korean	1950	China, North Korea, South Korea, United Nations: United States and 11 other countries
3	Soviet-Hungarian	1956	Hungary, Soviet Union
4	Sinai	1956	Egypt, France, Israel, United Kingdom
5	Sino-Indian	1962	China, India
6	Kashmir	1965	India, Pakistan
7	Vietnam	1965	Australia, North Vietnam, South Korea, South Vietnam, United States
8	Six-Day	1967	Egypt, Israel, Jordan, Syria
9	Soviet-Czech	1968	Czechoslovakia, Soviet Union
10	Football	1969	El Salvador, Honduras
11	Indo-Pakistani	1971	India, Pakistan
12	Yom Kippur	1973	Egypt, Israel, Syria
13	Cyprus	1974	Cyprus, Turkey
14	Ogaden	1977	Ethiopia, Somalia
15	Cambodian-Vietnamese	1978	Cambodia, China, Vietnam
16	Ugandan-Tanzanian	1978	Tanzania, Uganda
17	Afghanistan	1979	Afghanistan, Soviet Union
18	Persian Gulf	1980	Iran, Iraq
19	Angola	1981	Angola, Cuba, South Africa
20	Falklands	1982	Argentina, United Kingdom
21	Saharan	1983	Chad, Libya
22	Lebanon	1987	France, Israel, Lebanon, Syria, United States
23	Panama	1989	Panama, United States
24	Persian Gulf	1990	Iraq, United Nations: United States and 7 other countries
25	Yugoslavia	1990	Bosnia-Herzegovina, Croatia, Serbia
26	Peruvian-Ecuadorian	1995	Ecuador, Peru
27	Albania	1995	Albania, Yugoslavia (Serbia-Montenegro)
28	Rwanda	1995	Burundi, Rwanda
29	East Timor	1995	Indonesia, New Guinea insurgency
30	Cameroon	1996	Cameroon, Nigeria
31	Northern Iraq	1996	Iraq, Kurdish insurgency
32	Eritrea	1997	Eritrea, Yemen
33	Iraq	1998	Great Britain, Iraq, United States
34	Kosovo	1999	Albania, NATO, Yugoslavia
35	Democratic Republic of the Congo	1998	Angola, Chad, Dem. Rep. of Congo, Namibia, Uganda, Zimbabwe
36	Chechnya	1999	Chechnya, Russia
37	"War on Terrorism"	2001	Afghanistan (Taliban), al-Qaeda organization, Great Britain, United States, Others
38	Iraq	2003	Great Britain, Iraq, United States
39	Lebanon	2006	Israel, Lebanon, Hiz'ballah
40	Georgia	2008	Georgia, Russian Federation

[1] "Conflict" implies at least 1,000 battle deaths.
[2] "Belligerent" implies country supplied at least 5% of the combat troops in the conflict.

The Korean War and the Vietnam War dominated the post–World War II period in terms of international military conflict. But numerous smaller conflicts have taken place, with fewer numbers of belligerents and with fewer battle and related casualties. These smaller international conflicts have been mostly territorial conflicts, reflecting the continual readjustment of political boundaries and loyalties brought about by the end of colonial empires, and the dissolution of the Soviet Union. Many of these conflicts were not wars in the more traditional sense, in which two or more coun-

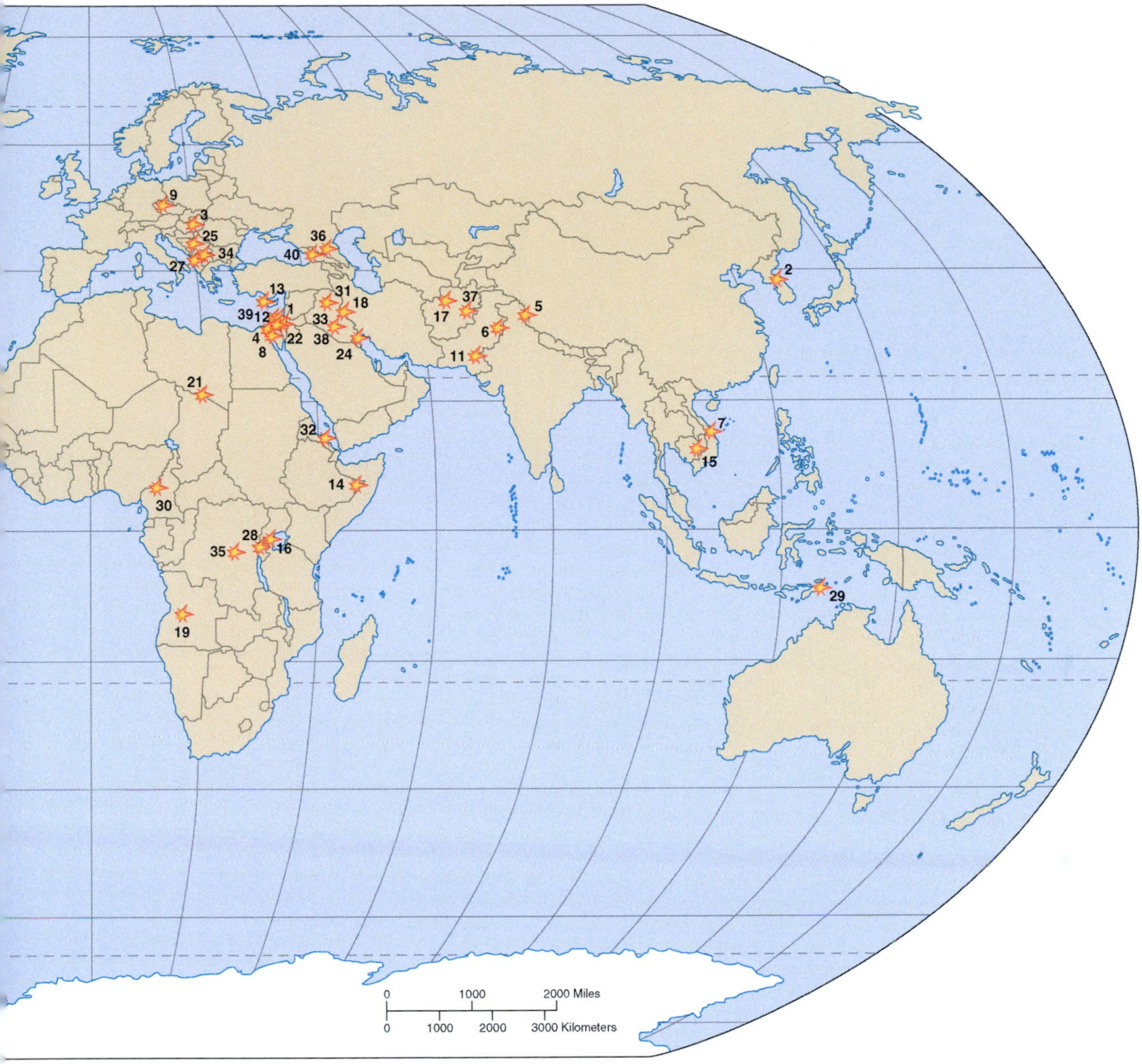

tries formally declare war on one another, severing diplomatic ties and devoting their entire national energies to the war effort. Rather, many of these conflicts were and are undeclared wars, sometimes fought between rival groups within the same country with outside support from other countries. The aftermath of the September 11, 2001, terrorist attacks on the United States indicate the dawn of yet another type of international conflict, namely a "war" fought between traditional nation-states and non-state actors.

Map 109 Political Realms: Regional Changes, 1945–2003

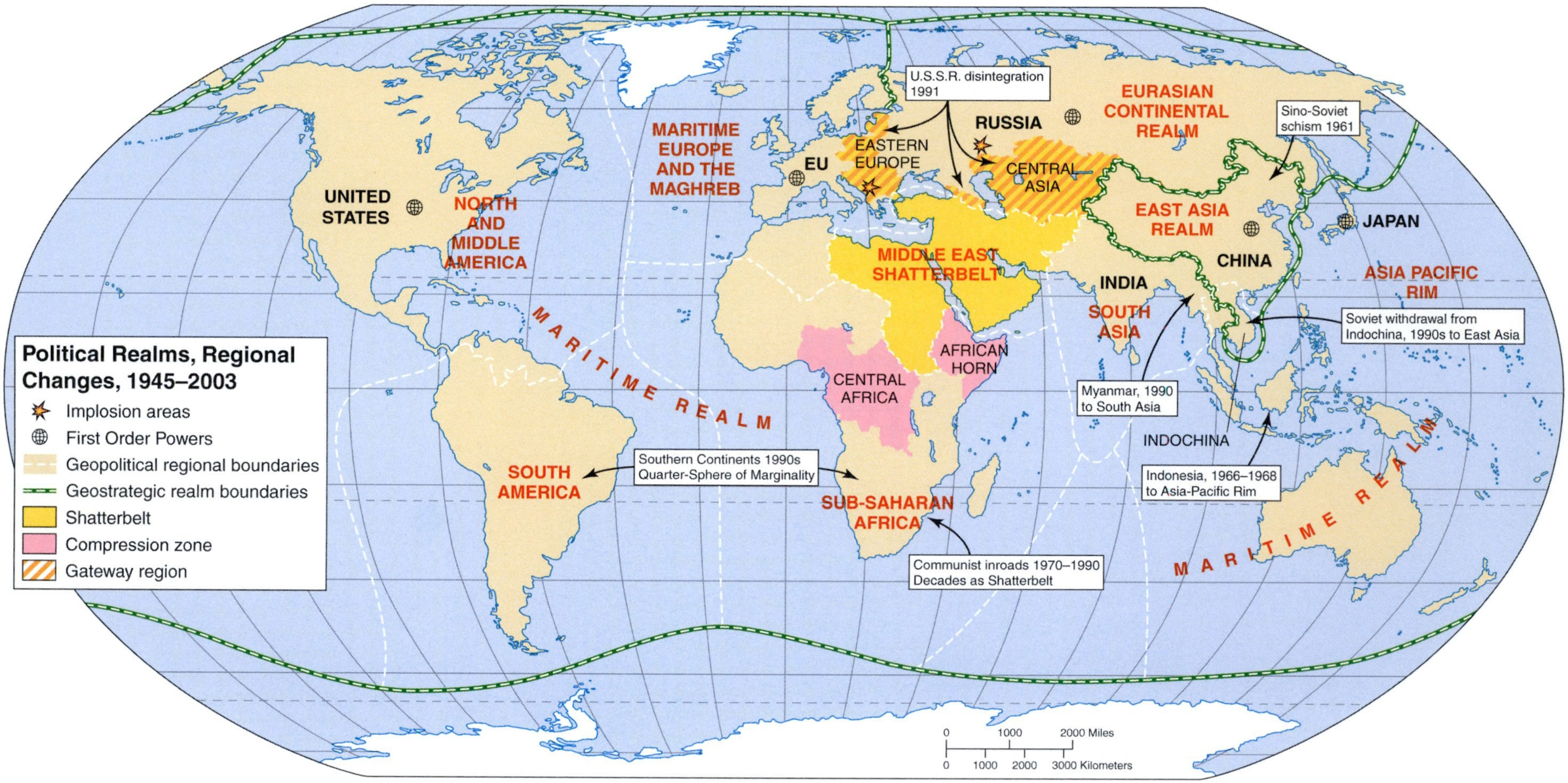

The Cold War following World War II shaped the major outlines of today's geopolitical relations. The Cold War included three phases. In the first, from 1945–1956, the Maritime Realm established a ring around the Continental Eurasian Realm in order to prevent its expansion. This phase included the Korean War (1950–1953), the Berlin Blockade (1948), the Truman doctrine and Marshall Plan (1947), and the founding of NATO (1949) and the Warsaw Pact (1955). Most of the world fell within one of the two realms: the Maritime (dominated by the United States) or the Eurasian Continental Realm (dominated by the Soviet Union). The Soviet Union sought to establish a ring of satellite states to protect it from a repeat of the invasions of World War II. The United States and other Maritime Realm states, in turn, sought to establish a ring of allies around the Continental Realm to prevent its expansion. South Asia was politically independent, but under pressure from both realms. During the second phase (1957–1979), Communist forces from the Continental Eurasian Realm penetrated deeply into the Maritime Realm. The Berlin Wall went up in 1961, Soviet missiles in Cuba ignited a crisis in 1962, and the United States became increasingly involved in the war in Vietnam (late 1960s). The Soviet Union sought increased political and military presence along important waterways including those in the Middle East, Southeast Asia, and the Caribbean. These regions became especially dangerous shatterbelts. The third phase (1980–1989) saw the retreat of Communist power from the Maritime Realm. China, after ten years of radical Communism and chaos of the Cultural Revolution (1966–1976), broke away from the Continental Eurasian Realm to establish a new East Asian realm. Soviet influence declined in the Middle East, Sub-Saharan Africa, and Latin America. In 1989 the Berlin Wall fell, and Eastern Europe began to establish democratic governments. In 1991 the Soviet Union broke apart as its constituent republics became independent states. Compression zones are areas of conflict, but they are not contested by major powers.

Map 110 The Geopolitical World at the Beginning of the Twenty-First Century

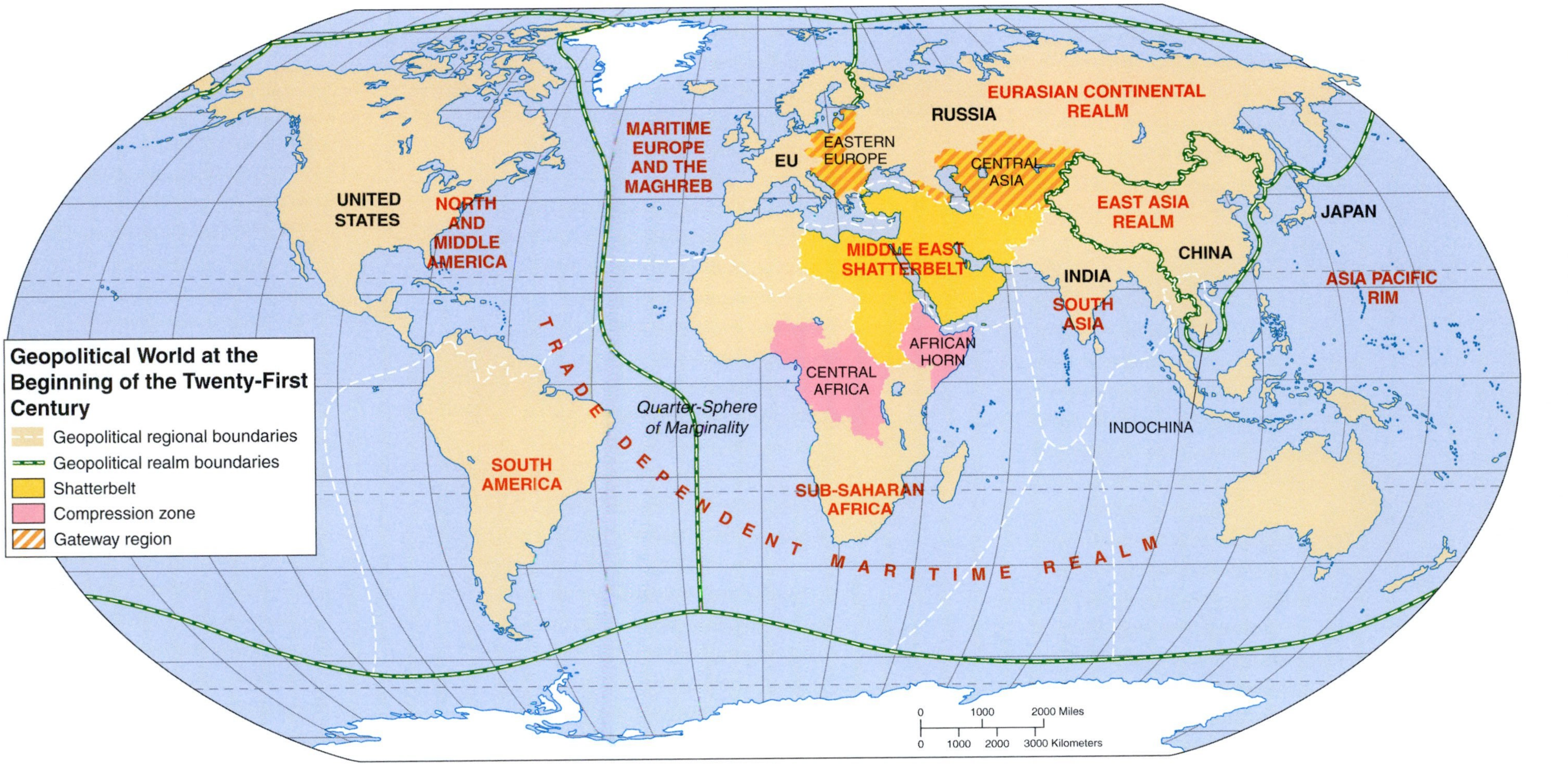

In the geostrategic structure of the world, the largest territorial units are realms. They are shaped by circulation patterns that link people, goods, and ideas. Realms are shaped by maritime and continental influences. Today's Atlantic and Pacific Trade-Dependent Maritime Realm has been shaped by international exchange over the oceans and their interior seas as mercantilism, capitalism, and industrialization gave rise to maritime-oriented states and to economic and political colonialism. The world's leading trading and economic powers are part of this realm. The Eurasian Continental Realm, centered around Russia, is inner-oriented, less influenced by outside economic or cultural forces, and politically closed, even after the fall of Communism. Expansion of NATO in Europe has increased its feeling of being "hemmed in." East Asia has mixed Maritime and Continental influences. China has traditionally been continental, but reforms that began in the late 1970s increased the importance of its maritime-oriented southern coasts. Even so, its trade volume is still low, and it maintains a hold on inland areas like Tibet and Xinjiang. Realms are subdivided into regions, some dependent on others, as South America is on North America. Regions located between powerful realms or regions may be shatterbelts (internally divided and caught up in competition between Great Powers) or gateways (facilitating the flow of ideas, goods, and people between regions).

Map 111 Democracy on the Rise

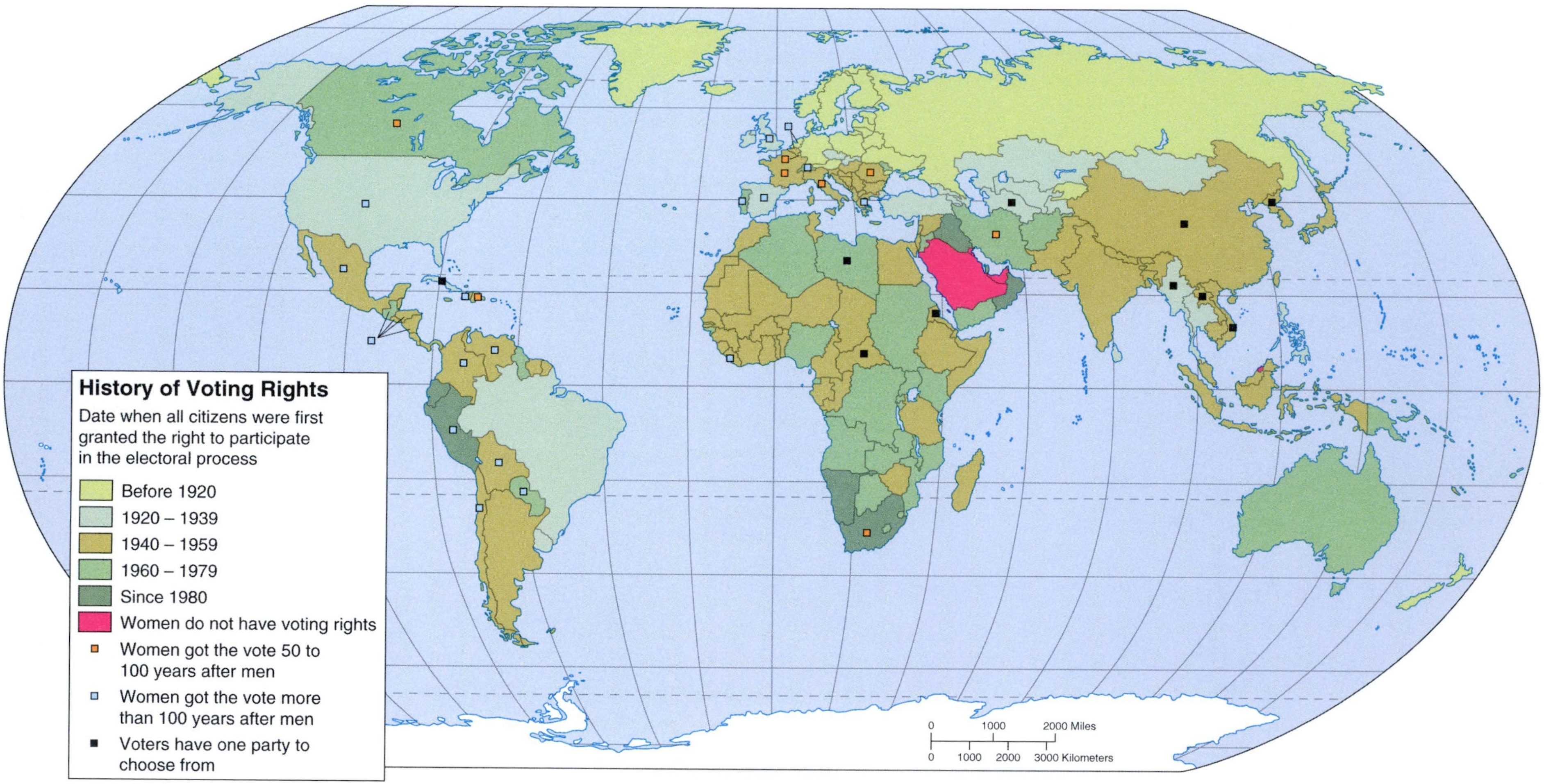

In 1800 most of the world's people were ruled by a monarch—sometimes with a constitution and sometimes with a parliament, but a monarch, nevertheless. Few monarchs remain and where they do, their role is largely that of a figurehead than of an actual ruler. For all the veneration given to, say, Queen Elizabeth II of the United Kingdom, whoever is prime minister in that government has enormously more power—and the prime minister is elected by a constituency. But elected leaders as opposed to hereditary rulers is not the only criterion for identifying a democracy. Suffrage or who possesses the right to vote ought to be universal (it is worth noting that in the United States—one of the world's oldest democracies—black men possessed the right to vote more than a half century before white women). Elections should be competitive: a true democracy can rarely exist without multiple political parties with equal access to the electorate. Elections must also be free from internal or external coercion or corruption. And, in true democracies, there are a number of other identifying components such as a free press, an advanced legal system in which laws rather than men rule, and a respect for human rights. By these more rigorous standards perhaps only two-thirds of the countries on this map are true, western-style liberal democracies. But that is a great deal better (unless, of course, you are a monarchist) than the situation in 1800. What explains the rise of democratic systems? Most experts point to such things as higher levels of economic development and greater literacy among populations. Even more important, however, is globalization and the shrinking of the world through increased communication systems. People who live in democracies tend to have certain advantages not possessed by those who do not. In a world of low or no barriers to interpersonal and interregional information flows, this fact is hard to hide. Isolated countries (North Korea is a case-in-point) are more likely to be autocratic than democratic and the future for truly democratic governance would seem bright.

Colombia: Colombia was one of three countries (along with Ecuador and Venezuela) that emerged from the collapse of the newly independent country of Gran Colombia in 1830 (the former Spanish Viceroyalty of New Granada). Although possessing a rich natural resource base and agricultural abundance, Colombia has been torn apart internally for more than four decades as the result of conflict between government forces and antigovernment insurgent groups and illegal paramilitary groups. The two largest insurgent groups are the Revolutionary Armed Forces of Colombia (FARC) and the National Liberation Army (ELN). The largest illegal paramilitary group is a semi-organized array of disparate paramilitary forces called the United Self-Defense Groups of Colombia (AUC). Both insurgency and paramilitary organizations are heavily funded by the illegal trade in cocaine, the most important (but untaxed) commodity in the Colombian economy. The insurgents do not have the military or popular support necessary to overthrow the government but continue to engage in attacks against government forces and civilians, and many areas of the country are under the control of guerillas who double as drug traffickers. The insurgency groups, with a more political agenda, and the paramilitary groups, who have a fairly simple agenda of controlling the cocaine trade, not only battle the government forces but also one another. Large swaths of internal Colombia are under the control of either insurgents or paramilitary drug lords. Over the last few years, the Colombian government has increased efforts to regain control of these areas, with some mild success. However, neighboring countries remain concerned about the probability of the violence spilling into their borders. Colombian-organized illegal narcotics, guerrilla, and paramilitary activities penetrate all of its neighbors' borders and have created a serious refugee crisis with over 300,000 persons having fled the country, mostly into neighboring states. In 2006 and 2007 a "fumigation" process sponsored and funded by the United States was implemented to reduce the acres in coca cultivation. This attempt at a reduction in the coca production system has been largely unsuccessful. Although active insurgency movements were somewhat diminished in 2008, the peripheral interior areas of the country remain unstable and are more controlled by political and economic opponents of the central government than they are by the official governing body of the state.

Congo: The war in the Democratic Republic of the Congo (formerly Zaire) has preoccupied the United Nations and African diplomats since 1999. Troops from Zimbabwe, Angola, Sudan, Chad, and Namibia joined with the Congo's President Laurent Kabila against his former allies Rwanda, Burundi, and Uganda, who each backed several separate Congolese rebel groups. The origins of the conflict originally lay in the overthrow of longtime dictator Mobutu Sese Seko by Kabila's army in May 1997 after a year of civil war. Kabila's failure to call elections or stabilize the country's economy led to further rounds of rebellion in the huge but fractious nation—rebellion supported by the economic and military assistance of neighboring Rwanda, Burundi, and Uganda. After Kabila's assassination in 2001 and the succession of his son, Joseph, to the presidency, accord seemed to have been reached, and the various conflicting parties agreed to withdraw troops in 2002. But in early 2003, new fighting flared along the country's eastern border, threatening a new and broadened war and the addition of more deaths to the 3.3 million since 1998. Diplomats called the conflict "Africa's first world war," and fears continue as late as 2008 that the Congo conflict could destabilize the entire southern half of the continent, leading to massive refugee flows and abject poverty. Although the stability of the central government of the Democratic Republic of the Congo seemed to increase in 2006 and 2007 with the successful election of a national assembly and a president, conflict and civil strife in the eastern border regions of the country continued to persist and even intensify in 2008.

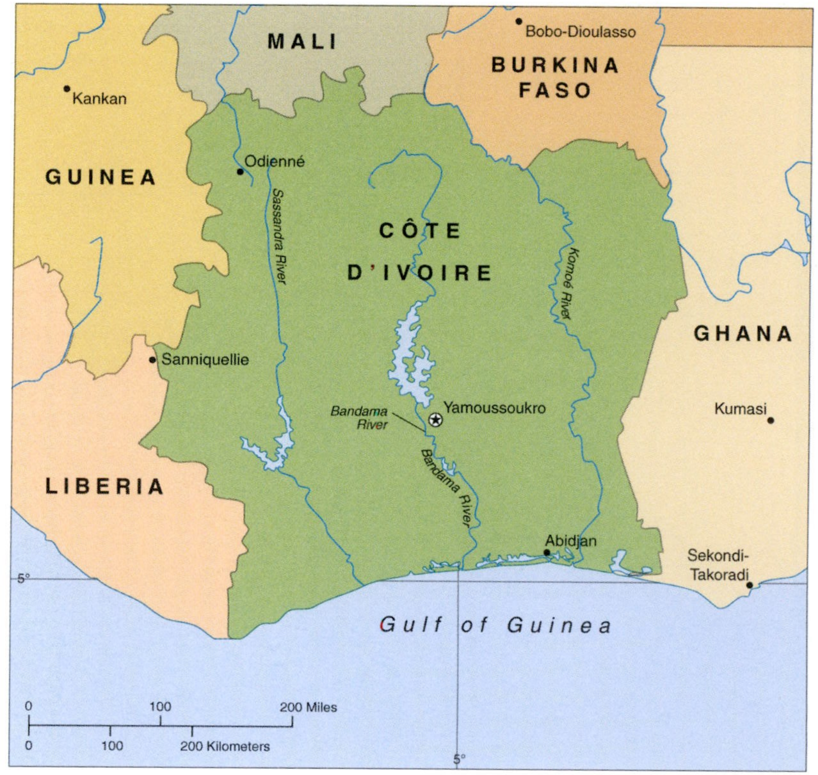

Côte d'Ivoire: Until relatively recently, its close ties to France following independence in 1960 and its development of cocoa production for export (along with significant foreign investment) made Côte d'Ivoire (Ivory Coast) one of the most prosperous of the tropical African states. Since 1999, however, political turmoil has disrupted economic development. In December 1999, the first military coup in Côte d'Ivoire's history overthrew the elected government. The leader of the coup rigged elections held in late 2000 and assumed "legitimate" power, but popular protest forced him to step aside in favor of the runner-up in the 2000 elections. Another coup, this one a failure, was launched by Ivorian dissidents and disaffected members of the military in September 2002. Rebel forces claimed the northern half of the country, and in January 2003 were granted ministerial positions in a unity government. For all practical purposes, Côte d'Ivoire has been since then in a state of civil war with the central government yet to exert control over the northern regions. In late 2006, anti-government protests intensified, leaving no apparent end in sight to a political impasse that has pitted the president against his prime minister. The international community and particularly the United Nations has been divided over how to deal with the continuing civil conflict in Côte d'Ivoire. A 2007 agreement between the president and the prime minister was reached to unify the country but both UN and French troops remain in the country to insure some stability.

Somalia: With the ouster of the government led by Mohamed Said Barre in January 1991, turmoil, factional fighting, and anarchy have followed in Somalia, with several separate governments arising in different parts of the country. The northern clans declared an independent Republic of Somaliland. Although not recognized by any government, it has maintained a stable existence, aided by the overwhelming dominance of a ruling clan and economic infrastucture left behind by British, Russian, and American military assistance programs. Puntland, the central portion of Somalia, from the Horn of Africa to the coast of the Indian Ocean and the border with Ethiopia, has been a self-governing autonomous state since 1998. The area of Southwestern Somalia is poorly organized and more conflict-ridden than any other part of the country, with much of that conflict centered on attempts to control the nominal Somali capital of Mogadishu and ongoing famine. UN humanitarian efforts were unable to either quell guerrilla activity or alleviate famine, and the UN withdrew in 1995. In 2004 a new government, the Transitional Federal Government (TFG), was created for the entire country. The president and parliament of the new government have not yet moved to Mogadishu, and discussions regarding the government's establishment are ongoing in Kenya. Numerous warlords and factions are still fighting for control of the capital city as well as for other southern regions. In 2006 and 2007 a push by central government forces, backed up by units of the Ethiopian regular army, succeeded in driving Islamic extremist forces out of the Mogadishu region. Ethiopian troops pulled out of Somalia in early 2009, but there is still no unification of the country. By any definition, Somalia is a "disordered" or "failed" state.

Sudan and Darfur Region: Since Sudan achieved independence from Great Britain in 1956, military regimes favoring Islamic-oriented governments have dominated national politics. These regimes have embroiled the country in a civil war for nearly all of the past half-century. These wars have been rooted in the attempts of northern economic, political, and social interests dominated by Muslims to control territories occupied by non-Muslim, non-Arab southern Sudanese such as the Dinka tribal groups. Since 1983, the war- and famine-related effects have resulted in more than 2 million deaths and over 4 million people displaced. The current ruling regime is a mixture of military elite and an Islamist party that came to power in a 1989 coup. Some northern opposition parties have made common cause with the southern rebels and entered the war as part of an anti-government alliance. Peace talks gained momentum in 2002–2003 with the signing of several accords, including a cease-fire agreement. However, conflicts have continued to persist in the three Darfur provinces of western Sudan adjacent to the border with Chad and the Central African Republic where government-backed Muslim militia have attacked and killed tens of thousands of non-Muslim tribal peoples. In 2005 and 2006, areas of conflict spilled over the borders of Sudan to involve both Chad and the Central African Republic. The Darfur region is relatively water-rich and forested in a country that is chiefly desert and is therefore desired by Muslim pastoral groups from the north for settlement purposes. International organizations have labeled the conflict in Darfur as "genocide" against the non-Muslim populations and conflict continues into 2009.

Zimbabwe: Relatively shortly after the former country of Rhodesia obtained its independence from the United Kingdom, it was reconstituted as the country of Zimbabwe in 1980. Its leader was Robert Mugabe, who has remained in power ever since. In 2000, Mugabe instituted an ill-advised land reform program, appropriating white-owned farms and giving them to tribal leaders. A country that had been one of Africa's leaders in agricultural production quickly deteriorated into conditions of abject poverty and famine. The United Nations has estimated that nearly half of Zimbabwe's population will be at starvation levels by the end of 2008. Nor does foreign aid do much to alleviate the problem since aid is appropriated and selectively distributed by the Mugabe government. An inflationary rate of nearly 11 million percent has brought the country to the brink of economic collapse. Elections held in March left Mugabe's chief rival, Morgan Tsvangirai, as the leading vote-getter but he did not receive enough votes to prevent a run-off election. Before that election could be held, Tsvangirai withdrew from the race, claiming (quite probably with some justification) that his supporters had been threatened with beatings, imprisonment, murder, and torture and that he did not wish to subject them to those dangers. Most recently, in September 2008, Mugabe ceded some power to Tsvangirai, and instituted—at least on paper—a power-sharing agreement in which he and Tsvangirai would combine their forces to try to lead the country out of its economic chaos and medical crisis created by a cholera epidemic. While Zimbabwe does not have the tribal conflicts that beset so many African nations (most Zimbabweans are Shona), the feelings between the supporters of the two rival political factions run deeply enough that, if the power-sharing arrangement does not work and the economy continues to deteriorate, the country could be plunged into a civil war.

Chechnya: The area in southern Russia known as the Caucasus Region is home to a large variety of non-Russian ethnic groups; many are Muslim and resent centuries of Russian domination and Soviet-era totalitarianism. After the Soviet Union disintegrated in 1991, several of these ethnic groups began agitating for more autonomy from Moscow or for outright independence. One of the more vocal groups with a history of opposition to Moscow's rule were the Chechens. The Chechens declared themselves a sovereign nation in 1991 and by 1994 relations between the breakaway government in Chechnya and the Russian government had drastically deteriorated. In December of that year, Russian forces attacked Chechnya, beginning the first of two (1994–96 and 1999–present) full-scale military conflicts that have also crept into the neighboring Russian autonomous area of Dagestan, itself largely Muslim. In the mid- and late 1990s Russia experienced several terrorist attacks in cities throughout the nation, which the Russian government attributed to Islamic extremists supporting Chechen independence. As a result, a second round of the conflict began in August 1999 with a full-scale Russian military assault on Dagestan and Chechnya. This assault is ongoing and continues to face intense resistance, with heavy casualties on both sides. In 2003 and 2004 Chechen rebels increased their pressure with urban terrorist activities in Russian cities, including Moscow. Although conflicts between Russian and Chechnyan forces decreased in intensity in 2005–2008, there were continuing terrorist incidents, including the bombing of a school building that killed numerous children.

Georgia: Along with Azerbaijan and Armenia, Georgia is one of the trans-Caucasus states that were former Soviet Socialist Republics, attaining their independence after the breakup of the U.S.S.R. Unlike Azerbaijan which is fairly homogenous in ethnicity and religion (Azerbaijani and Muslim) and Armenia which is also an ethnically (Armenian) and religiously (Christian) uniform area, Georgia is a crazy quilt of ethnic, linguistic, and religious groups. Such a lack of national uniformity does not bode well for the stability of a political unit and almost from its independence, Georgia has been beset by internal turmoil. Much of this has been centered in the north, in the province of South Ossetia, and in the west, in the province of Abkhazia. Both provinces have wanted, almost from the creation of Georgian independence, to achieve independence on their own. Matters came to a head in the summer of 2008 when the Russian Federation sent troops to augment already existing "peace-keeping" forces in the breakaway provinces of South Ossetia and Abkhazia. The Russians have no particular religious or ethnic links with populations in either South Ossetia or Abkhazia; the language of neither province is related to Russian and the populations are far from uniformly Russian Orthodox Christians in religion. Indeed, both the South Ossetian and Ahkhazian populations are more similar to the Chechens, a people who have given Russia no end of grief for the last decade and a half. So why did Russia "invade" Georgia? Partly to demonstrate that it could, perhaps, using an ill-advised Georgian military thrust into the breakaway South Ossetian region as an excuse. More important, the increased Russian occupation of areas of the Republic of Georgia signaled to the Georgians that the Russians would not take it kindly if Georgia continued its overtures to the West, particularly to the NATO alliance. Although probably not a significant enough flashpoint to precipitate a regional trans-Caucasian war, the Russian/Georgian conflict signals the beginning of what could be a period of unrest and conflict for the entire region. It may also signal the beginning of a period of increased military tension between the Russian Federation and the United States that some political and military experts are already calling "a new Cold War."

Israel and Its Neighbors: The modern state of Israel was created out of the former British Protectorate of Palestine, inhabited primarily by Muslim Arabs, after World War II. Conflict between Arabs and Israeli Jews has been a constant ever since. Much of the present tension revolves around the West Bank area, not part of the original Israeli state but taken from Jordan, an Arab country, in the Six-Day War of 1967. Many Palestinians had settled this part of Jordan after the creation of Israel and remain as a majority population in the West Bank region today. Israel has established many agricultural settlements within the region since 1967, angering Palestinian Arabs. For Israel, the West Bank is the region of ancient Judea and this region, won in battle, will not be ceded back to Palestinian Arabs without protracted or severe military action. The West Bank, inhabited by nearly 400,000 Israeli settlers and 4 million Palestinians, is also the location of most of the suicide bombings carried out by Islamic militant groups from 2001 to 2009. By early 2008, the Gaza Strip had emerged as the most critical flashpoint in the area. The Israeli government and the Palestinian Authority had agreed to resume peace talks with the goal being a peace agreement by the end of the year. But in late 2008 and early 2009, Israeli troops responded to rocket attacks by HAMAS, a leading Palestinian political party, by attacking Gaza in force. By January 2009, over 1000 Palestinians and Israelis had died in this latest conflict.

Lebanon-Israel: Following a devastating 1976–91 civil war between Lebanon's Muslim majority (about 53%) and Christian minority (about 5%), it was agreed to rebuild a political institution in which the Lebanese established a more equitable political system, giving Muslims a greater voice in the political process while institutionalizing sectarian divisions in the government. Lebanon's neighbor, Syria, intruded into Lebanese territory with military forces. In response, Israel occupied a good part of southern Lebanon until 2000 and then withdrew, leaving many areas of Lebanon under Syrian control. The passage of a United Nations resolution in 2004 called for Syria to withdraw from Lebanon and end its interference in Lebanese affairs, encouraging the Lebanese government to oppose Syria's presence. In February 2005, the assassination of a popular former prime minister led to the "Cedar Revolution," massive demonstrations in Beirut against the Syrian presence. In April 2005, Syria withdrew the remainder of its military forces from Lebanon. In spring 2005, Lebanon held its first free legislative elections since the end of the civil war, and the nation began rebuilding its formerly thriving economy. Most militias were disbanded, and the Lebanese Armed Forces extended authority over about two-thirds of the country. But Hizballah, a radical Shi'a organization listed by the U.S. State Department as a Foreign Terrorist Organization, retained its weapons and control over much of southern Lebanon. In July 2006, Hizballah militia kidnapped Israeli soldiers and launched rockets and mortars into northern Israel. Israel retaliated by blockading Lebanese ports and airports and by conducting airstrikes against Hizballah locations throughout Lebanon. Since Hizballah militia were located throughout urban areas, Israeli retaliation resulted in many civilian deaths. Most of the Israelis killed by Hizballah rockets were civilians. Nine hundred thousand Lebanese and three hundred thousand Israelis were displaced by warfare, and normal life was disrupted across all of Lebanon and northern Israel. A UN resolution, approved by both Lebanese and Israeli governments in August 2006, called for the disarming of Hizballah, Israeli withdrawal, and deployment of the Lebanese Army and an enlarged UN peace-keeping force in southern Lebanon. The situation remains tense into 2009 as Hizballah has not fully disarmed.

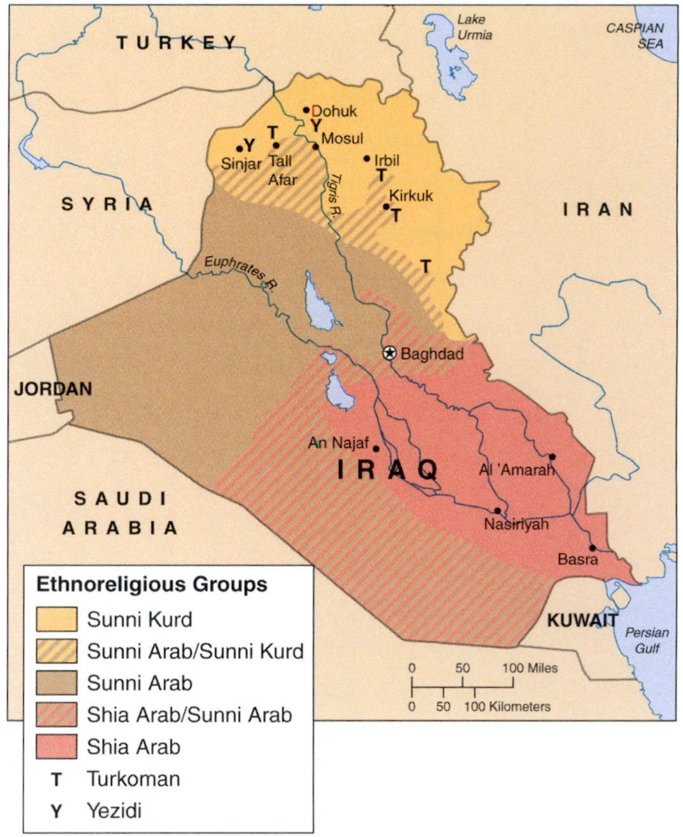

Iraq: Prior to the 1990–91 invasion of Kuwait by Iraq and the subsequent United Nations coalition's military expulsion of Iraq from its neighbor, Iraq was one of the most prosperous countries in the Middle East and the only one with full capacity to feed itself, even without the vast oil revenues generated by the country's immense reserves. Despite the inefficiencies of the Baathist dictatorship of Saddam Hussein, the country had a solid agricultural base and a burgeoning industry. The combination of military adventurism and conflict, in the form of a lengthy war with Iran and the ill-advised invasion of Kuwait, limited further economic development, however. Development was also problematic given the country's internal tensions between Arabic Sunni Muslims and Arabic Shiite Muslims, and between Arabs and Kurds and a few other minority populations in the northern parts of the country.

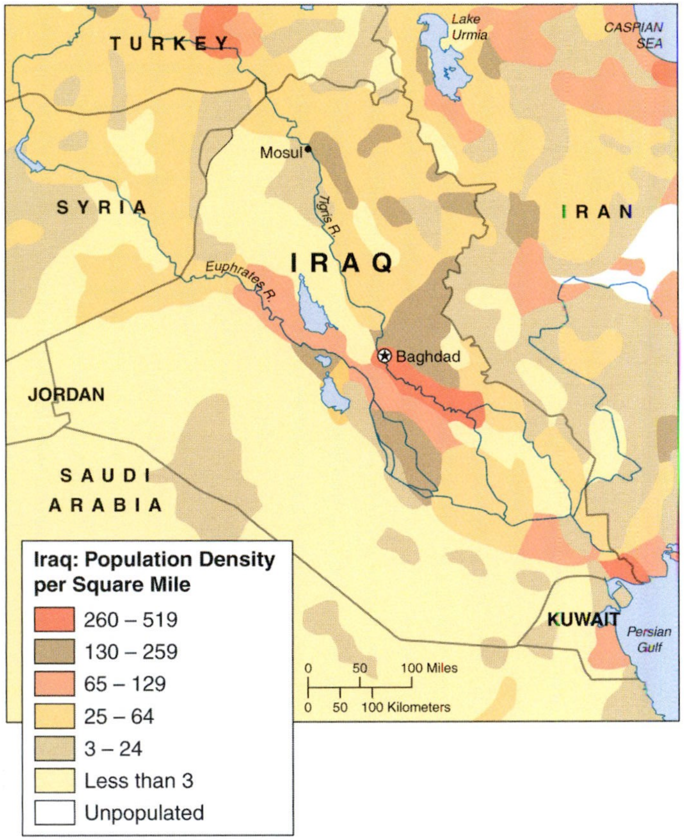

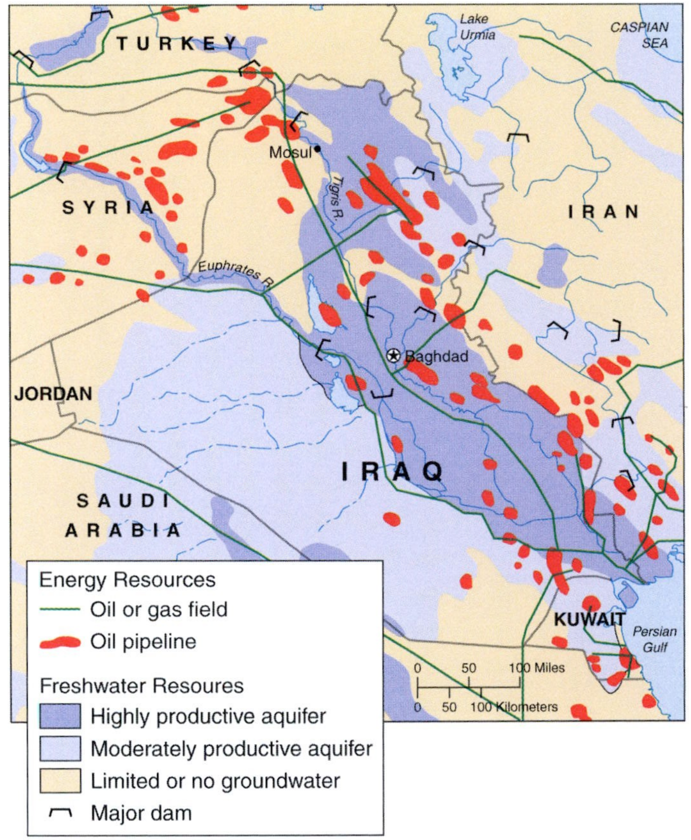

The most densely settled areas of Iraq are in the more well-watered portions of the country in the floodplains of the Tigris and Euphrates rivers, the historical center of Mesopotamian civilization since the fifth millennium B.C. To the west of the floodplain region lies the northeastern portions of the great Arabian desert, and population densities here are very low. Most of Iraq's huge oil reserves are located in the areas of greatest population density and agricultural production, between the rivers and north and east of the Tigris River. The combination of the Middle East's highest potential level of agricultural production and great oil wealth make Iraq unique among states of the region in terms of potential self-sufficiency. That potential is currently being hindered by the ongoing civil conflict between the Shiite majority and the Sunni minority populations.

	November 2004	November 2006	November 2008
Iraqi Civilian Deaths From War	2,650	3,475	500
Civilians Displaced, Net Rate (refugees and internally displaced; in thousands)	50	90	0
U.S./Other Foreign Forces (in thousands)	138 / 24	140 / 18	148 / 6
Iraqi Security Forces (in thousands)	114	323	558
Iraqi Security Forces in Top Two Readiness Tiers (in thousands)	5	220	380
U.S. Troop Deaths	137	69	12
Iraqi Security Force Deaths	65	123	27
U.S. Active-Duty Soldiers Having Served Multiple Deployments in Iraq/Afghanistan (percent)	5	20	31
Iraqi Trained Judges	300	1,200	1,180
Number of "Sons of Iraq" (in thousands)	0	5	100
"Sons of Iraq" Paid by Iraqi Government (in thousands)	0	0	54
Electricity Production (average gigawatts; official grid; prewar: 4)	3.2	3.7	5.1
Oil Production (in millions of barrels per day)	2.0 / 1.3	2.1 / 1.4	2.4 / 1.9
Unemployment Rate (percent)	35	33	30
Political Progress Achieved (out of 11 "Iraq index" criteria)	0	0.5	7

More favorable conditions — Less favorable conditions

Nearly 5 years after the U.S.-led invasion, Iraq is still in critical condition, although important strides have been made as the chart comparing 2004 and 2008 shows. Sectarian violence is still rampant, although death tolls have declined, and the price of Operation Iraqi Freedom will likely reach a staggering $3 trillion. As of 2008, the maps above show the state of Iraq nearly 5 years after the end of active military conflict between the Baathist forces of Saddam Hussein and the U.S.-led coalition forces. It is a country fractured by conflicting tribal and religious loyalties, beset by a continuing Sunni insurgency against the Shiite-dominated constitutional government. There are signs of hope, however, and some of the most vocal of the insurgents have toned down their rhetoric. Nevertheless, the threat remains that Iraq could descend more deeply into the spiral of all-out civil war after the probable pull-out of U.S. troops in 2009.

Kurdistan: Where Turkey, Iran, and Iraq meet in the high mountain region of the Tauros and Zagros mountains, a nation of 25 million people exists. This nation is "Kurdistan," but the Kurds, the occupants of this area for over 3,000 years, have no state, and receive much less attention than other stateless nations like the Palestinians. Following the 1991 Gulf War between Iraq and a U.S.-led coalition of European and Arabic states, the United Nations demarcated a Kurdish "security zone" in northern Iraq. From 1991 to 2003 the Security Zone was anything but secure as Iraqi militants from the south and Turks from the north infringed on Kurdish territory, and the internal militant extremist groups, such as the Kurdish Workers' Party, staged periodic attacks on rival villages. During the 2003 U.S.-led invasion of Iraq that eliminated the Baathist regime of Saddam Hussein, the Kurds played an important role in securing the northern portions of Iraq for the U.S.-British coalition and fought alongside American troops in expelling elements of the Iraqi army from cities like Mosul and Kirkuk. Rich in oil and history, Kurdistan will probably remain as a nation without a state, shared by Iraq, Turkey, and Iran—none of which is likely to give up substantial portions of territory for the establishment of a Kurdish state. By 2009, that portion of northern Iraq under Kurdish control was the most stable of that war-torn country.

Iran: Iran has been a bone in the throat of the United States and Israel since the overthrow of the pro-Western government of the Shah and the installation of a fundamentalist Islamic republic in 1979. The Iranians took a large number of American hostages following the political revolution and held them for over a year. During an eight-year war between Iran and its nearest neighbor, Iraq, the United States provided military aid to Iraq, further straining the relations between Iran and the West. Most recently, the statements of Iran's president, Mahmoud Amahdinejad, denying the Holocaust, threatening to "wipe Israel off the map," and promising to continue Iran's development of nuclear facilities that could lead to the development of nuclear weapons, have heightened distrust of Iran in the West. Described by President George W. Bush a number of years ago as a member of the "axis of evil" (the other states being Iraq and North Korea), Iran has an elected spokesman (Amahdinejad) whose rhetoric has simply ramped up a war of words between Iran and the United States that has the potential to erupt into open conflict. More likely, however, is that once Iran has developed facilities capable of producing weapons-grade plutonium, Israel will carry out the same type of preemptive strikes it has previously used on Iraq and Syria. Iran is a large and important country, poorly understood by the United States. With an ancient imperial tradition, Iran is the historical core of Shiah Islam (nearly 90% of Iranians are Shiite Muslims) and it possesses enormous reserves of oil and natural gas. As tensions mount between Iran, the United States, and its most important regional ally, Israel, open military conflict between Iran and the United States (and, conceivably, other NATO states as well) is far from unlikely.

Afghanistan: In the aftermath of the tragic September 11, 2001 terrorist attacks on the World Trade Center and the Pentagon, the United States (backed to varying degrees by its allies) has declared a massive and global "war on terrorism" and any states that may provide "safe harbor" to terrorists. To date, the most prominent target of this U.S. declaration of war has been the Taliban regime of Islamic extremists who controlled about 95 percent of the territory of the beleaguered nation of Afghanistan. International observers believe that the Taliban regime welcomed and provided a base for the al-Qaeda terrorist network dominated by Saudi expatriate and millionaire Osama bin Laden since the late 1990s. As a result of this intelligence, the U.S. and Britain pursued a daily bombardment of key al-Qaeda and Taliban installations inside Afghanistan for several months, with key logistical support in the form of air bases and supply depots provided by the government of Pakistan (in exchange for financial considerations and political support of the non-elected Pakistani government). U.S. and allied ground troops, aided by members of the Northern Alliance of Afghan rebels opposed to the Taliban regime, expelled the Taliban government in 2002. While now under home rule and with a duly elected government, Afghanistan still is plagued by warlords in remote areas of the country who refuse to recognize the legally constituted government. In addition, along the Afghanistan-Pakistan border, significant pockets of resistance from remnants of the former Taliban regime and from al-Qaeda forces are engaged in ongoing military conflict with American and Pakistani troops. In 2006 and 2007 a marked resurgence of Taliban military activity occurred along the border between Afghanistan and Pakistan, giving rise to fears that the central government in Kabul—not strong to begin with—was beginning to lose its grasp of the peripheral areas of the country. The Taliban resurgence grew throughout 2008 and into 2009, partly as the result of an ineffective and corrupt central government that, for all practical purposes, controls only the region of the capital city, Kabul. And that control is tenuous.

Pakistan: Pakistan's importance to the West and its potential as a flashpoint represent one of the most critical threats to world peace in the opening years of the twenty-first century. Pakistan has a population that places it in the top 10 countries in the world; it possesses both nuclear weapons and a delivery system (as does its nearest neighbor and chief antagonist, India); and it teeters on the brink of being either a western-style representative democracy or an Islamic fundamentalist state. Strategically, Pakistan lies at the western end of the core of the Muslim world (although large Muslim populations exist to the south and east in India, Malaysia, and Indonesia) and is immediately adjacent to U.S. military operations against the Taliban in Afghanistan. Indeed, it has been conjectured that al-Qaeda operatives (including the putative head of al-Qaeda, Osama bin Laden) are using the rugged terrain of the Pakistani-Afghan border as both hiding place and base of operations. Physically, Pakistan is an incredibly rugged country mixing a large river floodplain (the Indus) with high mountain country with peaks in excess of 25,000 feet in elevation in the northwest. Culturally, the country is a mixture of different linguistic and ethnic groups. The government has tried to encourage the use of Urdu as the national language but less than 10% of Pakistanis speak Urdu as their primary language. About the only source of unity in Pakistan is Sunni Islam, but even here, nearly 20% of the Pakistani population adheres to Shiah Islam. The long-serving president, Musharraf, who was deposed in August 2008, was a purported ally of the United States although his actions indicated otherwise. A new Pakistani chief-of-state has known leanings toward the west, as did his assassinated wife who probably would have, had it not been for her murder by (probably) Islamic extremists, kept Pakistan a regional ally of the United States. With its religious conflict between moderate and fundamentalist, with its ethnic and linguistic mix, with its difficult terrain which makes internal cohesion of the state problematic, Pakistan is a country that could, in the near future, become a center of military action, either internal or external.

Jammu and Kashmir: When Britain withdrew from South Asia in 1947, the former states of British India were asked to decide whether they wanted to become part of a new Hindu India or a Muslim Pakistan. In the state of Jammu and Kashmir, the rulers were Hindu and the majority population was Muslim. The maharajah (prince) of Kashmir opted to join India, but an uprising of the Muslim majority precipitated a war between India and Pakistan over control of this high mountain region. In 1949 a cease-fire line was established by the UN, leaving most of the territory of Jammu and Kashmir in Indian hands. Since then Pakistan and India have waged intermittent skirmishes over the disputed territory that holds the headwaters of the Indus River, a life-giving stream to desert Pakistan. In 1999 extremist Muslim groups demanding independence escalated the periodic battles into a full-fledged, if small, war between two of Asia's major powers—both possessing nuclear weapons. The specter of nuclear exchange caused both Pakistan and India to back down, and while the area remains disputed, military activity has quieted somewhat. There have been some suggestions that the terrorist attack on Mumbai (Bombay) in the summer of 2008 was fueled by tensions over Kashmir. That is less likely than the general simmering feud between Hindu and Muslim that has been part of South Asian politics since independence from Great Britain and partition into separate states in 1947.

Sri Lanka: The island state of Sri Lanka, historically known as Ceylon, is potentially one of the most agriculturally productive regions of Asia. Unfortunately for plans related to agricultural development, two quite different peoples have occupied the island country: The Buddhist Sinhalese originally from northern India and long the dominant population in Sri Lanka, and the minority Hindu Tamil, a Dravidian people from south India. Since independence from Britain, Sri Lankan governments have sought to "resettle" the Tamil population in south India, actions that finally precipitated an armed rebellion by Tamils against the Sinhalese-dominated government. The Tamils at present are demanding a complete separation of the state into two parts, with a Tamil homeland in the north and along the east coast. At one time viewed as an island paradise, Sri Lanka is now a troubled country with an uncertain future. A cease-fire between Sinhalese and Tamil fighters was brokered in 2001 by Norway but fell apart in late 2003 with the resumption of violence. In May 2004, Norwegian diplomats again made an attempt to negotiate a cease-fire but with questionable results. 2006 and 2007 brought a renewal of terrorist attacks instigated by the Tamil faction, bringing to an end the cease-fire brokered by Norway. Terrorist activities by the "Tamil Tigers" continued through 2008, even accelerating late in the year.

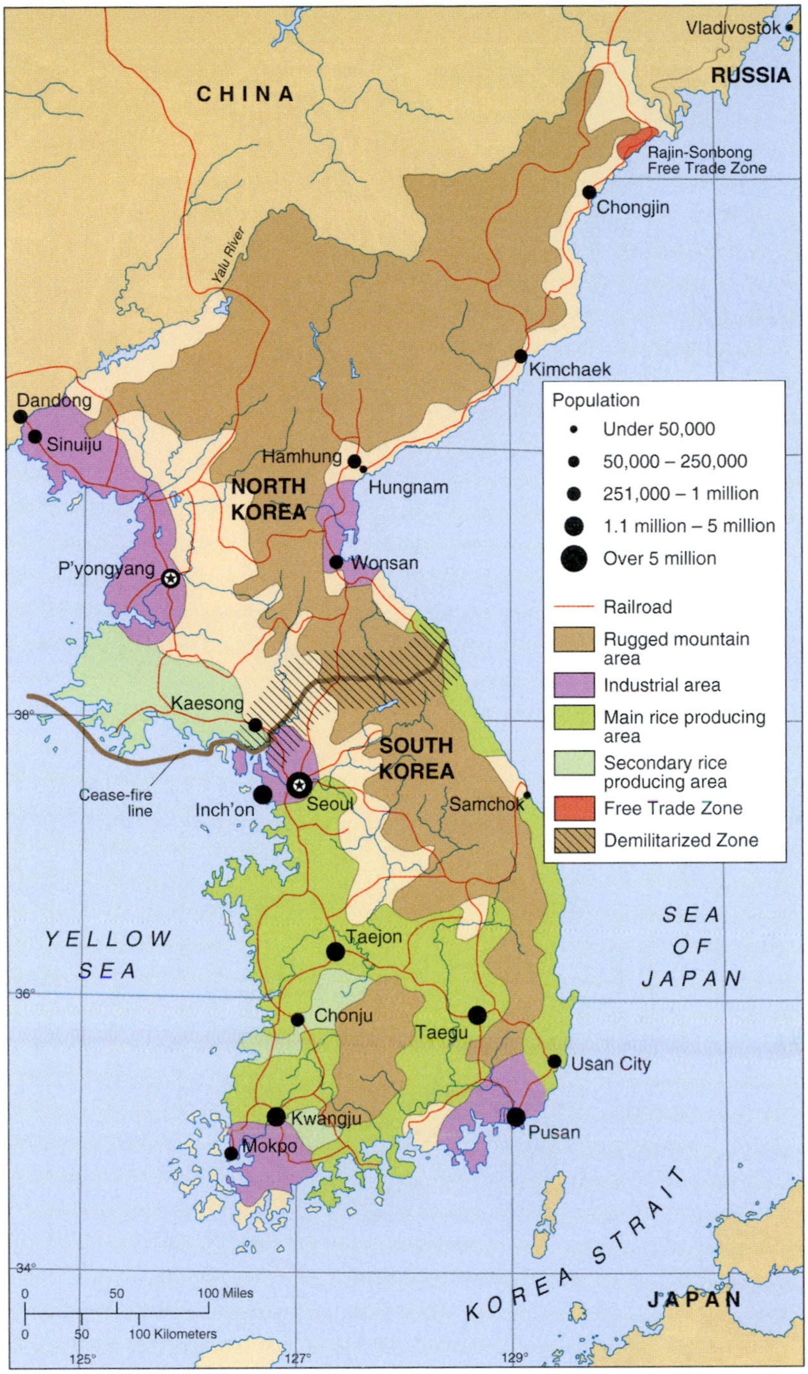

Korean Peninsula: Although active military conflict has not existed since the 1950s, the Korean peninsula remains an important flashpoint. Following the Japanese occupation during World War II, Korea was split into a Communist-dominated north, and an ostensibly republican south, the latter backed by the United States. Between 1950 and 1953 communist North Korea, allied with China, attempted unsuccessfully to reunite the entire peninsula under Communist rule. Since the end of the "Korean War," South Korea has flourished economically. North Korea, on the other hand, adopted a policy of diplomatic and economic "self-reliance," becoming one of the world's most authoritarian and isolated states. North Korea molded its political, economic, and military policies around the core ideological objective of eventual unification of Korea and retains that objective. Yet, North Korea so mismanaged and misallocated its resources that, by the mid-1990s, the country was unable to feed itself. It continues to expend resources to maintain one of the world's largest armies. North Korea also has active military programs, including long-range missile development as well as chemical and biological weapons programs, and the development of nuclear weapons with long-range missile delivery systems. All of these programs are of major concern to the international community. In 2003, North Korea announced that it would violate a 1994 agreement with the United States to freeze and ultimately dismantle its existing plutonium-based program. It also expelled monitors from the International Atomic Energy Agency (IAEA) and withdrew from the international Non-Proliferation Treaty. By 2006, North Korea announced the development of nuclear weapons and delivery systems designed to "protect" North Korea against American aggression. In 2007, the North Korean government, in exchange for the relaxing of sanctions against it, agreed to halt its attempt to build a nuclear arsenal, although the status of that agreement was unclear by 2008 and North Korea's nuclear ambitions and capability remain largely unknown.

Map 113 Nations with Nuclear Weapons

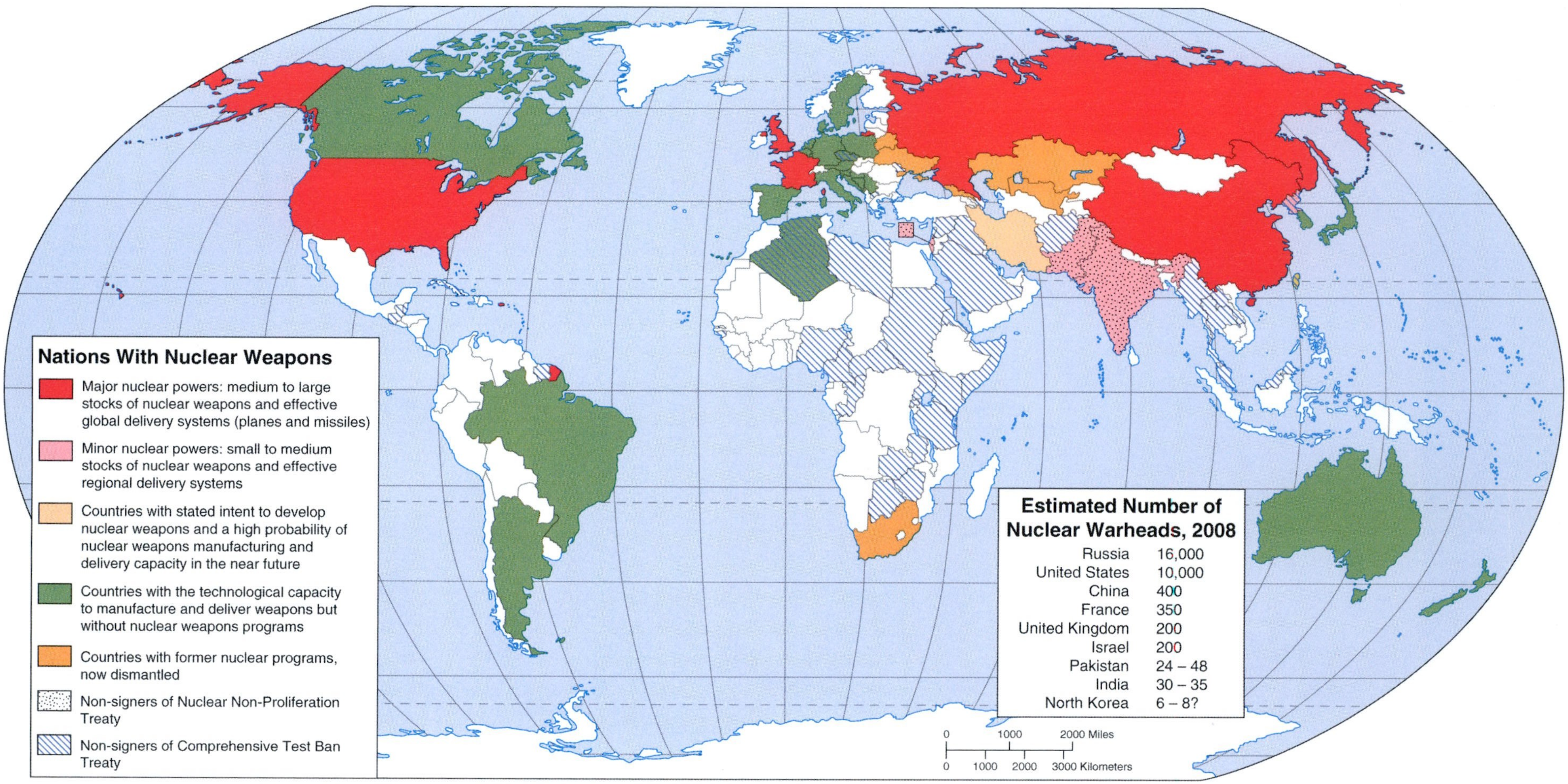

Since 1980, the number of countries possessing the capacity to manufacture and deliver nuclear weapons has grown dramatically, increasing the chances of accidental or intentional nuclear exchanges. In addition to the traditional nuclear powers of the United States, Russia, China, the United Kingdom, and France, must now be added Israel, India, and Pakistan as countries that, without possessing the large stocks of weapons of the major powers, nor the extensive delivery systems of the United States and Russia, still have effective regional (and possibly global) delivery systems and medium stocks of warheads. Countries such as Kazakhstan, Ukraine, Georgia, and Belarus that were created out of what had been the Soviet Union did have some nuclear capacity in the 1991–1995 period but have since had all nuclear weapons removed from their territories. North Korea has recently announced the re-suspension of its nuclear weapons programs, although it may possess a small stock of nuclear warheads along with the capacity to deliver those weapons regionally. Iran has announced nuclear ambitions and recently tested delivery systems. Until the overthrow of the Baathist regime of Saddam Hussein by a U.S.-led military coalition in 2003, Iraq also had nuclear ambitions. The proliferation of nuclear states threatens global security, and the objective of the Nuclear Non-Proliferation Treaty was to reduce the chances for expanding nuclear arsenals worldwide. This treaty has been partially successful in that a number of countries in the developed world certainly have the capacity to manufacture and deliver nuclear weapons but have chosen not to do so. These countries include Canada, European countries other than the United Kingdom and France, South Korea, Japan, Australia, and New Zealand, and Brazil and Argentina in South America. On the other side of the coin, the still-possible intent of North Korea to emerge as a nuclear power may force countries such as South Korea and Japan to re-think their positions as non-nuclear countries.

Map 114 Military Power

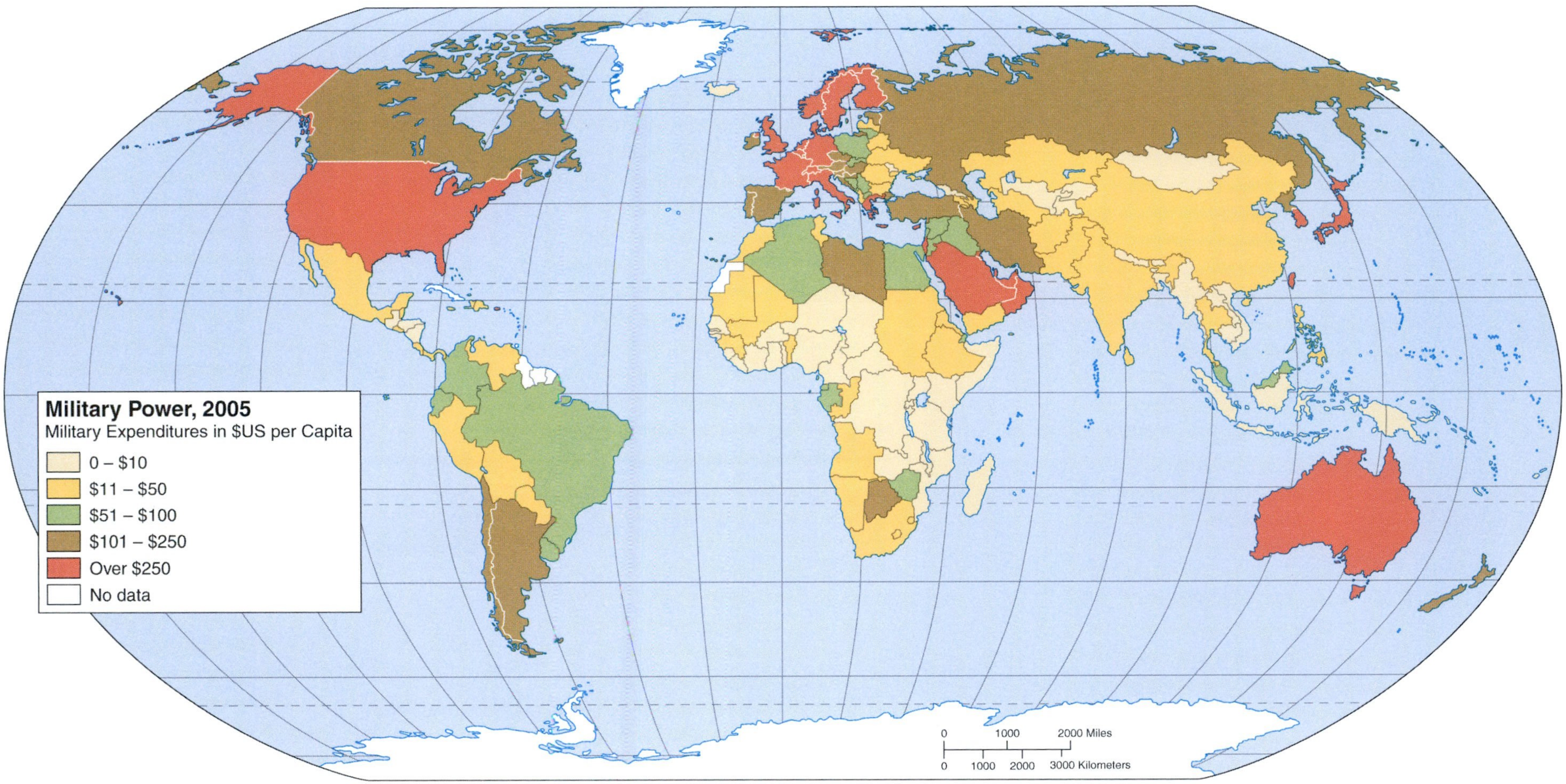

Military expenditures, measured in $US per capita, is often a good indicator of two things: a strong propensity toward defending "the Homeland" (few countries in the post–Cold War world actually think in terms of military aggression toward their neighbors), and an extensive network of political and military alliances that require significant military expenditures in the name of collective self-defense. The enormous per capita military expenditures of Saudi Arabia are explained in terms of the first category. Saudi Arabia has a limited ground force and almost no navy but one of the world's most advanced, well-trained, and well-equipped air forces. They view this as necessary to defend the homeland against attack from a neighboring country with which they might share political or, more likely, religious difference. Falling into the second category are those countries, such as the United States, and the other NATO allies that have obligation not only to themselves but to many others, including countries that are not members of NATO. The example of NATO troops and materiel being sent to Bosnia comes immediately to mind. Beyond these "normal" military expenditures are those of countries wishing to become regional powers through a large military force: Libya and Iran are excellent examples, although just during the last year, the expenditures by Libya have decreased and those of Iran have increased (later than 2005 data were not available for the map). Finally, there are large per-capita military expeditures in what might be categorized as "rogue" states like North Korea for which the money spent on one of the world's largest standing armies and the development of a nuclear weapons system seem to make little sense in light of the population living at the starvation level.

Map 115 International Terrorist Incidents, 2000–2006

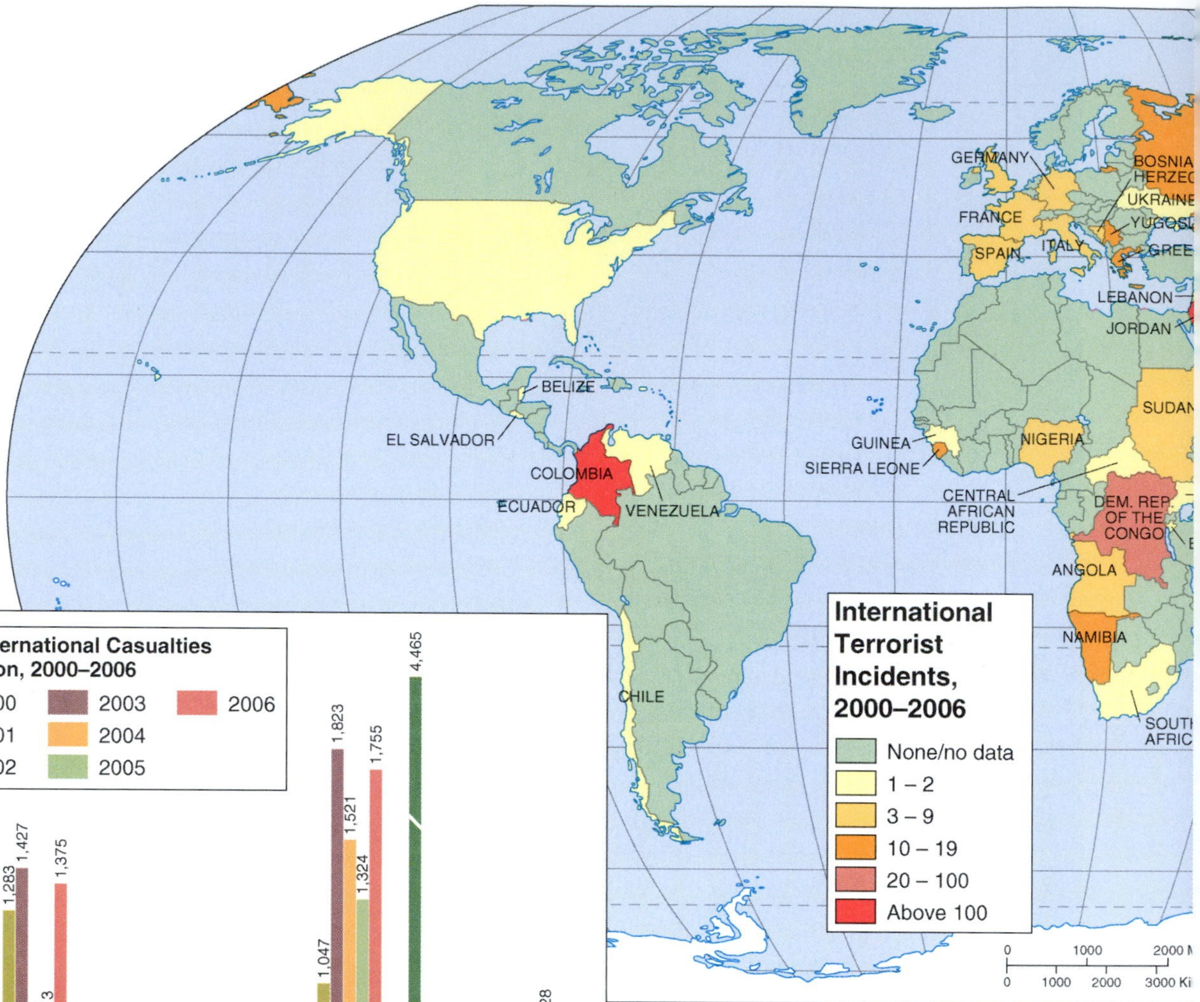

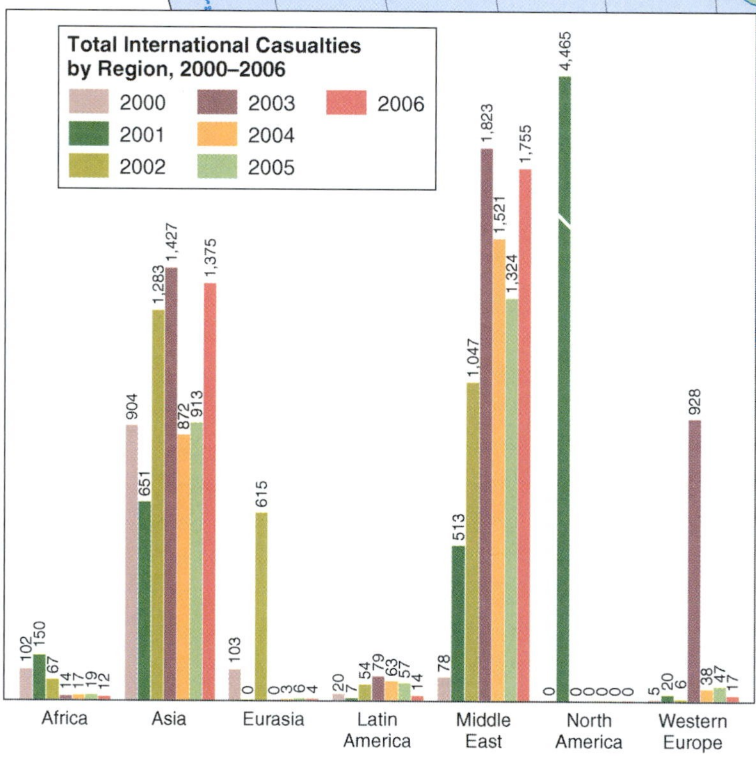

Americans have made a mantra of the saying "the world has changed" as a consequence of the terrorist attacks on the World Trade Center and the Pentagon on September 11, 2001. As the map above and the accompanying graphs of terrorist activities before and after 9/11/01 indicate, however, the world did not change, although the focus of a major terrorist attack shifted from Africa, Asia, and the Middle East to North America. Many other areas of the world have lived with terrorism and terrorist activity for years. In 2000 and 2001, despite the enormous losses in the United States in the 9/11 attacks, more lives were lost in Asia and Africa as a result of terrorism than were lost in North America. The world did not change, but Americans' perception of that world and their place in it has certainly changed. Events subsequent to 9/11 indicate a shift back to more "normal" patterns in North America from 2002 to the present, a time during which terrorist activities have been virtually absent. This is in sharp contrast to other parts of the world where terrorist activities have been on the upswing, the Madrid train bombing that killed nearly 200 people, and the London subway bombing that killed more than 50.

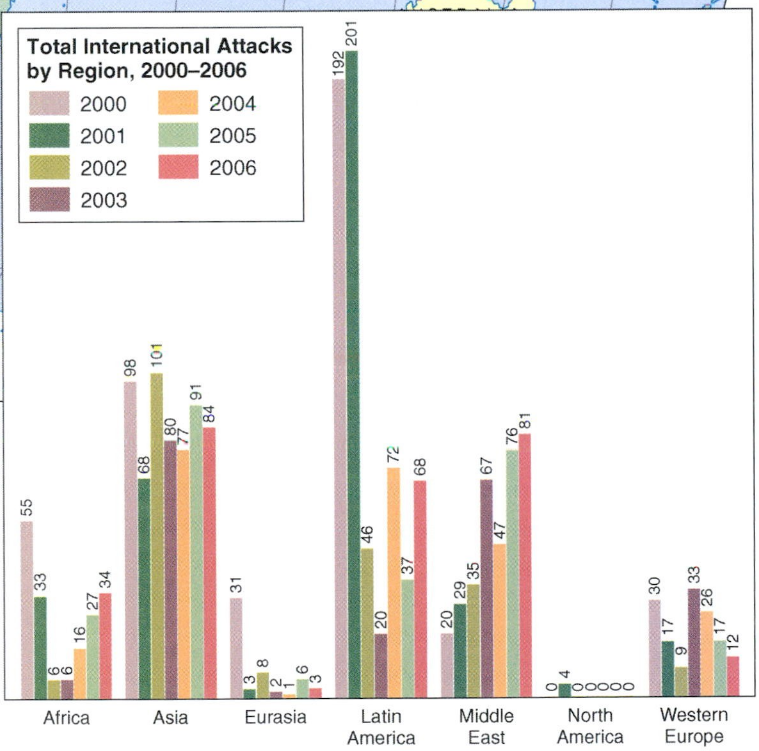

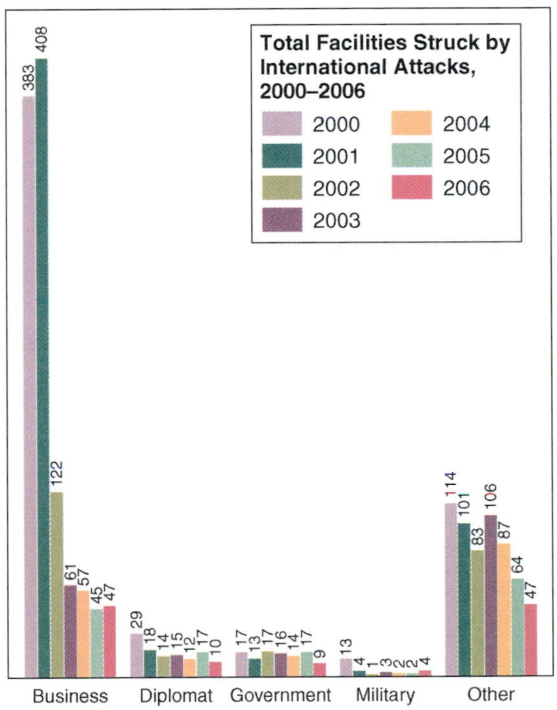

Two qualifiers need to be added regarding the data presented: the term "casualties" refers to dead and wounded (the number of the dead represents a relatively small percentage of total casualties by region); the term "terrorist attacks" does not include insurgency events in areas of military conflict or occupation and, therefore, does not include attacks carried out against U.S. forces or others in places like Afghanistan or Iraq where military action continued throughout the time frame of the data. And, as terrible as they are, terrorist attacks do not include the casualties of sectarian violence such as that between Iraqi Shiites and Sunnis or between paramilitary organizations such as Hizballah and the legal state of Israel. In late 2003, the manner of collecting data on terrorist incidents and casualties therefrom changed, as did the definition of what constituted an international terrorist event. Hence, data for 2004 through 2006 are not comparable to data from 2003 and earlier. And reliable data from 2007 on are simply not available, meaning that neither the map nor the chart reflect the deadly terrorist attack on Mumbai, India in late 2008.

Map 116 Distribution of Minority Populations

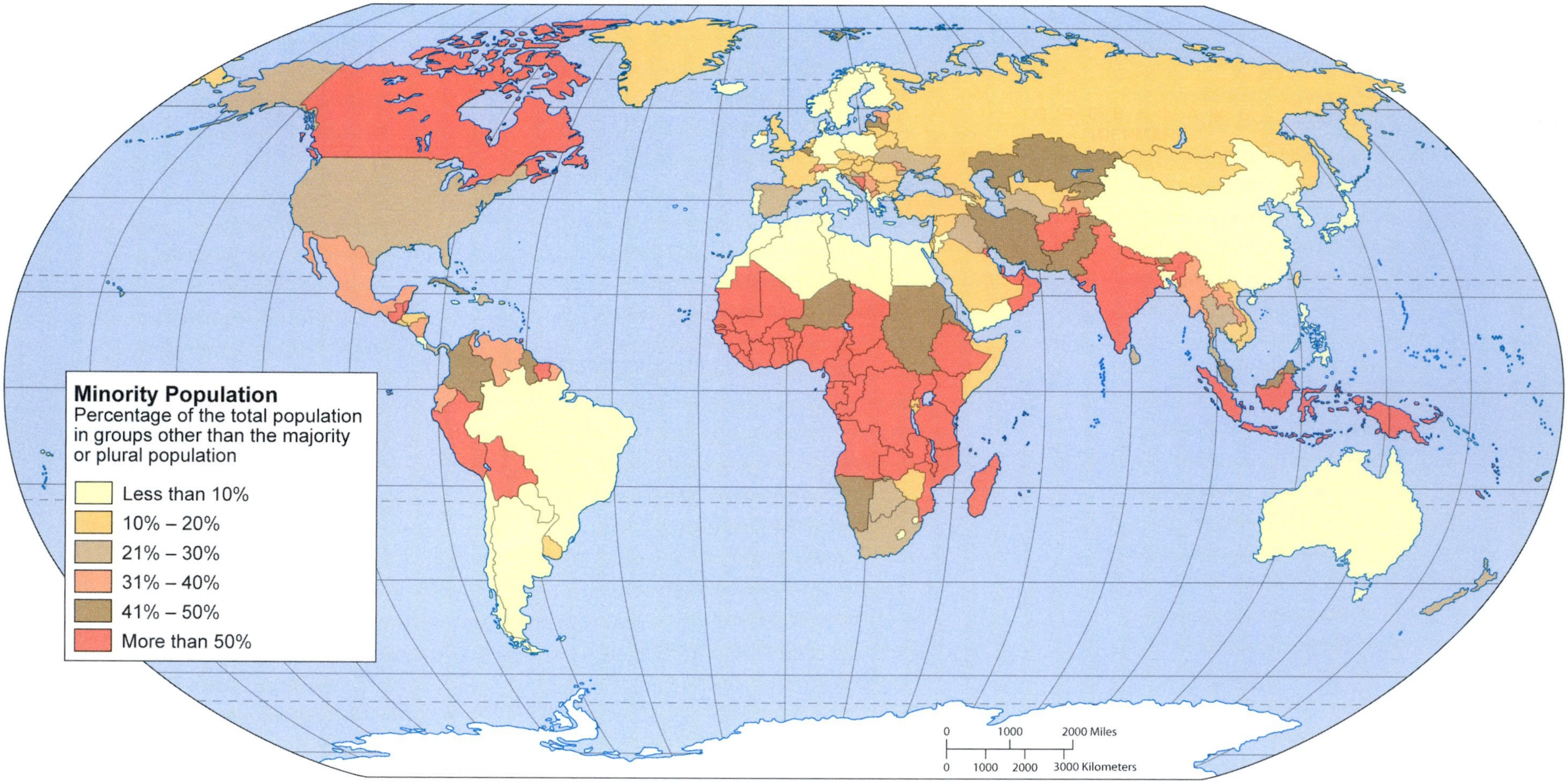

The presence of minority ethnic, national, or racial groups within a country's population can add a vibrant and dynamic mix to the whole. Plural societies with a high degree of cultural and ethnic diversity should, according to some social theorists, be among the world's most healthy. Unfortunately, the reality of the situation is quite different from theory or expectation. The presence of significant minority populations played an important role in the disintegration of the Soviet Union; the continuing existence of minority populations within the new states formed from former Soviet republics threatens the viability and stability of those young political units. In Africa, national boundaries were drawn by colonial powers without regard for the geographical distribution of ethnic groups, and the continuing tribal conflicts that have resulted hamper both economic and political development. Even in the most highly developed regions of the world, the presence of minority ethnic populations poses significant problems: witness the separatist movement in Canada, driven by the desire of some French-Canadians to be independent of the English majority, and the continuing ethnic conflict between Flemish-speaking and Walloon-speaking Belgians. This map, by arraying states on a scale of homogeneity to heterogeneity, indicates areas of existing and potential social and political strife.

Map 117 Marginalized Minorities: Declining Indigenous Populations

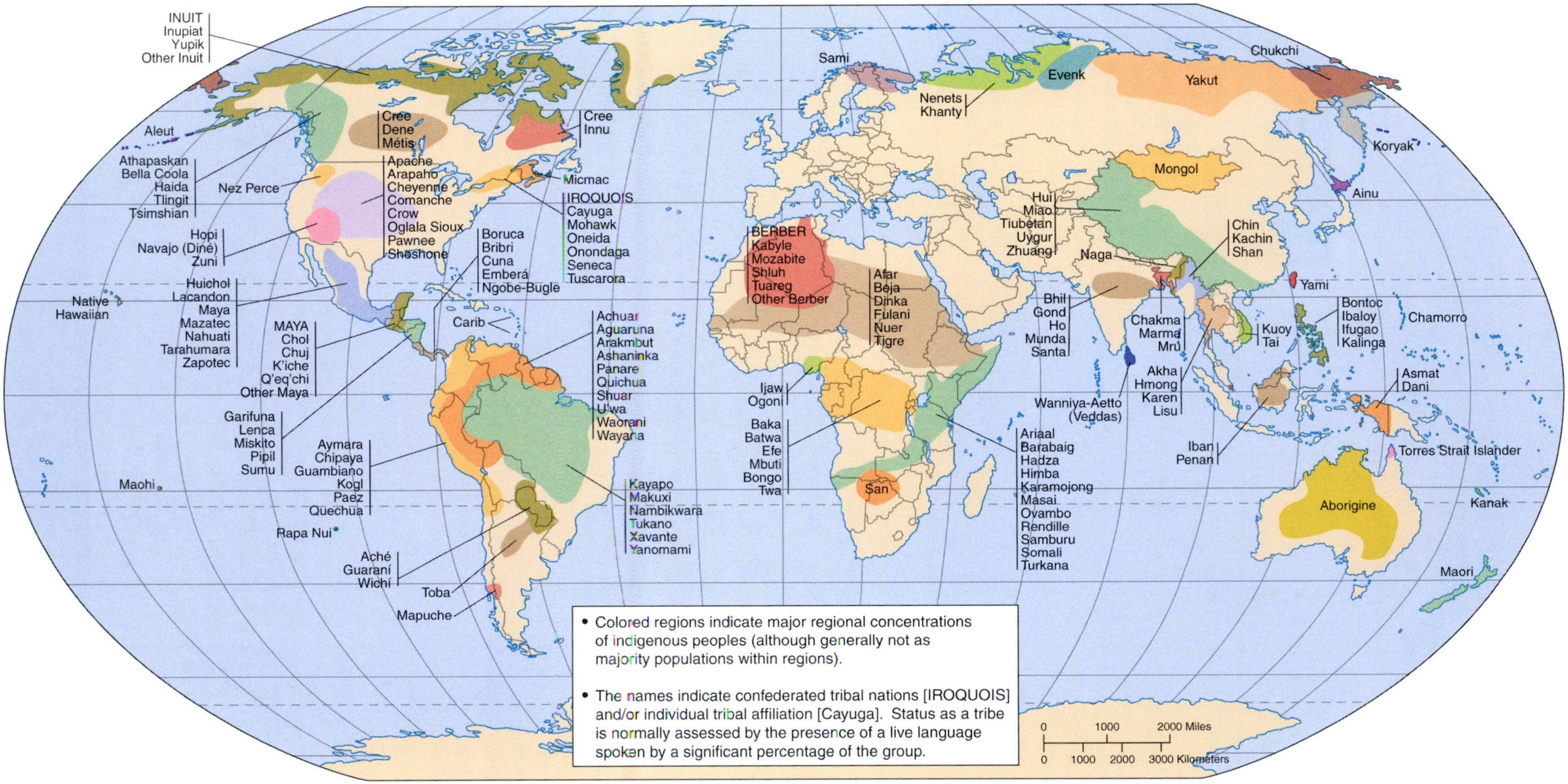

The world's indigenous (native) populations, those with the longest historical association with a geographic region, have nearly always been marginalized by colonialism, migration, economic development, or other external forces. Often confined to specific territories such as reservations or preserves, indigenous peoples have strived to preserve their languages, their cultures, their very identities—and often have sought to regain control of their ancestral lands as they have defined them. It has been estimated that, of the approximately 5,000 indigenous cultures remaining, fewer than half will survive through the end of the twenty-first century. There are many reasons for this: departure from the traditional life and homeland for urban areas and jobs; populations and population growth rates that have dropped below critical mass for maintenance; the scourge of substance abuse, particularly alcohol; disease rates that are higher among locationally marginal populations; because they are and have been marginal, they may not have had the same levels of exposure to certain diseases as more densely-crowded populations. The best historical example of this is the massive decline in American Indian populations (in both North and South America) after European colonization. American Indians simply did not have the built-in immunities against diseases such as smallpox, measles, chicken pox, or even the common cold as did Europeans and Africans and mortality rates were enormous among many populations—often exceeding 90%. A similar process occurred in the latter half of the twentieth century as the Amazon Basin was rapidly occupied by representatives of mixed Old World populations (European and African) and Old World diseases took a heavy toll among indigenous peoples of the Amazon. The loss of indigenous culture is not just an academic issue but one that bears upon the survival of humanity. For it is the indigenous cultures, living close to their natural environment, who may have the knowledge of local biology that could aid in the identification of substances that could assist in the eradication of major diseases among non-indigenous populations.

Map 118 The Political Geography of a Global Religion: The Islamic World

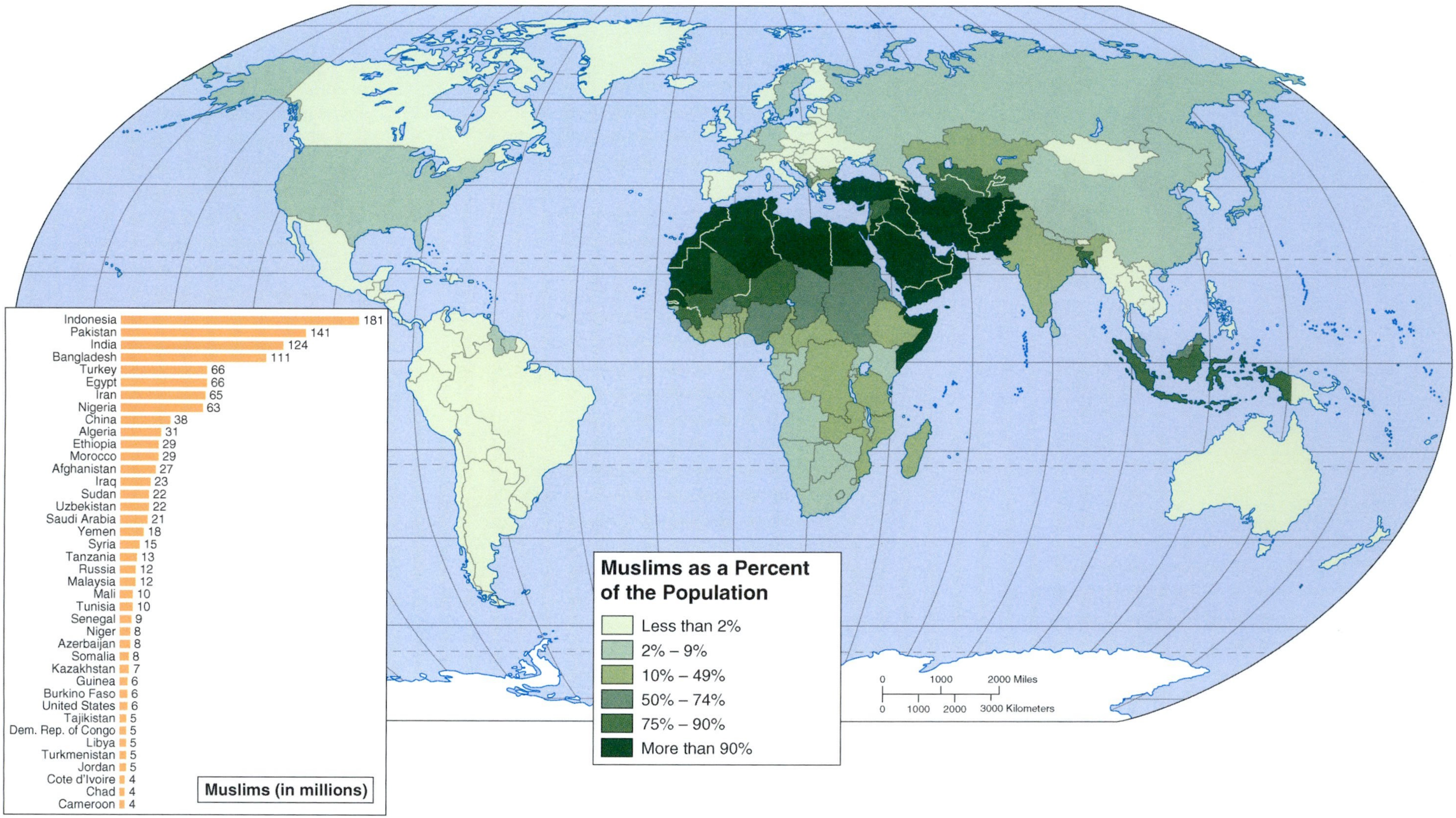

Islam, as a religion, does not promote conflict. The term **jihad**, often mistranslated to mean "holy war," in fact refers to the struggle to find God and to promote the faith. In spite of the beneficent nature of Islamic teachings, the tensions between Muslims and adherents of other faiths often flare into warfare. A comparison of this map with the map of international conflict will show a disproportionate number of wars in that portion of the world where Muslims are either majority or significant minority populations. The reasons for this are based more in the nature of government, cultures, and social structure, than in the tenets of the faith of Islam. Nevertheless, the spatial correlations cannot be ignored. Similarly, terrorist incidents falling considerably short of open armed warfare are spatially consistent with the distribution of Islam and even more consistent with the presence of Islamic fundamentalism or "Islamism," which tends to be less tolerant and more aggressive than the mainstream of the religion. Terrorism is also consistent with those areas where the legacy of colonialism or the persistent presence of non-Islamic cultures intrude into the Islamic world.

Map 119 World Refugee Population

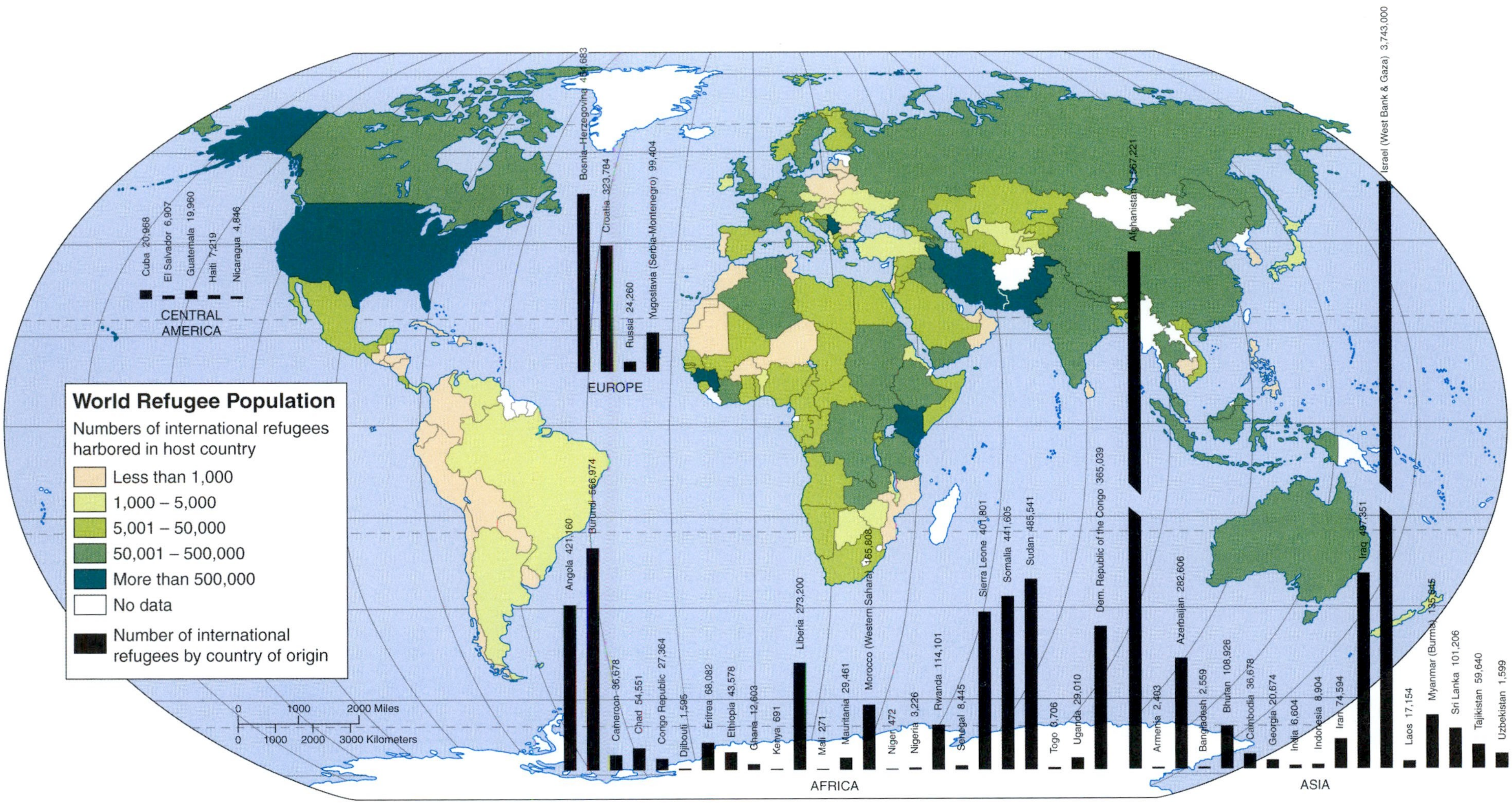

Refugees are persons who have been driven from their homes, normally by armed conflict, and have sought refuge by relocating. The most numerous refugees have traditionally been international refugees, who have crossed the political boundaries of their homelands into other countries. This refugee population is recognized by international agencies, and the countries of refuge are often rewarded financially by those agencies for their willingness to take in externally displaced persons. In recent years, largely because of an increase in civil wars, there have been growing numbers of internally displaced persons—those who leave their homes but stay within their country of origin. There are no rewards for harboring such internal refugee populations.

Map 120 Women's Rights

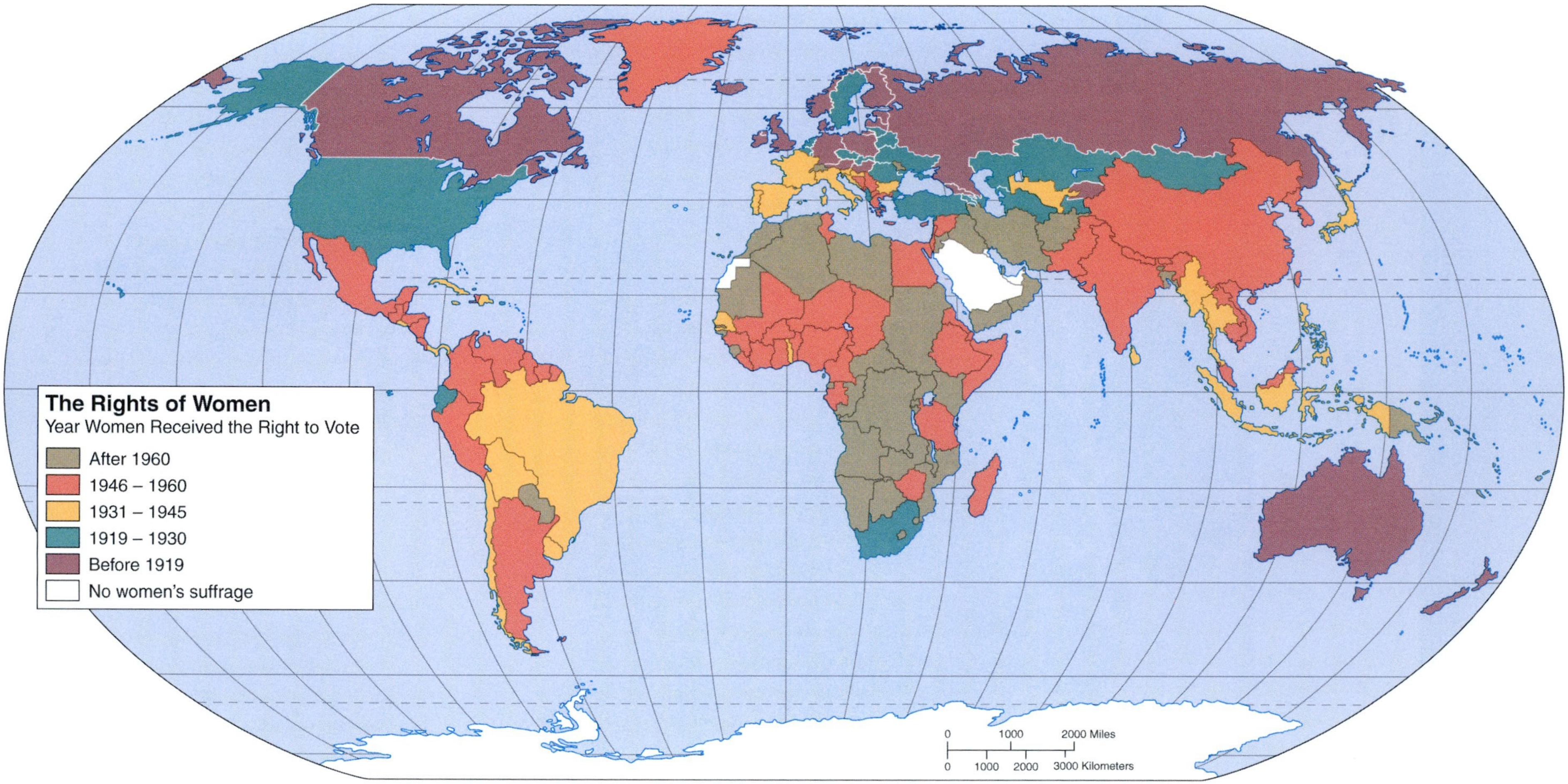

The "rights" referred to in this map refer primarily to the right to vote. But where women have the right to vote in free elections, the other fundamental rights tend to become available as well: the right to own property, the right to an education, the right to leave a domestic alliance without fear of retribution, or the right to be treated as a human being rather than property. But the time lag between women receiving the right to vote and their attainment of other fundamental human rights does not occur immediately or, in many cases, even relatively quickly. On the map, the most recent countries to grant suffrage to women are in Africa and Southwest Asia. In these regions, women still do not have access to many of the basic rights of what we would consider to be a civilized life. And, of course, there are still areas where women cannot vote: Kuwait, for example. Neither men nor women are allowed to vote in Brunei, Saudi Arabia, United Arab Emirates, or Western Sahara.

Map 121 Political and Civil Liberties, 2008

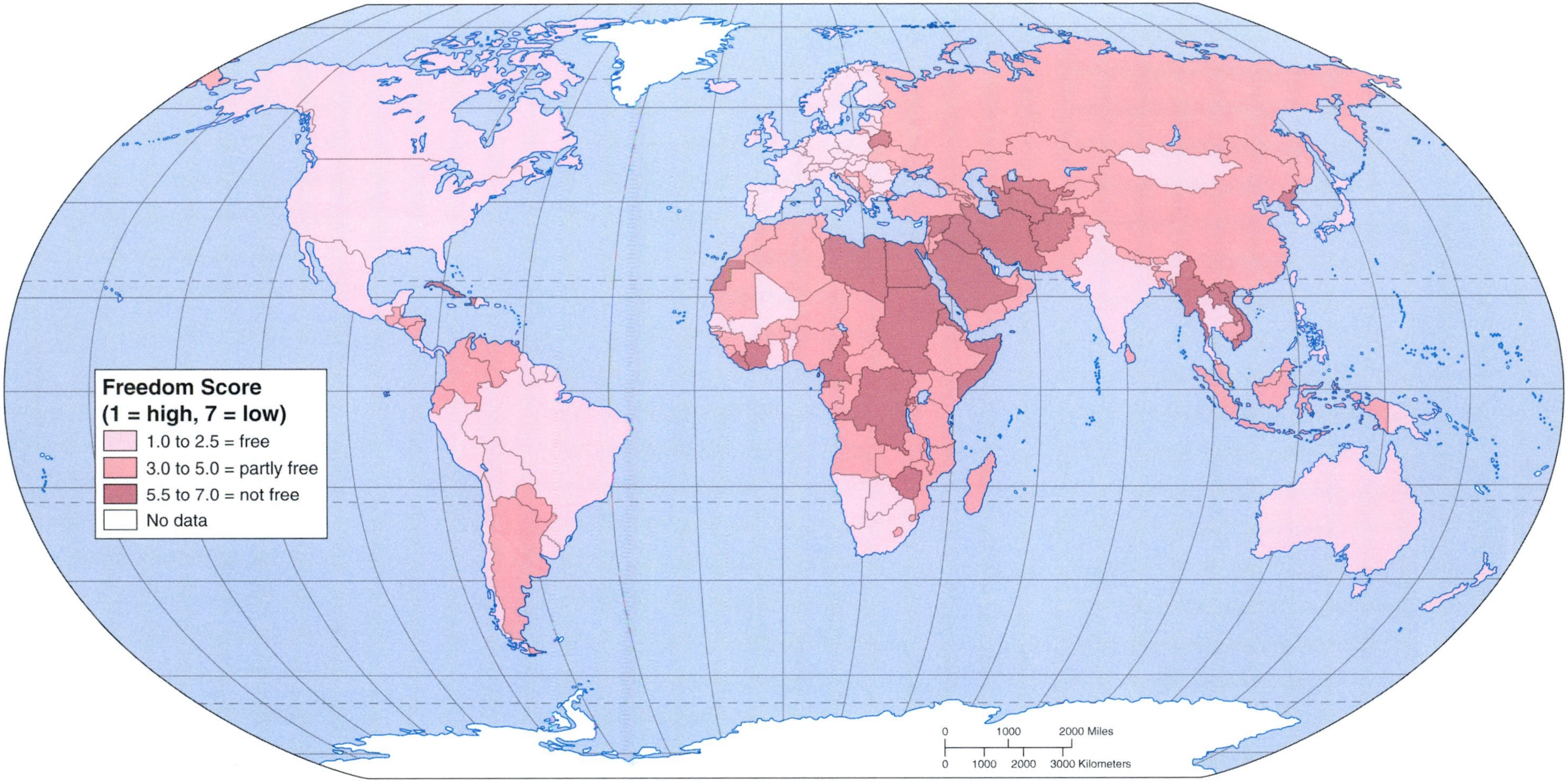

Although measures of political and civil liberty are somewhat difficult to obtain and assess, there are some generally accepted standards that can be evaluated: open elections and competitive political parties, the rule of law, freedoms of speech and press, judicial systems separate from other branches of government, and limits on the power of elected or appointed governmental officials. Interstingly, there appear to be correlations between "degrees of freedom" and such other characteristics of a state as per capita wealth, environmental quality, and healthy economic growth—characteristics that may be mutually contradictory. There is no empirical evidence of a causal link between democratic institutions and consumption; on the other hand, there is clear evidence of a positive relationship between wealth and consumption. Therefore, the three variables are closely correlated and should be used in assessing the nature of the state in any part of the world.

Map 122 Abuse of Public Trust

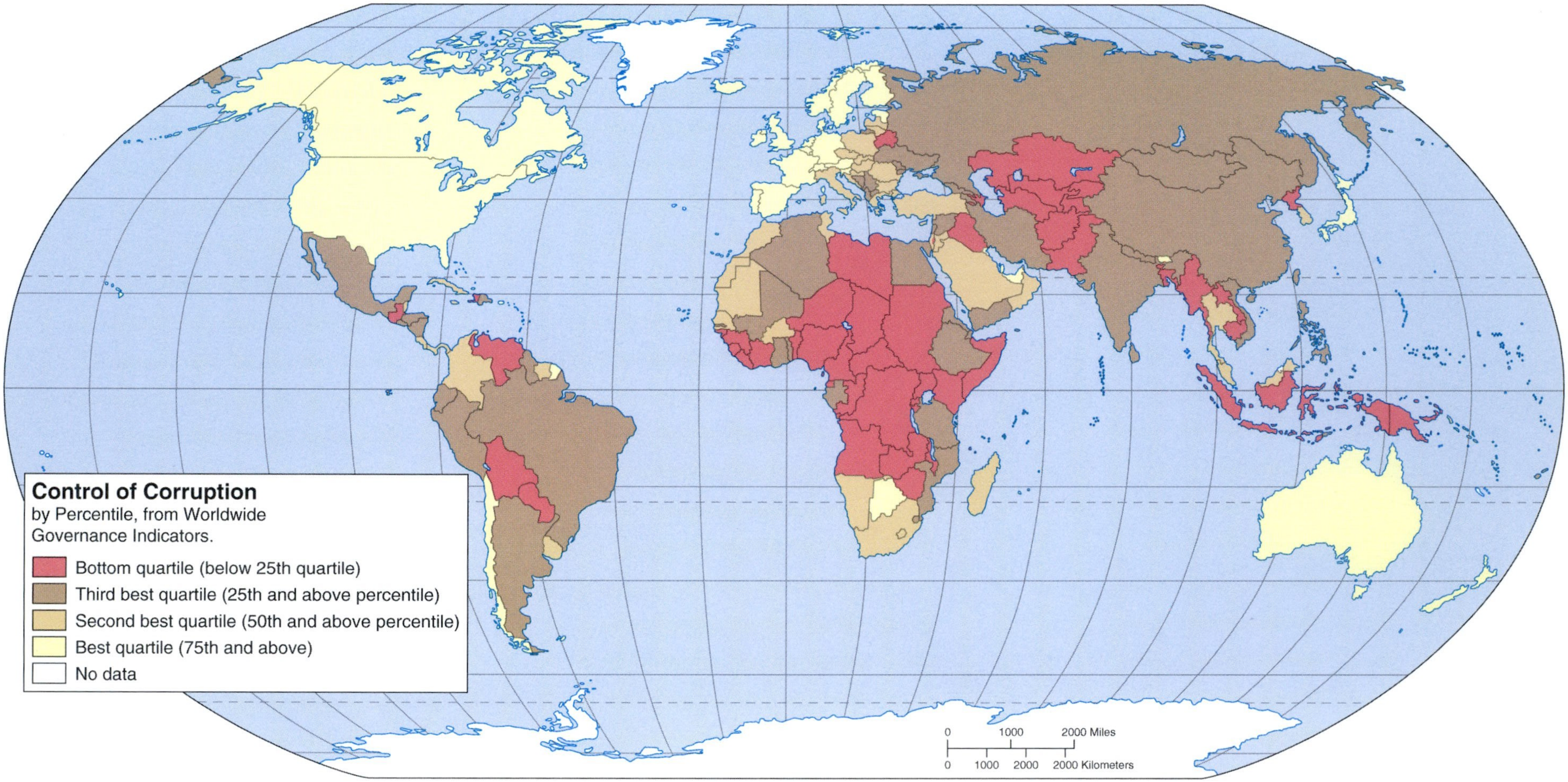

Abusing the public trust is simply another way of saying "corruption in government." In many parts of the world, corruption in the government is not an aberration but a way of life. Normally, although not always, governmental corruption is an indication of a weak and ineffective government, one that negatively affects such public welfare issues as public health, sanitation, education, and the provision of social services. It also tends to impact the cost of doing business and, thereby, drives away the foreign capital so badly needed in many African and Asian countries for economic development. Corruption is not automatic in poor countries, nor are rich countries free from it. But there is a general correlation between abuse of the public trust and lower levels of per capita income—excepting such countries as the Baltic states and Chile that have reached high standards of governance without joining the ranks of the wealthy countries. Studies by the World Bank have shown that countries that address issues of corruption and clean up the operations of their governments increase national incomes as much as four or five times. In those countries striving to attain governments that function according to a rule of law—rather than a rule of abusing the public trust—such important demographic measures as child mortality drop by as much as 75%. Clearly, good government and good business and higher incomes and better living conditions for the general public all go hand-in-hand.

Map 123 The Perception of Corruption

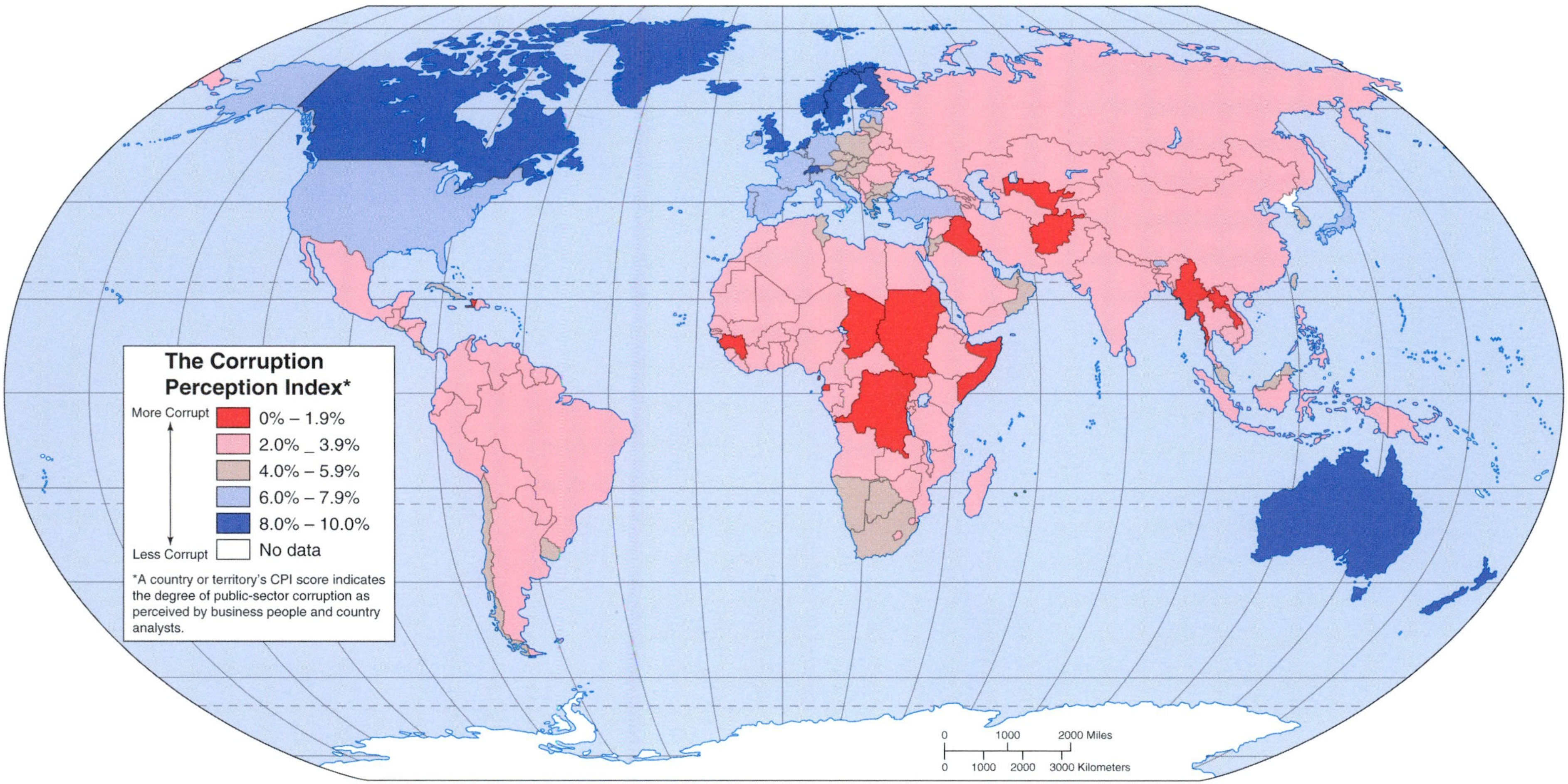

A country's Corruption Perception Index is based on analyses by business executives who work with a country's government or whose company does business within the country evaluated; in addition, the CPI is contributed to by analysts from press sources, international organizations, and others whose jobs are involved in the gathering of intelligence on foreign areas. A glance at the map tells us things that we might have easily intuited: Iraq, Afghanistan, Sudan, Somalia, Myanmar, and other disrupted or failed states or military dictatorships fall into the range of "more corrupt." At the same time, northern and western European governments with stable parliamentary democracies and high standards of living are, as we might expect, among the least corrupt. But why should the United States fall into a middle-range category, along with Mediterranean, Central, and Eastern European countries on the CPI scale? Have U.S. governmental actions and/or business dealings over the last few years reduced our international reputation? It would be tempting to think so. Non-U.S. analysts are not stupid and when such large percentages of American tax dollars devoted to the Iraq war end up in either American or Iraqi private hands, without discernible benefit, that fact registers with those analysts. The great Scottish poet Robert Burns wrote "Oh, would some Power the gift give us, to see ourselves as others see us" (translated from the Scottish). This may not be a bad plea for American officials to make. It is simply not good enough, in a country that supposedly governs itself by the rule of law, to say, "Well, we're not seen as corrupt as Somalia or Myanmar." It would also be self-defeating to dismiss these perceptions as unimportant. How a country is seen by others is often a very important factor in how that country is dealt with by others.

Map 124 A Decade of Risk of Civil Conflict 2005–2015

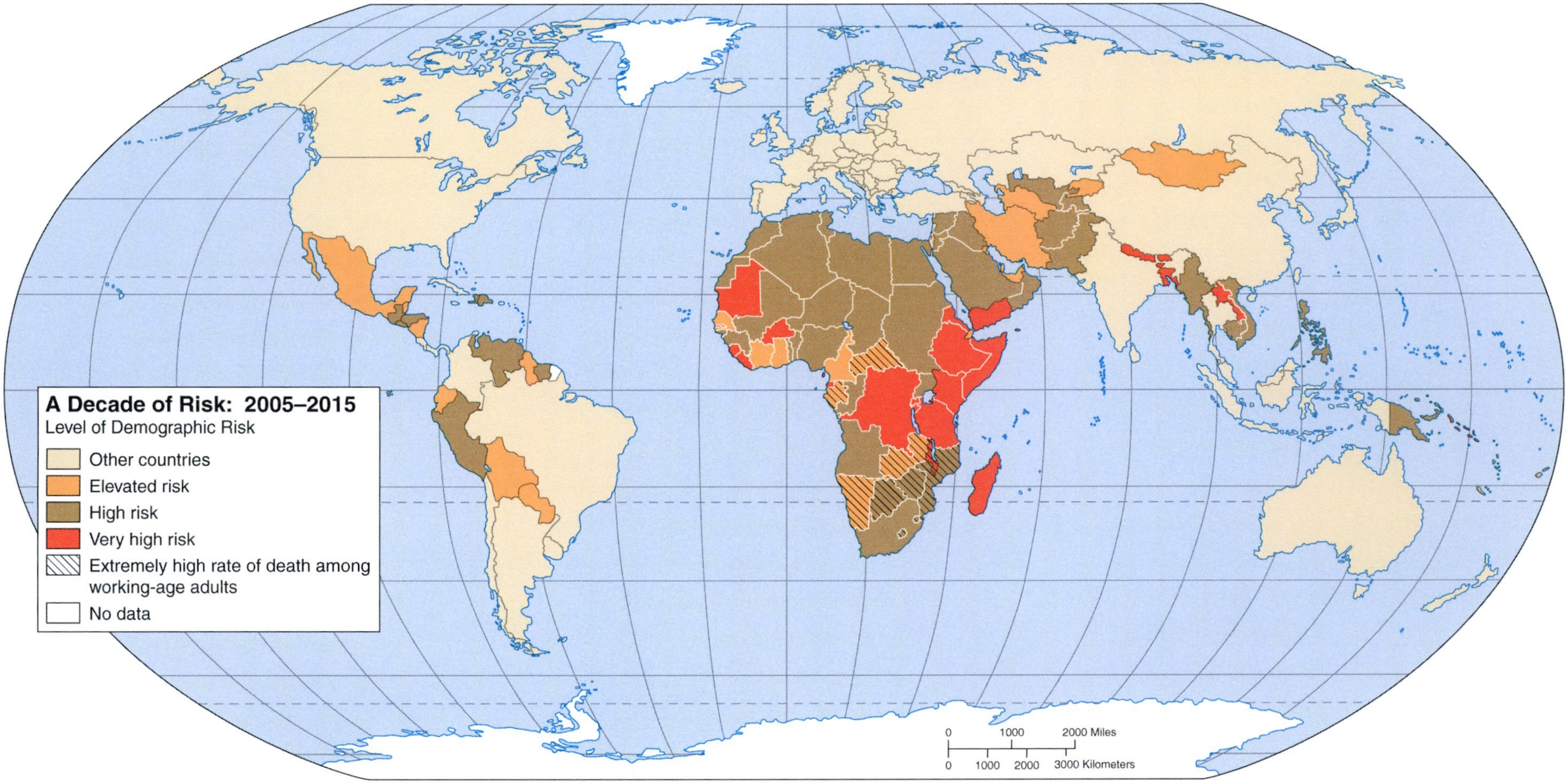

The point has been made earlier in this atlas (see maps 45–50) that demographic stresses enhance the risk of civil conflict, as opposed to international conflict. This map uses three demographic factors—percentage of a country's population between the ages of 15 and 29, rate of urban population growth, and per capita availability of cropland and fresh water—to assess the risk of domestic violence and civil conflict. It needs to be noted that the data are available on a national level and that geographically large countries such as Brazil, Russia, China, or India may well have areas of demographic stress within them that are masked by the portrayal of something other than subnational data. Nevertheless, the implications of the map are striking: Those parts of the world where civil conflict is in the highest risk category are almost all in Africa and Southwest Asia. We have already seen evidences of such civil conflict: the Hutu-Tutsi civil war of Rwanda and the Shiite-Sunni chaos of Iraq. We describe those as "tribal" or "sectarian" civil conflicts. But how much of the actual conflict is the result of a combination of demographic stresses?

Unit VII

World Regions

Map 125 North America: Political Divisions

ARCTIC OCEAN
GREENLAND (DENMARK)
ELLESMERE ISLAND
QUEEN ELIZABETH ISLANDS
Pt. Barrow
BROOKS RANGE
Arctic Circle
BEAUFORT SEA
BANKS ISLAND
Viscount Melville Sound
VICTORIA ISLAND
BOOTHIA PENINSULA
Cape York
Baffin Bay
BAFFIN ISLAND
Davis Strait
Mackenzie R.
Great Bear Lake
Mt. Logan 18,551 ft.
Mt. St. Elias 18,008 ft.
SOUTHAMPTON ISLAND
Hudson Strait
Cape Chidley
UNGAVA PENINSULA
Hudson Bay
SITKA ISLAND
Great Slave Lake
COAST MOUNTAINS
ROCKY
CORDILLERAN
Peace R.
Athabasca Lake
Athabasca R.
Reindeer Lake
Nelson R.
LABRADOR PENINSULA
Strait of Belle Isle
QUEEN CHARLOTTE ISLANDS
CANADA
Saskatchewan R.
James Bay
Cape Race
N. Saskatchewan R.
S. Saskatchewan R.
Fraser R.
Lake Winnipeg
St. Lawrence R.
Gulf of St. Lawrence
CAPE BRETON ISLAND
VANCOUVER ISLAND
LAURENTIAN HIGHLANDS
GASPÉ PENINSULA
Lake Nipigon
Lake of the Woods
Mt. Rainier 14,410 ft.
Missouri R.
Lake Superior
Montreal
Ottawa
Cape Sable
Columbia R.
Yellowstone R.
Red R.
Lake Huron
L. Ontario
Connecticut R.
Cape Cod
CASCADE RANGE
COAST RANGES
Mt. Shasta 14,162 ft.
Snake R.
Grand Teton 13,770'
UNITED STATES
Lake Michigan
Detroit
L. Erie
Hudson R.
New York
Philadelphia
Delaware R.
Sacramento R.
Great Salt Lake
Des Moines R.
Chicago
Wabash R.
Platte R.
GREAT BASIN
Ohio R.
APPALACHIAN MOUNTAINS
Washington, D.C.
Potomac R.
Chesapeake Bay
San Joaquin R.
SIERRA NEVADA
Colorado R.
Illinois R.
Mt. Whitney 14,494 ft.
Pike's Peak 14,110'
Arkansas R.
Roanoke R.
Cape Hatteras
BERMUDA
Los Angeles
MOUNTAIN HIGHLANDS
GREAT PLAINS
Rio Grande
Canadian R.
Red R.
Tennessee R.
Mississippi R.
Savannah R.
ATLANTIC OCEAN
GUADALUPE ISLAND
SIERRA MADRE OCCIDENTAL
SIERRA MADRE ORIENTAL
Lake Okeechobee
Miami
Straits of Florida
BAHAMAS
Gulf of California
Tropic of Cancer
Gulf of Mexico
Havana
CUBA
DOMINICAN REPUBLIC
HAITI
Cabo San Lucas
Yucatan Channel
WEST INDIES
PACIFIC OCEAN
MEXICO
Campeche Bay
Yucatan Peninsula
JAMAICA
ISLA REVILLAGIGEDO
Mexico City
Orizaba Peak 18,406 ft.
Popocatepetl 17,887 ft.
BELIZE
Gulf of Honduras
CARIBBEAN SEA
HONDURAS
GUATEMALA
NICARAGUA
EL SALVADOR
COSTA RICA
PANAMA
Gulf of Panama
ISLA DEL COCO
ISLA DE MALPELO
0 500 1000 Miles
0 500 1000 Kilometers
KAUAI
NIIHAU
OAHU
HAWAII
MOLOKAI
MAUI
LANAI
KAHOOLAWE
PACIFIC OCEAN
HAWAII
0 50 100 Miles
0 50 100 Kilometers
ST. LAWRENCE ISLAND
Bering Strait
Pt. Barrow
BROOKS RANGE
BERING SEA
ALASKA
Arctic Circle
Yukon R.
PRIBILOF ISLANDS
Mt. McKinley 20,320 ft.
Fairbanks
ALASKA RANGE
Bristol Bay
Anchorage
YUKON
Mt. Logan 18,551 ft.
ALEUTIAN ISLANDS
Mt. St. Elias 18,008 ft.
KODIAK ISLAND
ALASKA PENINSULA
Gulf of Alaska
Juneau
0 500 Miles
0 500 Kilometers
National capital
City
Elevation
Feet
3000
1500
300
0
Sea level
0
100
1000
Below sea level

Map 127a Environment and Economy: The Use of Land

The use of land in North America represents a balance between agriculture, resource extraction, and manufacturing that is unmatched. The United States, as the world's leading industrial power, is also the world's leader in commercial agricultural production. Canada, despite its small population, is a ranking producer of both agricultural and industrial products and Mexico has begun to emerge from its developing nation status to become an important industrial and agricultural nation as well. The countries of Middle America and the Caribbean are just beginning the transition from agriculture to modern industrial economies. Part of the basis for the high levels of economic productivity in North America is environmental: a superb blend of soil, climate, and raw materials. But just as important is the cultural and social mix of the plural societies of North America, a mix that historically aided the growth of the economic diversity necessary for developed economies.

Map 127b Population Density

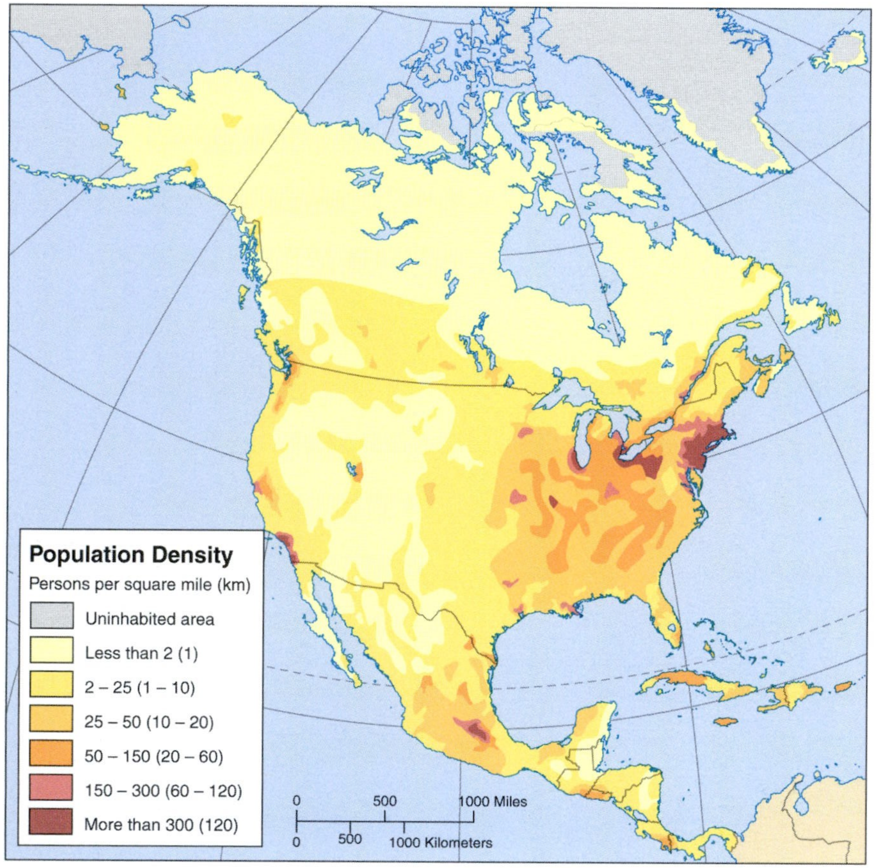

North America contains over 500 million people and the United States, with nearly 300 million inhabitants, is the third most populous country in the world after China and India. Most of the present North American population has roots in the Old World. The native populations of the Americas had little or no resistance to Old World diseases in 1500 and within a couple of centuries of their first contact with Europeans, most of the native peoples had either died out or preserved their genetic heritage by mixing with disease-resistant Old World populations. This left a North American population that is largely European, but with significant minorities resulting from the slave laborers imported from Africa and from the mixture of native Americans with Europeans and/or Africans. The density of that population is largely the consequence of environmental factors (good soil, the availability of water, the presence of other resources) and cultural/economic ones (agricultural production, urbanization, industrialization).

Map 127c Population Distribution: Clustering

Although population clustering is characteristic of highly economically developed regions, the Anglo-American population exhibits a remarkably clustered pattern. The primary reasons for this remarkable development are agricultural technology and affluence. Highly developed agricultural technology allows a small number of farmers using sophisticated machinery to grow and harvest enormous quantities of agricultural produce on large farms. In the United States, the world's leader in commercial agricultural production, only 2 percent of the population are farmers and that population is thinly distributed over wide areas. The vast majority of Americans live and work in city-suburb systems in which the widespread availability of private automobiles allows people to live considerable distances from where they work, keeping overall urban population densities relatively low but allowing for extensive urbanization—with cities large enough in area to be visible as population clusters on maps at this scale.

Map 127d Deforestation in North and Middle America

This map shows current forest cover adjacent to the estimated forest cover that would be present had there been no human intervention over the past several thousand years since the end of the last ice age, and assuming that climatic conditions would have been pretty much the same as they have been during that time. While in the popular imagination, it is the tropical forest regions of South America and Middle America that have been "deforested," the map shows otherwise. The greatest losses of forest cover have been in the now-agricultural regions of North America and the more temperate regions of South America (most in southern Brazil). Beside these forest losses in the temperate zones, the forest losses of the Amazon Basin seem relatively minor. This does not suggest that the current deforestation of Amazonia and other tropical forest regions is something about which we should not be concerned, but it does put that current deforestation in a more accurate historical context.

Map 128 Canada

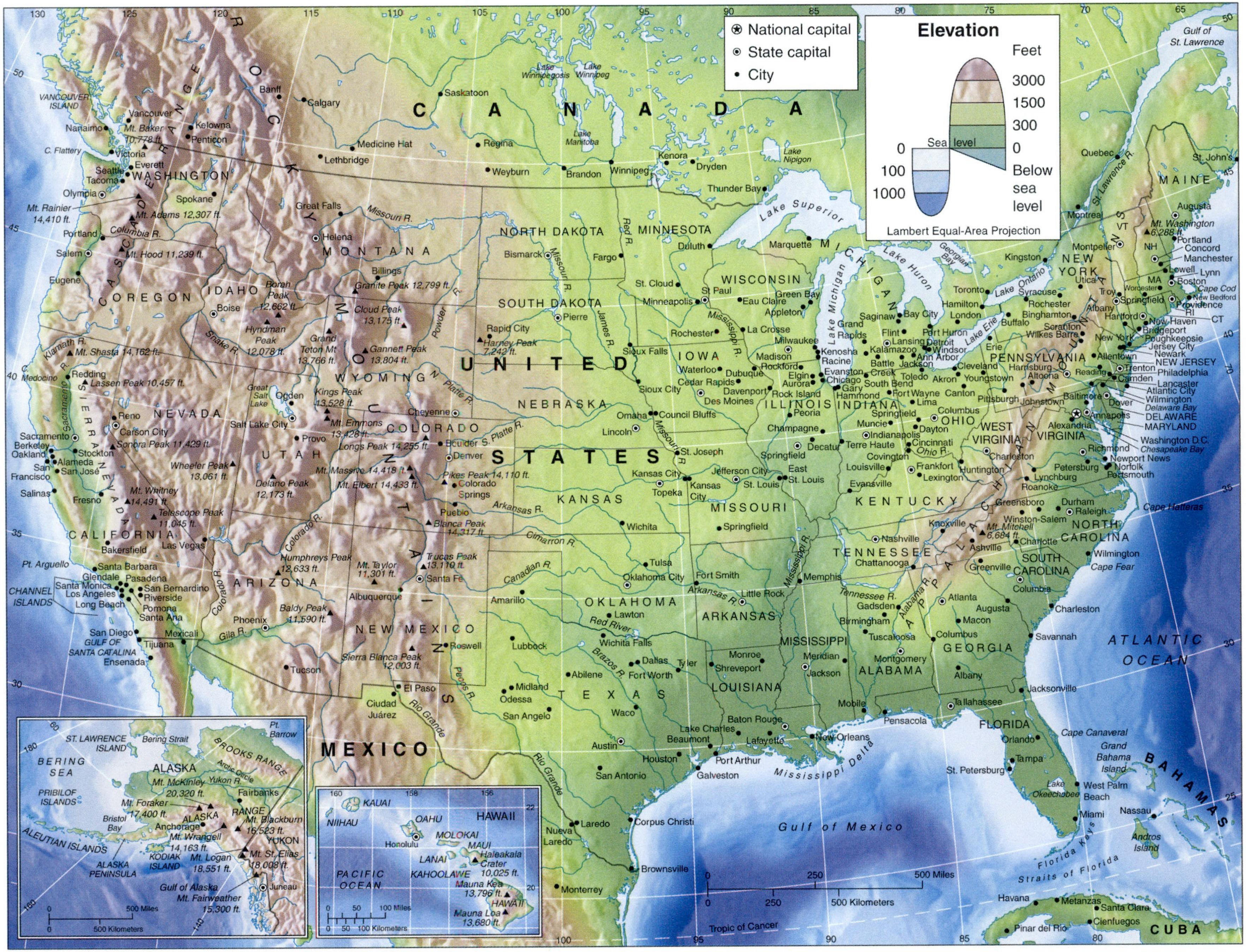
National capital
State capital
City
Elevation
Feet
3000
1500
300
0
Below sea level
Sea level
0
100
1000
Lambert Equal-Area Projection
C A N A D A
U N I T E D S T A T E S
M E X I C O
ATLANTIC OCEAN
Gulf of Mexico
BAHAMAS
CUBA
WASHINGTON
OREGON
CALIFORNIA
IDAHO
NEVADA
MONTANA
WYOMING
UTAH
ARIZONA
COLORADO
NEW MEXICO
NORTH DAKOTA
SOUTH DAKOTA
NEBRASKA
KANSAS
OKLAHOMA
TEXAS
MINNESOTA
IOWA
MISSOURI
ARKANSAS
LOUISIANA
WISCONSIN
ILLINOIS
MICHIGAN
INDIANA
OHIO
KENTUCKY
TENNESSEE
MISSISSIPPI
ALABAMA
GEORGIA
FLORIDA
SOUTH CAROLINA
NORTH CAROLINA
VIRGINIA
WEST VIRGINIA
PENNSYLVANIA
NEW YORK
MAINE
NEW JERSEY
DELAWARE
MARYLAND
Washington D.C.
ROCKY MOUNTAINS
CASCADE RANGE
SIERRA NEVADA
APPALACHIAN MOUNTAINS
Lake Superior
Lake Michigan
Lake Huron
Lake Erie
Lake Ontario
Mississippi R.
Missouri R.
Rio Grande
Colorado R.
Mt. Rainier 14,410 ft.
Mt. Whitney 14,491 ft.
Pikes Peak 14,110 ft.
Mt. Mitchell 6,684 ft.
Mt. Washington 6,288 ft.
Tropic of Cancer
500 Miles
500 Kilometers
ALASKA
BERING SEA
BROOKS RANGE
ALASKA RANGE
Mt. McKinley 20,320 ft.
Gulf of Alaska
ALEUTIAN ISLANDS
HAWAII
PACIFIC OCEAN
KAUAI
OAHU
MOLOKAI
MAUI
Honolulu
Mauna Kea 13,796 ft.
Mauna Loa 13,680 ft.

UNITED STATES OF AMERICA
MEXICO
ATLANTIC OCEAN
PACIFIC OCEAN
Gulf of Mexico
CARIBBEAN SEA
Gulf of California
Yucatan Channel
BAHAMA ISLANDS
CUBA
JAMAICA
CAYMAN ISLANDS (Br.)
ISLE OF YOUTH
ANDROS I.
BELIZE
GUATEMALA
HONDURAS
EL SALVADOR
NICARAGUA
COSTA RICA
PANAMA
BAJA CALIFORNIA
BAJA CALIFORNIA SUR
SONORA
CHIHUAHUA
COAHUILA
NUEVO LEON
TAMAULIPAS
SINALOA
DURANGO
ZACATECAS
SAN LUIS POTOSI
NAYARIT
AGUASCALIENTES
JALISCO
GUANAJUATO
QUERÉTARO
HILDAGO
COLIMA
MICHOACAN
MEXICO
TLAXCALA
MORELOS
GUERRERO
PUEBLA
VERACRUZ
OAXACA
TABASCO
CHIAPAS
CAMPECHE
YUCATAN
QUINTANA ROO
SIERRA MADRE OCCIDENTAL
SIERRA MADRE ORIENTAL
San Diego
Tijauana
Mexicali
Ensenada
Phoenix
El Paso
Ciudad Juárez
Hermosillo
Ciudad Obregón
Chihuahua
Culiacán
La Paz
Cabo San Lucas
Mazatlán
Gómez Palacio
Torreón
Saltillo
Monterrey
Nuevo Laredo
Matamoros
Durango
Zacatecas
San Luis Potosí
Ciudad Victoria
Tampico
Tepic
Aguascalientes
León
Guanajuato
Querétaro
Guadalajara
Celaya
Pachuca
Morelia
Mexico City
Toluca
Cuernavaca
Tlaxcala
Puebla
Xalapa
Veracruz
Orizaba
Colima
Chilpancingo de los Bravo
Oaxaca
Villahermosa
Tuxtla Gutiérrez
Campeche
Mérida
Ciudad Chetumal
Belize City
Belmopan
Guatemala
Tegucigalpa
Santa Ana
San Salvador
Managua
San Jose
Colón
Panama
Dallas
Houston
San Antonio
New Orleans
Memphis
Atlanta
Jacksonville
Tampa
Miami
Nassau
Guanabacoa
Havana
Marianao
Matanzas
Cárdenas
Santa Clara
Cienfuegos
Pinar del Río
Camagüey
Holguín
Manzanillo
Santiago de Cuba
Kingston
Canadian R.
Red R.
Pecos R.
Brazos R.
Mississippi R.
Rio Grande
Balsas R.
GUADALUPE ISLAND
CEDROS ISLAND
Punta Eugenie
Cape Corrientes
Tropic of Cancer
Lake Chapala
Campeche Bay
Terminos Lagoon
Yucatan Peninsula
Gulf of Honduras
Gulf of Tehuantepec
Lake Nicaragua
Gulf of Mosquitos
Gulf of Panama
Cerro de la Encantada 10,126 ft.
Cerro Pinacate 4,560 ft.
Vol. Las Tres Virgenes 6,547 ft.
Pico Cupula 5,000 ft.
Santa Genoveva 7,100 ft.
Nevado de Colima 13,993 ft.
Vol. Paricutín 9,213 ft.
Nevado de Toluca 14,409 ft.
Iztaccihuatl 17,159 ft.
Vol. Popocatépetl 17,930 ft.
Tacana 13,417 ft.
Pico Turquino 6,470 ft.
Mt. Denham 3,296 ft.
Cerro Chirripó 12,530 ft.
Vol. Barú 11,401 ft.
120
110
100
90
80
30
20
10
Elevation
Feet
3000
1500
300
0
Below sea level
Sea level
0
100
1000
National capital
State capital
City
0
250
500 Miles
0
250
500 Kilometers
Lambert Equal-Area Projection

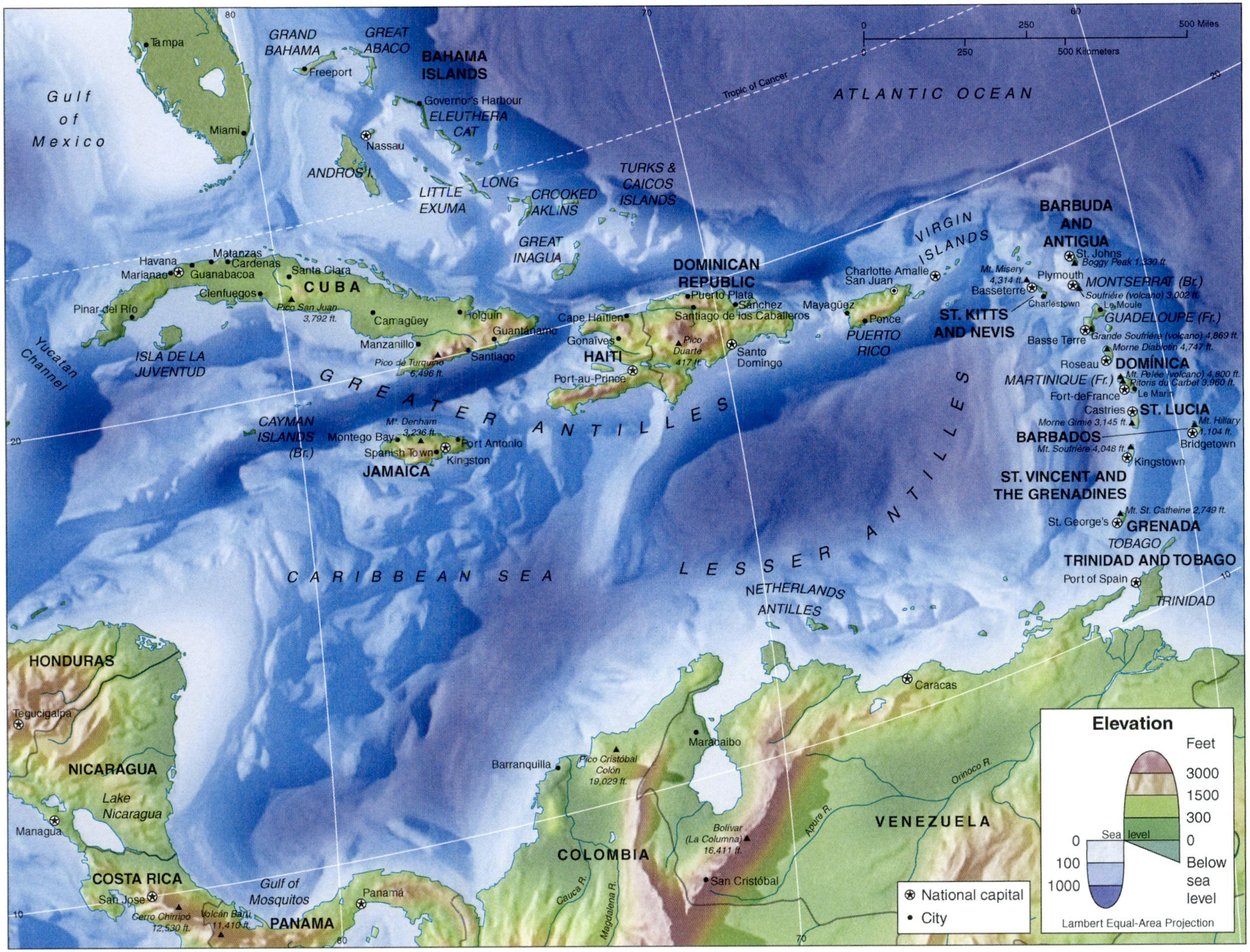
Gulf of Mexico
Tampa
Miami
GRAND BAHAMA
Freeport
GREAT ABACO
BAHAMA ISLANDS
Governor's Harbour
ELEUTHERA
CAT
Nassau
ANDROS I.
LITTLE EXUMA
LONG
CROOKED
AKLINS
TURKS & CAICOS ISLANDS
GREAT INAGUA
ATLANTIC OCEAN
Tropic of Cancer
Havana
Marianao
Matanzas
Cardenas
Guanabacoa
Santa Clara
CUBA
Cienfuegos
Pinar del Rio
Pico San Juan 3,792 ft.
Camagüey
Holguin
Guantánamo
Manzanillo
Santiago
Pico de Turquino 6,496 ft.
ISLA DE LA JUVENTUD
Yucatan Channel
DOMINICAN REPUBLIC
Puerto Plata
Sánchez
Santiago de los Caballeros
Cape Haitien
Gonaïves
HAITI
Pico Duarte 417 ft.
Santo Domingo
Port-au-Prince
GREATER ANTILLES
CAYMAN ISLANDS (Br.)
Montego Bay
Mt. Denham 3,236 ft.
Port Antonio
Spanish Town
Kingston
JAMAICA
VIRGIN ISLANDS
Charlotte Amalie
San Juan
Mayagüez
Ponce
PUERTO RICO
BARBUDA AND ANTIGUA
St. Johns
Boggy Peak 1,330 ft.
Mt. Misery 4,314 ft.
Plymouth
MONTSERRAT (Br.)
Soufrière (volcano) 3,002 ft.
Basseterre
Charlestown
Le Moule
ST. KITTS AND NEVIS
GUADELOUPE (Fr.)
Basse Terre
Grande Soufrière (volcano) 4,869 ft.
Morne Diablotin 4,747 ft.
Roseau
DOMINICA
MARTINIQUE (Fr.)
Mt. Pelée (volcano) 4,800 ft.
Pitons du Carbet 3,960 ft.
Fort-deFrance
Le Marin
Castries
ST. LUCIA
Morne Gimie 3,145 ft.
Mt. Hillary 1,104 ft.
BARBADOS
Bridgetown
Mt. Soufrière 4,048 ft.
Kingstown
ST. VINCENT AND THE GRENADINES
Mt. St. Catheine 2,749 ft.
St. George's
GRENADA
TOBAGO
TRINIDAD AND TOBAGO
Port of Spain
TRINIDAD
LESSER ANTILLES
NETHERLANDS ANTILLES
CARIBBEAN SEA
HONDURAS
Tegucigalpa
NICARAGUA
Lake Nicaragua
Managua
COSTA RICA
San José
Cerro Chirripó 12,530 ft.
Volcán Barú 11,410 ft.
Gulf of Mosquitos
PANAMA
Panamá
Barranquilla
Pico Cristóbal Colón 19,029 ft.
Maracaibo
COLOMBIA
Cauca R.
Magdalena R.
Bolívar (La Columna) 16,411 ft.
San Cristóbal
Apure R.
Orinoco R.
VENEZUELA
Caracas
0 250 500 Miles
0 250 500 Kilometers
National capital
City
Elevation
Feet
3000
1500
300
0
Sea level
Below sea level
0
100
1000
Lambert Equal-Area Projection
80
70
60
20
10

Map 132 South America: Political Divisions

CARIBBEAN SEA
COSTA RICA
PANAMA
Gulf of Panamá
Gulf of Venezuela
Cartagena
Gulf of Darién
Maracaibo
Caracas
Valencia
Port-of-Spain
BARBADOS
TRINIDAD and TOBAGO
Lake Maracaibo
Rio Magdalena
Rio Cauca
Rio Orinoco
VENEZUELA
Medellín
Santa Fe de Bogotá
Nevado del Tolima 17,110 ft.
LLANOS
COLOMBIA
GUIANA HIGHLANDS
Georgetown
GUYANA
Paramaribo
SURINAME
FRENCH GUIANA
Cayenne
NORTH ATLANTIC OCEAN
Rio Casiquiare
Rio Branco
Rio Negro
Quito
ECUADOR
Cotopaxi 19,347 ft.
Chimborazo 20,702 ft.
Guayaquil
Gulf of Guayaquil
Rio Putumayo
Rio Japurá
Rio Solimões
Manaus
Rio Amazon
Belém
Equator
Rio Tocantins
Fortaleza
Iquitos
Rio Marañón
Rio Juruá
Rio Purús
Rio Madeira
Rio Tapajós
Rio Xingu
Rio Aripuanã
Rio Roosevelt
Porto Velho
BRAZIL
Natal
Recife
Chiclayo
Trujillo
Nevado Huascarán 22,133 ft.
Rio Ucayali
PERU
ANDES MOUNTAINS
Rio São Francisco
Paulo Afonso Falls
SERRA DO PIAUÍ
Rio Araguaia
Callao
Lima
Cusco
Rio Beni
Rio Mamoré
Rio Guaporé
SERRA DO ESPINHAÇO
MATO GROSSO
Brasília
Lake Titicaca
La Paz
Misti 19,101 ft.
Arequipa
Nev. Illimani 20,741 ft.
BOLIVIA
Rio Paraguay
BRAZILIAN HIGHLANDS
Lake Poopó
Sucre
Potosí
Rio Paranaíba
Belo Horizonte
GRAN CHACO
Pico de Bandeira 9,843 ft.
Rio Paraná
PARAGUAY
Antofagasta
ATACAMA DESERT
Rio Pilcomayo
Asunción
Iguassú Falls
Santos
Tropic of Capricorn
Rio Bermejo
Rio Iguassú
Copiapó 19,947 ft.
Copiapó
PACIFIC OCEAN
Rio Salado
Rio Uruguay
Coquimbo
Porto Alegre
ARGENTINA
CHILE
URUGUAY
SOUTH ATLANTIC OCEAN
Cerro Aconcagua 22,881 ft.
Buenos Aires
Montevideo
Valparaíso
Santiago
La Plata
Rio de la Plata
JUAN FERNANDEZ ISLANDS
PAMPAS
Rio Colorado
Bahía Blanca
Rio Negro
Valdivia
Gulf of San Matías
Puerto Montt
CHILOÉ ISLAND
Rio Chubut
Comodoro Rivadavia
Gulf of San Jorge
LOS CHONOS ARCHIPELAGO
PATAGONIA
WELLINGTON ISLAND
FALKLAND ISLANDS (ISLAS MALVINAS)
HANOVER ISLAND
Strait of Magellan
DESOLACIÓN ISLAND
TIERRA DEL FUEGO
ISLA DE LOS ESTADOS
Mt. Sarmiento 8,100 ft.
Cape Horn
SOUTH GEORGIA
National capital
City
Elevation
Feet
3000
1500
300
0
Sea level
Below sea level
0
100
1000
0 250 500 Miles
0 250 500 Kilometers

South America is a region just beginning to emerge from a colonial-dependency economy in which raw materials flowed from the continent to more highly developed economic regions. With the exception of Brazil, Argentina, Chile, and Uruguay, most of the continent's countries still operate under the traditional mode of exporting raw materials in exchange for capital that tends to accumulate in the pockets of a small percentage of the population. The land use patterns of the continent are, therefore, still dominated by resource extraction and agriculture. A problem posed by these patterns is that little of the continent's land area is actually suitable for either commercial forestry or commercial crop agriculture without extremely high environmental costs. Much of the agriculture, then, is based on high value tropical crops that can be grown in small areas profitably, or on extensive livestock grazing. Even within the forested areas of the Amazon Basin where forest clearance is taking place at unprecedented rates, much of the land use that replaces forest is grazing.

Map 134b Population Density

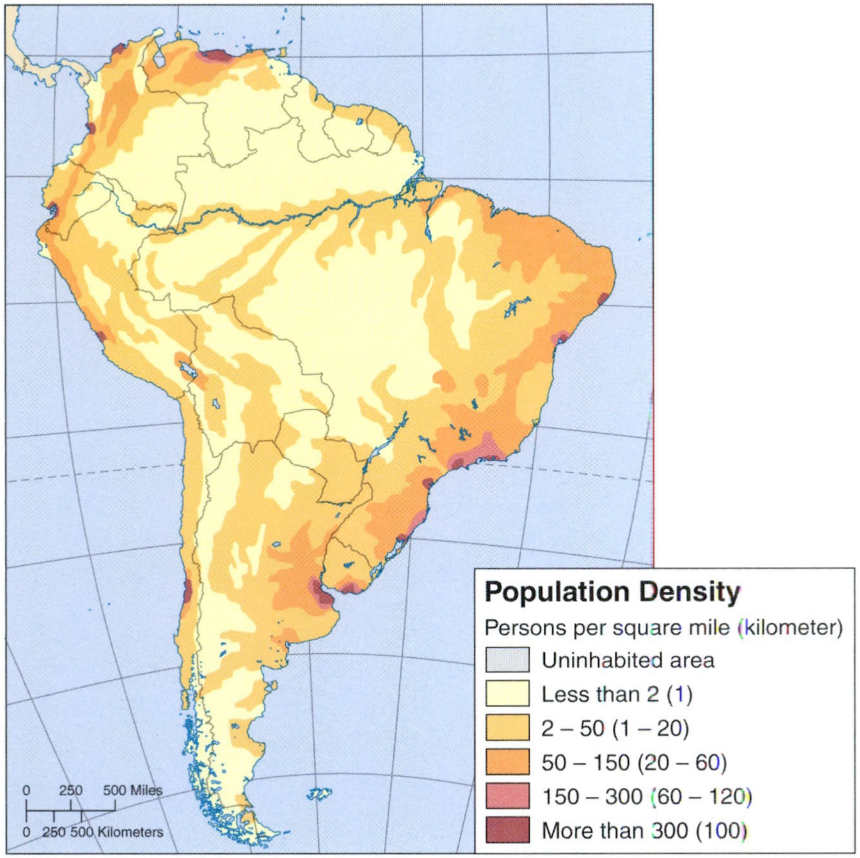

Since so much of interior South America is uninhabitable (the high Andes) or only sparsely populated (the interior of the Amazon Basin), the continent's population tends to be peripheral—approximately 90 percent of the continent's nearly 375 million people live within 150 miles of the sea. This population also tends to be heavily urbanized. Over 80 percent of South America's population lives in cities and the continent has three of the world's 15 largest cities—Rio de Janeiro, São Paulo, and Buenos Aires. São Paulo is the world's third largest urban agglomeration after Tokyo and New York. As in North America, most of the population of South America can trace at least part of its ancestry to the Old World. Throughout the Spanish-speaking parts of the continent the population is predominantly *mestizo* or mixed European and native South American. In Portuguese-speaking Brazil, in addition to a *mestizo* population, there is a significant admixture of African blood, the result of a large slave labor force imported from Africa to work the sugar, cotton, and other plantations of the colonial period.

Map 134c Environmental Problems

The drainage basin of the Amazon River and its tributaries, along with adjacent regions, is the world's largest remaining area of tropical forest. Much of the periphery of this vast forested region has already been cleared for farming and grazing, and the ax and chainsaw and flames are working their way steadily toward the interior. Tropical deforestation produces a loss of the biological diversity represented by the world's most biologically productive ecosystem, along with changes in soil and soil-water systems. South America has other environmental problems: *desertification* in which grassland and/or scrub vegetation is converted to desert through overgrazing or other unwise agricultural practices; soil *salinization* in which soils become increasingly salty as the consequence of the over-application of irrigation water; *coastal and estuarine pollution* resulting from unregulated or unchecked use of coastal waters for industrial, commercial, and transportation purposes; and *acid precipitation* resulting from the combination of airborne industrial wastes and automobile/truck exhausts with water vapor to produce dry or wet acidic fallout.

Map 134d Protection of Natural Areas

International conservation organizations are concerned that the protection of natural areas is an empty measure, as in many areas of the world, the park systems designated largely for the purposes of wildlife preservation have been subjected to population and resource pressures. In North and South America, recent government measures that seem directed against preservation have raised mounting fears that American protected areas could become, like their counterparts in many areas of Africa and Asia, simply parks on paper.

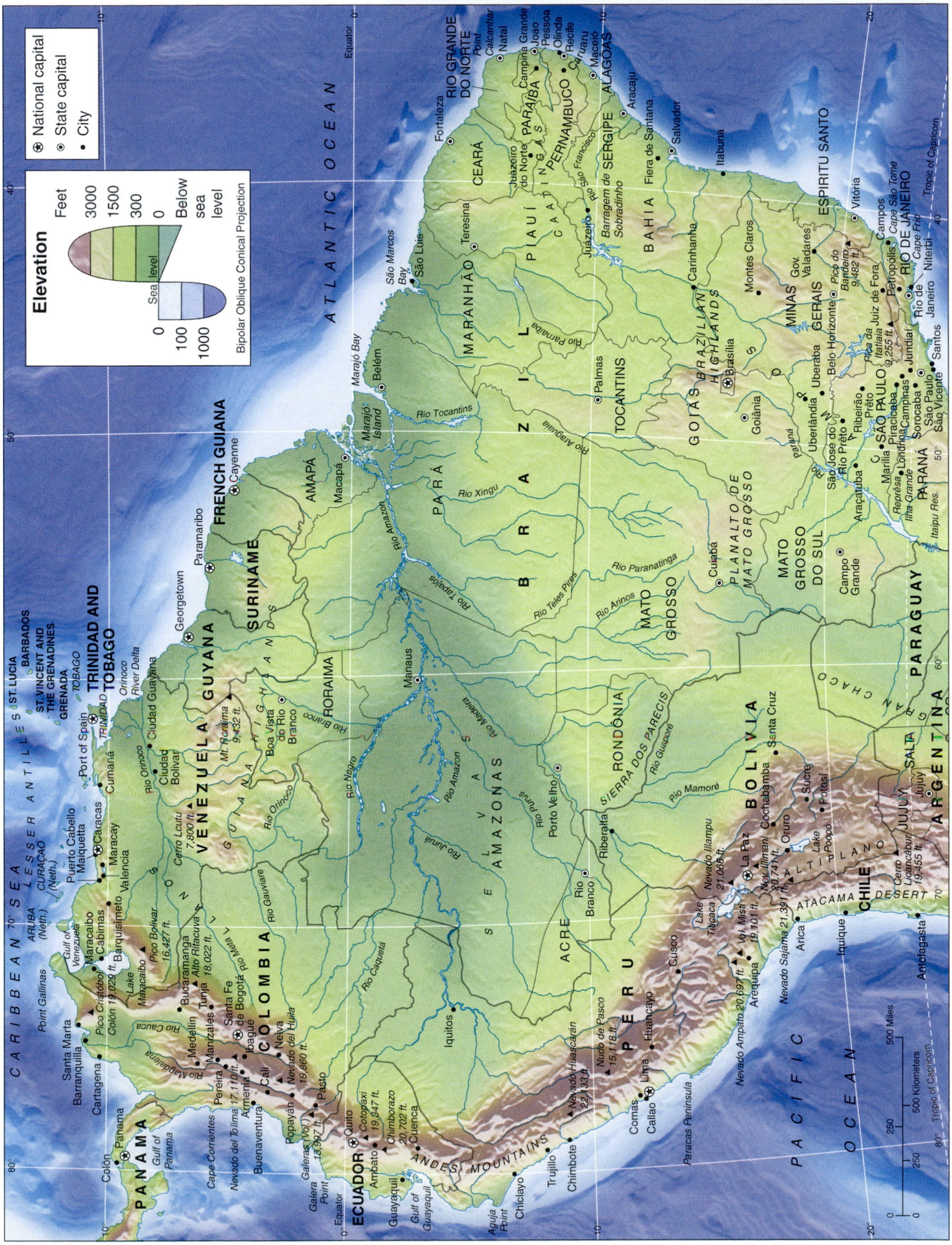
National capital
State capital
City
Elevation
Feet
3000
1500
300
0
Below sea level
Sea level
0
100
1000
Bipolar Oblique Conical Projection
CARIBBEAN SEA
LESSER ANTILLES
ATLANTIC OCEAN
PACIFIC OCEAN
Equator
Tropic of Capricorn
ARUBA (Neth.)
CURAÇAO (Neth.)
ST. LUCIA
BARBADOS
ST. VINCENT AND THE GRENADINES
GRENADA
TOBAGO
TRINIDAD
TRINIDAD AND TOBAGO
Port of Spain
Orinoco River Delta
PANAMA
Panama
Colón
Gulf of Panama
COLOMBIA
Barranquilla
Cartagena
Santa Marta
Point Gallinas
Gulf of Venezuela
Lake Maracaibo
Maracaibo
Cabimas
Pico Cristóbal Colón 19,029 ft.
Pico Bolívar 16,427 ft.
Bucaramanga
Medellín
Manizales
Pereira
Armenia
Ibagué
Tunja
Santa Fe de Bogotá
Alto Ritacuva 18,022 ft.
Neiva
Cali
Nevado del Huila 18,860 ft.
Nevado del Tolima 17,110 ft.
Popayán
Pasto
Buenaventura
Cape Corrientes
Galeras (Vol.) 13,997 ft.
Galera Point
Rio Magdalena
Rio Cauca
Rio Meta
Rio Guaviare
Rio Caquetá
VENEZUELA
Barquisimeto
Valencia
Maracay
Puerto Cabello
Maiquetía
Caracas
Cumaná
Ciudad Guayana
Ciudad Bolívar
Rio Orinoco
Cerro Lotu 7,800 ft.
Mt. Roraima 9,432 ft.
GUIANA HIGHLANDS
GUYANA
Georgetown
SURINAME
Paramaribo
FRENCH GUIANA
Cayenne
ECUADOR
Quito
Cotopaxi 19,347 ft.
Ambato
Chimborazo 20,702 ft.
Cuenca
Guayaquil
Gulf of Guayaquil
ANDES MOUNTAINS
PERU
Aguja Point
Chiclayo
Trujillo
Chimbote
Nevado Huascarán 22,133 ft.
Nudo de Pasco 15,118 ft.
Huancayo
Comas
Lima
Callao
Paracas Peninsula
Cusco
Iquitos
Nevado Ampato 20,637 ft.
Arequipa
Vol. Misti 19,101 ft.
Lake Titicaca
Nevado Illampu 21,066 ft.
BOLIVIA
La Paz
Nev. Illimani 20,741 ft.
Cochabamba
Oruro
Lake Poopó
Sucre
Potosí
Santa Cruz
ALTIPLANO
Nevado Sajama 21,391 ft.
Arica
ATACAMA DESERT
CHILE
Iquique
Antofagasta
Cerro Licancabur 19,455 ft.
ARGENTINA
JUJUY
Jujuy
SALTA
Salta
GRAN CHACO
PARAGUAY
BRAZIL
SELVAS
AMAZONAS
ACRE
Rio Branco
RONDÔNIA
Porto Velho
Riberalta
SIERRA DOS PARECIS
Rio Guaporé
Rio Mamoré
Rio Purus
Rio Juruá
Rio Amazon
Rio Negro
Rio Madeira
RORAIMA
Boa Vista do Rio Branco
Rio Branco
Manaus
Rio Tapajós
AMAPÁ
Macapá
Marajó Island
Marajó Bay
Belém
PARÁ
Rio Xingu
Rio Tocantins
MATO GROSSO
Rio Teles Pires
Rio Arinos
Rio Paranatinga
Cuiabá
PLANALTO DE MATO GROSSO
MATO GROSSO DO SUL
Campo Grande
TOCANTINS
Palmas
Rio Araguaia
GOIÁS
Goiânia
Brasília
BRAZILIAN HIGHLANDS
MARANHÃO
São Marcos Bay
São Luís
Teresina
PIAUÍ
Rio Parnaíba
CEARÁ
Fortaleza
RIO GRANDE DO NORTE
Point Calcanhar
Natal
PARAÍBA
Campina Grande
João Pessoa
PERNAMBUCO
Olinda
Recife
Caruaru
ALAGOAS
Maceió
SERGIPE
Aracaju
Juazeiro do Norte
CAATINGAS
Rio São Francisco
Juazeiro
Barragem de Sobradinho
BAHIA
Feira de Santana
Salvador
Itabuna
Carinhanha
Montes Claros
MINAS GERAIS
Gov. Valadares
Belo Horizonte
Uberaba
Uberlândia
Pico do Bandeira 9,482 ft.
ESPIRITU SANTO
Vitória
Campos
Cape São Tomé
RIO DE JANEIRO
Petrópolis
Juiz de Fora
Niterói
Rio de Janeiro
Cape Frio
Pico da Itatiaia 9,255 ft.
SÃO PAULO
Ribeirão Preto
São José do Rio Preto
Araçatuba
Marília
Piracicaba
Campinas
Jundiaí
Sorocaba
São Paulo
São Vicente
Santos
Londrina
PARANÁ
Represa Ilha Grande
Itaipu Res.
Rio Paraná
0
250
500 Miles
0
250
500 Kilometers

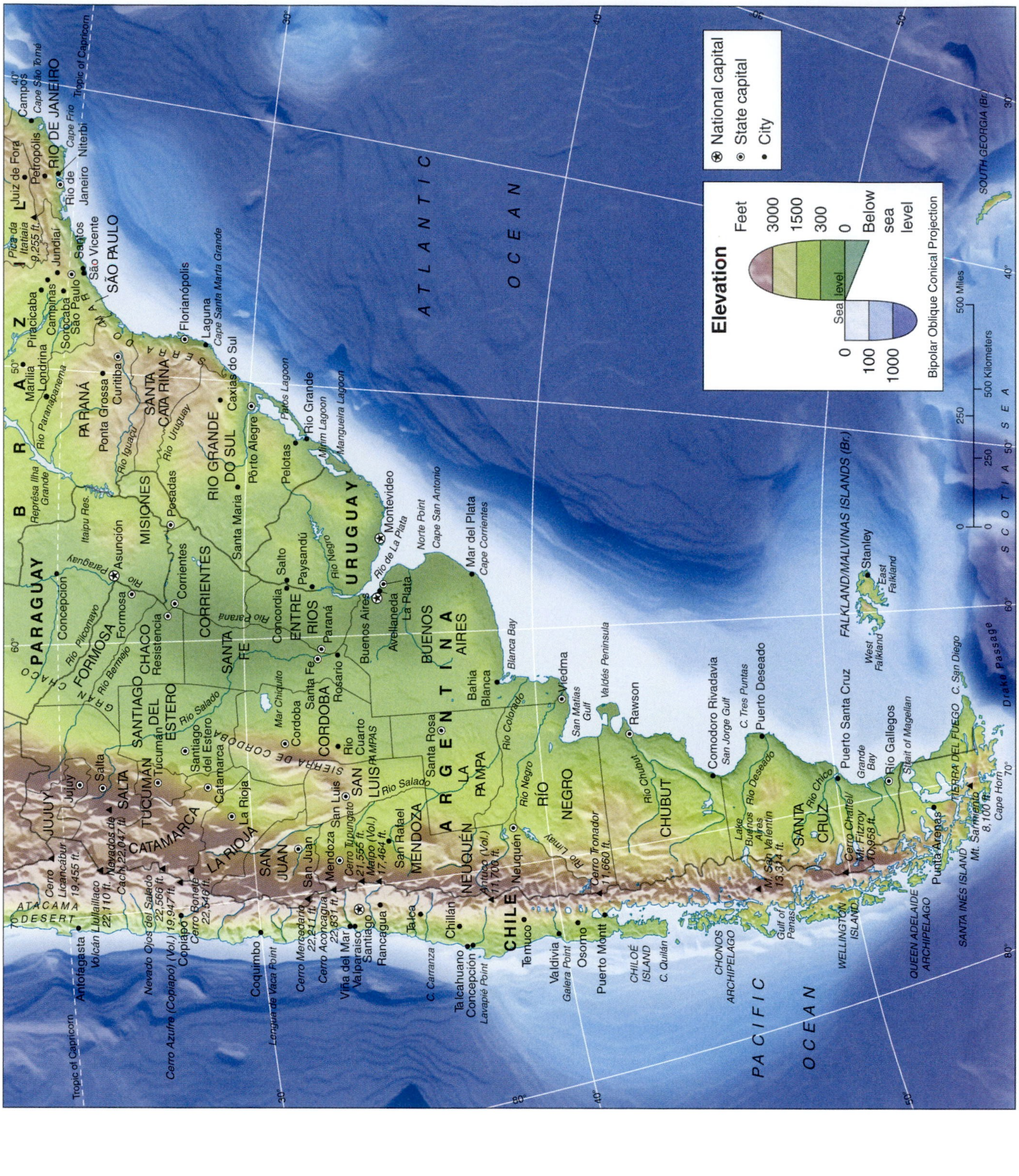

National capital
State capital
City
Elevation
Feet
3000
1500
300
0
Below sea level
Sea level
0
100
1000
Bipolar Oblique Conical Projection
0 250 500 Miles
0 250 500 Kilometers
ATLANTIC OCEAN
PACIFIC OCEAN
SCOTIA SEA
Tropic of Capricorn
BRAZIL
PARAGUAY
URUGUAY
ARGENTINA
CHILE
RIO DE JANEIRO
SÃO PAULO
PARANÁ
SANTA CATARINA
RIO GRANDE DO SUL
MISIONES
CORRIENTES
ENTRE RIOS
BUENOS AIRES
LA PAMPA
CORDOBA
SANTA FE
SANTIAGO DEL ESTERO
CHACO
FORMOSA
SALTA
JUJUY
TUCUMAN
CATAMARCA
LA RIOJA
SAN JUAN
SAN LUIS
MENDOZA
NEUQUÉN
RÍO NEGRO
CHUBUT
SANTA CRUZ
TIERRA DEL FUEGO
GRAN CHACO
PAMPAS
SIERRA DE CORDOBA
ATACAMA DESERT
Rio de Janeiro
Niterói
Cape Frio
Petrópolis
Juiz de Fora
Campos
Cape São Tomé
Pico da Itatiaia 9,255 ft.
Jundiaí
Santos
São Vicente
São Paulo
Campinas
Sorocaba
Piracicaba
Londrina
Marília
Curitiba
Ponta Grossa
Florianópolis
Laguna
Cape Santa Marta Grande
Caxias do Sul
Pôrto Alegre
Patos Lagoon
Rio Grande
Mirim Lagoon
Mangueira Lagoon
Pelotas
Santa Maria
Posadas
Represa Ilha Grande
Itaipu Res.
Rio Paranapanema
Rio Iguaçu
Rio Uruguay
Asunción
Concepción
Rio Paraguay
Rio Pilcomayo
Rio Bermejo
Formosa
Resistencia
Corrientes
Rio Paraná
Rio Salado
Salto
Paysandú
Rio Negro
Montevideo
Rio de La Plata
Norte Point
Cape San Antonio
Mar del Plata
Cape Corrientes
Concordia
Paraná
Santa Fe
Rosario
Buenos Aires
Avellaneda
La Plata
Bahía Blanca
Blanca Bay
Mar Chiquito
Córdoba
Río Cuarto
Santa Rosa
Santiago del Estero
Tucumán
Salta
Jujuy
Catamarca
La Rioja
San Juan
Mendoza
San Luis
San Rafael
Neuquén
Viedma
San Matías Gulf
Valdés Peninsula
Rawson
Río Colorado
Río Negro
Río Limay
Río Chubut
Comodoro Rivadavia
San Jorge Gulf
C. Tres Puntas
Puerto Deseado
Río Deseado
Río Chico
Puerto Santa Cruz
Grande Bay
Río Gallegos
Strait of Magellan
C. San Diego
Drake Passage
Cape Horn
Mt. Sarmiento 8,100 ft.
Punta Arenas
Lake Buenos Aires
Cerro Chaltel/Mt. Fitzroy 10,958 ft.
Mt. San Valentín 13,314 ft.
Cerro Tronador 11,660 ft.
Antuco (Vol.) 11,700 ft.
Cerro Tupungato 21,555 ft.
Maipo (Vol.) 17,464 ft.
Cerro Aconcagua 22,831 ft.
Cerro Mercedario 22,211 ft.
Cerro Bonete 22,546 ft.
Nevado Ojos del Salado 22,566 ft.
Cerro Azufre (Copiapó) (Vol.) 19,947 ft.
Nevados de Cachi 22,047 ft.
Volcán Llullaillaco 22,110 ft.
Cerro Licancabur 19,455 ft.
Antofagasta
Copiapó
Coquimbo
Lengua de Vaca Point
Viña del Mar
Valparaíso
Santiago
Rancagua
Talca
Chillán
C. Carranza
Talcahuano
Concepción
Lavapié Point
Temuco
Valdivia
Galera Point
Osorno
Puerto Montt
CHILOÉ ISLAND
C. Quilán
CHONOS ARCHIPELAGO
Gulf of Penas
WELLINGTON ISLAND
QUEEN ADELAIDE ARCHIPELAGO
SANTA INÉS ISLAND
FALKLAND/MALVINAS ISLANDS (Br.)
Stanley
East Falkland
West Falkland
SOUTH GEORGIA (Br.)

Map 137 Europe: Political Divisions

Elevation
Feet
3000
1500
300
0
Below sea level
0
100
1000
Sea level
National capital
City
0 250 500 Miles
0 250 500 Kilometers
GREENLAND
ARCTIC OCEAN
Denmark Strait
Arctic Circle
Reykjavik
ICELAND
FAEROE ISLANDS
SHETLAND ISLANDS
ORKNEY ISLANDS
UNITED KINGDOM
HEBRIDES
GRAMPIAN MTS.
NORTH SEA
Firth of Forth
North Channel
IRELAND
Dublin
IRISH SEA
Newcastle
Leeds
Liverpool
Manchester
Birmingham
St. George's Channel
Lands End
Thames R.
London
ISLES OF SCILLY
Strait of Dover
English Channel
NETHERLANDS
The Hague
Amsterdam
Rotterdam
Antwerp
BELGIUM
Brussels
Lille
Seine R.
Paris
LUX.
Luxembourg
FRANCE
Bay of Biscay
Cape Finisterre
SWITZ.
Bern
MASSIF CENTRAL
Rhone R.
JURA MTS.
VOSGES
ALPS
PYRENEES
MONACO
Monaco
Marseille
Gulf of Lyon
Gulf of Genoa
ANDORRA
Barcelona
BALEARIC IS.
PORTUGAL
Lisbon
Madrid
SPAIN
Cabo De Sao Vicente
Strait of Gibraltar
MADEIRA ISLANDS
CANARY ISLANDS
ATLANTIC OCEAN
NORWAY
Oslo
Vestfjord
Skagerrak
Kattegat
DENMARK
Copenhagen
SWEDEN
Lake Vanern
Lake Vattern
Stockholm
GOTLAND
OLAND
Gulf of Bothnia
BALTIC SEA
FINLAND
Helsinki
Gulf of Finland
Tallinn
ESTONIA
Gulf of Riga
LATVIA
Riga
LITHUANIA
Vilnius
RUSSIA
Hamburg
Weser R.
Elbe
Berlin
GERMANY
Bonn
Frankfurt
ERZGEBIRGE
Rhine R.
Stuttgart
Munich
Prague
CZECH REP.
POLAND
Warsaw
SLOVAKIA
Bratislava
Vienna
AUSTRIA
Budapest
HUNGARY
Ljubljana
SLOVENIA
Zagreb
CROATIA
Po
CORSICA
SARDINIA
ITALY
Rome
APENNINES
Naples
TYRRHENIAN SEA
SICILY
BOSNIA AND HERZEGOVINA
Sarajevo
DINARIC ALPS
MONTE-NEGRO
ADRIATIC SEA
Tirane
ALBANIA
SERBIA
Belgrade
KOSOVO
Skopje
MACEDONIA
ROMANIA
TRANSYLVANIAN ALPS
CARPATHIAN MTS.
Bucharest
Danube R.
BULGARIA
Sofia
GREECE
Athens
AEGEAN SEA
IONIAN SEA
MEDITERRANEAN SEA
MALTA
CRETE
WHITE SEA
Dvina R.
Lake Onega
Lake Ladoga
St. Petersburg
Lake Rybinsk
Nizhniy-Novgorod
Volga R.
Moscow
RUSSIA
URAL MOUNTAINS
Yekaterinburg
Chelyabinsk
Minsk
BELARUS
PRIPYAT MARSHES
Kiev
UKRAINE
Dnepr R.
Dnipropetrovs'k
Don R.
MOLDOVA
Chisinau
SEA OF AZOV
BLACK SEA
Bosporus
Istanbul
Volgograd
Volga R.
CASPIAN DEPRESSION
CASPIAN SEA
CAUCASUS
Tbilisi
GEORGIA
Yerevan
ARMENIA
Baku
AZERBAIJAN
KAZAKSTAN
ARAL SEA
Syr Darya
UZBEKISTAN
Amu Darya
KARAKUMY DESERT
TURKMENISTAN
Ashkhabad
ELBURZ MOUNTAINS
Tehran
IRAN
ZAGROS MOUNTAINS
TURKEY
Ankara
L.Van
Tigris R.
Euphrates R.
Baghdad
IRAQ
SYRIA
Damascus
Nicosia
CYPRUS
Beirut
LEBANON
ISRAEL
Tel-Aviv
Jerusalem
Amman
JORDAN
KUWAIT
Kuwait
Persian Gulf
BAHRAIN
Abu Dhabi
QATAR
UNITED ARAB EMIRATES
SAUDI ARABIA
Riyadh
RED SEA
Nile R.
Cairo
EGYPT
LIBYA
Tripoli
TUNISIA
Tunis
Algiers
ALGERIA
ATLAS MOUNTAINS
Rabat
MOROCCO
MAURITANIA

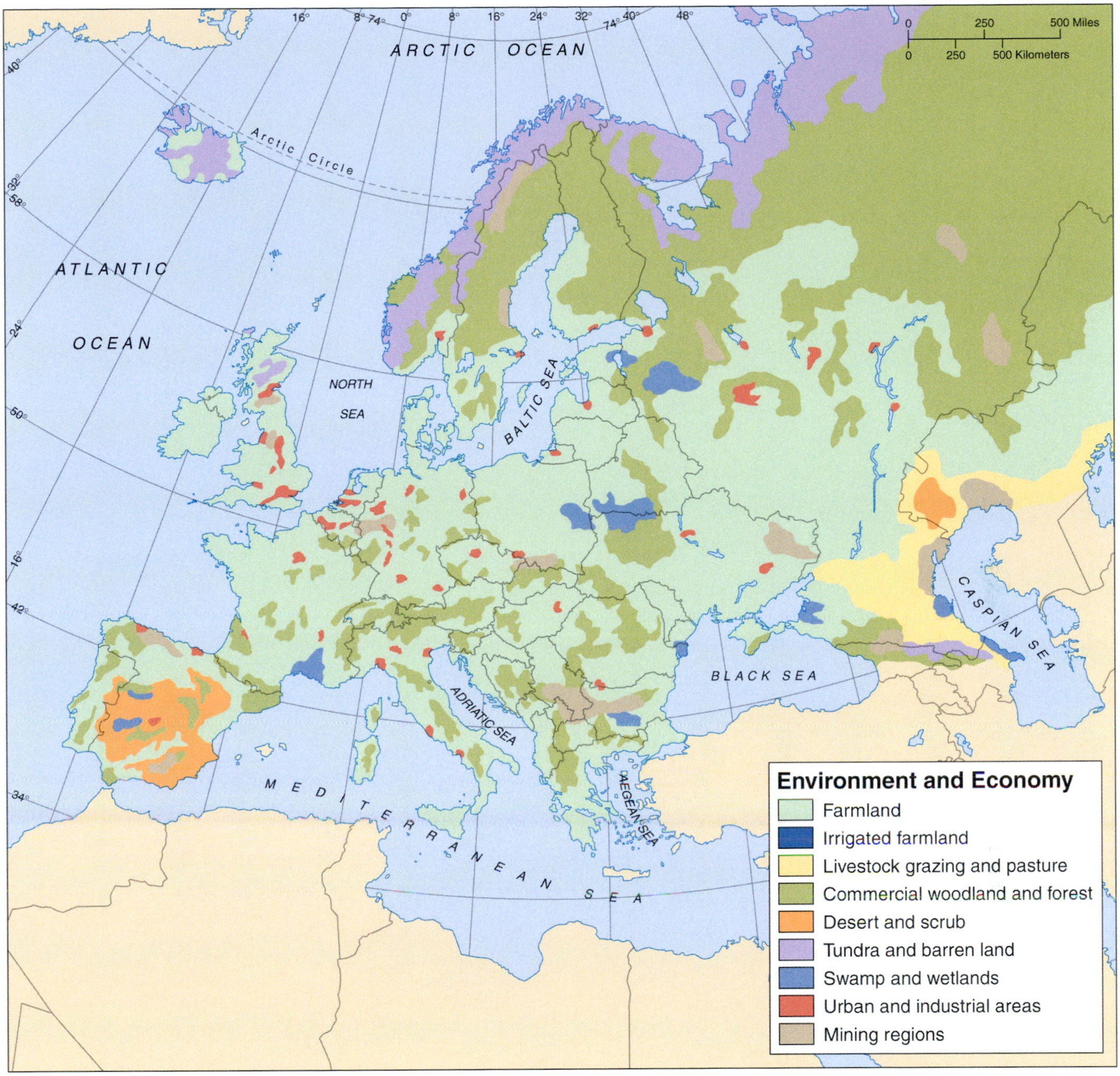

More than any other continent, Europe bears the imprint of human activity—mining, forestry, agriculture, industry, and urbanization. Virtually all of western and central Europe's natural forest vegetation is gone, lost to clearing for agriculture beginning in prehistory, to lumbering that began in earnest during the Middle Ages, or more recently, to disease and destruction brought about by acid precipitation. Only in the far north and the east do some natural stands remain. The region is the world's most heavily industrialized and the industrial areas on the map represent only the largest and most significant. Not shown are the industries that are found in virtually every small town and village and smaller city throughout the industrial countries for Europe. Europe also possesses abundant raw materials and a very productive agricultural base. The mineral resources have long been in a state of active exploitation and the mining regions shown on the map are, for the most part, old regions in upland areas that are somewhat less significant now than they may have been in the past. Agriculturally, the northern European plain is one of the world's great agricultural regions but most of Europe contains decent land for agriculture.

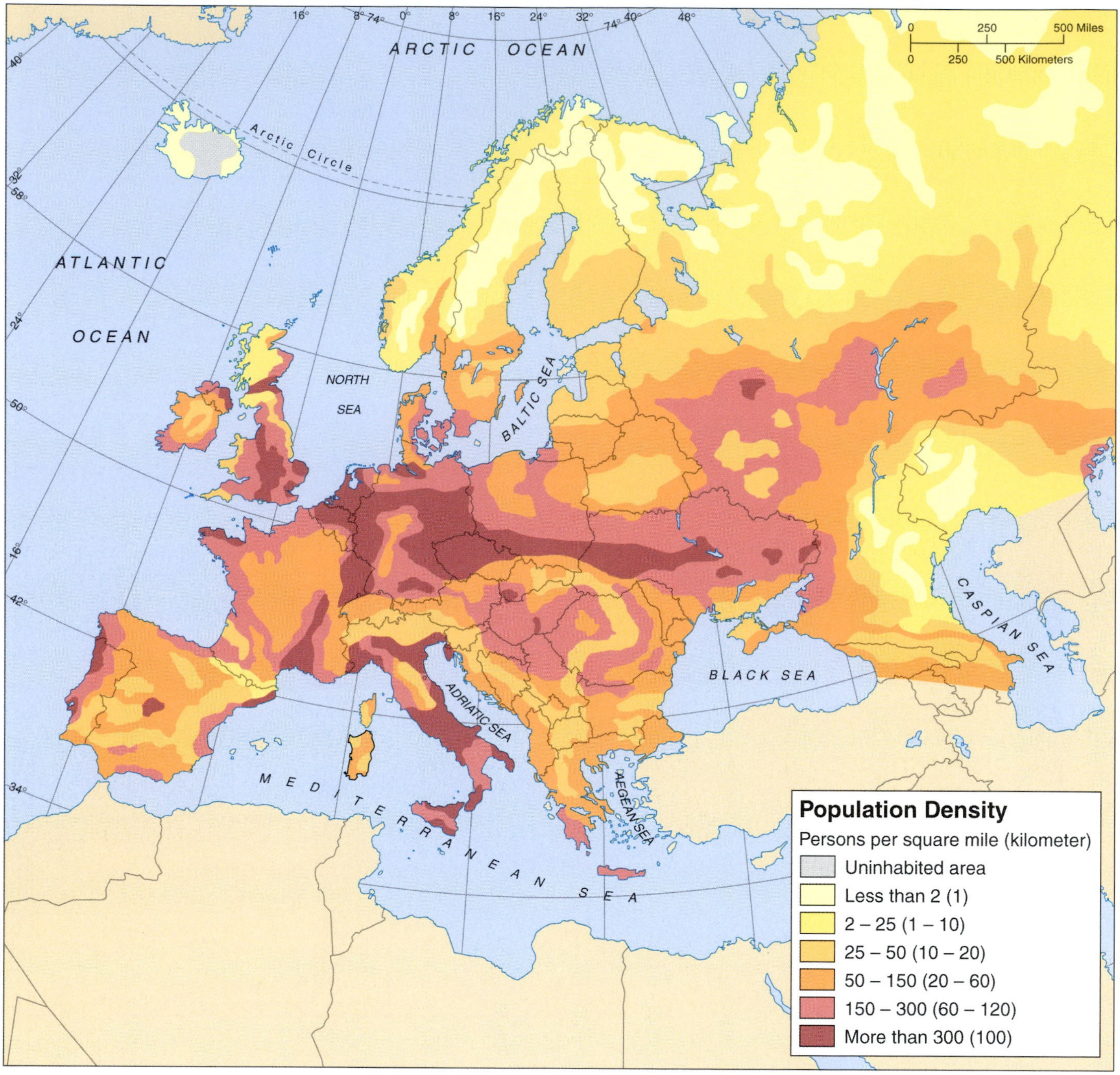

Europe is one of the most densely settled regions of the world with an overall population density nearing 200 persons per square mile (80 per square kilometer), the consequence of a high level of urbanization and an economic system that is heavily industrialized. Even in agricultural regions, the population density is high. Beyond high density, the two chief identifying marks of the European population are remarkable diversity and unusual dynamics. For a small part of the world, Europe has cultural and ethnic diversity that is rarely matched elsewhere; more than 60 languages are spoken in an area not much larger than the United States. The population dynamics of Europe show a mature population that has passed through the "Demographic Transition"—a remarkable increase and then decline in growth rates resulting from the rise of an urban-industrial society. Only in other heavily industrialized regions of the world are found the very small overall growth rates characteristic of Europe.

Map 139c European Political Boundaries, 1914–1948

In 1914, on the eve of the First World War, Europe was dominated by the United Kingdom and France in the west, the German Empire and the Austro-Hungarian Empire in central Europe, and the Russian Empire in the east. Battle lines for the conflict that began in 1914 were drawn when the United Kingdom, France, and the Russian Empire joined together as the Triple Entente. In the view of the Germans, this coalition was designed to encircle Germany and its Austrian ally, which, along with Italy, made up the Triple Alliance. The German and Austrian fears were heightened from 1912–14 when a Russian-sponsored "Balkan League" pushed the Ottoman Turkish Empire from Europe, leaving behind the weak and mutually antagonistic Balkan states Serbia and Montenegro. In August 1914, Germany and Austria-Hungary attacked in several directions and World War I began. Four years later, after massive loss of life and destruction, the central European empires were defeated. The victorious French, English, and Americans (who had entered the war in 1917) restructured the map of Europe in 1919, carving nine new states out of the remains of the German and Austro-Hungarian empires and the westernmost portions of the Russian Empire which, by the end of the war, was deep in the Revolution that deposed the czar and brought the Communists to power in a new Union of Soviet Socialist Republics.

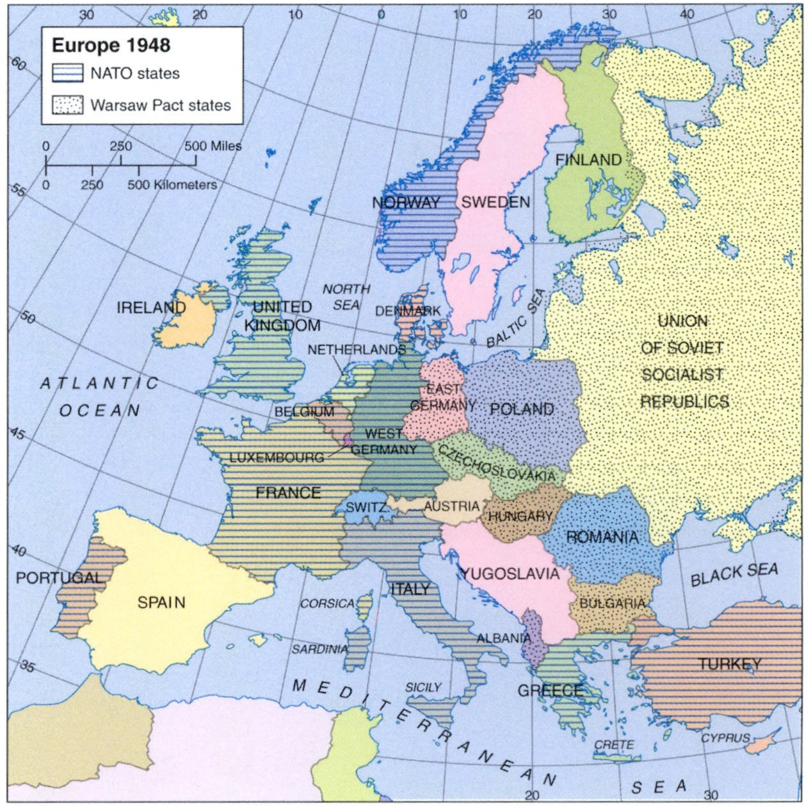

When the victorious Allies redrew the map of central and eastern Europe in 1919, they caused as many problems as they were trying to solve. The interval between the First and Second World Wars was really just a lull in a long war that halted temporarily in 1918 and erupted once again in 1939. Defeated Germany, resentful of the terms of the 1918 armistice and 1919 Treaty of Versailles and beset by massive inflation and unemployment at home, overthrew the Weimar republican government in 1933 and installed the National Socialist (Nazi) party led by Adolf Hitler in Berlin. Hitler quickly began making good on his promises to create a "thousand year realm" of German influence by annexing Austria and the Czech region of Czechoslovakia and allying Germany with a fellow fascist state in Mussolini's Italy. In September 1939 Germany launched the lightning-quick combined infantry, artillery, and armor attack known as **der Blitzkrieg** and took Poland to the east and, in quick succession, the Netherlands, Belgium, and France to the west. By 1943 the greater German Reich extended from the Russian Plain to the Atlantic and from the Black Sea to the Baltic. But the Axis powers of Germany and Italy could not withstand the greater resources and manpower of the combined United Kingdom-United States-USSR–led Allies and, in 1945, Allied armies occupied Germany. Once again, the lines of the central and eastern European map were redrawn. This time, a strengthened Soviet Union took back most of the territory the Russian Empire had lost at the end of the First World War. Germany was partitioned into four occupied sectors (English, French, American, and Russian) and later into two independent countries, the Federal Republic of Germany (West Germany) and the German Democratic Republic (East Germany). Although the Soviet Union's territory stopped at the Polish, Hungarian, Czechoslovakian, and Romanian borders, the eastern European countries (Poland, East Germany, Czechoslovakia, Hungary, Romania, Yugoslavia, Albania, and Bulgaria) became Communist between 1945 and 1948 and were separated from the West by the Iron Curtain.

Map 139d Political Changes, 1989–2008

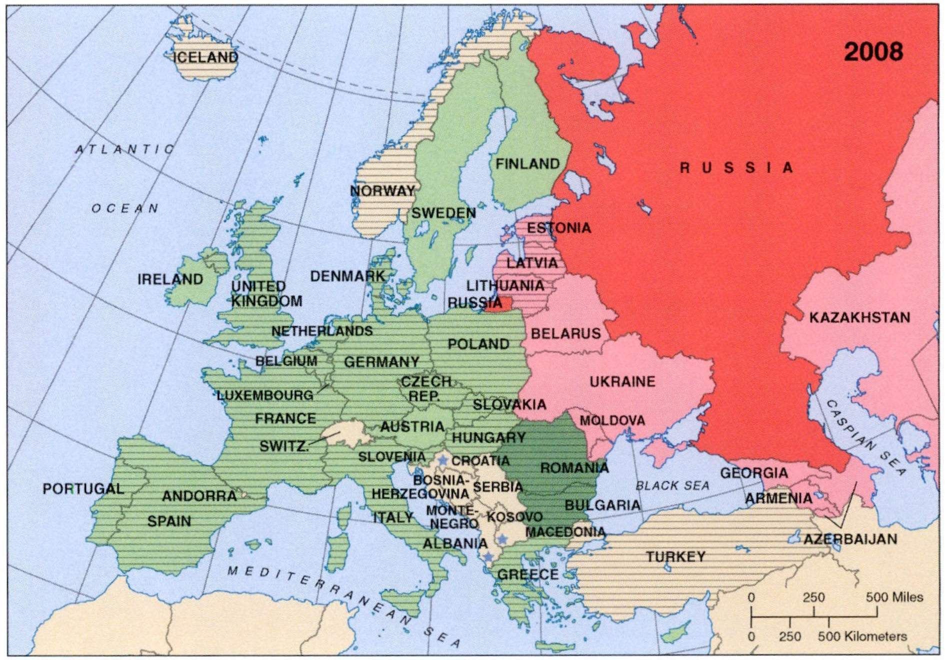

Europe: Political Changes 1989-2008

1989

- Union of Soviet Socialist Republics
- Warsaw Pact Countries (excluding the USSR)
- North Atlantic Treaty Organization Countries
- European Community (formerly the EEC)

2008

- Former Republics of the USSR, now independent countries
- Russian Federation
- NATO Countries 2005
- Associated with NATO / petitioned for entry
- European Union (formerly the EC)
- Joined European Union in 2007

During the last decade of the twentieth century, one of the most remarkable series of political geographic changes of the last 500 years took place. The bipolar "East-West" structure that had characterized Europe's political geography since the end of World War II altered in the space of a very few years. In the mid-1980s, as Soviet influence over eastern and central Europe weakened, those countries began to turn to the capitalist West. Between 1989, when the country of Hungary was the first Soviet satellite to open its borders to travel, and 1991, when the Soviet Union dissolved into 15 independent countries, abrupt change in political systems occurred. The result is a new map of Europe that includes a number of countries not present on the map of 1989. These countries have emerged as the result of reunification, separation, or independence from the old Soviet Union. The new political structure has been accompanied by growing economic cooperation. In May, 2004, ten new applicant countries (the Greek part of Cyprus, Czech Republic, Estonia, Hungary, Latvia, Lithuania, Malta, Poland, Slovakia, and Slovenia) were added to the European Union, increasing the economic strength of the EU, which now has its own currency (the euro) and is presently attempting to develop a constitution acceptable to the 25 member nations. The only remaining formal applicant country to EU membership is Turkey. Concerned about Turkish migration and the potential political problems of adding a country with a population of 60 million Muslims to the European Union, the majority of EU countries have thus far failed to act positively on Turkey's application for membership.

Elevation
Feet
3000
1500
300
0
Below sea level
Sea level
0
100
1000
Azimuthal Equidistant Projection
National capital
City
FAEROES
SHETLAND IS.
HEBRIDES
ORKNEY IS.
UNITED KINGDOM
SCOTLAND
NORTHERN IRELAND
IRELAND
IRISH SEA
WALES
ENGLAND
NORWAY
SWEDEN
NORTH SEA
DENMARK
Skagerrak
Kattegat
BALTIC SEA
NETHERLANDS
GERMANY
POLAND
BELGIUM
LUXEMBOURG
CZECH REPUBLIC
SLOVAKIA
FRANCE
AUSTRIA
HUNGARY
SWITZ.
LIECHTENSTEIN
SLOVENIA
CROATIA
BOSNIA-HERZEGOVINA
ALPS
ITALY
SAN MARINO
MONACO
ANDORRA
ADRIATIC SEA
CORSICA (France)
SARDINIA (Italy)
TYRRHENIAN SEA
SICILY
MALTA
IONIAN SEA
PORTUGAL
SPAIN
BALEARIC IS.
MINORCA
MAJORCA
IBIZA
ATLANTIC OCEAN
Bay of Biscay
English Channel
MASSIF CENTRAL
PYRENEES
MEDITERRANEAN SEA
Strait of Gibraltar
MOROCCO
ALGERIA
TUNISIA
MIDDLE ATLAS
SAHARAN ATLAS
Ben Nevis 4,406 ft.
Carrauntoohil 3,414 ft.
Snowdon 3,560 ft.
Scafell Pikes 3,210 ft.
Glittertinden 8,110 ft.
Helagsfjället 5,892 ft.
Sarfjället 4,190 ft.
Puy de Sancy 6,185 ft.
Plomb du Cantal 6,096 ft.
Mt. Perdido 11,000 ft.
Pico de Aneto 11,168 ft.
Pic de Montcalm 10,305 ft.
Mulhacén 11,424 ft.
La Sagra 7,999 ft.
Mt. Cinto 8,878 ft.
Punta la Marmora 6,016 ft.
Mt. Corno 9,554 ft.
Mt. Amaro 9,163 ft.
Vesuvio 4,190 ft.
Mt. Etna (volcano) 10,902 ft.
0
250
500 Miles
500 Kilometers
London
Paris
Madrid
Lisbon
Dublin
Rome
Berlin
Amsterdam
Brussels
Oslo
Stockholm
Copenhagen
Prague
Vienna
Bern
Algiers
Rabat
Tunis

Elevation
Feet
3000
1500
300
0
Below sea level
0
100
1000
Sea level
Azimuthal Equidistant Projection
National capital
City
500 Miles
500 Kilometers
NORWAY
SWEDEN
FINLAND
DENMARK
ESTONIA
LATVIA
LITHUANIA
RUSSIA (KALININGRAD OBLAST)
BELARUS
POLAND
GERMANY
CZECH REP.
SLOVAKIA
AUSTRIA
HUNGARY
SLOVENIA
CROATIA
BOSNIA-HERZEGOVINA
ITALY
ROMANIA
MOLDOVA
UKRAINE
BULGARIA
MACEDONIA
ALBANIA
GREECE
TURKEY
GEORGIA
ARMENIA
AZERBAIJAN
NAGORNO KARABACH
RUSSIA
KAZAKHSTAN
UZBEKISTAN
TURKMENISTAN
IRAN
IRAQ
SYRIA
LEBANON
CYPRUS
MALTA
BALTIC SEA
BLACK SEA
CASPIAN SEA
ARAL SEA
MEDITERRANEAN SEA
Sea of Azov
Gulf of Bothnia
Skagerrak
CAUCASUS MTS.
CARPATHIAN MOUNTAINS
TAURUS MOUNTAINS
ELBURZ MOUNTAINS
ZAGROS MOUNTAINS
ALPS
CRIMEA
SICILY
CRETE
RHODES
Lake Ladoga
Lake Peipus
Volga R.
Don R.
Dnieper R.
Ural R.
Danube R.
Tigris R.
Euphrates R.
Moscow
Warsaw
Minsk
Kyyiv
Bucharest
Budapest
Vienna
Prague
Berlin
Stockholm
Helsinki
Tallinn
Riga
Vilnius
Ankara
Tbilisi
Yerevan
Baku
Tehran
Baghdad
Damascus
Beirut
Nicosia
Athens
Sofiya
Belgrade
Zagreb
Ljubljana
Sarajevo
Tirane
Skopje
Chisinau
Copenhagen
Oslo
Rome
Ashkhabad

Elevation
Feet
3000
1500
300
0
Sea level
0
100
1000
Below sea level
Simple Conic Projection
National capital
City
0
100
200 Miles
0
100
200 Kilometers
Arctic Circle
BARENTS SEA
NORWEGIAN SEA
WHITE SEA
KOLA PENINSULA
NORTH SEA
BALTIC SEA
Gulf of Bothnia
Gulf of Finland
Gulf of Riga
Skagerrak
Kattegat
NORWAY
SWEDEN
FINLAND
RUSSIA
ESTONIA
LATVIA
LITHUANIA
BELARUS
POLAND
GERMANY
DENMARK
RUSSIA (KALININGRAD OBLAST)
LOFOTEN
VESTERÅLEN
HAILUOTO
ÅLAND
GOTLAND
ÖLAND
BORNHOLM
HIIUMAA
SAAREMAA
Lake Onega
Lake Ladoga
Lake Peipus
Vänern
Vättern
Nordkapp
Kebnekaise 6,962 ft.
Sulitelma 62,80 ft.
Sylarna 5,761 ft.
Helagsfjället 5,892 ft.
Snøhetta 7,500 ft.
Galdhøpiggen 8,097 ft.
Glittertinden 8,110 ft.
Sånfjället 4,190 ft.
Städjan 3,711 ft.
Oslo
Stockholm
Helsinki
Tallinn
Riga
Vilnius
Minsk
Copenhagen
Murmansk
Tromsø
Narvik
Bodø
Trondheim
Bergen
Stavanger
Kristiansand
Göteborg
Malmö
Uppsala
Luleå
Umeå
Oulu
Tampere
Turku
St. Petersburg
Petrozavodsk
Novgorod
Pskov
Kaliningrad
Klaipeda
Liepaja
Gdańsk
Daugava R.
Neman R.
Wisła (Vistula) R.
Elbe R.
Dnepr R.
Volkhov R.
Polisť R.
Lotta R.
Ponoy R.
Kem' R.
Skellefteälv R.
Umeälv R.
Västerdalälven R.
0°
3°
6°
9°
12°
15°
18°
21°
24°
27°
30°
33°
36°
39°
42°
45°
54°
57°
60°
63°
66°
69°

Map 143 Africa: Political Divisions

ATLANTIC OCEAN
PORTUGAL
SPAIN
FRANCE
ITALY
SERBIA
BULGARIA
MACEDONIA
ALBANIA
GREECE
BLACK SEA
RUSSIA
GEORGIA
ARMENIA
AZERBAIJAN
CASPIAN SEA
TURKEY
ASIA MINOR
TURKMENISTAN
IRAN
SYRIA
IRAQ
CYPRUS
LEBANON
ISRAEL
JORDAN
KUWAIT
Persian Gulf
BAHRAIN
QATAR
UNITED ARAB EMIRATES
SAUDI ARABIA
ARABIAN PENINSULA
SINAI PENINSULA
YEMEN
MEDITERRANEAN SEA
Strait of Gibraltar
Algiers
Tunis
TUNISIA
Rabat
MOROCCO
Jebel Toubkal 13,665 ft.
ATLAS MOUNTAINS
CANARY ISLANDS
GRAND ERG OCCIDENTAL
Gulf of Gabes
Tripoli
Khalij Bay
Banghazi
Alexandria
Cairo
Al Fayyum
Asyut
ARABIAN DESERT
Nile R.
ALGERIA
LIBYA
LIBYAN DESERT
EGYPT
Aswan
Al Khufra Oasis
RED SEA
Lake Nasser
NUBIAN DESERT
WESTERN SAHARA
Tropic of Cancer
Dakhla
ERG IGUIDI
AHAGGAR MTS
Mt. Tahat 9,541 ft.
EL DJOUF
Cape Blanc
Pic Tousidé 10,712 ft.
TIBESTI MASSIF
MAURITANIA
MALI
SAHARA
Nouakchott
Cape Vert
Senegal R.
Tombouctou
SENEGAL
Dakar
Gambie R.
GAMBIA
Banjul
SAHEL
Niger R.
NIGER
CHAD
Bur Sudan
ERITREA
Asmera
Omdurman
Khartoum
Al Fashir
DARFUR
Jabal Marrah 10,131 ft.
White Nile
Bab el Mandeb
Gulf of Aden
Gees Gwardafuy
DJIBOUTI
Djibouti
Ras Dashen Terara 15,158 ft.
Bissau
GUINEA-BISSAU
GUINEA
Conakry
Freetown
SIERRA LEONE
Monrovia
LIBERIA
Bamako
BURKINA FASO
Ouagadougou
Niamey
Black Volta R.
White Volta R.
IVORY COAST
GHANA
TOGO
BENIN
Kano
NIGERIA
Abuja
Ibadan
Lake Chad
N'Djamena
Logone R.
Chari R.
SUDAN
Blue Nile
Addis Ababa
Harer
Hargeysa
AS SUDD
Mountain Nile
ETHIOPIAN PLATEAU
ETHIOPIA
SOMALIA
Lake Volta
Lomé
Accra
Porto Novo
Lagos
Abidjan
Cape Palmas
Volta R.
Gulf of Guinea
Benue R.
CAMEROON
CENTRAL AFRICAN REPUBLIC
Douala
Malabo
Yaoundé
Bangui
Uele R.
UGANDA
Lake Rudolf
Jubba R.
EQUATORIAL GUINEA
SÃO TOMÉ AND PRINCIPE
Equator
Cape Lopez
Libreville
GABON
CONGO BASIN
RUWENZORI MTS.
Lake Albert
Ubangi R.
Congo R.
Kisangani
Kampala
Mt. Elgon 14,178 ft.
Mt. Kenya 17,058 ft.
KENYA
Mogadishu
Lake Edward
Lake Victoria
Nairobi
CONGO REPUBLIC
DEMOCRATIC REPUBLIC OF THE CONGO
RWANDA
Kigali
Brazzaville
Kinshasa
Kasai R.
Boma
Bujumbura
BURUNDI
Mt. Kilimanjaro 19,340 ft.
MASAI STEPPE
Mombasa
Dodoma
ZANZIBAR
INDIAN OCEAN
Lake Tanganyika
Dar es Salaam
ASCENSION ISLAND
ATLANTIC OCEAN
Cuango R.
Luanda
TANZANIA
ALDABRA ISLANDS
COSMOLEDO ISLANDS
Lake Nyasa
ANGOLA
Lubumbashi
Ruvuma R.
COMOROS
Moroni
Cape d'Ambre
Benguela
MALAWI
Lilongwe
Cubango R.
ZAMBIA
Namibe
Lusaka
Zambezi R.
Moçambique
ST. HELENA
Mozambique Channel
Cape Fria
Harare
Okavongo R.
Toamasino
Antananarivo
NAMIBIA
Okavango Swamp
ZIMBABWE
MOZAMBIQUE
MADAGASCAR
Windhoek
Limpopo R.
Tropic of Capricorn
Walvis Bay
BOTSWANA
Kalahari Desert
Gaborone
GREAT NAMALAND
Pretoria
Johannesburg
Maputo
Mbabane
Cape Ste. Marie
Vaal R.
Kimberly
SWAZILAND
Maseru
LESOTHO
DRAKENSBERG
Durban
SOUTH AFRICA
Cape Town
Cape of Good Hope
Cape Agulhas
Port Elizabeth
Elevation
Feet
3000
1500
300
0
Sea level
Below sea level
100
1000
National capital
City
0 250 500 750 1000 Miles
0 250 500 750 1000 Kilometers

Map 145a Environment and Economy

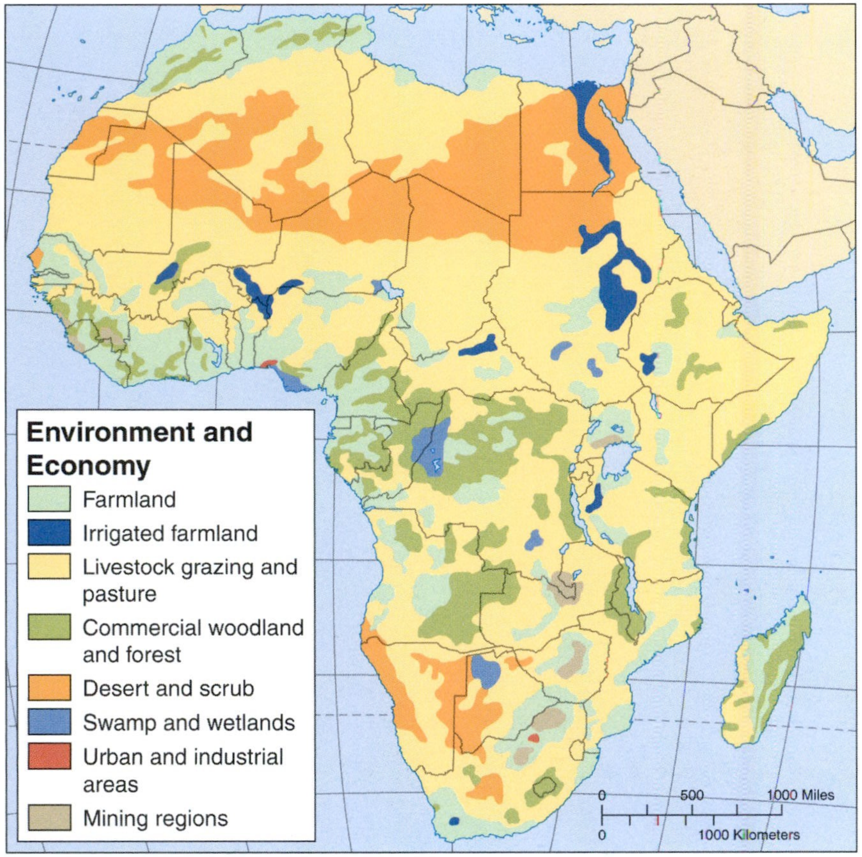

Africa's economic landscape is dominated by subsistence, or marginally-commercial agricultural activities and raw material extraction, engaging three-fourths of Africa's workers. Much of this grazing land is very poor desert scrub and bunch grass that is easily impacted by cattle, sheep, and goats. Growing human and livestock populations place enormous stress on this fragile support capacity and the result is desertification: the conversion of even the most minimal of grazing environments to a small quantity of land suitable for crop farming. Although the continent has approximately 20 percent of the world's total land area, the proportion of Africa's arable land is small. The agricultural environment is also uncertain; unpredictable precipitation and poor soils hamper crop agriculture.

Map 145b Population Density

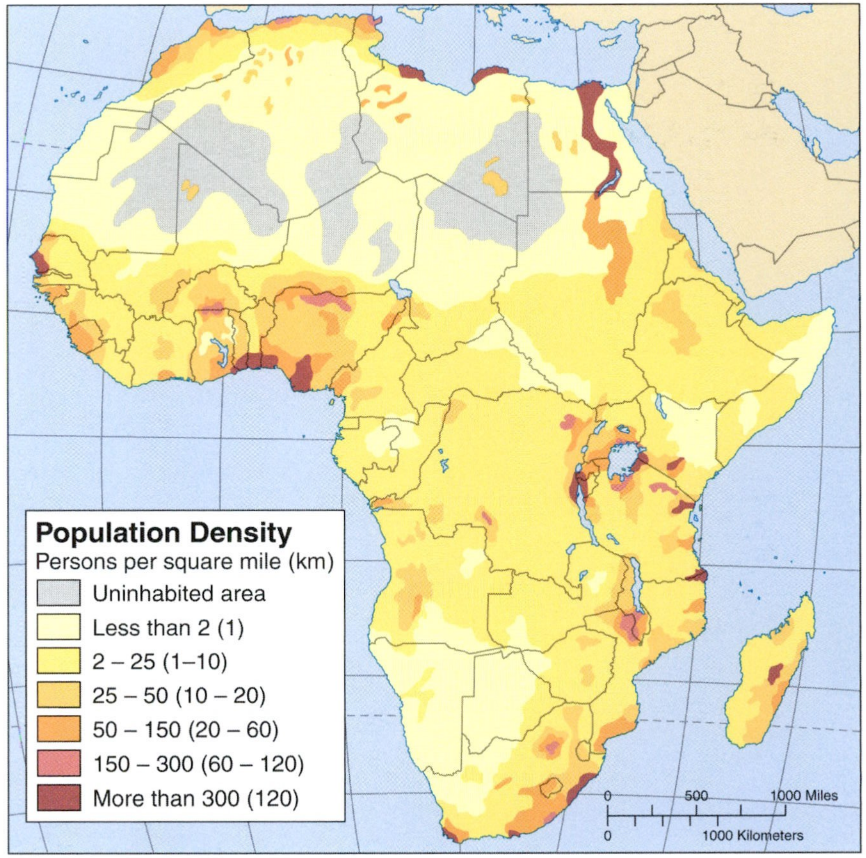

Over 900 million people occupy the African continent, approximately one-eighth of the world's population. In general, this population has two chief characteristics: a low level of quality of life and growth rates that are among the world's highest. On a continent beset by poverty, recurrent internal civil and tribal war, and a host of environmental problems, the populations of many African countries are nevertheless increasing at a rate above 3 percent per year. The bulk of the population is concentrated in relatively small areas of the Mediterranean coastal regions of the north, the bulge of West Africa, and the eastern coastal and highland zone stretching from South Africa to Kenya. Where populations in other parts of the world tend to avoid highland locations, in Africa highlands often tend to be the most densely settled regions: moister with better soils, freer of insect pests, and somewhat cooler.

Map 145c Colonial Patterns

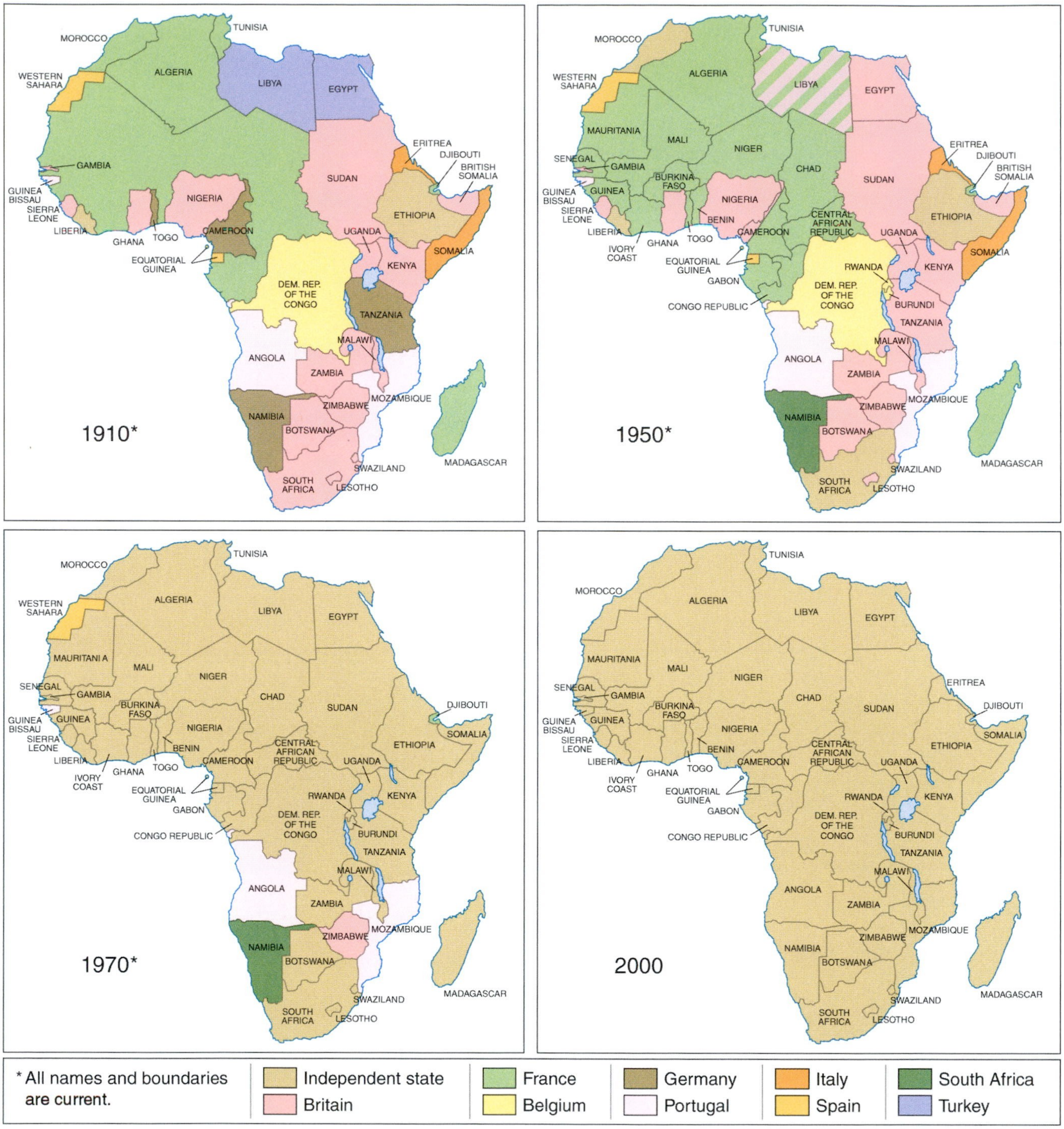

In few parts of the world has the transition from colonialism to independence been as abrupt as on the African continent. Most African states did not become colonies until the nineteenth century and did not become independent until the twentieth, nearly all of them after World War II. Much of the colonial power in Africa is social and economic. The African colony provided the mother country with raw materials in exchange for marginal economic returns, and many African countries still exist in this colonial dependency relationship. An even more important component of the colonial legacy of Europe in Africa is geopolitical. When the world's colonial powers joined at the Conference of Berlin in 1884, they divided up Africa to fit their own needs, drawing boundary lines on maps without regard for terrain or drainage features, or for tribal/ethnic linguistic, cultural, economic, or political borders. Traditional Africa was enormously disrupted by this process. After independence, African countries retained boundaries that are legacies of the colonial past and African countries today are beset by internal problems related to tribal and ethnic conflicts, the disruption of traditional migration patterns, and inefficient spatial structures of market and supply.

Map 145d African Cropland and Dryland Degradation

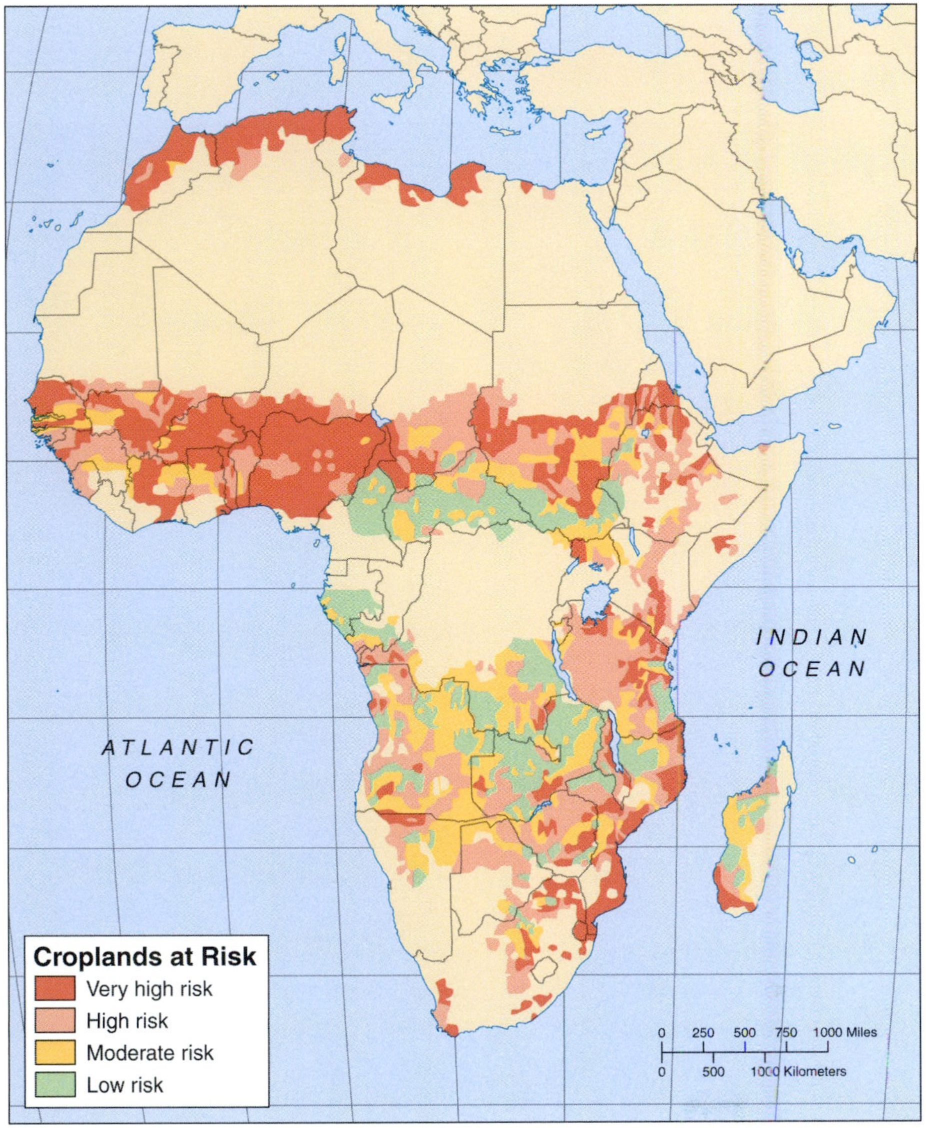

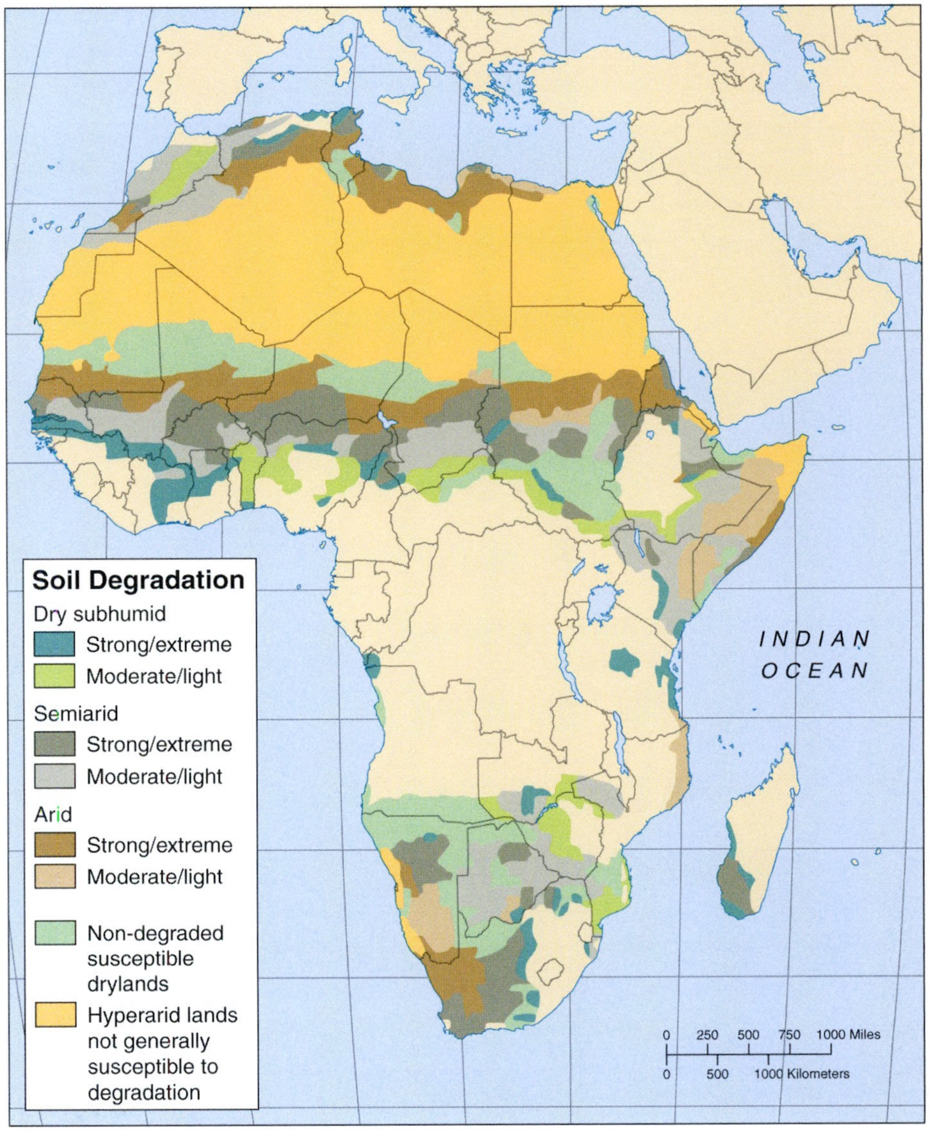

The economy of the African continent is largely agricultural and given an African population of nearly 800 million people, much of the agricultural environment is degraded. Two forms of degradation exist. The first of these is in cropland areas where susceptible tropical soils and high population densities have produced major cropland degradation, largely in the form of loss of fertility. Irrigation and the advent of artificial fertilizer have not helped soils to restore their natural chemical balances after generations of misuse. The second form of degradation is in the dryland areas where livestock grazing rather than cropping is the dominant agricultural form. Populations of domesticated stock that are too large for the carrying capacity of the environment, exacerbated by the view of livestock as wealth and worsened by increasing human populations, have all contributed to the conversion of semi-arid grasslands into desert.

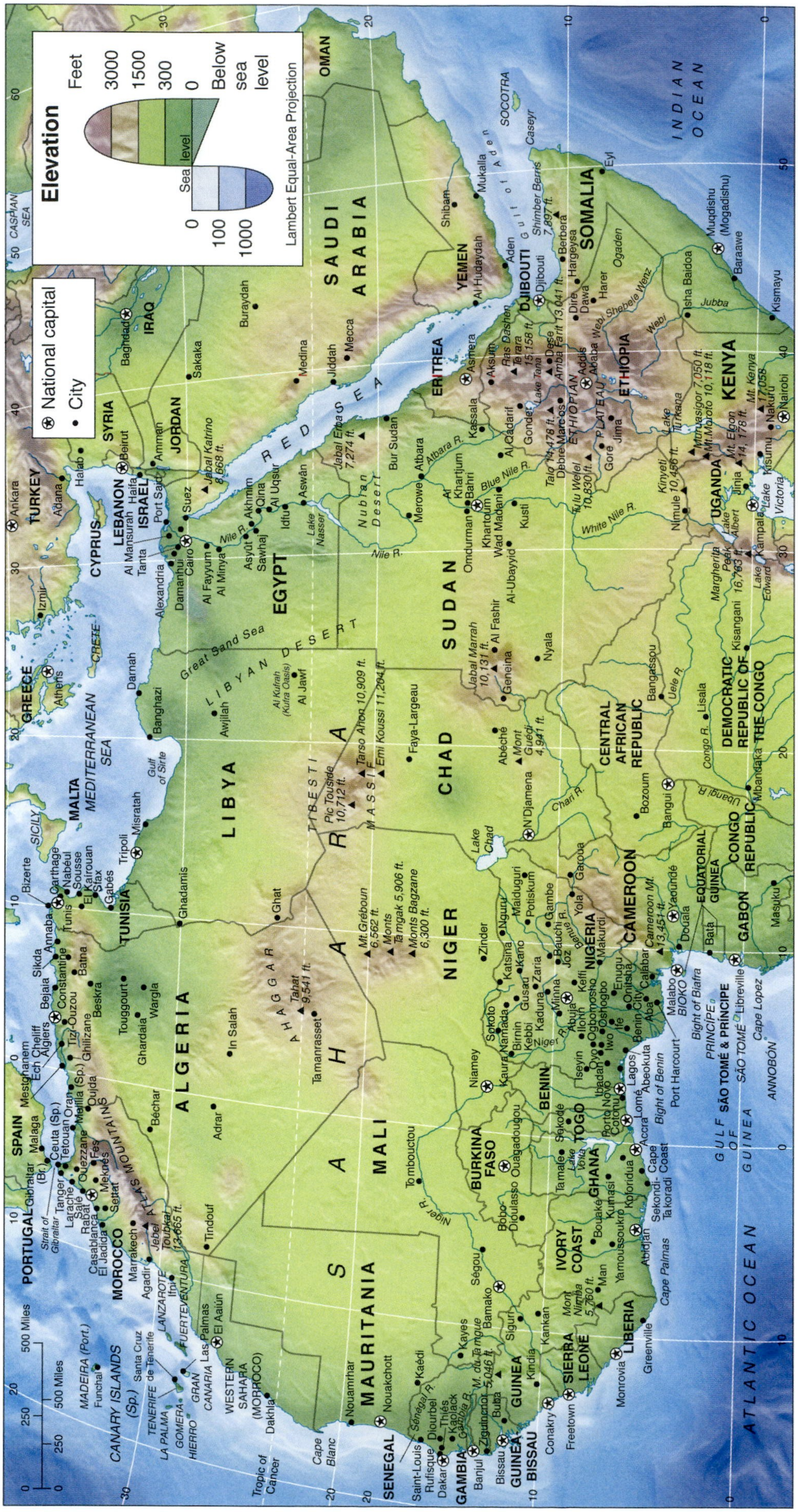
National capital
City
Elevation
Feet
3000
1500
300
0
Below sea level
Sea level
0
100
1000
Lambert Equal-Area Projection
MAURITANIA
MALI
NIGER
CHAD
SUDAN
EGYPT
LIBYA
ALGERIA
MOROCCO
TUNISIA
SENEGAL
GAMBIA
GUINEA-BISSAU
GUINEA
SIERRA LEONE
LIBERIA
IVORY COAST
GHANA
TOGO
BENIN
BURKINA FASO
NIGERIA
CAMEROON
CENTRAL AFRICAN REPUBLIC
EQUATORIAL GUINEA
GABON
CONGO REPUBLIC
DEMOCRATIC REPUBLIC OF THE CONGO
UGANDA
KENYA
ETHIOPIA
ERITREA
DJIBOUTI
SOMALIA
YEMEN
SAUDI ARABIA
OMAN
JORDAN
ISRAEL
LEBANON
SYRIA
IRAQ
TURKEY
CYPRUS
GREECE
MALTA
SPAIN
PORTUGAL
WESTERN SAHARA (MOROCCO)
CANARY ISLANDS (Sp.)
MEDITERRANEAN SEA
RED SEA
ATLANTIC OCEAN
INDIAN OCEAN
GULF OF GUINEA
SAHARA
LIBYAN DESERT
Tropic of Cancer

Map 147 Southern Africa

ATLANTIC OCEAN
ARCTIC OCEAN
ICELAND
GREENLAND
Arctic Circle
ALASKA (USA)
Bering Strait
WRANGELL ISLAND
ST. LAWRENCE ISLAND
BERING SEA
EAST SIBERIAN SEA
NEW SIBERIAN ISLANDS
LAPTEV SEA
SVALBARD
ZEMLYA FRANTSA IOSIFA
SEVERNAYA ZEMLYA
NOVAYA ZEMLYA
BARENTS SEA
KARA SEA
IRELAND
UNITED KINGDOM
London
NORTH SEA
NORWAY
Gulf of Bothnia
SWEDEN
FINLAND
DENMARK
Copenhagen
Stockholm
NETH.
BELG.
Paris
FRANCE
LUX.
GERMANY
Berlin
SWITZ.
ALPS
Danube R.
POLAND
BALTIC SEA
Prague
CZECH REP.
Warsaw
AUSTRIA
Vienna
SLOVENIA
SLOVAKIA
CROATIA
Budapest
HUNGARY
B-H
SERBIA
MONTENEGRO
ROMANIA
Rome
ITALY
KOSOVO
Sofiya
MACEDONIA
ALBANIA
BULGARIA
Bucharest
MOLDOVA
UKRAINE
Kiev
GREECE
St. Petersburg
ESTONIA
LATVIA
LITHUANIA
RUSSIA
Ladozhkoye Ozero
Onezhkoye Ozero
Arkhangelsk
Sev. Dvina
Dnepr R.
Moscow
Rostov
Nizhniy Novgorod
Don R.
Volga R.
Donetsk
Rostov na Donu
Samara
SEA OF AZOV
BLACK SEA
Istanbul
Ankara
TURKEY
TOROS DAGLARI
MEDITERRANEAN SEA
CYPRUS
GEORGIA
Tbilisi
Yerevan
ARMENIA
CAUCASUS MTS.
Astrakhan
CASPIAN SEA
Ural R.
URAL MOUNTAINS
Yekaterinburg
Irtysh R.
Ob R.
Omsk
WESTERN SIBERIAN LOWLAND
CENTRAL SIBERIAN PLATEAU
Podkammenaya Tunguska
Yenisey R.
Lena R.
Indigirka R.
Kolyma R.
VERKHOYANSKIY RANGE
GYDAN RANGE
KAMCHATKA RANGE
Okhotsk
SEA OF OKHOTSK
KURIL ISLANDS
DZHUGDZHUR MOUNTAINS
Yakutsk
RUSSIA
STANOVOY MOUNTAINS
Angara R.
Tomsk
Krasnoyarsk
Novosibirsk
Irkutsk
Lake Baikal
YABLONOVYY RANGE
SAYAN RANGE
ALTAI MTS.
Komsomolsk
Amur R.
SAKHALIN
Tatar Strait
Khabarovsk
SIKHOTE ALIN
HOKKAIDO
Hakodate
HONSHU
Vladivostok
Harbin
Changchun
GREATER KHINGAN RANGE
KIRGHIZ STEPPE
Ishim R.
Astana
ARAL SEA
KAZAKHSTAN
Syr Darya
Lake Balkash
UZBEKISTAN
Amu Darya
Uliastay
Ulan Bator
MONGOLIA
GOBI DESERT
Alexandria
LEBANON
ISRAEL
Beirut
Aleppo
SYRIA
Damascus
Tel Aviv
Jerusalem
Amman
JORDAN
EGYPT
Cairo
Tabriz
Baku
AZERBAIJAN
IRAQ
Euphrates R.
Tigris R.
Baghdad
ELBURZ MTS.
TURKMENISTAN
Tehran
Ashkabad
Mashad
Samarkand
Tashkent
Bishkek
Alma-Ata
Ürümqi
TIEN SHAN
KYRGYZSTAN
TARIM BASIN
Dushanbe
PAMIRS
Kashi
TAKLA MAKAN
TAJIKISTAN
ALTUN SHAN
KUNLUN SHAN
Nile R.
ARABIAN DESERT
AL HIJAZ
ASIR
Kuwait
KUWAIT
SAUDI ARABIA
ZAGROS MTS.
IRAN
PLATEAU OF IRAN
Shiraz
Persian Gulf
BAHRAIN
QATAR
Al Manamah
Ad Dawhah
Riyadh
Straits of Hormuz
AFGHANISTAN
Herat
Kabul
HINDU KUSH
KARAKORAM
Peshawar
Islamabad
Srinagar
Rawalpindi
Lahore
SULAIMAN RANGE
Indus R.
PAKISTAN
Karachi
UNITED ARAB EMIRATES
Abu Dhabi
Gulf of Oman
Muscat
OMAN
RUB AL KHALI
RED SEA
SUDAN
ERITREA
Bab el Mandeb
DJIBOUTI
Sanaa
YEMEN
Aden
Gulf of Aden
ETHIOPIA
SOMALIA
ARABIAN SEA
Equator
Huang
Lanzhou
PLATEAU OF TIBET
CHINA
Chengdu
Yangtze (Chang Jiang)
Chongqing
Brahmaputra R.
Lhasa
HIMALAYAS
GREAT INDIAN DESERT
Delhi
New Delhi
NEPAL
Kathmandu
BHUTAN
Thimphu
Kanpur
Ganges R.
Dhaka
Kolkata
Ahmadabad
Gulf of Kutch
INDIA
DECCAN PLATEAU
Gulf of Khambhat
Mumbai
Godavari R.
WESTERN GHATS
EASTERN GHATS
Hyderabad
Krishna R.
Bangalore
Madras
Calicut
Gulf of Mannar
Colombo
SRI LANKA
BANGLADESH
Bay of Bengal
Irrawaddy R.
Mandalay
MYANMAR (BURMA)
Salween R.
Rangoon
ANDAMAN ISLANDS
NICOBAR ISLANDS
Shenyang
Pyongyang
NORTH KOREA
Seoul
SOUTH KOREA
Korea Strait
Beijing
Tianjin
Qingdao
Nanjing
Wuhan
Shanghai
EAST CHINA SEA
Fuzhou
Taiwan Strait
Taipei
TAIWAN
NAN LING MTS.
Kunming
Guangzhou
Hong Kong
Luzon Strait
JAPAN
Tokyo
Yokohama
Kyoto
Osaka
Kobe
SHIKOKU
Kitayushu
KYUSHU
NANSEI SHOTO
PACIFIC OCEAN
LAOS
Vientiane
Hanoi
Gulf of Tonkin
Mekong R.
Hue
THAILAND
Moulmein
Bangkok
CAMBODIA
Phnom Penh
VIETNAM
Ho Chi Minh
Gulf of Thailand
SOUTH CHINA SEA
PHILIPPINE SEA
PHILIPPINES
LUZON
Manila
MINDORO
SAMAR
LEYTE
PANAY
NEGROS
PALAWAN
SULU SEA
SULU ISLANDS
MINDINAO
HALMAHERA
CELEBES SEA
Bandar Seri Begawan
BRUNEI
Straits of Malacca
Kuala Lumpur
MALAYSIA
SINGAPORE
BORNEO
CELEBES
INDIAN OCEAN
National capital
City
0 500 1000 Miles
0 500 1000 Kilometers
MADAGASCAR
THAILAND
Straits of Malacca
Bandar Seri Begawan
BRUNEI
SULU ISLANDS
CELEBES SEA
HALMAHERA
Equator
NEW GUINEA
PAPUA NEW GUINEA
Kuala Lumpur
MALAYSIA
SINGAPORE
BORNEO
CELEBES
SUMATRA
INDONESIA
EAST TIMOR
AUSTRALIA
INDIAN OCEAN
Jakarta
JAVA

National capital
City
ATLANTIC OCEAN
ARCTIC OCEAN
ICELAND
GREENLAND
SVALBARD
ZEMLYA FRANTSA IOSIFA
NOVAYA ZEMLYA
SEVERNAYA ZEMLYA
NEW SIBERIAN ISLANDS
WRANGELL ISLAND
ALASKA (USA)
Bering Strait
ST. LAWRENCE ISLAND
BERING SEA
EAST SIBERIAN SEA
LAPTEV SEA
KARA SEA
BARENTS SEA
Arctic Circle
IRELAND
UNITED KINGDOM
London
NORTH SEA
NORWAY
SWEDEN
Gulf of Bothnia
FINLAND
DENMARK
Copenhagen
NETH.
BELG.
Paris
FRANCE
LUX.
GERMANY
Berlin
Danube R.
SWITZ.
ALPS
POLAND
Prague
CZECH REP.
Warsaw
AUSTRIA
Vienna
SLOVAKIA
SLOVENIA
CROATIA
Budapest
HUNGARY
B-H
SERBIA
Rome
ITALY
KOSOVO
MONTENEGRO
ROMANIA
MACEDONIA
ALBANIA
Sofiya
BULGARIA
Bucharest
GREECE
MOLDOVA
UKRAINE
Kiev
BALTIC SEA
Stockholm
ESTONIA
LATVIA
LITHUANIA
RUSSIA
St.Petersburg
Ladozhkoye Ozero
Onezhkoye Ozero
Arkhangelsk
Sev. Dvina
Moscow
Rostov
Nizhniy Novgorod
Dnepr R.
Don R.
Volga R.
Donetsk
Rostov na Donu
SEA OF AZOV
BLACK SEA
Istanbul
Ankara
TURKEY
TOROS DAGLARI
Mt. Elbrus 18,510 ft.
CAUCASUS MTS.
GEORGIA
Tbilisi
ARMENIA
Yerevan
AZERBAIJAN
Baku
CASPIAN SEA
Astrakhan
Samara
Ural R.
URAL MOUNTAINS
Yekaterinburg
Irtysh R.
Ob R.
WESTERN SIBERIAN LOWLAND
Yenisey R.
CENTRAL SIBERIAN PLATEAU
Podkamennaya Tunguska
RUSSIA
Angara R.
Lena R.
VERKHOYANSKIY RANGE
Indigirka R.
Kolyma R.
GYDAN RANGE
KAMCHATKA RANGE
Klyuchevskaya 15,584 ft.
Yakutsk
Okhotsk
DZHUGDZHUR MOUNTAINS
SEA OF OKHOTSK
KURIL ISLANDS
SAKHALIN
STANOVOY MOUNTAINS
Komsomolsk
Amur R.
Tatar Strait
Khabarovsk
HOKKAIDO
Lake Baikal
YABLONOVYY RANGE
Irkutsk
Krasnoyarsk
Tomsk
Omsk
Novosibirsk
SAYAN RANGE
ALTAI MTS.
Ishim R.
KIRGHIZ
Astana
KAZAKHSTAN
ARAL SEA
Syr Darya
UZBEKISTAN
Amu Darya
Lake Balkash
Gora Kommunizma 24,590 ft.
Alma-Ata
TURKMENISTAN
Tashkent
Bishkek
TIEN SHAN
KYRGYZSTAN
Samarkand
Ashkabad
Dushanbe
TARIM BASIN
PAMIRS
TAKLA MAKAN
ALTUN SHAN
KUNLUN SHAN
TAJIKISTAN
AFGHANISTAN
Kabul
HINDU KUSH
Peshawar
Islamabad
Srinagar
K2 (Godwin-Austen) 28,250 ft.
Rawalpindi
Lahore
SULAIMAN RANGE
Indus R.
PLATEAU OF TIBET
HIMALAYAS
Mt. Everest 29,028 ft.
Nanda Devi 25,645 ft.
Lhasa
Brahmaputra R.
Delhi
New Delhi
NEPAL
Kathmandu
Kanchenjunga 28,208 ft.
BHUTAN
Thimphu
Ganges R.
GREAT INDIAN DESERT
PAKISTAN
Karachi
Gulf of Kutch
Ahmadabad
INDIA
Gulf of Khambhat
DECCAN PLATEAU
Mumbai
WESTERN GHATS
EASTERN GHATS
Godavari R.
Hyderabad
Krishna R.
Bangalore
Madras
Calicut
Gulf of Mannar
Colombo
SRI LANKA
Kolkata
Dhaka
BANGLADESH
Bay of Bengal
ANDAMAN ISLANDS
NICOBAR ISLANDS
MYANMAR (BURMA)
Mandalay
Irrawaddy R.
Salween R.
Rangoon
Moulmein
THAILAND
Bangkok
Gulf of Thailand
CAMBODIA
Phnom Penh
LAOS
Viangchan
Mekong R.
VIETNAM
Hanoi
Gulf of Tonkin
Hue
Ho Chi Minh
Straits of Malacca
Kuala Lumpur
MALAYSIA
SINGAPORE
BORNEO
BRUNEI
Bandar Seri Begawan
CELEBES SEA
CELEBES
SULU SEA
SULU ISLANDS
HALMAHERA
MINDANAO
NEGROS
PALAWAN
PANAY
LEYTE
SAMAR
MINDORO
Manila
LUZON
PHILIPPINES
PHILIPPINE SEA
Luzon Strait
SOUTH CHINA SEA
Hong Kong
Guangzhou
NAN LING MTS.
Gongga Shan 24,790 ft.
Phou Bia 9,249 ft.
Kunming
Chongqing
Chengdu
Yangtze (Chang Jiang)
CHINA
Lanzhou
Huang
Wuhan
Nanjing
Shanghai
EAST CHINA SEA
Fuzhou
Taiwan Strait
Taipei
TAIWAN
PACIFIC OCEAN
NANSEI SHOTO
Qingdao
Tianjin
Beijing
Shenyang
Pyongyang
NORTH KOREA
Seoul
SOUTH KOREA
Korea Strait
KYUSHU
SHIKOKU
Kitayushu
Osaka
Kobe
Kyoto
Yokohama
Tokyo
JAPAN
HONSHU
SEA OF JAPAN
Vladivostok
SIKHOTE ALIN
Harbin
Changchun
GREATER KHINGAN RANGE
MONGOLIA
Ulan Bator
GOBI DESERT
IRAN
Tehran
ELBURZ MTS.
Mashad
PLATEAU OF IRAN
ZAGROS MTS.
Tabriz
IRAQ
Baghdad
Euphrates R.
Tigris R.
Kuwait
KUWAIT
Shiraz
Persian Gulf
Straits of Hormuz
Al Manamah
BAHRAIN
QATAR
Ad Dawhah
Abu Dhabi
UNITED ARAB EMIRATES
Gulf of Oman
Muscat
OMAN
SAUDI ARABIA
ARABIAN DESERT
Riyadh
RUB AL KHALI
AL HIJAZ ASIR
YEMEN
Sanaa
Aden
Gulf of Aden
ARABIAN SEA
SYRIA
Aleppo
Damascus
LEBANON
Beirut
ISRAEL
Tel Aviv
Jerusalem
Amman
JORDAN
CYPRUS
EGYPT
Cairo
Alexandria
MEDITERRANEAN SEA
Nile R.
ARABIAN DESERT
RED SEA
SUDAN
ERITREA
Bab el Mandeb
DJIBOUTI
ETHIOPIA
SOMALIA
MADAGASCAR
INDIAN OCEAN
Equator
Elevation
Feet
3000
1500
300
0
Sea level
Below sea level
100
1000
0 500 1000 Miles
0 500 1000 Kilometers
THAILAND
Straits of Malacca
Kuala Lumpur
MALAYSIA
SINGAPORE
BRUNEI
Bandar Seri Begawan
BORNEO
SULU ISLANDS
CELEBES SEA
CELEBES
HALMAHERA
NEW GUINEA
PAPUA NEW GUINEA
SUMATRA
INDONESIA
EAST TIMOR
JAVA
Jakarta
AUSTRALIA
INDIAN OCEAN

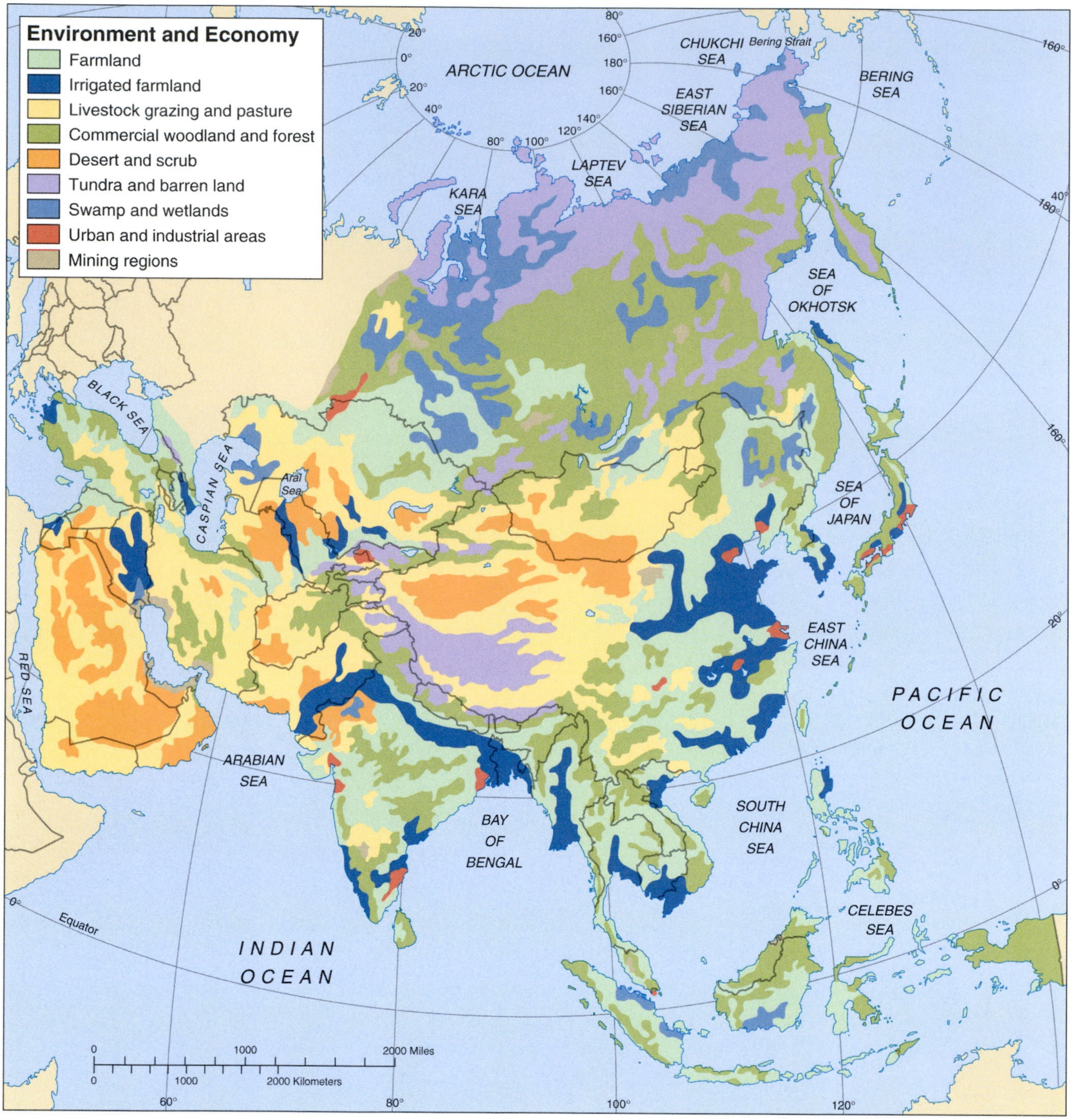

Asia is a land of extremes of land use with some of the world's most heavily industrialized regions, barren and empty areas, and productive and densely populated farm regions. Asia is a region of rapid industrial growth. Yet Asia remains an agricultural region with three out of every four workers engaged in agriculture. Asian commercial agriculture and intensive subsistence agriculture is characterized by irrigation. Some of Asia's irrigated lands are desert requiring additional water. But most of the Asian irrigated regions have sufficient precipitation for crop agriculture and irrigation is a way of coping with seasonal drought—the wet-and-dry cycle of the monsoon—often gaining more than one crop per year on irrigated farms. Agricultural yields per unit area in many areas of Asia are among the world's highest. Because the Asian population is so large and the demands for agricultural land so great, Asia is undergoing rapid deforestation and some areas of the continent have only small remnants of a once-abundant forest reserve.

Map 150b Population Density

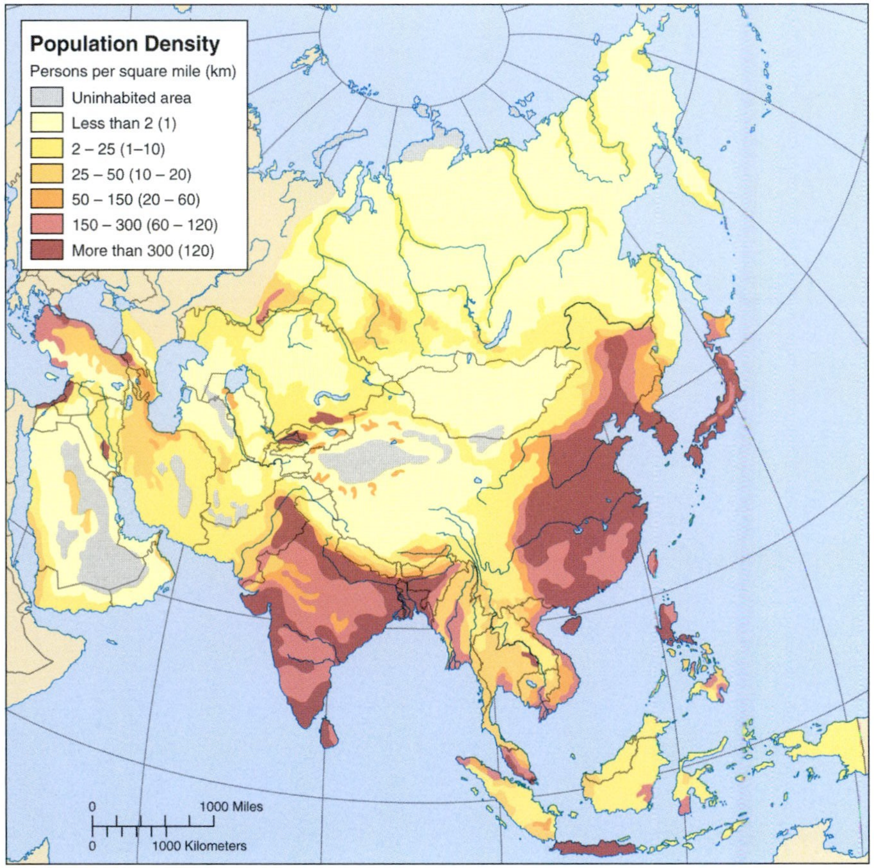

With one-third of the world's land area and nearly two-thirds of the world's population, Asia is more densely settled than any other region of the world. In some of the continent's farming regions, agricultural population density exceeds 2,000 persons per square mile. In some portions of the continent, particularly the Islamic areas of Central and Southwest Asia, this already large population is growing very rapidly with some countries having population growth rates above 3 percent per year and doubling times between 20 and 25 years. The populations of neither China nor India are growing particularly rapidly but since both countries have population bases that are enormous, the absolute number of Indians and Chinese added to the world's population each year is staggering. In spite of these massive populations, Asia also contains areas that are either completely uninhabited or have population densities that are as low as any on earth.

Map 150c Industrialization: Manufacturing and Resources

International economists have predicted that the next century will be one of Asian economic dominance as other Asian nations approach the economic levels of industrial giant Japan. Asian countries have begun to industrialize rapidly and have increased industrial production more than 100 percent in the last decade. But nothing guarantees industrial output; it must master the great distances separating critical raw material and production sites from the locations of the markets for them. Such a mastery is achieved only by the development of efficient transportation systems. Usually that means water travel and here Asia is remarkably deficient when compared with Europe and North America. Nevertheless, the mixed prospects for Asian economic growth into the next century are a great deal better than would have been predicted a decade ago.

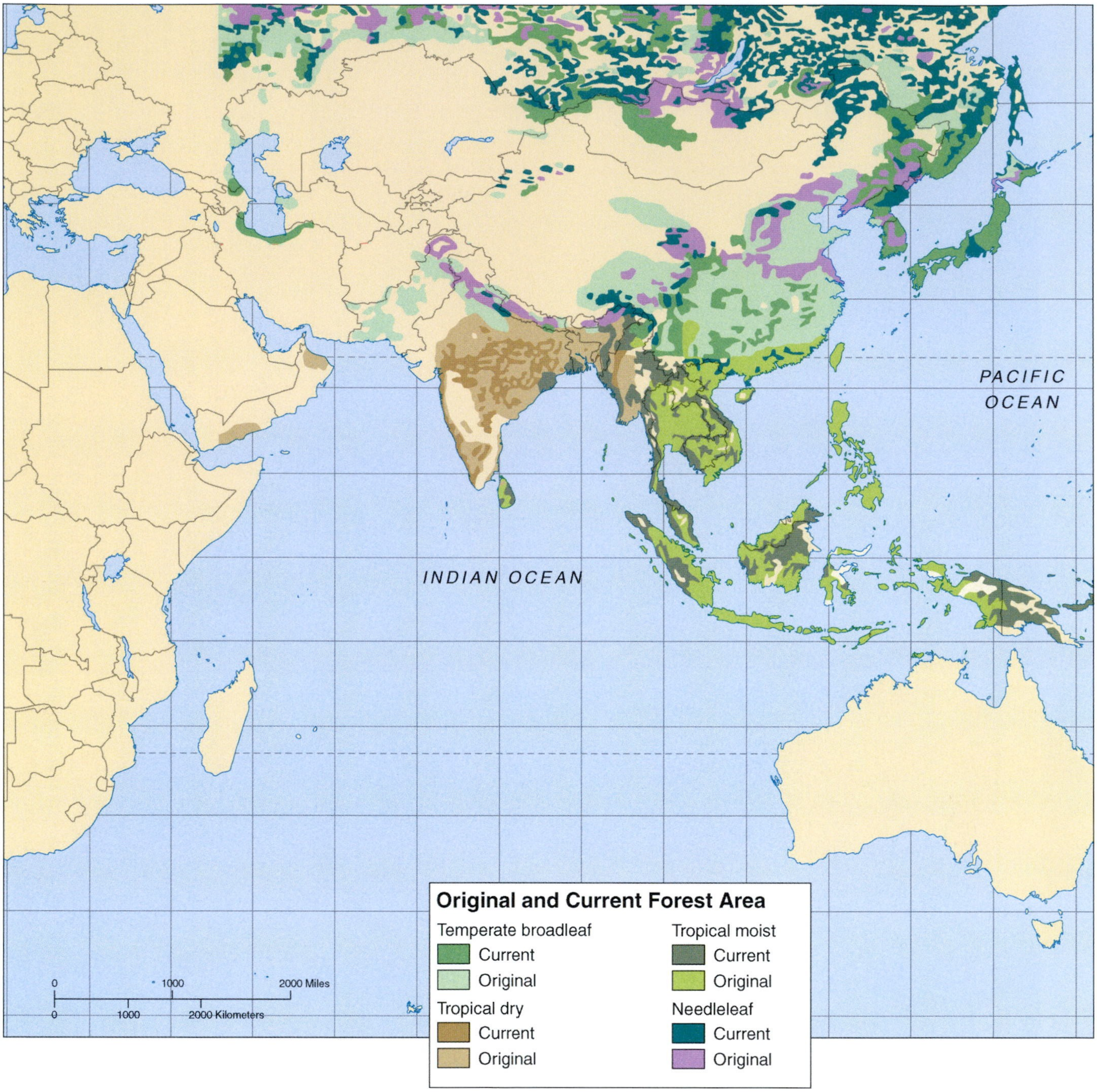

In no part of the world has deforestation produced such dramatic changes in the environment as in Asia. Here, the world's oldest continuous agricultural civilizations have placed enormous demands upon the vegetative and soil environments to the extent that precious little natural forest cover remains in the world's two most populous countries, China and India. While Southeast Asia has been historically less denuded than the major culture centers to its north and west, recent decades have seen significant forest loss to commercial timber operations, stimulated by the high-market value of tropical hardwoods. Also, a fairly recent trend has been the massive deforestation in the northeastern parts of Russia where economic expansion over the last century has stimulated the rapid loss of softwood forests for construction lumber and for paper production.

Map 151 Northern Asia

BLACK SEA
CASPIAN SEA
MEDITERRANEAN SEA
RED SEA
Persian Gulf
Gulf of Oman
ARABIAN SEA
Gulf of Aden
Gulf of Suez
Strait of Hormuz
Bab al Mandab
TURKEY
ARMENIA
AZERBAIJAN
TURKMENISTAN
AFGHANISTAN
IRAN
IRAQ
SYRIA
LEBANON
ISRAEL
JORDAN
CYPRUS
KUWAIT
QATAR
BAHRAIN
UNITED ARAB EMIRATES
OMAN
SAUDI ARABIA
YEMEN
EGYPT
SUDAN
ERITREA
DJIBOUTI
ETHIOPIA
SOMALIA
Istanbul
Zonguldak
Kastamonu
Samsun
Giresun
Trabzon
Bursa
Çankiri
Ankara
Merzifon
Çorum
Mitilini
Kuthaya
Bergama
Eskisehir
Kirsehir
Yozgat
Sivas
Erzincan
Erzurum
Kars
Yerevan
Izmir
Aydin
Afyon
Lake Tuz
Kayseri
Erciyes Dagi 12,848 ft.
Mt. Ararat 16,854 ft.
Elazig
Malatya
Khvoy
Van
Lake Van
Isparta
Konya
Antalya
Kahramanmaras
Tarsus
Adana
Gaziantep
Siverek
Diyarbakir
Mardin
Icel
Iskenderun
Hatay
Sanliurfa
Aleppo
Al Ladhiqiyah
Hamah
Nicosia
Limassol
Tripoli
Hims
Dayr az Zawr
Beirut
Damascus
Haifa
As Suwayda
Tel Aviv
Jerusalem
Gaza
Amman
Alexandria
Port Said
Cairo
Suez
SINAI
Elat
QATTÂRA DEPRESSION
SYRIAN DESERT
Ganca
Baku
Krasnovodsk
Nebit-Dag
KARAKUMY
Ashkhabad
Mary
Lankaran
Ardabil
AZERBAIJAN
Tabriz
Bandar-e Anzali
Rasht
Bojnurd
KOPET MTS.
Binalud 11,208 ft.
Mashhad
Meymaneh
Orumiyeh
Lake Urumiyeh
Mianeh
Zanjan
Qazvin
Babol
Gorgan
ELBURZ MOUNTAINS
Emamshahr
Neyshabur
Qolleh-ye Damavand 18,386 ft.
Tehran
Herat
Al Mawsil
Irbil
As Sulaymaniyah
Kirkuk
Sanandaj
Bakhtaran
Hamadan
Arak
DASH-E-KAVIR
Mileh Tharthar
Baghdad
Borujerd
Kashan
Birjand
Esfahan
Karbala
An Najaf
An Nasiriyah
Masjed Soleyman
Qomsheh
Yazd
DASHT-E LUT
ZAGROS MOUNTAINS
Ahvaz
Kalar 14,100 ft.
Al Basrah
Khorramshahr
Abadan
HAWR AL HAMMAR
Kerman
Zahedan
Al Jawf
Sakakah
Kazerun
Shiraz
Kuwait
Bushehr
Jahram
Jabal Katrina 8,668 ft.
AN NAFUD
Jabal Sharr 6,398 ft.
Ha'il
AD DAHNA
Bandar Abbas
Bandar e Lengeh
Al Qatif
Al Manamah
Ad Dammam
Jask
Qena
Nile R.
Buraydah
Unayzah
Shaqra
Doha
Dubai
Al Hufuf
Abu Dhabi
Jabal Radwah 5,906 ft.
Al Madinah
HIJAZ
Riyadh
Muscat
Tropic of Cancer
LIBYAN DESERT
Aswân
Jabal ash Sham 9,957 ft.
Sur
NUBIAN DESERT
Jiddah
Mecca
At Taif
Jabal Ibrahim 8,500 ft.
MASIRAH
Port Sudan
Suakin
ASIR
TIHAMAH
RUB' AL KHALI
Ra's al Madrakah
Abha
Salalah
FARASAN ISLANDS
Sa'dah
Atbara R.
Hadur Shu'ayb 12,008 ft.
DAHLAK ARCHIP.
Tarim
HADRAMAWT
Ra's Fartak
Khartoum
Kassala
Asmera
Mits'iwa
San'a
Al Hudaydah
Jabal Rema 10,720 ft.
Al Mukalla
Wad Medani
Blue Nile
White Nile
SOCOTRA
'ABD AL KURI
Gonder
Aden
Lake Tana
Djibouti
Raas Caseyr
Debre Mark'os
Berbera
Xaafuun
Dire Dawa
Hargeysa
Harer
Addis Ababa
Bandarbeyla
0 150 300 Miles
0 150 300 Kilometers
National capital
City
Elevation
Feet
3000
1500
300
0
Sea level
Below sea level
100
1000
Simple Conic Projection

Map 152a The Middle East: Territorial Changes, 1918–2004

The Middle East, encompassing the northeastern part of Africa and southwestern Asia, has experienced a turbulent history. In the last century alone, many of the region's countries have gone from being ruled by the Turkish Ottoman Empire, to being dependencies of Great Britain or France, to being independent. Having experienced the Crusades and colonial domination by European powers, the region's predominantly Islamic countries are now resentful of interference in the region's affairs by countries with a European and/or Christian heritage. The tension between Israel (settled largely in the late nineteenth and twentieth centuries by Jews of predominantly European background) and its neighbors is a matter of European–Middle Eastern cultural stress as well as a religious conflict between Islamic Arab culture and Judaism. The political boundaries on the map, like those throughout most of Africa, are the invention of European colonial powers and often do not take into account the pre-existing lines of tribal control or authority. In Iraq, for example, three distinct cultural areas exist: Shiite Muslim (Arabic), Sunni Muslim (Arabic), and Kurd (also Islamic but distinctly not Arabic). These three cultures occupying the territory of a state makes stability in that tortured region a distinct problem for the future.

Map 152b South-Central Eurasia: An Ethnolinguistic Crazy Quilt

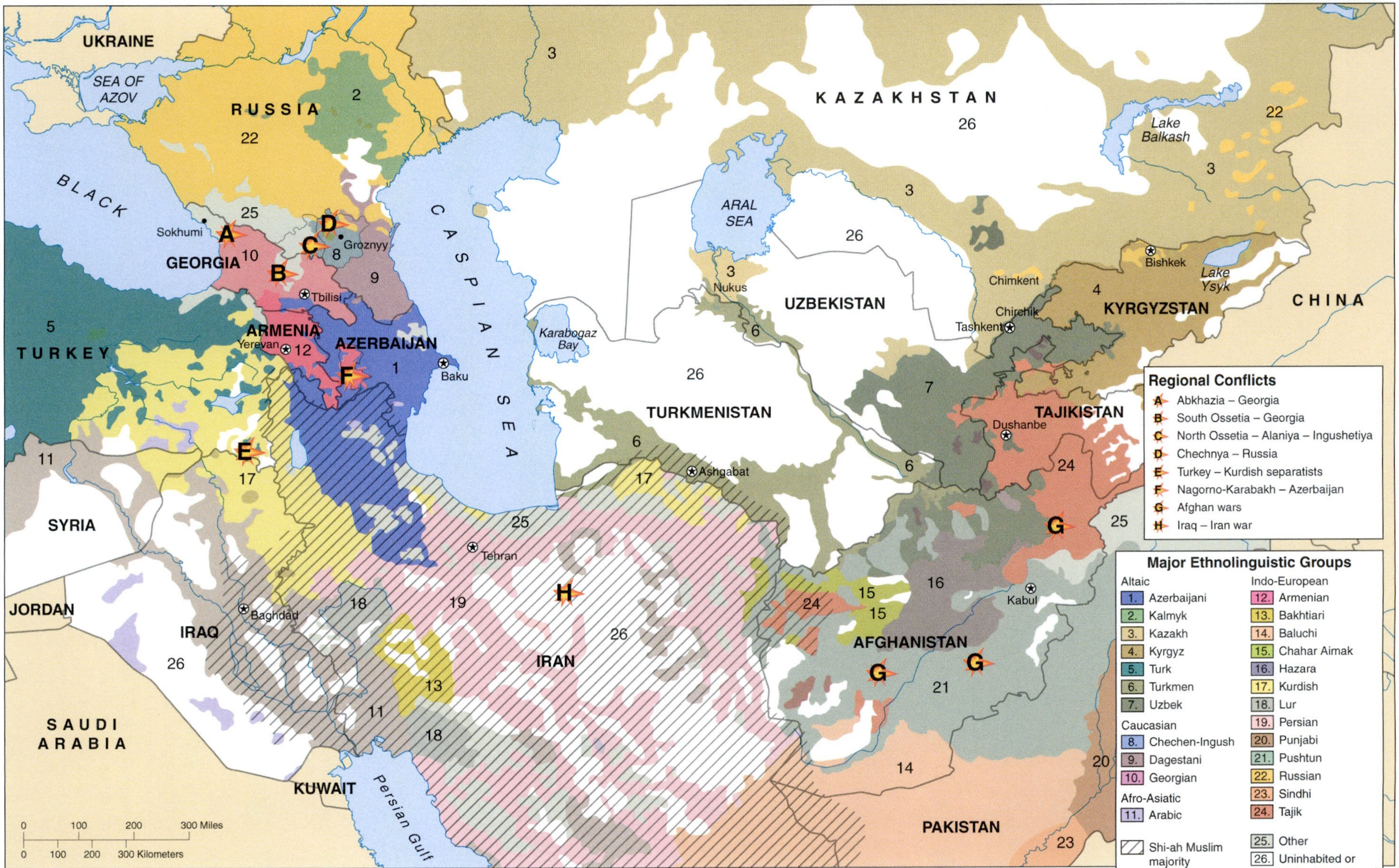

South-Central Eurasia: The area of south-central Eurasia represents what is perhaps the world's most volatile area in terms of potential military conflict. A crazy quilt of ethnic and linguistic groups, this region contains politically-defined "states" hampered by the mixture of languages and ethnic populations within their borders. For some countries, such as Afghanistan, the ethnolinguistic mix is so historically fixed as to render any attempts at modern state building nearly hopeless. This vast area—although nearly universally Muslim—is also split between the two primary Islamic sects, Sunni and Shi'ah, with several smaller sectarian divisions as well. The division of one of the world's global religions into two principal factions occurred in the seventh century over a theological issue. But the separation between Sunni and Shi'ah Muslims is now a political issue—a mutual antipathy that periodically flares into civil unrest and conflict that goes far beyond the bounds of religious debate. The area is rich in natural resources, but the mixture of ethnicity, language, and religion has and will continue to produce the human conflicts that inhibit human development.

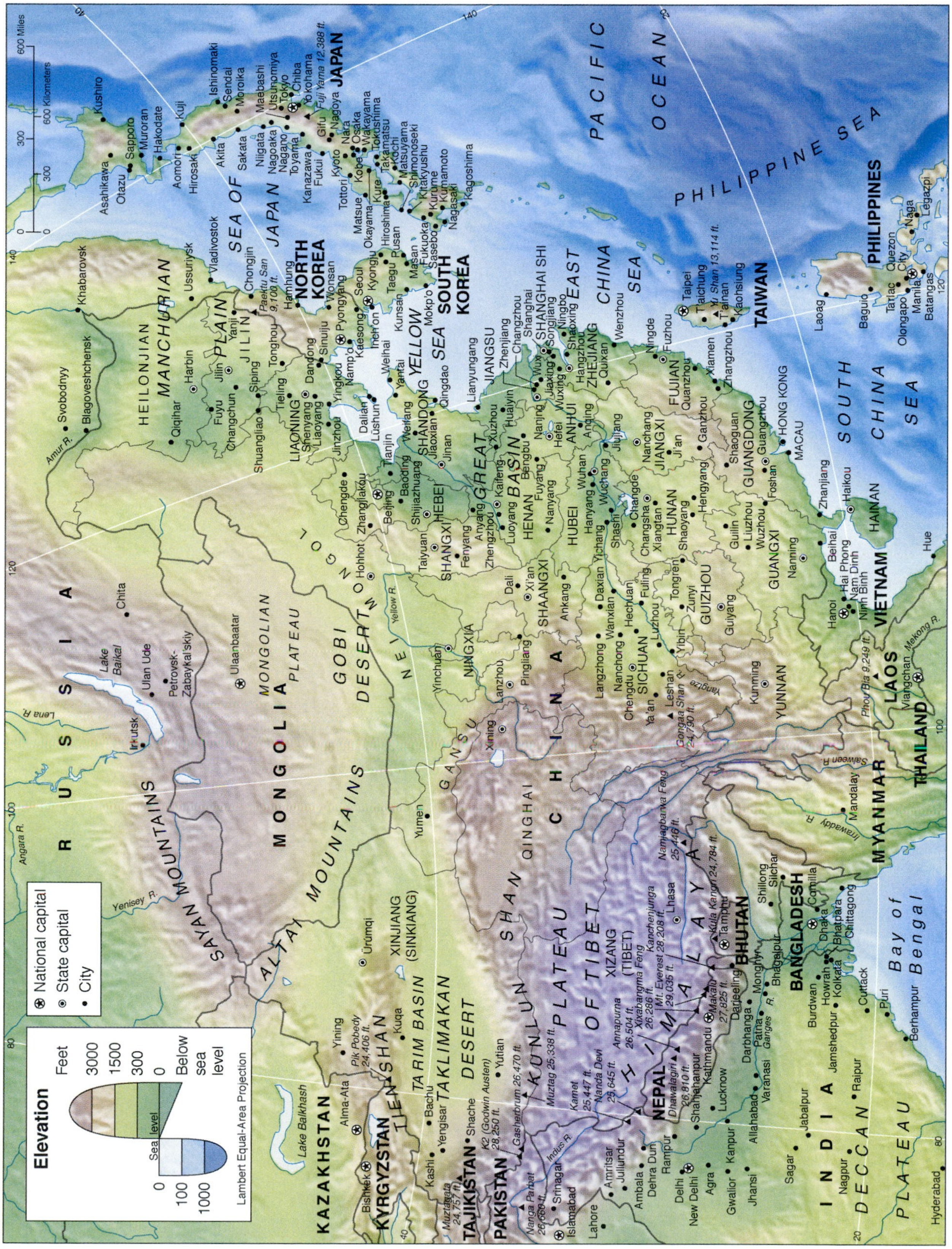
Elevation
Feet
3000
1500
300
0
Below sea level
Sea level
0
100
1000
Lambert Equal-Area Projection
National capital
State capital
City
0
300
600 Miles
0
300
600 Kilometers
RUSSIA
MONGOLIA
CHINA
KAZAKHSTAN
KYRGYZSTAN
TAJIKISTAN
PAKISTAN
INDIA
NEPAL
BHUTAN
BANGLADESH
MYANMAR
THAILAND
LAOS
VIETNAM
NORTH KOREA
SOUTH KOREA
JAPAN
TAIWAN
PHILIPPINES
HONG KONG
MACAU
SAYAN MOUNTAINS
ALTAY MOUNTAINS
MONGOLIAN PLATEAU
GOBI DESERT
TIEN SHAN
TARIM BASIN
TAKLIMAKAN DESERT
KUNLUN SHAN
PLATEAU OF TIBET
HIMALAYA
QINGHAI
GANSU
NEI MONGOL
MANCHURIAN PLAIN
GREAT BASIN
DECCAN PLATEAU
SEA OF JAPAN
YELLOW SEA
EAST CHINA SEA
SOUTH CHINA SEA
PHILIPPINE SEA
PACIFIC OCEAN
Bay of Bengal
Lake Balkhash
Lake Baikal
XINJIANG (SINKIANG)
XIZANG (TIBET)
HEILONGJIAN
JILIN
LIAONING
HEBEI
SHANDONG
SHANXI
SHAANXI
NINGXIA
HENAN
HUBEI
ANHUI
JIANGSU
ZHEJIANG
JIANGXI
FUJIAN
HUNAN
GUIZHOU
SICHUAN
YUNNAN
GUANGXI
GUANGDONG
HAINAN
SHANGHAI SHI
Beijing
Shanghai
Ulaanbaatar
Seoul
Pyongyang
Tokyo
Taipei
Manila
Hanoi
Dhaka
Kathmandu
Thimphu
New Delhi
Islamabad
Bishkek
Alma-Ata
Viangchan
Mt. Everest 29,035 ft.
K2 (Godwin Austen) 28,250 ft.
Fuji Yama 12,388 ft.

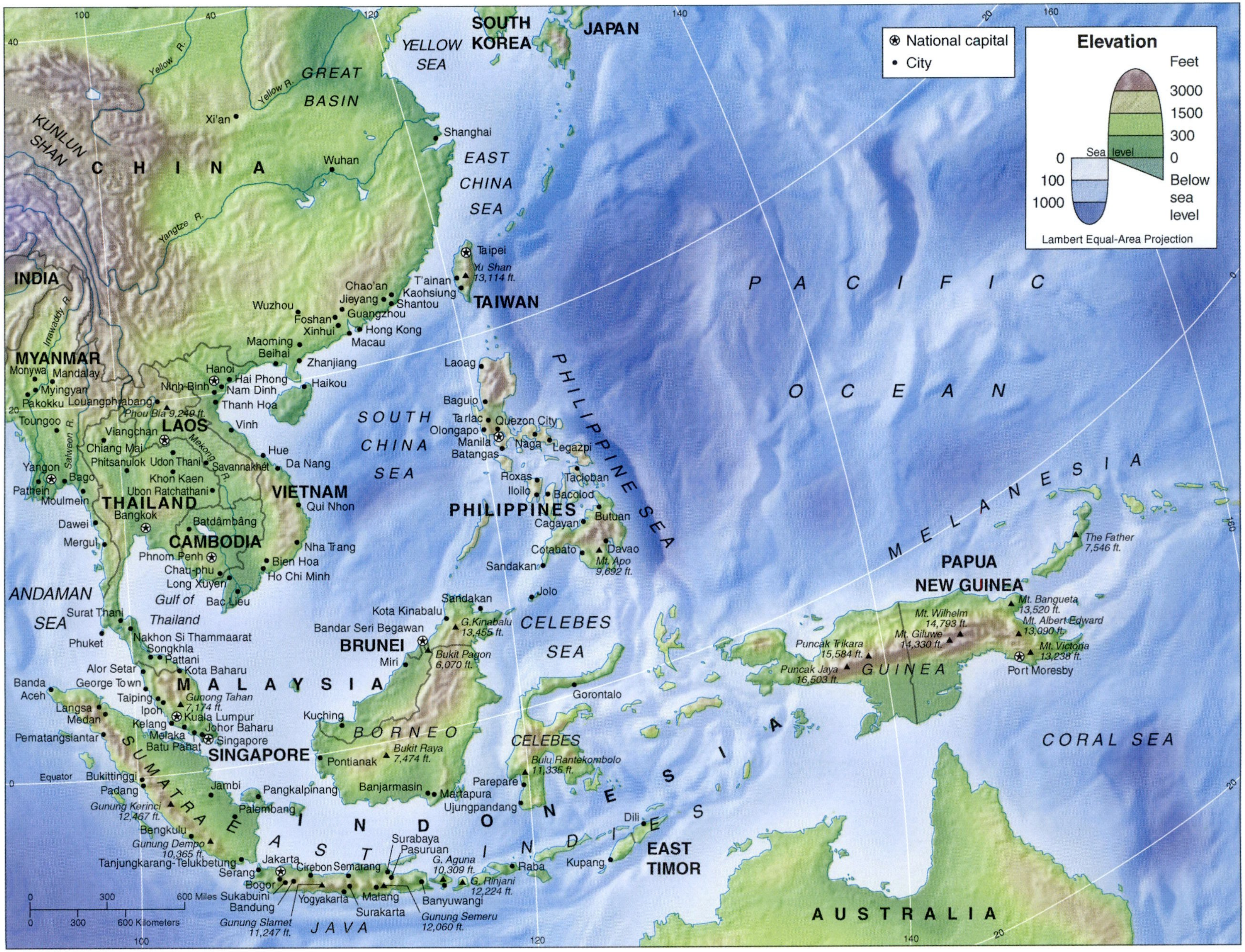
National capital
City
Elevation
Feet
3000
1500
300
0
Below sea level
Sea level
0
100
1000
Lambert Equal-Area Projection
CHINA
KUNLUN SHAN
GREAT BASIN
Yellow R.
Yangtze R.
Irawaddy R.
Salween R.
Mekong R.
Xi'an
Wuhan
Shanghai
YELLOW SEA
SOUTH KOREA
JAPAN
EAST CHINA SEA
Taipei
Yu Shan 13,114 ft.
T'ainan
Kaohsiung
TAIWAN
Chao'an
Jieyang
Shantou
Guangzhou
Wuzhou
Foshan
Xinhui
Hong Kong
Macau
Maoming
Beihai
Zhanjiang
Haikou
INDIA
MYANMAR
Monywa
Mandalay
Myingyan
Pakokku
Toungoo
Yangon
Bago
Pathein
Moulmein
Dawei
Mergui
Louangphrabang
Phou Bia 9,249 ft.
LAOS
Viangchan
Chiang Mai
Phitsanulok
Udon Thani
Savannakhet
Khon Kaen
Ubon Ratchathani
THAILAND
Bangkok
Hanoi
Ninh Binh
Hai Phong
Nam Dinh
Thanh Hoa
Vinh
Hue
Da Nang
VIETNAM
Qui Nhon
Nha Trang
Bien Hoa
Ho Chi Minh
CAMBODIA
Batdâmbâng
Phnom Penh
Chau-phu
Long Xuyen
Bac Lieu
Gulf of Thailand
ANDAMAN SEA
Surat Thani
Phuket
Nakhon Si Thammaarat
Songkhla
Pattani
Kota Baharu
Alor Setar
George Town
Taiping
Ipoh
Gunong Tahan 7,174 ft.
Kuala Lumpur
Kelang
Melaka
Batu Pahat
Johor Baharu
Singapore
SINGAPORE
MALAYSIA
Banda Aceh
Langsa
Medan
Pematangsiantar
Equator
Bukittinggi
Padang
Gunung Kerinci 12,467 ft.
SUMATRA
Jambi
Palembang
Pangkalpinang
Bengkulu
Gunung Dempo 10,365 ft.
Tanjungkarang-Telukbetung
Serang
Jakarta
Bogor
Sukabuini
Bandung
Cirebon
Semarang
Gunung Slamet 11,247 ft.
Yogyakarta
Surakarta
Malang
JAVA
Surabaya
Pasuruan
G. Aguna 10,309 ft.
Banyuwangi
Gunung Semeru 12,060 ft.
G. Rinjani 12,224 ft.
Raba
SOUTH CHINA SEA
Laoag
Baguio
Tarlac
Quezon City
Olongapo
Manila
Batangas
Naga
Legazpi
Roxas
Iloilo
Tacloban
Bacolod
PHILIPPINES
Butuan
Cagayan
Cotabato
Davao
Mt. Apo 9,692 ft.
Sandakan
Jolo
PHILIPPINE SEA
Kota Kinabalu
Sandakan
G. Kinabalu 13,455 ft.
Bandar Seri Begawan
BRUNEI
Miri
Bukit Pagon 6,070 ft.
Kuching
BORNEO
Bukit Raya 7,474 ft.
Pontianak
Banjarmasin
Martapura
CELEBES SEA
Gorontalo
CELEBES
Bulu Rantekombolo 11,335 ft.
Parepare
Ujungpandang
INDONESIA
Dili
EAST TIMOR
Kupang
PACIFIC OCEAN
MELANESIA
PAPUA NEW GUINEA
The Father 7,546 ft.
Mt. Wilhelm 14,793 ft.
Mt. Giluwe 14,330 ft.
Mt. Bangueta 13,520 ft.
Mt. Albert Edward 13,090 ft.
Mt. Victoria 13,238 ft.
Port Moresby
Puncak Trikara 15,584 ft.
Puncak Jaya 16,503 ft.
GUINEA
CORAL SEA
AUSTRALIA
0 300 600 Miles
0 300 600 Kilometers
100
120
140
160
40
20
0

INDONESIA
PAPUA NEW GUINEA
SOLOMON ISLANDS
EAST TIMOR
VANUATU
FIJI
NEW CALEDONIA
AUSTRALIA
WESTERN AUSTRALIA
NORTHERN TERRITORY
SOUTH AUSTRALIA
QUEENSLAND
NEW SOUTH WALES
VICTORIA
AUSTRALIAN CAPITAL TERRITORY
TASMANIA
NEW ZEALAND
NORTH ISLAND
SOUTH ISLAND
PACIFIC OCEAN
INDIAN OCEAN
CORAL SEA
TASMAN SEA
ARAFURA SEA
TIMOR SEA
SUNDA ISLANDS
BORNEO
CELEBES
BURU
CERAM
JAVA
SUMBAWA
FLORES
SUMBA
NEW GUINEA
BISMARCK ARCHIPELAGO
ADMIRALTY ISLANDS
NEW HANOVER
NEW IRELAND
NEW BRITAIN
TROBRIAND ISLANDS
WOODLARK ISLAND
D'ENTRECASTEAUX ISLANDS
NEW GEORGIA ISLAND
GUADALCANAL
SAN CRISTOBAL
SANTA CRUZ ISLANDS
ESPIRITU SANTO
NEW HEBRIDES
MALEKULA
Equator
Tropic of Capricorn
Surabaya
Ujung Pandang
Jayapura
Merauke
Port Moresby
Honiara
Port Vila
Suva
Noumea
Darwin
Katherine
Wyndham
Derby
Victoria River Downs
Tennant Creek
Alice Springs
Charlotte Waters
Oodnadatta
Hughes
Woomera
Kalgoorlie-Boulder
Perth
Fremantle
Albany
Carnarvon
Port Lincoln
Adelaide
Cairns
Townsville
Rockhampton
Brisbane
Newcastle
Sydney
Wollongong
Canberra
Melbourne
Ballarat
Geelong
Hobart
Auckland
Wellington
Lower Hutt
Christchurch
Dunedin
Cape van Diemen
Cape Londonderry
Cape Leveque
Roebuck Bay
North West Cape
Cape Farquhar
Steep Point
Geographe Bay
Cape Naturaliste
West Cape Howe
Cape Arnhem
Gulf of Carpentaria
Cape York
Torres Strait
Gulf of Papua
South Cape
Cape Melville
Halifax Bay
Repulse Bay
Botany Bay
Cape Howe
Cape Otway
Bass Strait
Great Australian Bight
North Cape
East Cape
Cape Farewell
Cook Strait
Southwest Cape
ARNHEM LAND
KING LEOPOLD RANGES
GREAT SANDY DESERT
GIBSON DESERT
GREAT VICTORIA DESERT
HAMERSLEY RANGE
MACDONNELL RANGES
MUSGRAVE RANGES
EVERARD RANGES
STUART RANGE
SIMPSON DESERT
NULLARBOR PLAIN
SWANLAND
DARLING RANGE
FLINDERS RA.
GREAT ARTESIAN BASIN
GREY RANGE
GREAT DIVIDING RANGE
GREGORY RA.
GREAT BARRIER REEF
BLUE MTS.
SNOWY MTS.
SOUTHERN ALPS
PEGUNUNGAN VAN REES
PEGUNUNGAN MAOKE
BISMARCK RA.
OWEN STANLEY RA.
Sepik R.
Fly R.
Daly R.
Victoria R.
Roper R.
Fitzroy R.
DeGrey R.
Ashburton R.
Gascoyne R.
Murchison R.
Swan R.
Mitchell R.
Gilbert R.
Burdekin R.
Belyando R.
Barwon R.
Namoi R.
Macquarie
Darling R.
Lachlan R.
Murrumbidgee R.
Murray R.
Lake Eyre
Lake Everard
Lake Gairdner
Lake Torrens
National capital
State capital
City
0 250 500 Miles
0 250 500 Kilometers

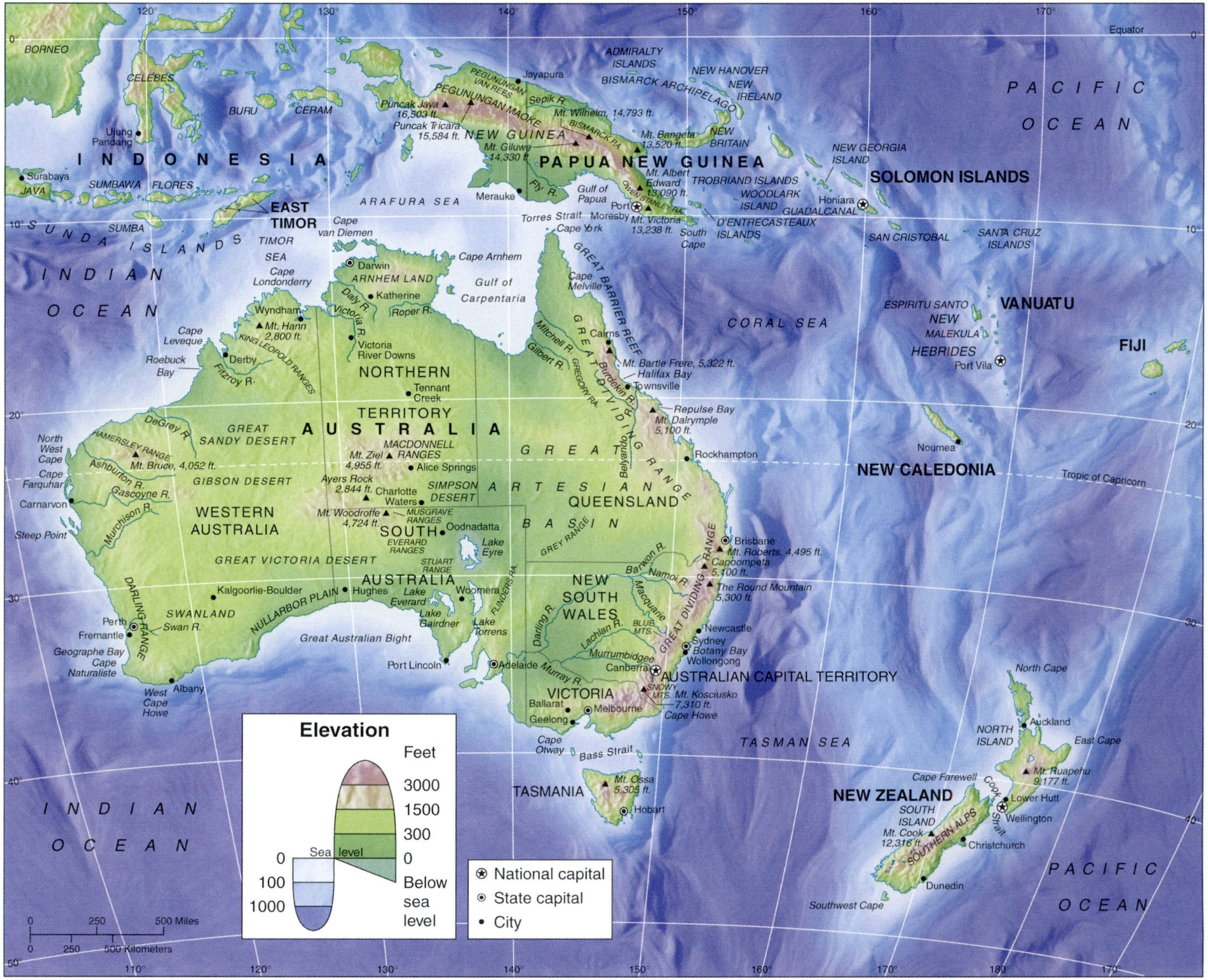
PACIFIC OCEAN
INDIAN OCEAN
INDONESIA
BORNEO
CELEBES
BURU
CERAM
Ujung Pandang
Surabaya
JAVA
SUMBAWA
FLORES
SUMBA
SUNDA ISLANDS
EAST TIMOR
TIMOR SEA
ARAFURA SEA
Cape van Diemen
Cape Londonderry
Jayapura
PEGUNUNGAN VAN REES
PEGUNUNGAN MAOKE
Sepik R.
Puncak Jaya 16,503 ft.
Puncak Trikora 15,584 ft.
NEW GUINEA
Mt. Giluwe 14,330 ft.
Mt. Wilhelm, 14,793 ft.
BISMARCK RA.
Mt. Bangeta 13,520 ft.
PAPUA NEW GUINEA
Fly R.
Merauke
Gulf of Papua
Port Moresby
OWEN STANLEY RA.
Mt. Albert Edward 13,090 ft.
Mt. Victoria 13,238 ft.
South Cape
Torres Strait
Cape York
ADMIRALTY ISLANDS
BISMARCK ARCHIPELAGO
NEW HANOVER
NEW IRELAND
NEW BRITAIN
TROBRIAND ISLANDS
WOODLARK ISLAND
D'ENTRECASTEAUX ISLANDS
NEW GEORGIA ISLAND
SOLOMON ISLANDS
Honiara
GUADALCANAL
SAN CRISTOBAL
SANTA CRUZ ISLANDS
ESPIRITU SANTO
NEW HEBRIDES
MALEKULA
VANUATU
Port Vila
FIJI
CORAL SEA
Noumea
NEW CALEDONIA
Tropic of Capricorn
Equator
Darwin
ARNHEM LAND
Cape Arnhem
Gulf of Carpentaria
Katherine
Daly R.
Victoria R.
Roper R.
Victoria River Downs
Wyndham
Mt. Hann 2,800 ft.
KING LEOPOLD RANGES
Cape Leveque
Derby
Roebuck Bay
Fitzroy R.
NORTHERN TERRITORY
Tennant Creek
AUSTRALIA
GREAT SANDY DESERT
DeGrey R.
HAMERSLEY RANGE
Mt. Bruce, 4,052 ft.
North West Cape
Cape Farquhar
Carnarvon
Ashburton R.
Gascoyne R.
Murchison R.
Steep Point
GIBSON DESERT
WESTERN AUSTRALIA
MACDONNELL RANGES
Mt. Ziel 4,955 ft.
Alice Springs
Ayers Rock 2,844 ft.
Charlotte Waters
SIMPSON DESERT
Mt. Woodroffe 4,724 ft.
MUSGRAVE RANGES
Oodnadatta
SOUTH AUSTRALIA
EVERARD RANGES
STUART RANGE
Lake Eyre
GREAT VICTORIA DESERT
Kalgoorlie-Boulder
NULLARBOR PLAIN
Hughes
Lake Everard
Lake Gairdner
Woomera
Lake Torrens
FLINDERS RA.
DARLING RANGE
SWANLAND
Perth
Swan R.
Fremantle
Geographe Bay
Cape Naturaliste
Albany
West Cape Howe
Great Australian Bight
Port Lincoln
Adelaide
Cape Melville
GREAT BARRIER REEF
Cairns
Mitchell R.
Gilbert R.
GREGORY RA.
GREAT DIVIDING RANGE
Burdekin R.
Mt. Bartle Frere, 5,322 ft.
Halifax Bay
Townsville
Repulse Bay
Mt. Dalrymple 5,100 ft.
Belyando R.
Rockhampton
GREAT ARTESIAN BASIN
QUEENSLAND
GREY RANGE
Brisbane
Mt. Roberts, 4,495 ft.
Capoompeta 5,100 ft.
The Round Mountain 5,300 ft.
Barwon R.
Namoi R.
NEW SOUTH WALES
Macquarie R.
Darling R.
Lachlan R.
BLUE MTS.
Newcastle
Sydney
Botany Bay
Wollongong
Murrumbidgee R.
Canberra
AUSTRALIAN CAPITAL TERRITORY
Murray R.
SNOWY MTS.
Mt. Kosciusko 7,310 ft.
Cape Howe
VICTORIA
Ballarat
Melbourne
Geelong
Cape Otway
Bass Strait
TASMANIA
Mt. Ossa 5,305 ft.
Hobart
TASMAN SEA
North Cape
Auckland
NORTH ISLAND
East Cape
Mt. Ruapehu 9,177 ft.
Cape Farewell
Cook Strait
Lower Hutt
Wellington
NEW ZEALAND
SOUTH ISLAND
Mt. Cook 12,316 ft.
SOUTHERN ALPS
Christchurch
Dunedin
Southwest Cape
Elevation
Feet
3000
1500
300
0
Sea level
Below sea level
0
100
1000
National capital
State capital
City
0 250 500 Miles
0 250 500 Kilometers
0°
10°
20°
30°
40°
50°
110°
120°
130°
140°
150°
160°
170°
180°

Map 157a Environment and Economy

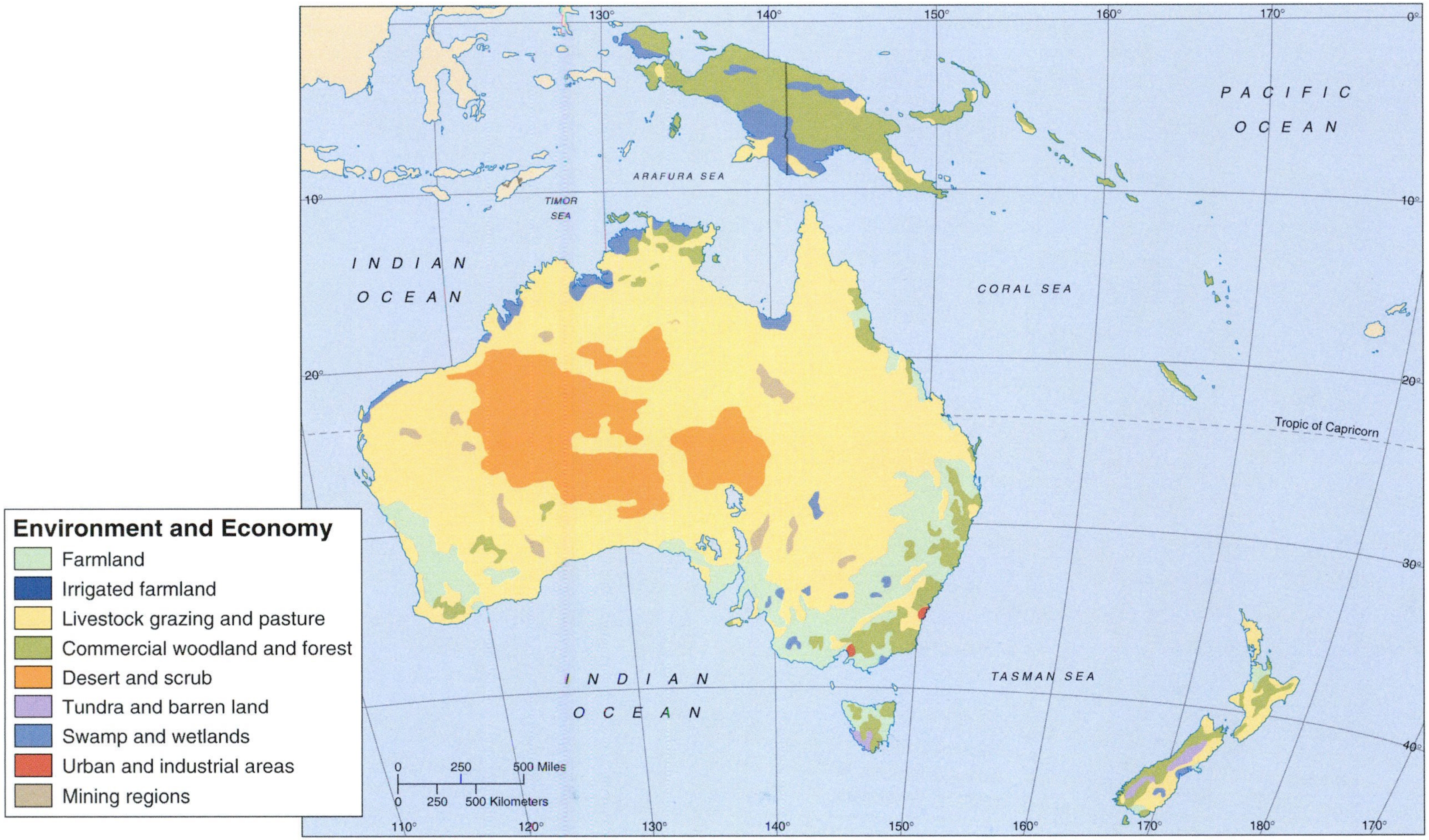

Australasia is dominated by the world's smallest and most uniform continent. Flat, dry and mostly hot, Australia has the simplest of land use patterns: where rainfall exists so does agricultural activity. Two agricultural patterns dominate the map: livestock grazing, primarily sheep, and wheat farming, although some sugar cane production exists in the north and some cotton is grown elsewhere. Only about 6 percent of the continent consists of arable land so the areas of wheat farming, dominant as they may be in the context of Australian agriculture, are small. Australia also supports a healthy mineral resource economy, with iron and copper and precious metals making up the bulk of the extraction. Elsewhere in the region, tropical forests dominate Papua New Guinea, with some subsistence agriculture and livestock. New Zealand's temperate climate with abundant precipitation supports a productive livestock industry and little else besides tourism—which is an important economic element throughout the remainder of the region as well.

Map 157b Population Density

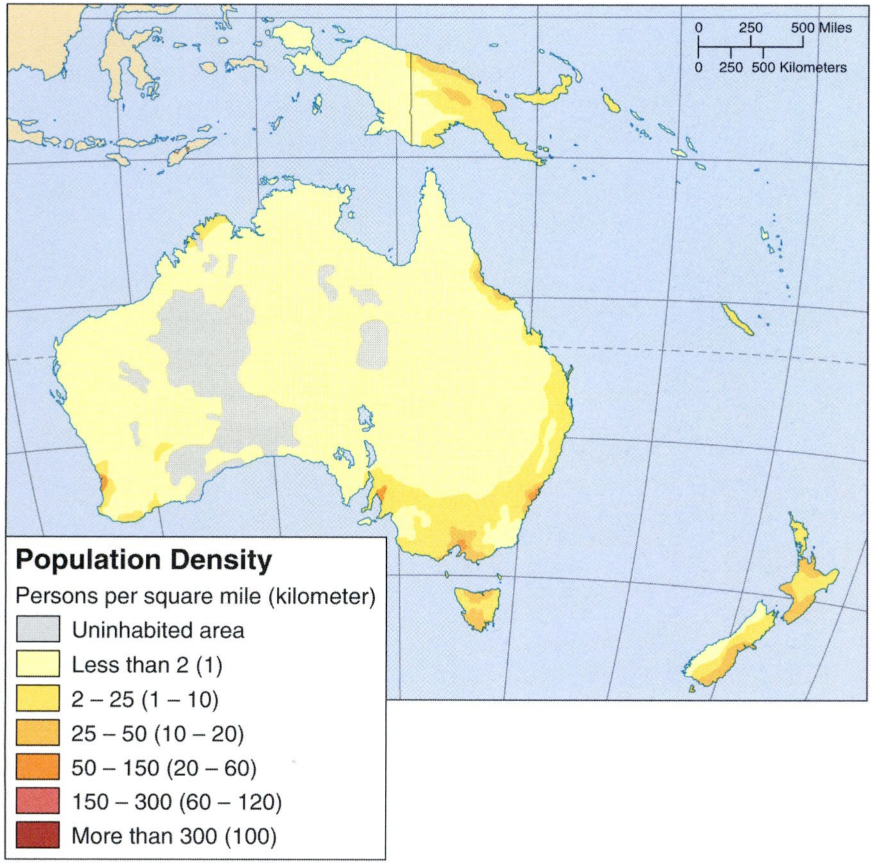

The region's small population is remarkably diverse. To the north are the Melanesian New Guinea peoples, while Europeans dominate the populations of Australia and New Zealand, both of which have significant indigenous populations. Throughout most of the smaller island groups, the bulk of the population is Melanesian but with a scattering of Europeans. The distribution of population is extremely uneven with New Guinea, southeastern Australia, and New Zealand supporting the bulk of the region's people while the remainder of the region—meaning nearly all of Australia—is either sparsely populated or devoid of population altogether. Nowhere do population densities reach the levels they do in other major regions of the world and densities of 50 persons per square mile are the highest to be found. Population location is dependent on precipitation and population growth patterns are culturally variable.

Map 157c Climatic Patterns

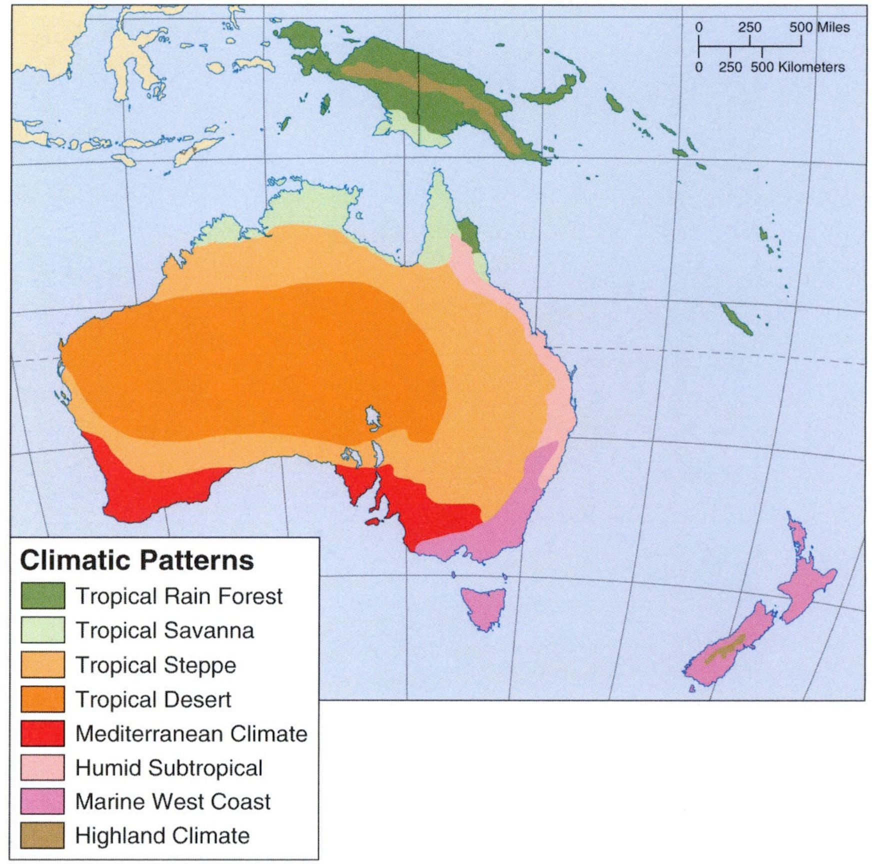

Because of its nearly uniform surface, with only a few low and scattered uplands, Australian climate is a consequence only of the two great climate controls—latitudinal position and location relative to continental margin and interior. The continent bestrides the 30th parallel of latitude and its climatic pattern is dominated by the subtropical high pressure system with dry air masses that are responsible for the existence of great deserts. Toward the equator, the desert grades into steppe, savanna, and tropical forest as the subtropical high gives way to equatorial low pressure and abundant precipitation. Toward the pole, arid land fades into more well-watered steppe grasslands, the Mediterranean type climate of the southern margins of the continent, and the marine west coast climate of the Australian southeast and New Zealand. This latter climate is where most of the region's people live.

Map 158 The Pacific Rim

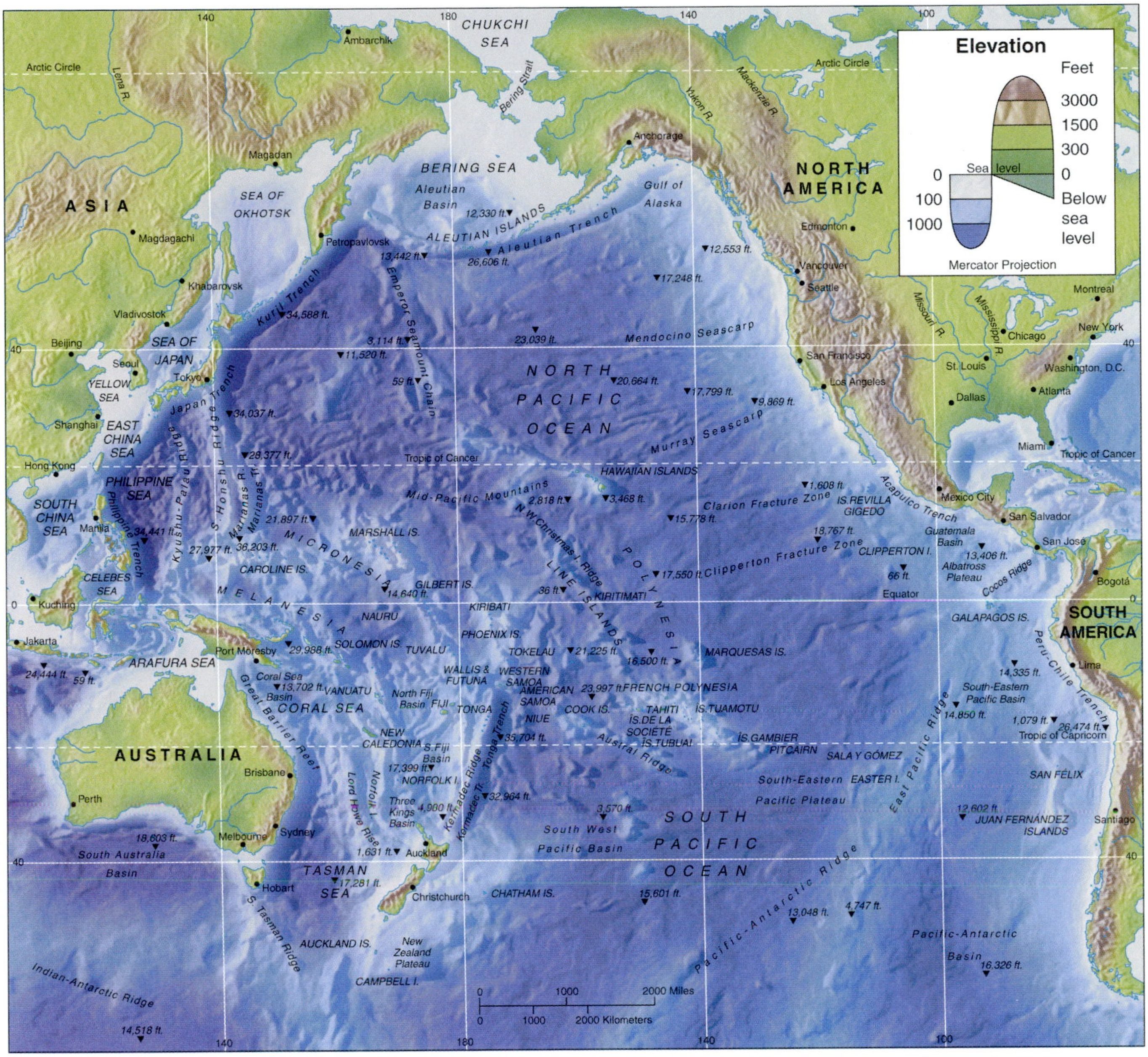

Elevation
Feet
3000
1500
300
0
Sea level
0
100
1000
Below sea level
Mercator projection
Baffin Bay
7,008 ft.
Gunnbjørn Fjeld 12,139 ft.
Denmark Strait
Mount Forel 11,024 ft.
NORWEGIAN SEA
Norwegian Basin
Arctic Circle
Hvannadalshnúkur 6,952 ft.
Reykjavik
Hekla 4,892 ft.
Iceland-Faroe Rise
FAEROE ISLANDS
SHETLAND IS.
9,433 ft.
C.Chidley
Cape Farewell
LABRADOR SEA
Reykjanes Ridge
2,398 ft.
9,869 ft.
ROCKALL
Oceanic Bank
Rockall Deep
11,195 ft.
Oslo
Stockholm
299 ft.
BALTIC SEA
Glasgow
2,425 ft.
NORTH SEA
Copenhagen
Berlin
London
EUROPE
NORTH ATLANTIC OCEAN
Goose Bay
Winnipeg
NORTH AMERICA
NEWFOUNDLAND
656 ft.
C.Race
39 ft.
Grand Banks
Newfoundland Basin
2,218 ft.
217 ft.
Land's End
Brussels
Prague
Paris
Pointe de Raz
Danube R.
Budapest
Minneapolis
Montreal
C.Sable
SABLE I.
20,700 ft.
Newfoundland Rise
200 ft.
North East Atlantic Basin
15,398 ft.
Bordeaux
C.Finisterre
Marseille
3,881 ft.
Rome
Missouri R.
Chicago
C.Cod
New York
2,881 ft.
Washington, D.C.
St. Louis
Mississippi R.
Madrid
9,403 ft.
11,910 ft.
Lisbon
AZORES
Mid-Atlantic Ridge
North American Basin
486 ft.
Gibraltar
Tunis
MEDITERRANEAN SEA
16,802 ft.
Dallas
Atlanta
C.Fear
20,995 ft.
BERMUDA
MADEIRA
17,921 ft.
Canary Basin
CANARY ISLANDS
Gulf of Mexico
13,200 ft.
Miami
Tropic of Cancer
Str. of Florida
Havana
Sargasso Sea
22,951 ft.
Nares Deep
883 ft.
20,647 ft.
856 ft.
Tropic of Cancer
4,689 ft.
Puerto Rico Trench
West Indies
28,374 ft.
53 ft.
Cayman Trench
25,218 ft.
Belmopan
Cape Verde Plateau
10,381 ft.
AFRICA
CARIBBEAN SEA
17,783 ft.
14,263 ft.
23,295 ft.
CAPE VERDE IS.
C. Verde
Dakar
Niger R.
Tegucigalpa
Cape Verde Basin
Kano
San José
Caracas
2,398 ft.
Freetown
Georgetown
Guyana Basin
Monrovia
Abidjan
Lagos
Accra
Bogotá
C.Orange
Cocos Ridge
387 ft.
ST. PAUL ROCKS
19,817 ft.
Sierra Leone Basin
C.Palmas
BIOKO
PRINCIPE
Equator
Quito
Equator
Romanche Gap
25,454 ft.
Guinea Basin
GULF OF GUINEA
SÃO TOMÉ
Port Gentil
Zaire R.
GALAPAGOS IS.
Manaus
R.Amazon
ROCAS
FERNANDO DE NORONHA
18,895 ft.
7,645 ft.
Kinshasa
C. São Roque
Recife
21,448 ft.
ASCENSION I.
Luanda
SOUTH AMERICA
276 ft.
18,288 ft.
Lima
Salvador
20,076 ft.
Angola Basin
ST. HELENA
19,850 ft.
Brasília
La Paz
Brazil Basin
Peru-Chile Trench
Nazca Ridge
TRINDADE
MARTIN VAZ IS.
Tropic of Capricorn
SOUTH ATLANTIC OCEAN
C.Frio
Walvis Bay
I.SAN FELIX
I.SAN AMBROSIA
Asunción
18,879 ft.
Bromley Plateau
South-Eastern Atlantic Basin
Walvis Ridge
1,686 ft.
1,788 ft.
Cape Basin
Rio Grande Rise
R.Paraná
Santiago
IS.JUAN FERNANDEZ
Buenos Aires
Montevideo
Cape Town
C.Agulhas
C.Corrientes
17,278 ft.
TRISTAN DA CUNHA
GOUGH I.
5,023 ft.
1,349 ft.
Discovery Tablemount
Argentine Basin
358 ft.
C.Blanco
20,382 ft.
18,164 ft.
Atlantic-Indian Ridge
7,579 ft.
FALKLAND IS.
Scotia Ridge
S. GEORGIA
S. Sandwich Tr.
BOUVET I.
Cape Horn
S. SANDWICH IS.
Drake Passage
SCOTIA SEA
27,652 ft.
0
1000
2000 Miles
0
1000
2000 Kilometers
Atlantic-Indian Antarctic Basin
80°
60°
40°
20°
0°
20°
60°
40°
20°
0°
20°
40°
60°

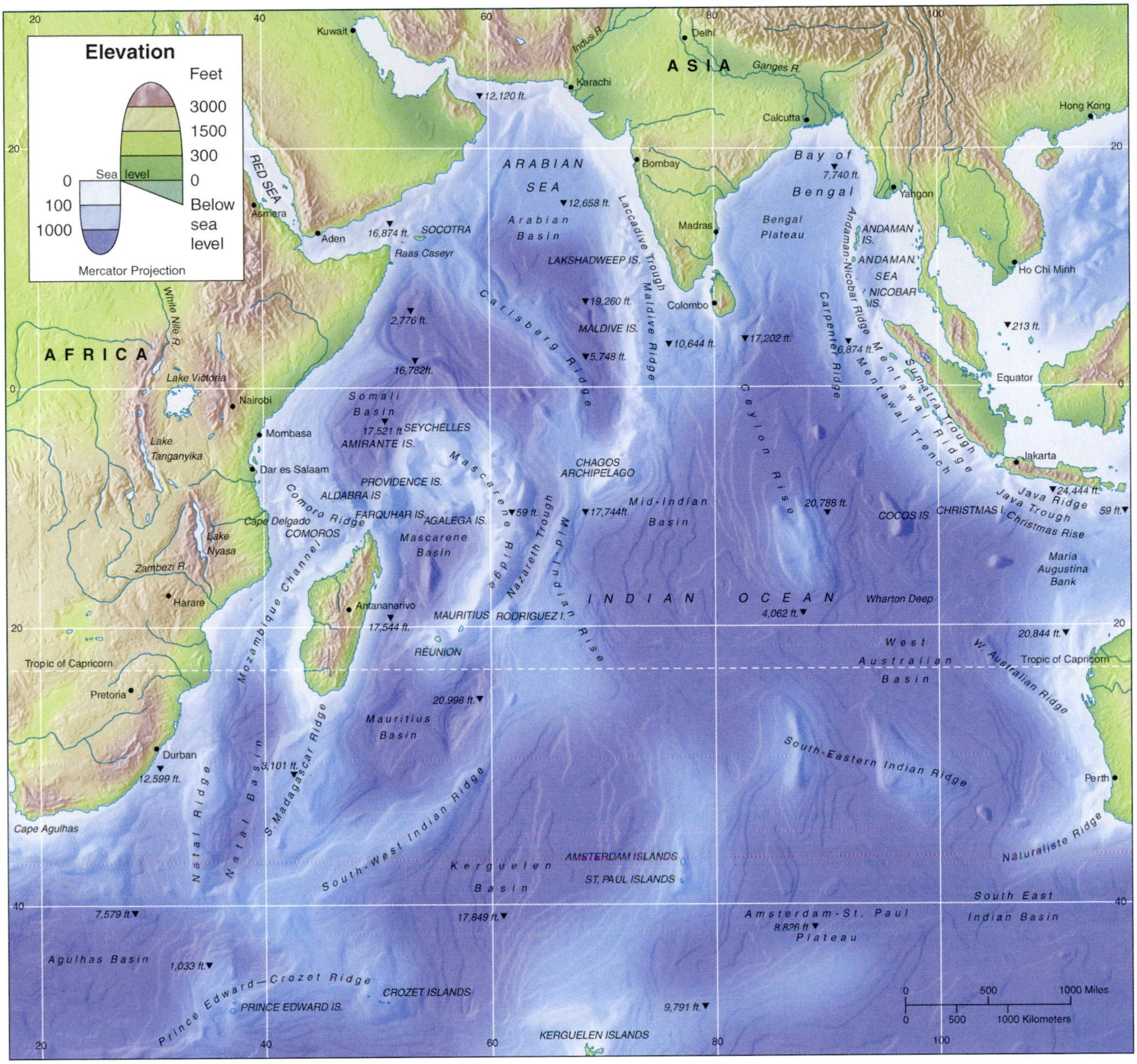

Elevation
Feet
3000
1500
300
0
Sea level
Below sea level
100
1000
Mercator Projection
ASIA
AFRICA
ARABIAN SEA
Bay of Bengal
INDIAN OCEAN
RED SEA
ANDAMAN SEA
Kuwait
Delhi
Karachi
Indus R.
Ganges R.
Calcutta
Hong Kong
Bombay
Yangon
Madras
Colombo
Ho Chi Minh
Aden
Asmera
SOCOTRA
Raas Caseyr
LAKSHADWEEP IS.
MALDIVE IS.
ANDAMAN IS.
NICOBAR IS.
Equator
Jakarta
Nairobi
Mombasa
Dar es Salaam
Lake Victoria
Lake Tanganyika
Lake Nyasa
White Nile R.
Zambezi R.
Harare
Pretoria
Durban
Cape Agulhas
Cape Delgado
COMOROS
Antananarivo
Tropic of Capricorn
SEYCHELLES
AMIRANTE IS.
PROVIDENCE IS.
ALDABRA IS.
FARQUHAR IS.
AGALEGA IS.
MAURITIUS
RODRIGUEZ I.
RÉUNION
CHAGOS ARCHIPELAGO
COCOS IS.
CHRISTMAS I.
Perth
AMSTERDAM ISLANDS
ST. PAUL ISLANDS
CROZET ISLANDS
PRINCE EDWARD IS.
KERGUELEN ISLANDS
Arabian Basin
Somali Basin
Mascarene Basin
Mascarene Ridge
Mid-Indian Basin
Mid-Indian Rise
Nazareth Trough
Carlsberg Ridge
Laccadive Trough
Maldive Ridge
Ceylon Rise
Bengal Plateau
Carpenter Ridge
Andaman-Nicobar Ridge
Mentawai Ridge
Mentawai Trench
Sumatra Trough
Java Ridge
Java Trough
Christmas Rise
Maria Augustina Bank
Wharton Deep
West Australian Basin
W. Australian Ridge
South-Eastern Indian Ridge
Naturaliste Ridge
South East Indian Basin
Amsterdam-St. Paul Plateau
Kerguelen Basin
South-West Indian Ridge
Mauritius Basin
S. Madagascar Ridge
Mozambique Channel
Comoro Ridge
Natal Basin
Natal Ridge
Agulhas Basin
Prince Edward—Crozet Ridge
12,120 ft.
12,658 ft.
16,874 ft.
2,776 ft.
16,782ft.
17,521 ft.
19,260 ft.
5,748 ft.
10,644 ft.
17,202 ft.
7,740 ft.
6,874 ft.
213 ft.
24,444 ft.
59 ft.
20,788 ft.
17,744ft.
4,062 ft.
17,544 ft.
20,844 ft.
20,998 ft.
3,101 ft.
12,599 ft.
17,849 ft.
8,826 ft
7,579 ft.
1,033 ft.
9,791 ft.
0 500 1000 Miles
0 500 1000 Kilometers
20
40
60
80
100
0

Map 161 The Arctic

Map 162 Antarctica

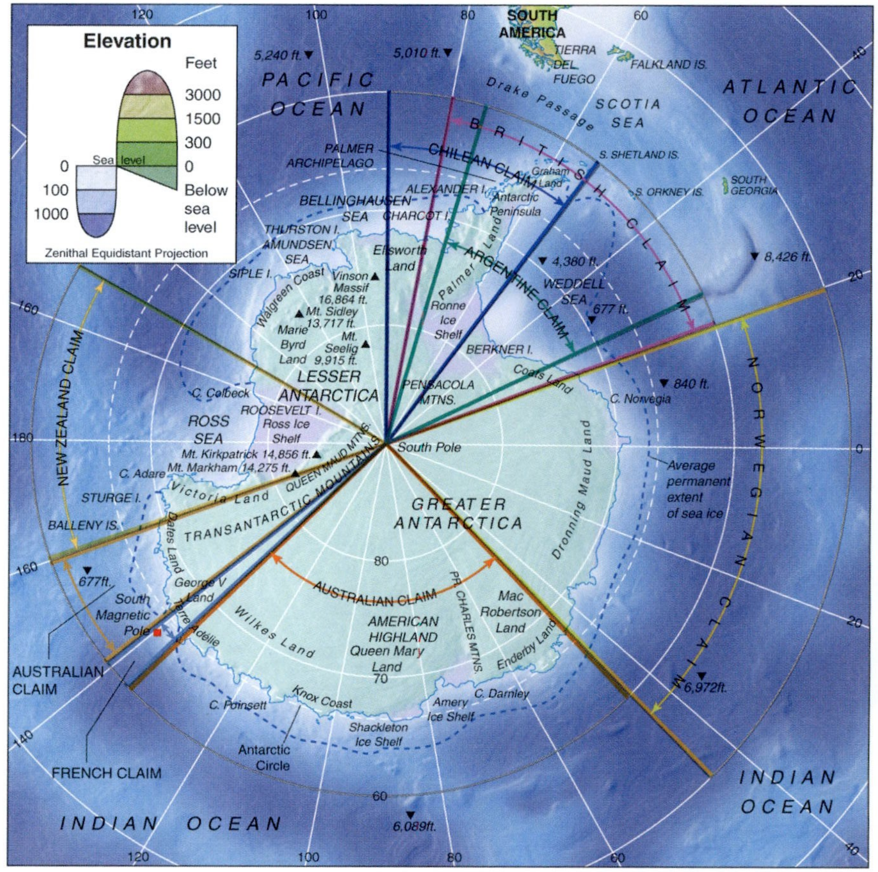

Unit VIII

Tables

Table A
World Countries: Area, Population, and Population Density, 2008

COUNTRY	AREA (MI.²)	AREA (KM.²)	POPULATION (ESTIMATED 12/2007)	DENSITY (POP/MI.²)	DENSITY (POP/KM.²)
Afghanistan (est. 2005)	251,826	652,090	30,328,497	120.43	46.51
Albania	11,100	28,750	3,172,155	285.78	110.34
Algeria	919,595	2,381,740	33,351,137	36.27	14.00
Angola	481,354	1,246,700	16,557,050	34.40	13.28
Antigua and Barbuda	171	440	84,097	491.80	191.13
Argentina	1,073,400	2,780,400	39,134,297	36.46	14.08
Armenia	11,506	29,800	3,009,549	261.56	100.99
Australia	2,966,155	7,741,220	20,701,488	6.98	2.67
Austria	32,377	83,860	8,280,900	255.76	98.75
Azerbaijan	33,436	86,600	8,484,400	253.75	97.97
Bahamas	5,382	13,880	327,279	60.81	23.58
Bahrain	267	710	738,913	2,767.46	1,040.72
Bangladesh	55,598	144,000	155,990,777	2,805.69	1,083.27
Barbados	166	430	292,930	1,764.64	681.23
Belarus	80,155	207,600	9,732,500	121.42	46.88
Belgium	11,783	30,510	10,540,900	894.59	345.49
Belize	8,866	22,960	297,613	33.57	12.96
Benin	43,475	112,620	8,759,655	201.49	77.78
Bhutan	18,200	47,000	648,766	35.65	13.80
Bolivia	424,165	1,098,580	9,353,846	22.05	8.51
Bosnia-Herzegovina	19,776	51,210	3,926,406	198.54	76.67
Botswana	231,803	581,730	1,858,163	8.02	3.19
Brazil	3,286,488	8,514,880	189,322,987	57.61	22.23
Brunei	2,228	5,770	381,952	171.43	66.20
Bulgaria	42,823	110,990	7,692,600	179.64	69.31
Burkina Faso	105,869	274,000	14,358,500	135.63	52.40
Burundi	10,745	27,830	8,173,070	760.64	293.68
Cambodia	69,898	181,040	14,196,611	203.10	78.42
Cameroon	183,569	475,440	18,174,696	99.01	38.23
Canada	3,849,674	9,970,610	32,649,000	8.48	3.27
Cape Verde	1,557	4,030	518,562	333.05	128.68
Central African Republic	240,535	622,980	4,037,747	16.79	6.48
Chad	495,755	1,284,000	9,748,931	19.66	7.59
Chile	292,259	756,630	16,432,674	56.23	21.72
China	3,705,392	9,598,050	1,311,797,692	354.02	136.67
Colombia	439,734	1,138,910	45,558,450	103.60	40.00
Comoros	838	2,230	613,606	732.23	275.16
Congo, Dem. Rep.	905,564	2,344,860	60,643,890	66.97	25.86
Congo, Rep.	132,047	342,000	3,689,299	27.94	10.79
Costa Rica	19,730	51,100	4,398,770	222.95	86.08
Côte d'Ivoire	124,502	322,460	18,914,474	151.92	58.66
Croatia	21,824	56,540	4,441,300	203.51	78.55
Cuba	42,804	110,860	11,266,701	263.22	101.63
Cyprus	3,571	9,250	771,200	215.96	83.37
Czech Republic	30,387	78,870	10,270,000	337.97	130.21

Table A (*continued*) World Countries: Area, Population, and Population Density, 2008					
COUNTRY	AREA		POPULATION	DENSITY	
	(MI.²)	(KM.²)	(ESTIMATED 12/2007)	(POP/MI.²)	(POP/KM.²)
Denmark	16,629	43,090	5,436,600	326.93	126.17
Djibouti	8,494	23,200	818,508	96.36	35.28
Dominica	290	750	72,396	249.64	96.53
Dominican Republic	18,815	48,730	9,614,670	511.01	197.30
Ecuador	109,484	283,560	13,201,995	120.58	46.56
Egypt	386,662	1,001,450	74,166,496	191.81	74.06
El Salvador	8,124	21,040	6,762,319	832.39	321.40
Equatorial Guinea	10,831	28,050	495,639	45.76	17.67
Eritrea	46,842	117,600	4,692,115	100.17	39.90
Estonia	17,413	45,230	1,342,300	77.09	29.68
Ethiopia	435,184	1,104,300	77,154,000	177.29	69.87
Fiji	7,054	18,270	833,330	118.14	45.61
Finland	130,127	338,150	5,266,300	40.47	15.57
France	176,460	551,500	61,256,600	347.14	111.07
Gabon	103,347	267,670	1,310,820	12.68	4.90
Gambia	4,363	11,300	1,663,031	381.17	147.17
Georgia	26,911	69,700	4,432,981	164.73	63.60
Germany	137,803	357,030	82,374,900	597.77	230.72
Ghana	92,098	238,540	23,008,443	249.83	96.46
Greece	50,942	131,960	11,147,100	218.82	84.47
Grenada	131	340	108,148	825.56	318.08
Guatemala	42,042	108,890	13,028,572	309.89	119.65
Guinea	94,926	245,860	9,181,335	96.72	37.34
Guinea-Bissau	13,948	36,120	1,645,529	117.98	45.56
Guyana	83,000	214,970	739,065	8.90	3.44
Haiti	10,714	27,750	9,445,947	881.65	340.39
Honduras	43,277	112,090	6,968,687	161.03	62.17
Hong Kong	422	1,092	6,857,100	16,249.05	6,279.40
Hungary	35,920	93,030	10,067,200	280.27	108.21
Iceland	39,768	103,000	301,800	7.59	2.93
India	1,269,340	3,287,260	1,109,811,147	874.32	337.61
Indonesia	741,097	1,904,570	223,041,632	300.96	117.11
Iran	636,294	1,648,200	70,097,913	110.17	42.53
Iraq (est. 2004)	168,754	438,320	24,699,540	146.36	56.47
Ireland	27,137	70,270	4,267,900	157.27	60.74
Israel	8,019	22,140	7,048,600	878.99	318.36
Italy	116,305	301,340	58,842,800	505.94	195.27
Jamaica	4,244	10,990	2,667,300	628.49	242.70
Japan	145,882	377,890	127,756,000	875.75	338.08
Jordan	35,445	89,210	5,537,600	156.23	62.07
Kazakhstan	1,049,156	2,724,900	15,308,084	14.59	5.62
Kenya	224,961	580,370	36,553,490	162.49	62.98
Kiribati	277	730	100,202	361.74	137.26
Korea, North	46,540	120,540	23,707,548	509.40	196.68
Korea, South	38,023	99,260	48,418,077	1,273.39	487.79

Table A (*continued*)
World Countries: Area, Population, and Population Density, 2008

COUNTRY	AREA		POPULATION	DENSITY	
	(MI.²)	(KM.²)	(ESTIMATED 12/2007)	(POP/MI.²)	(POP/KM.²)
Kuwait	6,880	17,820	2,599,444	377.83	145.87
Kyrgyzstan	76,641	199,900	5,192,100	67.75	25.97
Laos	91,429	236,800	5,759,402	62.99	24.32
Latvia	24,749	64,600	2,287,500	92.43	35.41
Lebanon	4,015	10,400	4,055,301	1,010.04	389.93
Lesotho	11,720	30,350	1,994,888	170.21	65.73
Liberia	43,000	111,370	3,578,925	83.23	32.14
Libya	679,362	1,759,540	6,038,643	8.89	3.43
Liechtenstein	62	160	34,900	562.90	218.12
Lithuania	25,174	65,300	3,394,500	134.84	51.98
Luxembourg	998	2,586	461,900	462.83	178.62
Macedonia	9,781	25,710	2,036,376	208.20	79.21
Madagascar	226,658	587,040	19,159,010	84.53	32.64
Malawi	45,747	118,480	13,570,713	296.65	114.54
Malaysia	127,317	329,750	26,113,731	205.11	79.19
Maldives	115	300	300,292	2,611.23	1,000.97
Mali	478,767	1,240,190	11,968,376	25.00	9.65
Malta	124	320	406,000	3,274.19	1,268.75
Marshall Islands	70	181	65,383	934.04	360.63
Mauritania	397,954	1,025,520	3,043,639	7.65	2.97
Mauritius	718	2,040	1,253,434	1,745.73	614.43
Mexico	761,603	1,958,200	104,221,361	136.84	53.22
Micronesia	271	702	110,617	408.18	157.57
Moldova	13,012	33,840	3,832,709	294.55	113.26
Monaco	1	2	32,600	26,942.15	16,717.95
Mongolia	604,427	1,566,500	2,584,655	4.28	1.65
Montenegro	5,395	14,026	601,022	111.40	42.85
Morocco	171,750	446,550	30,496,553	177.56	68.29
Mozambique	309,494	801,590	20,971,449	67.76	26.16
Myanmar	261,969	676,580	48,379,206	184.68	71.51
Namibia	318,259	824,290	2,046,555	6.43	2.48
Nepal	54,363	147,180	27,641,362	508.46	187.81
Netherlands	14,413	41,530	16,340,200	1,133.71	393.46
New Zealand	103,738	270,530	4,184,600	40.34	15.47
Nicaragua	49,998	130,000	5,532,364	110.65	42.56
Niger	489,191	1,267,000	13,736,722	28.08	10.84
Nigeria	356,669	923,770	144,719,953	405.75	156.66
Norway	125,182	323,760	4,660,100	37.23	14.39
Oman	82,030	309,500	2,546,325	31.04	8.23
Pakistan	310,402	796,100	159,002,039	512.25	199.73
Palau	177	460	20,200	114.12	43.91
Panama	30,193	75,520	3,287,538	108.88	43.53
Papua New Guinea	178,259	462,840	6,201,692	34.79	13.40
Paraguay	157,048	406,750	6,016,000	38.31	14.79
Peru	496,225	1,285,220	27,588,577	55.60	21.47

Table A (*continued*) World Countries: Area, Population, and Population Density, 2008					
	AREA		POPULATION	DENSITY	
COUNTRY	(MI.²)	(KM.²)	(ESTIMATED 12/2007)	(POP/MI.²)	(POP/KM.²)
Philippines	115,831	300,000	86,263,713	744.74	287.55
Poland	120,728	312,690	38,129,400	315.83	121.94
Portugal	35,552	91,980	10,589,300	297.85	115.13
Qatar	4,247	11,000	821,313	193.39	74.66
Romania	91,699	238,390	21,590,400	235.45	90.57
Russia	6,592,745	17,075,400	142,500,000	21.61	8.35
Rwanda	10,169	26,340	9,464,241	930.70	359.31
Samoa	1,104	2,840	185,366	167.90	65.27
San Marino	23	61	28,600	1,243.48	472.73
São Tomé and Príncipe	372	960	155,126	417.01	161.59
Saudi Arabia	756,982	2,149,690	23,678,849	31.28	11.02
Senegal	75,749	196,720	12,072,479	159.37	61.37
Serbia	297,987	77,474	7,439,000	24.96	96.02
Seychelles	175	450	84,600	483.43	188.00
Sierra Leone	27,699	71,740	5,742,694	207.32	80.05
Singapore	244	680	4,483,900	18,376.64	6,593.97
Slovakia	18,859	48,845	5,390,400	285.83	110.36
Slovenia	7,836	20,250	2,006,800	256.10	99.10
Solomon Islands	10,985	28,900	484,022	44.06	16.75
Somalia	246,201	637,660	8,445,397	34.30	13.24
South Africa	471,444	1,219,090	47,391,025	100.52	38.87
Spain	194,885	505,990	44,121,300	226.40	87.20
Sri Lanka	25,332	65,610	19,886,000	785.02	303.09
Sudan	967,500	2,505,810	37,707,483	38.97	15.05
Suriname	63,039	163,270	455,273	7.22	2.79
Swaziland	6,704	17,360	1,137,915	169.74	65.55
Sweden	173,732	449,960	9,083,800	52.29	20.19
Switzerland	15,943	41,290	7,491,200	469.87	181.43
Syria	71,498	185,180	19,407,558	271.44	104.80
Taiwan	13,892	35,980	22,942,926	1,651.52	637.66
Tajikistan	55,251	143,100	6,639,837	120.18	46.40
Tanzania	364,900	945,090	39,458,709	108.14	41.75
Thailand	198,456	513,120	63,443,950	319.69	123.64
Timor-Leste	5,741	14,870	1,029,195	179.27	69.21
Togo	21,925	56,790	6,410,428	292.38	112.88
Tonga	290	750	99,811	344.18	133.08
Trinidad and Tobago	1,980	5,130	1,328,432	670.93	258.95
Tunisia	63,170	163,610	10,128,100	160.33	61.90
Turkey	301,382	774,820	72,975,000	242.13	94.18
Turkmenistan	188,456	488,100	4,899,456	26.00	10.04
Uganda	93,135	241,040	29,898,598	321.02	124.04
Ukraine	233,090	603,700	46,787,750	200.73	77.50
United Arab Emirates	31,696	83,600	4,248,476	134.04	50.82
United Kingdom	94,525	242,910	60,550,100	640.57	249.27
United States	3,717,797	9,629,090	299,398,000	80.53	31.09

Table A (*continued*)
World Countries: Area, Population, and Population Density, 2008

COUNTRY	AREA		POPULATION	DENSITY	
	(MI.^{2})	(KM.^{2})	(ESTIMATED 12/2007)	(POP/MI.^{2})	(POP/KM.^{2})
Uruguay	68,039	176,220	3,314,466	48.71	18.81
Uzbekistan	172,742	447,400	26,540,312	153.64	59.32
Vanuatu	5,699	12,190	220,772	38.74	18.11
Venezuela	352,145	912,050	27,020,920	76.73	29.63
Vietnam	127,243	331,690	84,108,100	661.00	253.57
West Bank and Gaza	2,401	6,220	3,774,671	1,572.12	606.86
Yemen	203,850	527,970	21,732,247	106.61	41.16
Zambia	290,586	752,610	11,696,160	40.25	15.54
Zimbabwe	150,803	390,760	13,228,191	87.72	33.85
World	51,715,141	133,941,600	6,531,056,406	126.29	48.76

Source: *World Development Indicators 2008* (World Bank, 2008)

Table B
World Countries: Form of Government, Capital City, Major Languages

COUNTRY	GOVERNMENT	CAPITAL	MAJOR LANGUAGES
Afghanistan	Transitional muliti-party republic, verging on failed state	Kabul	Dari, Pashtu, Uzbek, Turkmen
Albania	Emerging multi-party democracy	Tiranë	Albanian, Greek
Algeria	Republic	Algiers	Arabic, Berber, dialects, French
Andorra	Parliamentary democracy	Andorra	Catalán, French, Spanish, Portuguese
Angola	Multi-party republic	Luanda	Portugese; Bantu and other African
Antigua and Barbuda	Constitutional parliamentary democracy	St. John's	English, local dialects
Argentina	Federal republic	Buenos Aires	Spanish, English, Italian, German, other
Armenia	Republic	Yerevan	Armenian, Russian, other
Australia	Federal parliamentary democracy	Canberra	English, indigenous
Austria	Federal republic	Vienna	German
Azerbaijan	Republic	Baku	Azerbaijani, Russian, Armenian, other
Bahamas	Parliamentary democracy; independent Commonwealth	Nassau	English, Creole
Bahrain	Constitutional hereditary monarchy	Al Manamah	Arabic, English, Farsi, Urdu
Bangladesh	Parliamentary democracy	Dhaka	Bangla, English
Barbados	Parliamentary democracy	Bridgetown	English
Belarus	Republic in name; dictatorship in fact	Minsk	Byelorussian, Russian, other
Belgium	Constitutional monarchy	Brussels	Dutch (Flemish), French, German
Belize	Parliamentary democracy	Belmopan	English, Spanish, Garifuna, Mayan
Benin	Multi-party republic	Porto-Novo	French, Fon, Yoruba
Bhutan	Transitional constitutional monarchy	Timphu	Dzongkha, Tibetan, Nepalese
Bolivia	Republic	La Paz, Sucre	Spanish, Quechua, Aymara
Bosnia-Herzegovina	Federal democratic republic	Sarajevo	Croatian, Serbian, Bosnian
Botswana	Parliamentary republic	Gaborone	English, Setswana
Brazil	Federal republic	Brasilia	Portugese, Spanish, English, French
Brunei	Constitutional monarchy	Bandar Seri Begawan	Malay, English, Chinese
Bulgaria	Parliamentary democracy	Sofia	Bulgarian
Burkina Faso	Parliamentary republic	Ouagadougou	French, indigenous
Burundi	Republic	Bujumbura	French, Kirundi, Swahili
Cambodia	Multi-party democracy under a constitutional monarchy	Phnom Penh	Khmer, French, English
Cameroon	Multi-party republic	Yaoundé	English, French, indigenous
Canada	Federal parliamentary	Ottawa	English, French, other
Cape Verde	Republic	Cidade de Praia	Portugese, Crioulu
Central African Republic	Republic	Bangui	French, Sangho
Chad	Republic	N'Djamena	French, Arabic, Sara, other indigenous
Chile	Republic	Santiago	Spanish
China	Single-party communist state	Beijing	Various Chinese dialects
Colombia	Republic/insurgency	Bogotá	Spanish
Comoros	Republic	Moroni	Arabic, French, Shikomoro
Congo, Dem. Rep.	Republic/transitional from military dictatorship/insurgencies	Kinshasa	French, Lingala, Kingwana, Kikongo, Tshilluba
Congo, Rep.	Multi-party republic	Brazzaville	French, Lingala, Monokutuba, Kikongo
Costa Rica	Democratic republic	San José	Spanish
Côte d'Ivoire	Multi-party republic/insurgencies	Abidjan, Yamoussoukro	French, indigenous
Croatia	Parliamentary democracy	Zagreb	Croatian
Cuba	Single-party communist state	Havana	Spanish

Table B (*continued*)
World Countries: Form of Government, Capital City, Major Languages

COUNTRY	GOVERNMENT	CAPITAL	MAJOR LANGUAGES
Cyprus	Republic	Nicosia	Greek, Turkish, English
Czech Republic	Parliamentary democracy	Prague	Czech
Denmark	Constitutional monarchy	Copenhagen	Danish, Faroese, Greenlandic, German
Djibouti	Republic	Djibouti	French, Somali, Afar, Arabic
Dominica	Parliamentary democracy, republic within Commonwealth	Roseau	English, French
Dominican Republic	Republic	Santo Domingo	Spanish
East Timor	Republic	Dili	Tetum, Portuguese, Indonesian, English
Ecuador	Republic	Quito	Spanish, Quechua, indigenous
Egypt	Republic	Cairo	Arabic
El Salvador	Republic	San Salvador	Spanish, Nahua
Equatorial Guinea	Republic	Malabo	Spanish, French, indigenous, English
Eritrea	Transitional government	Asmara	Afar, Amharic, Arabic, Tigre, other indigenous
Estonia	Parliamentary republic	Tallinn	Estonian, Russian, Ukranian, Finnish
Ethiopia	Federal republic	Addis Ababa	Amharic, Tigrinya, Orominga, Somali, Arabic, English
Fiji	Republic	Suva	English, Fijian, Hindustani
Finland	Republic	Helsinki	Finnish, Swedish
France	Republic	Paris	French
Gabon	Multi-party republic	Libreville	French, Fang, indigenous
Gambia	Multi-party democratic republic	Banjul	English, Mandinka, Wolof, Fula
Georgia	Republic	Tbilisi	Georgian, Russian, Armenian
Germany	Federal republic	Berlin	German
Ghana	Parliamentary democracy	Accra	English, indigenous
Greece	Parliamentary republic	Athens	Greek
Grenada	Parliamentary democracy	St. George's	English, French
Guatemala	Republic	Guatemala City	Spanish, Quiche, Cakchiquel, other indigenous
Guinea	Republic	Conakry	French, indigenous
Guinea-Bissau	Multi-party republic	Bissau	Portugese, Crioulo, indigenous
Guyana	Republic (within Commonwealth)	Georgetown	English, Creole, Hindi, Urdu, indigenous
Haiti	Republic	Port-au-Prince	Creole, French
Honduras	Republic	Tegucigalpa	Spanish, indigenous
Hungary	Parliamentary democracy	Budapest	Hungarian
Iceland	Republic	Reykjavík	Icelandic
India	Federal republic	New Delhi	English, Hindi, 14 other official
Indonesia	Republic	Jakarta	Bahasa Indonesian, English, Dutch, Javanese
Iran	Theocratic republic	Tehran	Farsi, Turkish, Kurdish
Iraq	Transitional parliamentary democracy/ insurgenicies/U.S. occupation	Baghdad	Arabic, Kurdish, Assyrian, Armenian
Ireland	Republic	Dublin	English, Irish Gaelic
Israel	Parliamentary democracy/insurgency	Jerusalem	Hebrew, Arabic, English
Italy	Republic	Rome	Italian
Jamaica	Parliamentary democracy	Kingston	English, Creole
Japan	Constitutional monarchy	Tokyo	Japanese
Jordan	Constitutional monarchy	Amman	Arabic

Table B (*continued*) World Countries: Form of Government, Capital City, Major Languages			
COUNTRY	GOVERNMENT	CAPITAL	MAJOR LANGUAGES
Kazakhstan	Republic	Astana	Kazakh, Russian
Kenya	Republic	Nairobi	English, Swahili, indigenous
Kiribati	Republic	Tarawa	English, I-Kiribati
Korea, North	Single-party communist state	Pyongyang	Korean
Korea, South	Republic	Seoul	Korean
Kuwait	Constitutional monarchy	Kuwait	Arabic, English
Kyrgyzstan	Republic	Bishkek	Kirghiz, Russian
Laos	Single-party communist state	Vientiane	Lao, French, English, indigenous
Latvia	Parliamentary democracy	Riga	Latvian, Lithuanian, Russian
Lebanon	Republic/insurgency	Beirut	Arabic, French, Armenian, English
Lesotho	Constitutional monarchy	Maseru	English, Sesotho, Zulu, Xhosa
Liberia	Republic/insurgency	Monrovia	English, indigenous
Libya	Single party/military dictatorship	Tripoli	Arabic
Liechtenstein	Constitutional monarchy	Vaduz	German
Lithuania	Parliamentary democracy	Vilnius	Lithuanian, Russian, Polish
Luxembourg	Constitutional monarchy	Luxembourg	French, Luxembourgian, German
Macedonia	Parliamentary democracy	Skopje	Macedonian, Albanian, Turkish, Serbo-Croatian
Madagascar	Republic	Antananarivo	Malagasy, French
Malawi	Multi-party democracy	Lilongwe	Chichewa, English, Tombuka
Malaysia	Constitutional monarchy	Kuala Lumpur	Malay, Chinese, English, indigenous
Maldives	Republic	Male	Dhivehi
Mali	Republic	Bamako	French, Bambara, indigenous
Malta	Parliamentary democracy	Valletta	English, Maltese
Marshall Islands	Constitutional government (free association with U.S.)	Majuro	English, Polynesian dialects, Japanese
Mauritania	Republic	Nouakchott	Arabic, Wolof, Pular, French, Solinke
Mauritius	Parliamentary democracy	Port Louis	English, Creole, French, Hindi, Urdu, Bojpoori, Hakka
Mexico	Federal republic	Mexico City	Spanish, indigenous
Micronesia	Constitutional government (free association with U.S.)	Palikir	English, Trukese, Pohnpeian, Yapese, others
Moldova	Republic	Chisinau	Moldavian, Russian, Gagauz
Monaco	Constitutional monarchy	Monaco	French, English, Italian, Monegasque
Mongolia	Republic	Ulaanbaatar	Khalkha Mongol, Turkic, Russian
Montenegro	Republic	Podgorica	Serbian, Bosnian, Albanian, Croatian
Morocco	Constitutional monarchy	Rabat	Arabic, Berber dialects, French
Mozambique	Republic	Maputo	Portuguese, indigenous
Myanmar	Military regime	Rangoon	Burmese, indigenous
Namibia	Republic	Windhoek	Afrikaans, English, German, indigenous
Nauru	Republic	Yaren district	Nauruan, English
Nepal	Constitutional monarchy	Kathmandu	Nepali, indigenous
Netherlands	Constitutional monarchy	Amsterdam	Dutch
New Zealand	Parliamentary democracy	Wellington	English, Maori
Nicaragua	Republic	Managua	Spanish, English, indigenous
Niger	Republic; transitional from military	Niamey	French, Hausa, Djerma
Nigeria	Federal republic/insurgencies	Abuja	English, Hausa, Fulani, Yorbua, Ibo

Table B (*continued*)
World Countries: Form of Government, Capital City, Major Languages

COUNTRY	GOVERNMENT	CAPITAL	MAJOR LANGUAGES
Norway	Constitutional monarchy	Oslo	Norwegian, Sami
Oman	Monarchy	Muscat	Arabic, English, Baluchi, Urdu
Pakistan	Federal republic	Islamabad	Punjabi, Sindhi, Siraiki, Pashtu, Urbu, English, others
Palau	Constitutional government (free association with U.S.)	Koror	English, Palauan, Sonsolorese, Tobi, Angaur, Japanese
Panama	Constitutional democracy	Panama	Spanish, English
Papua New Guinea	Parliamentary democracy	Port Moresby	Various indigenous, English, Motu
Paraguay	Constitutional republic	Asunción	Spanish, Guarani
Peru	Republic	Lima	Quechua, Spanish, Aymara
Philippines	Republic	Manila	English, Filipino, 8 major dialects
Poland	Republic	Warsaw	Polish
Portugal	Parliamentary democracy	Lisbon	Portuguese
Qatar	Traditional monarchy	Doha	Arabic, English
Romania	Republic	Bucharest	Romanian, Hungarian, German
Russia	Federation	Moscow	Russian, numerous other
Rwanda	Republic	Kigali	French, Kinyarwanda, English, Kiswahili
St. Kitts and Nevis	Constitutional monarchy	Basseterre	English
St. Lucia	Parliamentary democracy	Castries	English, French
St. Vincent and the Grenadines	Parliamentary monarchy independent within Commonwealth	Kingstown	English, French
Samoa	Constitutional monarchy	Apia	Samoan, English
San Marino	Republic	San Marino	Italian
São Tomé and Príncipe	Republic	São Tome	Portuguese
Saudi Arabia	Monarchy	Riyadh	Arabic
Senegal	Republic	Dakar	French, Wolof, indigenous
Serbia	Republic	Belgrade	Serbian, Albanian
Seychelles	Republic	Victoria	English, French, Creole
Sierra Leone	Constitutional democracy/insurgency	Freetown	English, Krio, Mende, Temne
Singapore	Parliamentary republic	Singapore	Chinese, English, Malay,Tamil
Slovakia	Parliamentary democracy	Bratislava	Slovak, Hungarian
Slovenia	Republic	Ljubljana	Slovenian, Serbo-Croatian, other
Solomon Islands	Parliamentary democracy	Honiara	English, indigenous
Somalia	No single permanent national government/failed state	Mogadishu	Arabic, Somali, English, Italian
South Africa	Republic	Pretoria	Afrikaans, English, Zulu, Xhosa, other
Spain	Parliamentary monarchy	Madrid	Castilian Spanish, Catalan, Galician, Basque
Sri Lanka	Republic/insurgency	Colombo	English, Sinhala, Tamil
Sudan	Provisional military	Khartoum	Arabic, Nubian, others
Suriname	Constitutional democracy	Paramaribo	Dutch, Sranang Tongo, English, Hindustani, Javanese
Swaziland	Monarchy within Commonwealth	Mbabane	English, siSwati
Sweden	Constitutional monarchy	Stockholm	Swedish
Switzerland	Federal republic	Bern	German, French, Italian, Romansch
Syria	Republic (under military regime)	Damascus	Arabic, Kurdish, Armenian, Aramaic
Taiwan	Multi-party democracy	Taipei	Mandarin Chinese, Taiwanese, Hakka
Tajikistan	Republic	Dushanbe	Tajik, Russian

Table B (*continued*)
World Countries: Form of Government, Capital City, Major Languages

COUNTRY	GOVERNMENT	CAPITAL	MAJOR LANGUAGES
Tanzania	Republic	Dar es Salaam	Kiswahili, English, Arabic, indigenous
Thailand	Constitutional monarchy	Bangkok	Thai, English
Togo	Republic/transitional	Lomé	French, indigenous
Tonga	Constitutional monarchy	Nuku'alofa	Tongan, English
Trinidad and Tobago	Parliamentary democracy	Port-of-Spain	English, Hindi, French, Spanish, Chinese
Tunisia	Republic	Tunis	Arabic, French
Turkey	Parliamentary republic	Ankara	Turkish, Kurdish, Arabic, Armenian, Greek
Turkmenistan	Republic	Ashkhabad	Turkmen, Russian, Uzbek, other
Tuvalu	Constitutional monarchy	Funafuti	Tuvaluan, English, Samoan, Kiribati
Uganda	Republic	Kampala	English, Luganda, Swahili, Arabic, indigenous
Ukraine	Republic	Kiev	Ukranian, Russian, Romanian, Polish, Hungarian
United Arab Emirates	Federated monarchy	Abu Dhabi	Arabic, English, Farsi, Hindi, Urdu
United Kingdom	Constitutional monarchy	London	English, Welsh, Scottish Gaelic
United States	Federal republic	Washington	English, Spanish
Uruguay	Republic	Montevideo	Spanish, Portunol/Brazilero
Uzbekistan	Republic	Tashkent	Uzbek, Russian, Kazakh, Tajik, other
Vanuatu	Republic	Port-Vila	English, French, Bislama (pidgin)
Venezuela	Federal republic	Caracas	Spanish, indigenous
Vietnam	Single-party communist state	Hanoi	Vietnamese, French, Chinese, English, Khmer
Yemen	Republic	San`aa	Arabic
Zambia	Republic	Lusaka	English, Tonga, Lozi, other indigenous
Zimbabwe	Parliamentary democracy/insurgency	Harare	English, Shona, Sindebele, other

Source: *The World Factbook* 2008 (CIA, Washington, DC, 2008).

Notes: Unless indicated otherwise, republics are multi-party. "Theocratic" normally refers to fundamentalist Islamic rule. "Transitional" governments are those still in the process of change from a previous form (eg. Single-party communist state to multi-party republic or parliamentary democracy). Those states within which powerful insurgent groups disrupt the normal functioning of the government have been identified as having insurgencies.

Table C
World Countries: Basic Economic Indicators

	GROSS NATIONAL INCOME (GNI) 2006		PURCHASING POWER PARITY (GNI) 2006			AVERAGE ANNUAL % GROWTH IN GDP			STRUCTURE OF ECONOMIC OUTPUT (GDP) 2006 (VALUE ADDED IN % OF GDP)			
COUNTRY	TOTAL ($US BILLIONS)	PER CAPITA ($US)	TOTAL ($INT BILLIONS)	PER CAPITA ($INT)	RANK	1995	2000	2006	AGRI-CULTURE	INDUSTRY	MANUFAC-TURING	SERVICES
Afghanistan	8.4	—	23.94	—	—	—	—	—	36.10	24.48	14.93	39.42
Albania	9.27	2,930	19.03	6,000	92	10.18	7.34	4.39	22.80	21.50	—	55.70
Algeria	107.32	3,030	198.02	5,940	93	1.82	0.74	1.47	8.46	61.46	5.65	30.09
Angola	39.66	1,970	64.48	3,890	114	7.24	0.41	15.26	8.90	69.67	4.25	21.44
Antigua and Barbuda	0.91	11,050	1.27	15,130	43	-6.63	1.09	10.10	—	22.86	2.02	73.52
Argentina	208.66	5,150	456.79	11,670	55	-4.07	-1.84	7.39	8.43	35.62	22.27	55.95
Armenia	6.49	1,920	14.91	4,950	103	9.09	6.52	13.61	19.64	43.61	16.79	36.75
Australia	745.17	35,860	702.54	33,940	15	2.87	0.74	1.01	—	27.97	11.00	68.95
Austria	321.91	39,750	298.44	36,040	12	1.60	3.24	2.53	1.72	30.90	19.40	67.38
Azerbaijan	17.17	1,840	46.08	5,430	97	-12.81	10.20	33.03	7.42	70.11	5.58	22.47
Bahamas	—	15,290	—	—	—	-1.49	3.51	—	—	—	—	—
Bahrain	—	10,740	15.55	23,930	30	1.02	2.97	—	—	—	—	—
Bangladesh	65.95	450	191.93	1,230	153	2.72	3.91	4.78	19.61	27.91	17.21	52.48
Barbados	—	8,480	—	—	—	0.94	1.89	—	3.68	17.97	7.13	78.34
Belarus	36.84	3,470	94.44	9,700	66	-10.11	6.12	10.43	9.30	42.01	32.78	48.69
Belgium	397.8	38,460	356.95	33,860	16	2.17	3.48	2.55	1.01	24.29	17.08	74.70
Belize	1.1	3,740	2.11	7,080	81	-2.23	9.24	3.52	13.97	21.00	12.27	65.03
Benin	4.71	530	10.91	1,250	151	1.10	2.61	0.90	32.20	13.42	7.51	54.38
Bermuda	—	—	—	—	—	3.72	—	—	—	—	—	—
Bhutan	0.94	1,430	2.60	4,000	113	7.97	4.29	6.50	22.34	37.90	7.37	39.77
Bolivia	10.8	1,100	35.63	3,810	117	2.35	0.42	2.71	13.97	34.15	14.72	51.88
Bosnia-Herzegovina	12.82	3,230	26.64	6,780	85	24.90	2.91	5.70	10.42	24.69	11.89	64.90
Botswana	9.94	5,570	21.79	11,730	54	1.87	6.51	0.93	1.96	53.12	3.61	44.92
Brazil	1,037.97	4,710	1,647.50	8,700	72	2.84	2.77	2.35	5.15	30.90	18.41	63.95
Brunei	11.56	26,930	19.06	49,900	3	1.75	0.43	2.91	0.70	73.39	10.48	25.91
Bulgaria	31.47	3,990	79.00	10,270	61	3.29	7.34	6.74	8.54	31.44	18.59	60.02
Burkina Faso	6.16	440	16.19	1,130	157	2.68	-1.20	3.24	30.67	20.61	12.55	40.88
Burundi	0.87	100	2.62	320	177	-9.19	-2.78	1.08	34.85	20.01	8.83	45.14
Cambodia	6.97	490	22.06	1,550	147	3.43	6.61	8.95	30.10	26.22	18.59	43.68
Cameroon	18.07	990	37.44	2,060	135	0.63	1.77	1.59	19.87	33.18	18.09	46.95
Canada	1,256.52	36,650	1,184.44	36,280	10	1.96	4.31	1.74	—	—	—	—
Cape Verde	1.1	2,130	1.34	2,590	128	4.93	4.14	3.69	9.09	16.86	4.78	74.05
Central African Republic	1.49	350	2.94	690	168	4.39	0.32	2.30	55.83	15.47	7.52	28.70
Chad	5.18	450	12.26	1,170	154	-1.96	-4.34	-2.62	20.52	54.82	5.34	24.67
Chile	126.44	6,810	185.62	11,300	57	8.83	3.21	3.10	4.11	47.68	13.54	48.21
China	2,656.44	2,000	6,119.05	4,660	109	9.70	7.64	10.08	11.71	48.37	33.48	39.91
Colombia	147.42	3,120	279.23	6,130	89	3.32	1.25	5.35	12.01	35.59	16.67	52.40
Comoros	0.4	660	0.70	1,140	155	1.45	-1.21	-1.64	45.18	11.81	4.19	37.99
Congo, Dem. Rep.	8.14	130	16.24	270	178	-2.30	-9.05	1.79	45.67	27.73	6.49	26.60
Congo, Rep.	2.28	550	5.95	1,860	137	1.10	4.82	4.11	4.21	73.48	4.87	22.32

Table C (*continued*)
World Countries: Basic Economic Indicators

COUNTRY	GROSS NATIONAL INCOME (GNI) 2006		PURCHASING POWER PARITY (GNI) 2006			AVERAGE ANNUAL % GROWTH IN GDP			STRUCTURE OF ECONOMIC OUTPUT (GDP) 2006 (VALUE ADDED IN % OF GDP)			
	TOTAL ($US BILLIONS)	PER CAPITA ($US)	TOTAL ($INT BILLIONS)	PER CAPITA ($INT)	RANK	1995	2000	2006	AGRI-CULTURE	INDUSTRY	MANUFAC-TURING	SERVICES
Costa Rica	21.43	4,980	40.56	9,220	69	1.39	-0.51	6.39	8.83	29.40	21.79	61.78
Côte d'Ivoire	16.76	880	29.79	1,580	144	4.01	-5.79	-0.91	22.68	26.32	18.26	50.99
Croatia	41.54	9,310	61.50	13,850	50	6.34	3.94	4.81	7.44	31.63	20.61	60.92
Cuba	—	—	—	—	—	1.96	5.30	—	—	—	—	—
Cyprus	17.79	23,270	19.33	25,060	28	4.19	3.94	2.24	—	—	—	—
Czech Republic	135.3	12,790	214.89	20,920	35	6.00	3.75	5.69	2.69	39.01	27.15	58.30
Denmark	279.18	52,110	196.73	36,190	11	2.61	3.18	2.81	1.65	25.98	14.18	72.37
Djibouti	0.85	1,060	1.79	2,180	133	-5.54	-2.38	3.02	3.51	16.38	2.52	80.11
Dominica	—	3,200	0.39	5,420	98	1.20	0.97	3.43	18.54	23.78	8.14	57.68
Dominican Republic	30.11	2,910	53.32	5,550	95	2.77	6.33	9.03	12.05	26.19	13.96	61.76
Ecuador	39.45	2,910	89.89	6,810	84	-0.13	1.41	2.79	6.71	34.58	9.00	58.71
Egypt	107.22	1,360	366.46	4,940	104	2.72	3.46	4.95	14.07	38.43	16.55	47.50
El Salvador	18.13	2,680	37.90	5,610	94	4.27	0.39	2.74	10.93	29.36	22.41	59.71
Equatorial Guinea	5.24	8,510	8.24	16,620	42	11.52	10.82	-7.76	2.72	94.32	8.59	2.96
Eritrea	1.08	190	3.18	680	170	1.98	-16.23	-4.47	17.47	22.99	8.69	59.54
Estonia	15.65	11,400	24.29	18,090	38	6.42	8.34	11.71	3.17	29.07	16.76	67.76
Ethiopia	13.28	170	48.95	630	172	2.93	2.73	6.19	47.32	13.50	5.30	39.19
Fiji	3.07	3,720	3.71	4,450	110	1.28	-2.42	2.94	14.98	25.78	14.94	59.24
Finland	211.57	41,360	174.66	33,170	19	3.54	4.79	5.05	2.64	32.45	23.06	64.91
France	2,265.51	36,560	1,974.93	32,240	22	1.79	3.43	1.35	2.05	20.75	12.37	77.20
French Polynesia	3.45	16,070	—	—	—	-1.39	2.22	—	—	—	—	—
Gabon	7.51	5,360	14.65	11,180	58	2.28	-3.86	-0.37	4.94	61.17	4.10	33.89
Gambia	0.5	290	1.84	1,110	159	-2.76	1.98	1.61	32.65	13.11	5.17	54.25
Georgia	7.49	1,580	17.18	3,880	115	4.51	3.00	10.40	12.97	24.90	12.75	62.14
Germany	2,929.33	36,810	2,692.34	32,680	21	1.73	3.06	2.89	0.97	29.97	23.24	69.06
Ghana	12.81	510	28.44	1,240	152	1.42	1.34	4.01	37.39	25.38	8.47	37.24
Greece	303.39	27,390	344.08	30,870	23	1.32	4.15	3.94	3.28	20.78	9.53	75.94
Grenada	0.38	3,670	0.73	7,160	80	1.53	6.84	-0.83	6.73	28.99	5.92	64.28
Guatemala	34.95	2,590	66.72	5,120	99	2.55	1.18	1.94	22.17	19.11	12.47	58.72
Guinea	3.27	400	10.41	1,130	157	1.16	-0.06	0.82	12.94	37.48	3.69	49.58
Guinea-Bissau	0.29	190	0.76	460	176	1.25	4.46	1.12	61.76	11.46	7.17	26.79
Guyana	0.86	1,150	2.52	3,410	123	4.77	-1.31	4.86	31.03	24.50	8.03	44.47
Haiti	4.33	430	10.07	1,070	160	-5.74	-1.32	0.68	—	—	—	—
Honduras	8.93	1,270	23.86	3,420	122	1.59	3.64	3.98	13.78	31.14	19.73	55.08
Hong Kong	190.46	29,040	268.78	39,200	7	1.86	9.23	6.07	0.06	9.31	3.42	90.63
Hungary	104.82	10,870	170.80	16,970	40	1.63	5.47	4.10	4.17	30.14	22.60	65.69
Iceland	14.86	49,960	10.18	33,740	17	-0.63	3.05	0.90	5.83	23.71	10.09	70.46
India	907.22	820	2,726.25	2,460	129	5.67	2.31	7.70	17.53	27.89	16.28	54.58
Indonesia	349.03	1,420	737.20	3,310	124	6.87	3.55	4.31	12.90	47.05	28.05	40.06
Iran	215.57	2,930	686.93	9,800	65	1.02	3.43	3.06	10.40	44.60	11.76	45.00
Iraq	—	—	—	—	—	—	—	—	—	—	—	—

Table C (*continued*)
World Countries: Basic Economic Indicators

COUNTRY	GROSS NATIONAL INCOME (GNI) 2006		PURCHASING POWER PARITY (GNI) 2006			AVERAGE ANNUAL % GROWTH IN GDP			STRUCTURE OF ECONOMIC OUTPUT (GDP) 2006 (VALUE ADDED IN % OF GDP)			
	TOTAL ($US BILLIONS)	PER CAPITA ($US)	TOTAL ($INT BILLIONS)	PER CAPITA ($INT)	RANK	1995	2000	2006	AGRI-CULTURE	INDUSTRY	MANUFAC-TURING	SERVICES
Ireland	189.84	44,830	148.21	34,730	13	8.48	7.93	3.01	2.08	36.07	24.53	61.85
Israel	138.98	20,170	168.06	23,840	31	3.87	5.88	3.21	—	—	—	—
Italy	1,845.92	31,990	1,704.89	28,970	24	2.82	3.53	1.46	2.07	26.57	18.13	71.37
Jamaica	9.34	3,560	18.81	7,050	82	1.52	0.02	2.01	5.90	32.80	12.68	61.29
Japan	4,490.96	38,630	4,195.92	32,840	20	1.57	2.68	2.21	1.50	29.88	21.05	68.61
Jordan	14.68	2,650	26.68	4,820	105	2.79	1.70	3.29	3.10	29.51	19.17	67.39
Kazakhstan	71.69	3,870	133.20	8,700	72	-6.58	10.13	9.54	5.88	42.10	12.44	52.02
Kenya	22.85	580	53.77	1,470	149	1.43	-1.96	3.34	27.09	18.82	11.46	54.09
Kiribati	0.12	1,240	0.62	6,230	88	3.36	-0.72	4.53	7.08	6.56	0.85	86.36
Korea, North	—	—	—	—	—	—	—	—	—	—	—	—
Korea, South	888.01	17,690	1,113.02	22,990	32	7.62	7.58	4.72	3.25	39.60	27.84	57.15
Kuwait	44.42	16,790	83.37	38,070	8	—	0.73	—	—	—	—	—
Kyrgyzstan	2.78	500	9.30	1,790	140	-6.45	4.36	1.71	32.99	20.10	12.88	46.91
Laos	3.03	500	10.05	1,740	142	4.31	3.82	5.78	42.01	32.46	20.86	25.53
Latvia	19.44	8,100	33.94	14,840	44	0.35	7.73	12.56	3.68	21.48	11.84	74.84
Lebanon	22.39	5,580	38.92	9,600	67	3.52	0.27	-1.10	6.70	23.72	11.39	69.59
Lesotho	1.87	980	3.60	1,810	139	2.90	0.96	6.42	16.30	43.22	17.93	40.48
Liberia	0.49	130	0.94	260	179	-7.53	18.95	3.67	65.97	15.78	12.42	18.24
Libya	50.36	7,290	70.24	11,630	56	—	-0.89	3.49	—	—	—	—
Lithuania	27.51	7,930	49.38	14,550	45	4.03	4.85	8.29	5.31	35.28	17.64	59.41
Luxembourg	33.39	71,240	28.12	60,870	1	0.02	6.96	5.02	0.35	14.63	8.84	85.02
Macao, China	6.2	14,250	9.19	20,840	36	2.26	4.30	15.52	0.00	18.74	3.87	85.10
Macedonia	6.21	3,070	15.98	7,850	77	-1.60	4.15	2.89	12.97	29.32	18.61	57.71
Madagascar	5.42	280	16.58	870	165	-1.28	1.73	2.06	27.54	15.25	13.40	57.21
Malawi	3.13	230	9.38	690	168	15.03	-1.36	4.69	34.15	19.70	13.59	46.14
Malaysia	146.11	5,620	317.44	12,160	52	7.08	6.42	4.03	8.71	49.94	29.79	41.35
Maldives	0.88	3,010	1.42	4,740	107	—	2.63	21.48	—	—	—	—
Mali	5.52	460	11.92	1,000	162	3.44	0.33	2.16	36.91	24.03	3.08	39.05
Malta	6.16	15,310	8.52	20,990	34	5.04	5.75	2.76	—	—	—	—
Marshall Islands	0.19	2,980	0.53	8,040	76	0.73	-2.75	-0.33	—	—	—	—
Mauritania	2.77	760	5.98	1,970	136	6.82	-1.06	8.74	13.13	47.81	5.05	39.06
Mauritius	6.39	5,430	13.34	10,640	60	3.18	3.01	2.70	5.56	26.87	19.13	67.57
Mexico	826.02	7,830	1,249.19	11,990	53	-7.86	5.10	3.63	3.87	26.74	18.04	69.39
Micronesia	0.26	2,390	0.67	6,070	91	1.26	8.49	-1.20	—	—	—	—
Moldova	3.76	1,080	10.20	2,660	127	-0.93	3.47	5.15	18.10	15.09	14.47	66.81
Monaco	—	—	—	—	—	—	—	—	—	—	—	—
Mongolia	3.05	1,000	7.26	2,810	125	4.84	-0.31	7.27	21.85	42.29	3.89	35.86
Montegro	2.46	4,130	5.37	8,930	70	—	2.87	17.52	7.80	17.60	12.60	74.60
Morocco	64.47	2,160	117.69	3,860	116	-8.13	0.44	6.75	15.68	27.77	16.47	56.54
Mozambique	6.14	310	13.92	660	171	-0.12	-1.41	5.71	28.31	25.93	15.54	45.76
Myanmar	—	—	23.24	510	175	5.51	8.65	—	—	—	—	—
Namibia	6.49	3,210	9.75	4,770	106	1.22	1.36	1.55	10.95	30.56	13.92	58.49

Table C (*continued*)
World Countries: Basic Economic Indicators

COUNTRY	GROSS NATIONAL INCOME (GNI) 2006		PURCHASING POWER PARITY (GNI) 2006			AVERAGE ANNUAL % GROWTH IN GDP			STRUCTURE OF ECONOMIC OUTPUT (GDP) 2006 (VALUE ADDED IN % OF GDP)			
	TOTAL ($US BILLIONS)	PER CAPITA ($US)	TOTAL ($INT BILLIONS)	PER CAPITA ($INT)	RANK	1995	2000	2006	AGRI-CULTURE	INDUSTRY	MANUFAC-TURING	SERVICES
Nepal	9.01	320	27.81	1,010	161	0.75	3.73	0.76	34.36	16.34	7.68	49.31
Netherlands	687.35	43,050	620.00	37,940	9	2.59	3.15	2.72	2.25	24.55	14.06	73.19
New Caledonia	2.68	14,020	—	—	—	2.71	1.13	—	—	—	—	—
New Zealand	105.46	26,750	107.74	25,750	27	2.67	1.54	0.67	—	—	—	—
Nicaragua	5.18	930	15.07	2,720	126	3.58	2.47	2.38	19.68	29.52	18.49	50.80
Niger	3.65	270	8.60	630	172	-0.97	-4.89	1.20	—	—	—	—
Nigeria	100.75	620	203.69	1,410	150	-0.34	2.67	2.75	23.35	56.78	—	19.87
Norway	334.89	68,440	233.33	50,070	2	3.63	2.54	2.09	1.56	44.87	9.40	53.57
Oman	19.11	6,720	34.74	14,460	46	2.05	3.95	—	—	—	—	—
Pakistan	129.34	800	382.81	2,410	130	2.41	1.78	4.75	19.39	27.20	19.47	53.41
Palau	0.16	7,990	0.29	14,340	48	—	—	5.18	3.13	19.00	0.38	76.87
Panama	16.05	5,000	28.57	8,690	74	-0.30	0.75	6.28	8.17	18.59	8.48	73.24
Papua New Guinea	5.07	740	10.11	1,630	143	-5.86	-3.75	0.42	—	—	—	—
Paraguay	9.3	1,410	24.32	4,040	112	3.03	-5.33	2.22	20.97	18.34	11.79	60.69
Peru	84.63	2,980	179.17	6,490	86	6.80	1.58	6.52	6.86	37.88	16.51	55.26
Philippines	128.04	1,390	296.22	3,430	121	2.40	3.77	3.37	14.18	31.63	22.90	54.19
Poland	325.36	8,210	543.37	14,250	49	6.88	4.81	6.23	4.54	31.52	18.79	63.94
Portugal	186.98	17,850	211.32	19,960	37	3.98	3.40	0.93	2.79	24.98	—	72.23
Puerto Rico	41.44	10,560	—	—	—	3.79	1.66	—	—	—	—	—
Qatar	8.38	—	—	—	—	—	—	—	—	—	—	—
Romania	118.37	4,830	219.22	10,150	62	7.40	2.17	7.92	10.51	37.94	25.50	51.55
Russia	958.38	5,770	1,814.89	12,740	51	-4.02	10.00	7.19	4.88	39.35	19.40	55.76
Rwanda	2.47	250	6.93	730	167	37.57	-0.72	2.74	41.00	21.20	8.49	37.81
Samoa	0.42	2,270	0.94	5,090	101	5.52	6.01	1.46	12.15	27.37	13.92	60.48
San Marino	1.36	45,130	—	—	—	—	—	3.53	—	—	—	—
São Tomé and Príncipe	0.12	800	0.23	1,490	148	—	—	5.27	16.99	20.75	6.44	62.26
Saudi Arabia	349.14	13,980	527.95	22,300	33	-2.42	2.52	1.83	3.01	65.03	9.48	31.96
Senegal	9.02	760	18.80	1,560	145	2.58	0.52	-0.26	15.87	23.01	13.93	61.13
Serbia	31.59	4,030	69.31	9,320	68	—	4.85	5.77	12.67	25.53	—	61.81
Seychelles	0.73	8,870	1.21	14,360	47	-2.64	3.33	3.18	3.00	25.55	15.25	71.45
Sierra Leone	1.41	240	3.53	610	174	-8.29	0.64	4.45	46.38	24.99	—	28.63
Singapore	127.98	28,730	194.14	43,300	5	4.92	8.17	4.46	0.09	34.74	29.19	65.17
Slovakia	52.96	9,610	91.95	17,060	39	5.52	0.84	8.20	3.58	31.62	19.84	64.80
Slovenia	36.72	18,660	48.11	23,970	29	3.59	3.92	4.87	2.32	34.59	24.61	63.09
Solomon Islands	0.34	690	0.90	1,850	138	4.00	-16.58	3.56	—	—	—	—
Somalia	—	—	—	—	—	—	—	—	—	—	—	—
South Africa	249.87	5,390	421.73	8,900	71	0.91	1.61	3.88	2.72	30.85	18.22	66.43
Spain	1,205.42	27,340	1,244.17	28,200	25	2.52	4.17	2.15	3.13	29.70	15.49	67.17
Sri Lanka	26.85	1,310	74.19	3,730	118	4.07	4.27	6.18	16.46	27.07	13.93	56.47
St. Kitts and Nevis	0.3	6,490	0.45	10,110	63	4.14	-1.90	4.95	3.00	27.62	9.71	69.38

Table C (*continued*)
World Countries: Basic Economic Indicators

COUNTRY	GROSS NATIONAL INCOME (GNI) 2006		PURCHASING POWER PARITY (GNI) 2006			AVERAGE ANNUAL % GROWTH IN GDP			STRUCTURE OF ECONOMIC OUTPUT (GDP) 2006 (VALUE ADDED IN % OF GDP)			
	TOTAL ($US BILLIONS)	PER CAPITA ($US)	TOTAL ($INT BILLIONS)	PER CAPITA ($INT)	RANK	1995	2000	2006	AGRI-CULTURE	INDUSTRY	MANUFAC-TURING	SERVICES
St. Lucia	0.61	3,910	1.10	7,040	83	1.34	-1.66	3.74	3.88	18.79	5.47	77.33
St. Vincent and the Grenadines	0.32	2,740	0.58	5,040	102	0.43	1.28	—	8.15	24.70	5.70	67.15
Sudan	34.54	800	67.19	1,780	141	3.28	5.97	9.41	32.29	28.52	6.15	39.19
Suriname	2.07	4,210	3.51	7,720	78	0.34	-0.99	5.11	5.16	35.75	14.19	59.09
Swaziland	2.66	2,400	5.35	4,700	108	0.58	-0.45	1.46	10.94	45.59	36.77	43.47
Sweden	385.17	43,530	311.71	34,310	14	3.31	4.20	3.50	1.44	29.01	—	69.55
Switzerland	417.65	58,050	305.90	40,840	6	-0.29	2.97	2.45	1.29	28.36	—	70.35
Syria	32.47	1,560	79.70	4,110	111	3.05	0.18	2.32	18.34	32.18	7.09	49.48
Tajikistan	2.72	390	10.33	1,560	145	-13.65	6.97	5.56	24.79	27.44	19.33	47.77
Tanzania	12.62	350	38.77	980	163	0.59	2.56	3.31	45.30	17.37	6.91	37.33
Thailand	202.12	3,050	472.23	7,440	79	8.01	3.76	4.29	10.70	44.62	35.05	44.68
Timor-Leste	0.85	840	5.25	5,100	100	—	13.12	-6.73	32.19	12.75	2.56	55.03
Togo	2.18	350	4.91	770	166	4.67	-4.13	1.31	43.63	23.96	10.14	32.41
Tonga	0.23	2,250	0.55	5,470	96	4.03	5.24	0.92	—	—	—	—
Trinidad and Tobago	17.2	12,500	22.32	16,800	41	3.26	5.71	11.59	0.63	61.66	5.84	37.71
Tunisia	28.66	2,970	65.72	6,490	86	0.73	3.52	4.17	11.35	28.38	17.42	60.27
Turkey	402.33	5,400	613.68	8,410	75	5.24	5.56	4.78	9.67	26.84	22.16	63.49
Turkmenistan	9.87	650	—	—	—	-9.23	17.08	—	—	—	—	—
Uganda	9.28	300	26.31	880	164	7.95	2.49	2.08	32.31	18.44	9.06	49.25
Ukraine	104.75	1,940	285.95	6,110	90	-11.50	6.97	7.83	8.74	34.57	20.71	56.68
United Arab Emirates	74.38	19,270	87.04	26,810	26	-0.74	-1.94	—	2.25	55.72	14.10	42.03
United Kingdom	2,417.09	40,560	2,037.23	33,650	18	2.76	2.05	2.21	0.93	24.09	13.61	74.98
United States	13,195.67	44,710	13,195.67	44,070	4	1.32	2.52	1.91	1.19	22.84	14.42	75.97
Uruguay	18.8	5,310	32.93	9,940	64	-2.17	-1.80	6.72	9.18	32.41	23.22	58.41
Uzbekistan	17.16	610	58.12	2,190	132	-2.67	2.75	5.79	26.14	27.40	10.76	46.46
Vanuatu	0.36	1,690	0.77	3,480	120	-2.49	0.59	4.58	15.62	9.17	3.43	81.71
Venezuela	180.35	6,070	296.38	10,970	59	1.78	1.79	8.52	—	—	—	—
Vietnam	59.67	700	194.44	2,310	131	7.59	6.62	6.88	20.36	41.56	21.25	38.08
West Bank and Gaza	4.19	1,580	14.05	3,720	119	1.93	-9.50	-2.61	—	—	—	—
Yemen	17.61	760	45.47	2,090	134	7.14	1.37	0.3	—	—	—	—
Zambia	9.73	630	13.35	1,140	155	-5.26	1.33	4.22	21.82	32.91	11.1	45.27
Zimbabwe	7.15	450	—	—	—	-1.77	-8.87	—	19.13	23.92	13.52	58.95
World	47,673.39	1,732	59,080.87	11,214	—	2.03	2.51	4.21	14.53	29.80	13.66	54.45

Table D
World Countries: Population Growth, 1950–2025

COUNTRY	POPULATION (THOUSANDS)			AVERAGE ANNUAL POPULATION CHANGE (PERCENT)		AVERAGE ANNUAL INCREMENT TO THE POPULATION (MID-YEAR POPULATION, IN THOUSANDS)		
	1950	2000	2025[A]	1975–1980	2001–2015[A]	1985–1990	1995–2000	2005–2010
WORLD	2,518,629.0	6,070,581.0	7,851,455.0	1.7	2.5	85,831.0	—	—
AFRICA								
Algeria	8,753.0	30,245.0	42,429.0	3.1	1.5	631.8	566.0	539.3
Angola	4,131.0	12,386.0	19,268.0	2.7	2.6	130.2	187.2	267.5
Benin	2,046.0	6,222.0	11,120.0	2.5	2.4	135.6	184.8	204.6
Botswana	419.0	1,725.0	1,614.0	3.5	0.5	43.3	21.5	-15.4
Burkina Faso	3,960.0	11,905.0	24,527.0	2.5	2.1	233.8	307.0	360.3
Burundi	2,456.0	6,267.0	12,328.0	2.3	1.7	95.2	123.0	166.5
Cameroon	4,466.0	15,127.0	20,831.0	2.8	1.7	326.5	370.9	374.8
Central African Republic	1,314.0	3,715.0	5,193.0	2.3	1.5	57.5	62.0	60.0
Chad	2,658.0	7,861.0	15,770.0	2.1	2.9	171.2	259.8	340.4
Congo, Dem. Rep.	12,184.0	48,571.0	95,448.0	3.0	2.6	1,146.2	1,234.4	1,907.8
Congo, Rep.	808.0	3,447.0	6,750.0	2.9	2.7	56.3	62.4	67.2
Côte d'Ivoire	2,775.0	15,827.0	22,140.0	3.9	1.6	411.1	350.2	394.2
Egypt	21,834.0	67,784.0	103,165.0	2.4	1.5	1,318.5	1,199.9	1,125.0
Equatorial Guinea	226.0	456.0	812.0	-0.7	—	8.7	11.1	13.6
Eritrea	1,140.0	3,712.0	7,261.0	2.6	2.3	35.4	134.9	153.2
Ethiopia	18,434.0	65,590.0	116,006.0	2.4	2.1	1,530.7	1,670.9	1,848.4
Gabon	469.0	1,258.0	1,915.0	3.1	2.2	11.6	13.9	8.6
Gambia	294.0	1,312.0	2,177.0	3.1	2.0	33.0	42.3	48.0
Ghana	4,900.0	19,593.0	30,618.0	1.9	1.6	435.4	380.2	285.2
Guinea	2,550.0	8,117.0	13,704.0	1.5	1.9	173.7	63.1	193.9
Guinea-Bissau	505.0	1,367.0	2,774.0	4.7	2.2	22.1	28.4	35.1
Kenya	6,265.0	30,549.0	39,917.0	3.8	1.4	723.5	604.9	206.8
Lesotho	734.0	1,785.0	1,608.0	2.5	0.8	41.5	39.6	11.5
Liberia	824.0	2,943.0	6,081.0	3.1	2.3	-2.9	236.3	107.3
Libya	1,029.0	5,237.0	7,785.0	4.4	1.9	92.8	92.2	136.3
Madagascar	4,230.0	15,970.0	30,249.0	2.5	2.5	308.1	433.2	590.5
Malawi	2,881.0	11,370.0	18,245.0	3.3	1.8	416.6	169.9	103.8
Mali	3,520.0	11,904.0	25,679.0	2.1	2.1	164.0	305.1	390.7
Mauritania	825.0	2,645.0	4,973.0	2.5	2.3	47.5	65.2	94.9
Morocco	8,953.0	29,108.0	40,721.0	2.3	1.4	565.7	535.1	515.0
Mozambique	6,442.0	17,861.0	25,350.0	2.8	1.6	78.7	359.2	75.1
Namibia	511.0	1,894.0	2,350.0	2.7	1.2	58.6	33.5	7.5
Niger	2,500.0	10,742.0	25,722.0	3.2	2.8	207.6	259.7	322.0
Nigeria	29,790.0	114,746.0	192,115.0	2.8	1.9	2,530.7	3,225.5	3,161.6
Rwanda	2,162.0	7,724.0	12,509.0	3.3	1.6	187.9	311.1	49.1
Senegal	2,500.0	9,393.0	15,663.0	2.8	2.0	191.6	278.9	336.2
Sierra Leone	1,944.0	4,415.0	7,593.0	2.0	1.9	106.3	143.8	162.5
Somalia	2,264.0	8,720.0	20,978.0	7.0	3.1	45.8	192.4	266.1
South Africa	13,683.0	44,000.0	42,962.0	2.2	0.4	943.4	383.4	-419.6
Sudan	9,190.0	31,437.0	47,536.0	3.1	2.0	634.6	902.5	1,059.5

Table D (*Continued*)
World Countries: Population Growth, 1950–2025

COUNTRY	POPULATION (THOUSANDS)			AVERAGE ANNUAL POPULATION CHANGE (PERCENT)		AVERAGE ANNUAL INCREMENT TO THE POPULATION (MID-YEAR POPULATION, IN THOUSANDS)		
	1950	2000	2025[A]	1975–1980	2001–2015[A]	1985–1990	1995–2000	2005–2010
Tanzania	7,886.0	34,837.0	53,435.0	3.1	1.7	799.7	832.1	970.9
Togo	1,329.0	4,562.0	7,551.0	2.7	1.9	122.0	160.9	115.3
Tunisia	3,530.0	9,586.0	12,843.0	2.6	1.3	168.9	124.3	104.8
Uganda	5,310.0	23,487.0	54,883.0	3.2	2.4	590.9	598.7	877.5
Zambia	2,440.0	10,419.0	14,401.0	3.4	1.2	210.9	174.1	190.9
Zimbabwe	2,744.0	12,650.0	12,857.0	3.0	0.6	308.9	78.1	-58.6
NORTH AND CENTRAL AMERICA								
Belize	69.0	240.0	356.0	1.7	—	5.0	6.3	7.2
Canada	13,737.0	30,769.0	36,128.0	1.2	0.6	369.8	331.8	289.5
Costa Rica	966.0	3,929.0	5,621.0	3.0	1.4	76.6	65.4	58.0
Cuba	5,850.0	11,202.0	11,479.0	0.9	0.3	93.2	48.4	37.3
Dominican Republic	2,353.0	8,353.0	10,955.0	2.4	1.3	141.1	136.4	147.1
El Salvador	1,951.0	6,209.0	8,418.0	2.1	1.6	87.1	110.8	117.7
Guatemala	2,969.0	11,428.0	19,456.0	2.5	2.4	255.9	317.6	366.3
Haiti	3,261.0	8,005.0	10,670.0	2.1	1.7	111.8	89.1	114.6
Honduras	1,380.0	6,457.0	10,115.0	3.4	2.1	117.2	151.1	134.9
Jamaica	1,403.0	2,580.0	3,263.0	1.2	1.1	18.3	16.8	23.9
Mexico	27,737.0	98,933.0	129,866.0	2.7	1.4	1,594.3	1,572.4	1,425.0
Nicaragua	1,134.0	5,073.0	8,318.0	3.1	2.1	91.0	107.6	100.9
Panama	860.0	2,950.0	4,290.0	2.5	1.3	44.8	39.8	32.5
Trinidad and Tobago	636.0	1,289.0	1,340.0	1.3	0.8	6.5	-4.9	-6.1
United States	157,813.0	285,003.0	358,030.0	0.9	0.8	2,296.3	2,503.8	2,429.2
SOUTH AMERICA								
Argentina	17,150.0	37,074.0	47,043.0	1.5	1.0	445.4	427.4	401.1
Bolivia	2,714.0	8,317.0	12,495.0	2.4	1.8	127.7	155.2	128.3
Brazil	53,975.0	171,796.0	216,372.0	2.4	1.1	2,756.3	1,975.6	1,285.4
Chile	6,082.0	15,224.0	19,651.0	1.5	1.0	212.2	189.7	148.3
Colombia	12,568.0	4,120.0	58,157.0	2.3	1.3	636.0	681.0	630.9
Ecuador	3,387.0	12,420.0	16,704.0	2.8	1.5	262.0	264.1	257.2
Guyana	423.0	759.0	724.0	0.7	—	-3.2	-3.5	4.1
Paraguay	1,488.0	5,470.0	9,173.0	3.2	2.1	113.5	141.6	162.8
Peru	7,632.0	25,952.0	35,622.0	2.7	1.3	472.9	491.4	432.4
Suriname	215.0	425.0	486.0	-0.5	—	3.9	3.3	1.4
Uruguay	2,239.0	3,342.0	3,875.0	0.6	0.6	19.4	23.7	26.7
Venezuela	5,094.0	24,277.0	31,189.0	3.4	1.5	465.5	397.3	351.7
ASIA								
Afghanistan	8,151.0	21,391.0	44,940.0	0.9	2.5	170.3	879.9	724.7
Armenia	1,354.0	3,112.0	2,866.0	1.8	0.3	-0.7	-13.8	7.6
Azerbaijan	2,896.0	8,157.0	10,222.0	1.6	0.7	103.6	23.6	61.8
Bangladesh	41,783.0	137,952.0	208,268.0	2.8	1.6	2,028.8	2,001.0	2,119.6
Bhutan	734.0	2,063.0	3,701.0	2.3	—	34.7	42.4	48.8
Cambodia	4,346.0	13,147.0	21,899.0	-1.8	1.5	313.1	271.6	314.6

Table D (*Continued*)
World Countries: Population Growth, 1950–2025

COUNTRY	POPULATION (THOUSANDS)			AVERAGE ANNUAL POPULATION CHANGE (PERCENT)		AVERAGE ANNUAL INCREMENT TO THE POPULATION (MID-YEAR POPULATION, IN THOUSANDS)		
	1950	2000	2025[A]	1975–1980	2001–2015[A]	1985–1990	1995–2000	2005–2010
China[B]	556,924.0	1,282,472.0	1,454,141.0	1.5	0.6	16,833.4	11,408.3	8,726.8
Georgia	3,527.0	5,262.0	4,429.0	0.7	-0.7	49.9	-53.5	-14.4
India	357,561.0	1,016,938.0	1,369,284.0	2.1	1.2	16,448.0	16,317.4	15,140.5
Indonesia	79,538.0	211,559.0	270,113.0	2.1	1.1	3,283.4	3,702.8	3,388.7
Iran	16,913.0	664,423.0	90,927.0	3.3	1.6	1,632.8	818.3	1,000.3
Iraq	5,158.0	23,224.0	41,707.0	3.3	1.9	488.2	623.7	719.5
Israel	1,258.0	6,042.0	8,598.0	2.3	1.5	87.4	107.5	73.6
Japan	83,625.0	127,034.0	123,444.0	0.9	-0.2	556.6	252.5	-30.4
Jordan	472.0	5,035.0	8,116.0	2.3	2.2	126.9	159.4	145.2
Kazakhstan	6,703.0	15,640.0	15,388.0	1.1	0.1	148.5	-42.0	85.9
Korea, North	10,815.0	22,268.0	24,665.0	1.6	0.6	307.4	27.2	4,294.4
Korea, South	18,859.0	46,835.0	50,165.0	1.6	0.4	943.5	459.2	322.0
Kuwait	152.0	2,247.0	3,930.0	6.2	2.1	81.8	70.6	90.4
Kyrgyzstan	1,740.0	4,921.0	6,484.0	1.9	1.1	76.8	30.0	80.9
Laos	1,755.0	5,279.0	8,635.0	1.2	2.2	110.7	130.3	155.3
Lebanon	1,443.0	3,478.0	4,554.0	-0.7	1.2	11.8	48.7	46.0
Malaysia	6,110.0	23,001.0	33,479.0	2.3	1.5	391.7	436.4	438.2
Mongolia	761.0	2,500.0	3,368.0	2.8	1.3	62.1	37.4	44.4
Myanmar	17,832.0	47,544.0	59,760.0	2.1	1.0	452.8	317.4	162.3
Nepal	8,643.0	23,518.0	37,831.0	2.5	2.0	457.5	559.0	616.3
Oman	456.0	2,609.0	4,785.0	5.0	2.2	58.3	80.5	104.3
Pakistan	39,659.0	142,654.0	249,766.0	2.6	2.2	2,984.4	2,984.8	2,936.8
Philippines	1,996.0	75,711.0	108,589.0	2.3	1.6	1,450.5	1,658.1	1,666.1
Saudi Arabia	3,201.0	22,147.0	39,751.0	5.6	2.9	527.8	678.3	922.2
Singapore	1,022.0	4,016.0	4,905.0	1.3	1.1	56.1	134.2	169.1
Sri Lanka	7,483.0	18,595.0	21,464.0	1.7	1.1	234.4	186.9	153.5
Syria	3,495.0	16,560.0	26,979.0	3.1	2.1	391.1	399.2	431.5
Tajikistan	1,532.0	6,089.0	8,193.0	2.8	1.5	149.0	115.3	168.7
Thailand	19,626.0	60,925.0	73,869.0	2.4	0.6	755.5	598.8	468.5
Turkey	21,484.0	68,281.0	88,995.0	2.1	1.1	1,083.1	895.5	732.4
Turkmenistan	1,211.0	4,643.0	6,549.0	2.5	1.1	85.4	83.3	95.8
United Arab Emirates	70.0	2,820.0	3,944.0	14.0	1.8	76.1	38.6	40.0
Uzbekistan	6,314.0	24,913.0	33,774.0	2.6	1.3	473.1	381.7	485.8
Vietnam	27,369.0	78,137.0	104,649.0	2.2	1.2	1,321.7	1,178.3	1,123.4
Yemen	4,316.0	18,017.0	43,204.0	3.2	3.0	436.3	524.0	782.1
EUROPE								
Albania	1,215.0	3,113.0	3,629.0	1.9	1.0	60.3	50.7	34.4
Austria	6,935.0	8,102.0	7,979.0	-0.1	-0.1	32.1	17.8	11.3
Belarus	7,745.0	10,034.0	8,950.0	0.6	-0.5	46.7	-7.5	-1.4
Belgium	8,639.0	10,251.0	10,516.0	0.1	0.0	22.2	20.9	5.3
Bosnia-Herzegovina	2,661.0	3,977.0	4,183.0	0.9	0.5	29.7	96.0	15.5
Bulgaria	7,251.0	8,099.0	6,609.0	0.3	-0.7	-9.9	-95.1	-74.3

Table D (*Continued*)
World Countries: Population Growth, 1950–2025

COUNTRY	POPULATION (THOUSANDS)			AVERAGE ANNUAL POPULATION CHANGE (PERCENT)		AVERAGE ANNUAL INCREMENT TO THE POPULATION (MID-YEAR POPULATION, IN THOUSANDS)		
	1950	2000	2025[A]	1975–1980	2001–2015[A]	1985–1990	1995–2000	2005–2010
Croatia	3,850.0	4,446.0	4,088.0	0.5	-0.3	10.1	-34.6	11.2
Czech Republic	8,925.0	10,269.0	9,806.0	0.6	-0.2	-0.1	-10.6	-14.7
Denmark	4,271.0	5,322.0	5,469.0	0.2	0.1	5.5	20.8	12.0
Estonia	1,101.0	1,369.0	1,017.0	0.6	-0.5	7.0	-10.5	-4.9
Finland	4,009.0	5,177.0	5,289.0	0.3	0.1	16.9	12.3	4.8
France	41,829.0	59,296.0	64,165.0	0.4	0.3	312.8	236.0	142.8
Germany	68,376.0	82,282.0	81,959.0	-0.1	-0.2	339.1	229.8	152.4
Greece	7,566.0	10,903.0	10,707.0	1.3	0.1	44.5	22.4	11.0
Hungary	9,338.0	10,012.0	8,865.0	0.3	-0.6	-55.4	-31.4	-31.1
Iceland	143.0	282.0	325.0	0.9	—	2.7	1.8	1.1
Ireland	2,969.0	3,819.0	4,668.0	1.4	0.8	-6.4	37.2	31.9
Italy	47,104.0	57,536.0	52,939.0	0.4	-0.4	4.0	74.2	-67.5
Latvia	1,949.0	2,373.0	1,857.0	0.4	-0.7	12.3	-23.5	-12.9
Lithuania	2,567.0	3,501.0	3,035.0	0.7	-0.2	22.2	-10.4	-3.6
Macedonia	1,230.0	2,024.0	2,199.0	1.4	0.4	6.9	11.0	7.2
Moldova	2,341.0	4,283.0	4,096.0	0.9	-0.2	49.9	-5.8	16.0
Netherlands	10,114.0	15,898.0	17,123.0	0.7	0.4	92.0	86.6	62.5
Norway	3,265.0	4,473.0	4,859.0	0.4	0.4	17.9	24.4	18.2
Poland	24,824.0	38,671.0	37,337.0	0.9	0.0	178.7	8.5	11.2
Portugal	8,405.0	10,016.0	9,834.0	1.4	-0.1	5.1	15.9	9.6
Romania	16,311.0	22,480.0	20,806.0	0.9	-0.3	69.0	-56.3	-49.8
Russia	102,192.0	145,612.0	124,428.0	0.6	-0.5	820.8	-422.7	-281.7
Serbia and Montenegro[C]	7,131.0	10,555.0	10,230.0	0.9	0.1	21.0	-5.2	-4.6
Slovakia	3,463.0	5,391.0	5,397.0	1.0	0.0	23.6	9.3	6.1
Slovenia	1,473.0	1,990.0	1,859.0	1.0	-0.2	4.6	3.6	1.2
Spain	28,009.0	40,752.0	40,369.0	1.1	0.0	163.2	49.1	-2.8
Sweden	7,014.0	8,856.0	9,055.0	0.3	0.0	40.5	9.5	0.5
Switzerland	4,694.0	7,173.0	6,801.0	-0.1	-0.1	54.8	19.2	7.9
Ukraine	37,298.0	49,688.0	40,775.0	0.4	-0.7	142.7	-432.6	-264.4
United Kingdom	49,816.0	58,689.0	63,275.0	0.0	0.0	189.0	178.9	94.6
OCEANIA								
Australia	8,219.0	19,153.0	23,205.0	0.9	0.7	246.8	209.7	167.0
Fiji	289.0	814.0	965.0	1.9	—	7.8	11.3	12.8
New Zealand	1,908.0	3,784.0	4,379.0	0.2	0.5	12.3	50.8	38.5
Papua New Guinea	1,798.0	5,334.0	8,443.0	2.5	1.9	89.3	116.4	125.1
Solomon Islands	90.0	437.0	783.0	3.5	—	11.0	13.8	14.3

Source: United Nations Population Division and International Labour Organisation. World Resources Institute, World Resources 2000–2001; U.S. Bureau of the Census International Database (2000).

[A] Data include projections based on 1990 base year population data

[B] Includes Hong Kong and Macao

[C] Separate data are not available for Montenegro and Serbia in this longitudinal analysis

Table E
World Countries: Basic Demographic Data, 1990–2006

COUNTRY	CRUDE BIRTH RATE (BIRTHS PER 1,000 POPULATION)		LIFE EXPECTANCY AT BIRTH (YEARS)		LIFE EXPECTANCY OF FEMALES AND MALES				TOTAL FERTILITY RATE		PERCENTAGE OF POPULATION IN SPECIFIC AGE GROUPS					
					FEMALES		MALES				1990			2006		
	1990	2006	1990	2006	1990	2006	1990	2006	1990	2006	0–14	15–64	65 AND OLDER	0–14	15–64	65 AND OLDER
Afghanistan	48.58	—	41.52	—	41.64	—	41.40	—	6.90	—	44.08	52.99	2.93	—	—	—
Albania	23.87	16.45	71.89	76.34	74.91	79.57	69.02	73.26	2.9	1.37	32.78	61.9	5.33	25.5	65.83	8.67
Algeria	31.18	20.79	67.02	71.99	68.3	73.41	65.81	70.63	4.59	2.41	43.03	53.36	3.61	28.94	66.48	4.58
Angola	52.72	47.57	40.35	42.36	42.06	43.96	38.71	40.84	7.14	6.5	47.29	50.17	2.54	46.28	51.31	2.42
Antigua and Barbuda	20.1	—	73.75	—	76.35	—	71.28	—	—	—	—	—	—	—	—	—
Argentina	21.63	17.62	71.67	75.03	75.33	78.87	68.19	71.37	2.96	2.27	30.73	60.33	8.94	26.12	63.63	10.26
Armenia	20.51	12.27	68.46	71.6	71.52	75.03	65.55	68.32	2.46	1.3	30.37	64.01	5.62	20.01	67.9	12.09
Australia	15.4	12.9	76.99	81	80.16	83.41	73.98	78.7	1.9	1.8	21.9	66.95	11.15	19.28	67.44	13.28
Austria	11.6	9.2	75.53	79.84	78.89	82.68	72.33	77.13	1.45	1.41	17.43	67.63	14.94	15.56	68.03	16.41
Azerbaijan	25.9	17.8	70.8	72.33	74.8	75.2	67	69.6	2.74	2.3	34.26	61.5	4.23	24.24	68.53	7.23
Bahamas	23.66	17.14	69.54	72.91	73.28	75.77	65.97	70.19	2.61	2.04	32.51	62.96	4.53	27.08	66.47	6.45
Bahrain	27.97	17.57	72.16	75.66	74.14	77.28	70.28	74.11	3.65	2.33	31.46	66.31	2.23	25.81	71.08	3.11
Bangladesh	34.85	25.42	54.79	63.66	55.15	64.58	54.45	62.78	4.32	2.91	40.74	56.25	3.01	34.72	61.67	3.61
Barbados	15.6	11.22	74.47	76.77	77.06	79.62	72	74.06	1.66	1.5	23.95	66.62	9.43	18.48	72.27	9.26
Belarus	13.9	9.49	70.84	68.59	75.6	74.48	66.3	62.98	1.91	1.29	23.07	66.28	10.65	15.26	70.43	14.31
Belgium	12.4	11.5	75.97	79.48	79.4	82.4	72.7	76.7	1.62	1.73	18.06	66.99	14.95	16.9	65.78	17.32
Belize	34.82	26.35	72.41	71.9	73.73	74.31	71.16	69.6	4.49	2.9	44.27	51.47	4.26	37.11	58.73	4.16
Benin	47.24	40.58	53.21	56.2	54.53	57.35	51.95	55.1	6.74	5.51	47.13	49.99	2.87	43.96	53.34	2.7
Bermuda	14.6	13.02	—	78.58	—	81.34	69.89	75.95	—	1.75	—	—	—	—	—	—
Bhutan	37.61	19.26	53.09	65.26	54.69	67.04	51.57	63.57	5.71	2.34	42.58	54.12	3.29	31.72	63.57	4.71
Bolivia	36.22	27.93	58.94	65.18	60.73	67.36	57.23	63.1	4.88	3.59	40.93	55.31	3.75	37.67	57.75	4.58
Bosnia-Herzegovina	14.53	8.91	72.05	74.57	74.86	77.28	69.38	71.99	1.68	1.18	24.19	69.7	6.11	17.32	68.63	14.05
Botswana	34.42	25.09	62.99	49.78	65.56	50.04	60.55	49.54	4.64	2.95	45.01	52.28	2.71	35.09	61.5	3.42
Brazil	24.09	19.46	66.59	72.08	70.44	75.84	62.93	68.51	2.8	2.27	35.3	60.3	4.4	27.59	66.15	6.26
Brunei	27.77	21.94	74.17	77.13	76.39	79.54	72.05	74.84	3.2	2.34	34.49	62.82	2.69	29.21	67.61	3.18
Bulgaria	12.1	9.4	71.64	72.61	75.15	76.3	68.3	69.1	1.81	1.38	20.32	66.54	13.14	13.57	69.18	17.25
Burkina Faso	48.93	44.41	49.98	51.87	51.61	53.46	48.43	50.36	7.27	6.08	47.59	48.42	3.99	45.99	50.96	3.06
Burundi	47.19	46.51	46.07	49.05	47.95	50.49	44.28	47.67	6.8	6.8	45.42	51.51	3.07	44.72	52.7	2.57
Cambodia	42.84	26.64	54.94	58.93	56.61	61.4	53.35	56.58	5.73	3.27	44.66	52.6	2.73	36.71	60.11	3.18
Cameroon	41.68	35.2	54.61	50.28	56.25	50.73	53.05	49.85	5.86	4.43	45.36	51.1	3.54	41.47	54.99	3.53
Canada	15	10.69	77.38	80.36	80.65	82.78	74.26	78.05	1.83	1.49	20.69	68.04	11.27	17.33	69.4	13.27
Cape Verde	38.06	29.3	65.29	71	68	74.16	62.7	67.99	5.36	3.45	47.61	48.08	4.31	38.99	56.8	4.21
Central African Republic	41.86	36.5	49.81	44.38	52.47	45.82	47.28	43.01	5.64	4.65	43.44	52.58	3.98	42.48	53.66	3.85
Chad	47.54	45.86	51.45	50.6	53.22	51.99	49.76	49.27	6.67	6.26	45.91	50.68	3.41	46.22	50.86	2.93
Chile	22.55	15.18	73.66	78.29	76.77	81.38	70.71	75.35	2.59	1.96	29.91	64.01	6.08	24.29	67.41	8.3
China	21.06	11.9	68.87	72	70.47	73.86	67.36	70.23	2.1	1.78	27.72	66.84	5.44	21.08	71.13	7.79
Colombia	26.77	19.22	68.4	72.59	72.49	76.38	64.5	68.99	3.03	2.27	36.18	59.57	4.25	29.8	64.97	5.23
Comoros	45.8	34.6	55.97	63.24	57.4	64.38	54.6	62.15	5.8	4.03	46.99	50.57	2.43	41.76	55.54	2.7

Table E (*continued*)
World Countries: Basic Demographic Data, 1990–2006

COUNTRY	CRUDE BIRTH RATE (BIRTHS PER 1,000 POPULATION)		LIFE EXPECTANCY AT BIRTH (YEARS)		LIFE EXPECTANCY OF FEMALES AND MALES				TOTAL FERTILITY RATE		PERCENTAGE OF POPULATION IN SPECIFIC AGE GROUPS					
					FEMALES		MALES				1990			2006		
	1990	2006	1990	2006	1990	2006	1990	2006	1990	2006	0–14	15–64	65 AND OLDER	0–14	15–64	65 AND OLDER
Congo, Dem. Rep.	48.61	44	46.28	46.12	48.26	47.46	44.39	44.84	6.7	6.3	46.89	50.37	2.74	47.26	50.14	2.6
Congo, Rep.	38.93	35.5	56.87	54.79	59.45	56.12	54.41	53.52	5.34	4.55	44.18	52.62	3.2	41.88	54.92	3.21
Costa Rica	26.38	18.04	75.8	78.66	78.17	81.1	73.55	76.32	3.12	2.14	35.9	59.21	4.89	27.76	66.31	5.93
Côte d'Ivoire	44.29	35.74	53.3	48.08	55.55	49	51.16	47.21	6.53	4.58	46.09	51.31	2.6	41.39	55.41	3.2
Croatia	11.7	9.3	72.17	75.82	75.93	79.3	68.59	72.5	1.63	1.38	20.5	68.16	11.34	15.27	67.41	17.32
Cuba	16.16	10.66	74.73	78.04	76.67	80.14	72.89	76.03	1.73	1.52	22.48	68.62	8.9	18.87	69.72	11.41
Cyprus	18.3	11.2	76.3	79.29	78.6	81.7	74.1	77	2.42	1.38	25.82	63.26	10.93	19.37	68.33	12.3
Czech Republic	12.7	10.2	71.38	76.48	75.42	79.67	67.54	73.45	1.89	1.33	21.42	66.03	12.55	14.46	71.2	14.34
Denmark	12.3	12	74.81	78.1	77.73	80.43	72.02	75.88	1.67	1.85	17.03	67.37	15.6	18.66	65.98	15.35
Djibouti	41.38	29.22	50.99	54.48	52.54	55.73	49.52	53.28	6.07	4.06	44.03	53.58	2.39	37.87	59.1	3.04
Dominica	23	—	73.16	—	75	—	71.4	—	2.7	—	—	—	—	—	—	—
Dominican Republic	28.84	23.82	68.01	72.03	70.56	75.2	65.57	69.01	3.31	2.84	37.74	58.46	3.8	33.23	61.11	5.66
Ecuador	28.9	21.43	69.03	74.83	71.53	77.86	66.65	71.96	3.64	2.63	38.95	56.89	4.15	32.24	61.71	6.05
Egypt	31.47	24.47	62.16	71.01	63.62	73.31	60.76	68.83	4.28	2.95	40.86	55.28	3.86	32.99	62.14	4.87
El Salvador	30	23.28	65.61	71.54	69.83	74.68	61.59	68.55	3.67	2.72	40.79	54.99	4.22	33.71	60.71	5.58
Equatorial Guinea	41.86	38.79	46.79	51.1	48.44	52.36	45.23	49.9	5.89	5.41	41.11	54.62	4.27	42.31	53.61	4.08
Eritrea	40.6	39.52	48.57	57.31	50.83	59.77	46.41	54.98	6.22	5.14	46.32	51.05	2.62	42.91	54.75	2.34
Estonia	14.2	11	69.48	72.57	74.6	78.14	64.6	67.27	2.04	1.55	22.18	66.27	11.55	14.95	68.37	16.68
Ethiopia	47.16	38.72	47.55	52.48	49.12	53.82	46.05	51.2	6.79	5.33	45.77	51.52	2.71	44.15	52.94	2.91
Fiji	28.32	21.52	66.69	68.59	68.84	70.89	64.64	66.4	3.4	2.8	37.86	59.03	3.11	32.56	63.12	4.32
Finland	13.1	11.1	74.81	79.23	78.88	82.83	70.94	75.8	1.78	1.84	19.32	67.3	13.38	17.2	66.71	16.1
France	13.4	12.8	76.75	80.56	80.93	84.11	72.76	77.17	1.78	1.98	20.25	65.77	13.99	18.34	65.35	16.31
French Polynesia	27.17	18.54	69.55	73.97	72.14	76.59	67.07	71.47	3.26	2.29	35.32	61.44	3.24	26.95	67.72	5.34
Gabon	35.56	26.13	60.94	56.74	62.59	57.21	59.36	56.28	4.7	3.12	40.62	53.75	5.63	35.39	59.97	4.65
Gambia	43.33	35.57	51.33	59.15	52.39	60.13	50.31	58.21	6	4.79	43.08	54.13	2.79	41	55.22	3.78
Georgia	15.9	10.86	70.25	70.72	74.18	74.69	66.5	66.95	2.08	1.35	24.6	66.07	9.33	18.35	67.25	14.4
Germany	11.4	8.2	75.21	79.13	78.58	82	72	76.4	1.45	1.34	16.08	68.96	14.96	14.15	66.63	19.23
Ghana	39.2	30.16	56.86	59.7	58.1	60.14	55.68	59.27	5.71	3.95	44.61	52.36	3.03	38.6	57.74	3.66
Greece	10.1	9.7	76.94	79.41	79.5	81.95	74.5	77	1.4	1.38	19.25	67.06	13.69	14.16	67.39	18.45
Grenada	22.64	19.27	—	—	—	—	—	—	—	2.44	38.89	53.25	7.86	33.47	59.72	6.81
Guam	25.95	18.9	72.18	75.36	74.55	77.76	69.92	73.08	3.12	2.58	30.17	65.96	3.87	29.03	64.3	6.68
Guatemala	39.18	33.75	62.51	69.91	65.54	73.52	59.63	66.48	5.55	4.24	45.39	51.27	3.34	42.88	52.8	4.32
Guinea	46.45	40.22	47.37	55.52	48.84	57.15	45.98	53.96	6.63	5.52	44.38	52.46	3.16	43.26	53.66	3.09
Guinea-Bissau	49.71	49.65	42.46	46.19	44.15	47.78	40.85	44.67	7.1	7.08	45.15	51.43	3.42	47.58	49.43	3
Guyana	24.87	17.91	62.3	66.28	65.65	69.24	59.12	63.46	2.61	2.35	36.63	58.77	4.6	31.15	62.98	5.87
Haiti	36.96	28.3	54.52	60.32	56.04	62.2	53.07	58.53	5.37	3.64	43.16	52.92	3.92	37.54	58.28	4.19

Table E (*continued*)
World Countries: Basic Demographic Data, 1990–2006

COUNTRY	CRUDE BIRTH RATE (BIRTHS PER 1,000 POPULATION)		LIFE EXPECTANCY AT BIRTH (YEARS)		LIFE EXPECTANCY OF FEMALES AND MALES				TOTAL FERTILITY RATE		PERCENTAGE OF POPULATION IN SPECIFIC AGE GROUPS					
					FEMALES		MALES				1990			2006		
	1990	2006	1990	2006	1990	2006	1990	2006	1990	2006	0–14	15–64	65 AND OLDER	0–14	15–64	65 AND OLDER
Honduras	38.16	28.34	65.88	69.89	68.4	73.47	63.48	66.48	5.1	3.39	45.45	51.23	3.32	39.42	56.41	4.17
Hong Kong	12	9.6	77.38	81.63	80.3	84.6	74.6	78.8	1.27	0.97	21.5	70.04	8.46	14.76	73.17	12.07
Hungary	12.1	9.9	69.32	73.09	73.7	77.35	65.14	69.03	1.84	1.35	20.23	66.42	13.35	15.5	69.14	15.36
Iceland	18.7	13.9	76.61	81.17	78.4	83	74.9	79.43	2.31	2.07	24.95	64.41	10.64	21.82	66.35	11.83
India	30.2	23.5	59.13	64.47	59.22	65.95	59.04	63.06	3.8	2.54	37.78	58.28	3.94	32.54	62.42	5.05
Indonesia	25.4	19.64	61.71	68.16	63.5	70.04	60	66.37	3.1	2.23	35.82	60.4	3.78	28.02	66.34	5.64
Iran	33.94	18.3	64.8	70.65	65.95	72.28	63.71	69.1	4.84	2.06	44.69	51.78	3.53	27.76	67.77	4.47
Iraq	39.23	—	61.75	—	63.34	—	60.24	—	5.88	—	44.61	52.31	3.08	—	—	—
Ireland	15.1	15.1	74.58	79.39	77.4	81.8	71.9	77.1	2.12	1.88	27.33	61.32	11.35	20.73	68.16	11.11
Israel	22.2	21	76.61	80.02	78.4	82	74.9	78.14	2.82	2.74	31.29	59.62	9.08	27.87	62.04	10.09
Italy	10	9.7	76.86	81.08	80.26	83.98	73.62	78.32	1.26	1.35	15.87	68.81	15.32	13.94	66.12	19.94
Jamaica	25.2	17.04	71.5	71.12	73.32	72.74	69.76	69.59	2.94	2.34	35.16	57.47	7.36	31.31	61.23	7.45
Japan	10	8.7	78.84	82.32	81.91	85.81	75.91	79	1.54	1.32	18.4	69.61	11.99	13.78	65.97	20.25
Jordan	36.16	29.1	67.52	72.2	68.84	73.83	66.26	70.66	5.43	3.21	46.82	50.01	3.17	36.53	60.2	3.26
Kazakhstan	21.7	19.7	68.34	66.16	73.1	72	63.8	60.6	2.72	2.12	31.47	62.68	5.85	23.88	68.17	7.95
Kenya	41.16	39.21	59.38	53.44	61.44	54.55	57.42	52.39	5.84	4.97	48.87	48.47	2.66	42.63	54.71	2.66
Kiribati	32.18	—	56.8	—	59.1	—	54.6	—	4.04	—	—	—	—	—	—	—
Korea, North	20.65	13.61	69.92	66.97	73.56	69.16	66.46	64.88	2.39	1.86	26.17	69.15	4.67	23.65	67.55	8.81
Korea, South	15.4	9.2	71.29	78.5	75.5	82.03	67.29	75.14	1.59	1.13	25.84	69.16	5	18.12	72.04	9.84
Kuwait	24.8	20.91	74.95	77.66	76.88	79.84	73.12	75.59	3.5	2.33	36.62	62.17	1.21	23.57	74.56	1.86
Kyrgyzstan	29.3	23.3	68.3	67.7	72.6	72.1	64.2	63.5	3.69	2.41	37.57	57.43	5	30.39	63.84	5.77
Laos	42.49	27.12	54.58	63.86	55.84	65.26	53.38	62.53	6.08	3.29	44.67	51.94	3.39	38.95	57.54	3.51
Latvia	14.2	9.5	69.27	70.86	74.6	76.5	64.2	65.49	2.02	1.36	21.44	66.73	11.83	14	69.18	16.81
Lebanon	25.56	18.39	68.8	71.78	71.16	73.99	66.56	69.68	3.13	2.23	35.73	59.07	5.21	28.23	64.48	7.29
Lesotho	36.25	29.48	58.56	42.93	60.49	42.94	56.72	42.92	4.87	3.46	44.94	50.76	4.3	40.14	55.13	4.73
Liberia	49.92	49.62	42.89	45.27	44.08	46.2	41.75	44.38	6.9	6.78	47	50.56	2.44	47.01	50.8	2.18
Libya	26.62	23.53	68.2	73.98	70.58	76.63	65.94	71.46	4.72	2.79	43.24	54.24	2.51	30.24	65.86	3.9
Lithuania	15.3	9.1	71.16	71.04	76.17	77.06	66.39	65.31	2.03	1.3	22.55	66.53	10.92	16.2	68.27	15.53
Luxembourg	12.9	12.1	74.67	79.18	77.73	82.3	71.76	76.2	1.62	1.7	17.35	69.23	13.42	18.39	67.46	14.15
Macao	18.96	7.51	76.84	80.47	79.28	82.68	74.52	78.36	1.75	0.89	25.67	67.82	6.51	15.16	77.3	7.54
Macedonia	17.09	11.11	71.34	73.99	73.55	76.43	69.22	71.66	1.95	1.45	26.15	66.41	7.44	19.24	69.49	11.27
Madagascar	44.25	36.94	51.09	58.99	52.33	60.8	49.91	57.27	6.19	4.88	44.98	51.99	3.03	43.56	53.28	3.16
Malawi	49.85	41.34	48.61	47.61	50	47.89	47.29	47.35	6.92	5.68	46.72	50.6	2.68	47.04	49.95	3.01
Malaysia	29.68	21.06	70.29	74.05	72.47	76.45	68.22	71.76	3.68	2.65	37.4	58.94	3.66	30.95	64.61	4.44
Maldives	38.7	23.14	60.51	67.92	59.19	68.69	61.77	67.19	5.97	2.67	46.53	50.18	3.28	32.94	63.22	3.84
Mali	51.6	48.2	47.62	53.78	49.52	56.07	45.81	51.61	7.41	6.55	47.66	48.61	3.73	47.63	48.79	3.58
Malta	15.2	9.4	75.5	78.55	77.85	80.7	73.26	76.5	2.05	1.4	23.46	66.09	10.45	16.97	69.59	13.45
Mauritania	39.54	33.06	57.53	63.75	59.52	65.59	55.63	61.99	5.79	4.46	43.8	52.88	3.32	40.06	56.34	3.6
Mauritius	21.3	14.7	69.4	73.17	73.4	76.64	65.6	69.87	2.32	1.99	29.67	64.97	5.36	24.01	69.3	6.69
Mexico	28.85	19.02	70.88	74.47	73.84	77.24	68.06	71.84	3.44	2.17	38.63	57.13	4.24	30.22	63.8	5.98

Table E (*continued*)
World Countries: Basic Demographic Data, 1990–2006

COUNTRY	CRUDE BIRTH RATE (BIRTHS PER 1,000 POPULATION)		LIFE EXPECTANCY AT BIRTH (YEARS)		LIFE EXPECTANCY OF FEMALES AND MALES				TOTAL FERTILITY RATE		PERCENTAGE OF POPULATION IN SPECIFIC AGE GROUPS					
					FEMALES		MALES				1990			2006		
	1990	2006	1990	2006	1990	2006	1990	2006	1990	2006	0–14	15–64	65 AND OLDER	0–14	15–64	65 AND OLDER
Micronesia	33.55	26.7	66.25	68.31	66.82	69.1	65.72	67.56	4.96	3.81	44.11	52.32	3.57	38.32	57.89	3.79
Moldova	18.1	11.43	67.28	68.53	70.82	72.27	63.91	64.96	2.32	1.22	28.25	63.46	8.29	19.35	69.52	11.13
Mongolia	30.92	18.3	62.66	67.17	64	68.72	61.38	65.7	4.03	2.29	41.67	54.29	4.04	27.97	68.07	3.96
Montenegro	11.38	13.48	75.33	74.42	78.27	76.71	72.54	72.24	1.98	1.64	26.29	65.38	8.33	19.36	66.42	14.22
Morocco	29.01	22.41	64.32	70.7	66.06	72.92	62.66	68.58	3.97	2.41	39.8	56.39	3.81	29.73	64.98	5.28
Mozambique	43.55	40.29	43.67	42.46	45.25	43	42.17	41.94	6.2	5.19	46.58	50.22	3.21	44.31	52.48	3.21
Myanmar	26.78	18.46	59.01	61.65	61.12	64.88	57	58.57	3.38	2.1	35.62	59.47	4.91	26.74	67.65	5.61
Namibia	40.63	26.02	61.52	52.5	63.66	52.98	59.48	52.04	5.71	3.27	44.54	52.13	3.34	38.31	58.19	3.5
Nepal	38.41	28.51	54.45	63.23	54.23	63.72	54.66	62.76	5.11	3.1	41.86	54.75	3.4	38.46	57.84	3.7
Netherlands	13.2	11.3	76.88	79.7	80.11	81.9	73.8	77.6	1.62	1.71	18.24	68.92	12.84	18.26	67.39	14.35
Netherlands Antilles	19.52	12.87	74.45	74.92	77.56	78.74	71.48	71.27	2.29	1.89	26.2	66.7	7.1	21.95	68.19	9.86
New Caledonia	25.6	16.52	70.49	75.33	73.2	78.15	67.9	72.65	3.19	2.3	32.29	63.15	4.57	26.62	66.54	6.84
New Zealand	17.5	14.29	75.38	79.93	78.4	82.01	72.5	77.95	2.18	2.05	23.38	65.49	11.13	21.19	66.53	12.28
Nicaragua	36.58	25.15	64.5	72.48	67.42	75.58	61.72	69.52	4.7	2.8	45.96	50.86	3.18	37.21	58.74	4.05
Niger	55.72	49.4	46.51	56.42	45.28	55.51	47.68	57.29	7.88	7	48.3	49.01	2.68	47.98	48.84	3.18
Nigeria	47.08	40.43	47.24	46.78	48.57	47.33	45.98	46.26	6.71	5.43	46.14	50.94	2.92	44.1	52.97	2.94
Norway	14.4	12.1	76.54	80.33	79.8	82.66	73.43	78.12	1.93	1.9	18.94	64.75	16.31	19.43	65.9	14.68
Oman	37.06	22.36	69.94	75.51	71.54	77.14	68.42	73.95	6.5	3.14	44.13	53.95	1.93	33.14	64.12	2.74
Pakistan	41.4	25.56	59.1	65.21	60	65.76	58.24	64.68	5.84	3.92	44.03	52.66	3.31	36.37	59.69	3.94
Panama	25.94	21.19	72.47	75.4	75.23	78.06	69.84	72.87	3	2.59	35.18	59.88	4.94	30.08	63.81	6.11
Papua New Guinea	36.82	30.45	54.96	57.32	57.75	60.27	52.3	54.51	4.81	3.89	41.72	56.1	2.17	40.33	57.29	2.38
Paraguay	35.08	25.24	68.14	71.65	70.45	73.93	65.94	69.48	4.69	3.16	41.41	54.57	4.02	35.35	59.8	4.85
Peru	30.02	21.13	65.79	71.12	68.23	73.71	63.47	68.64	3.86	2.55	38.21	57.83	3.96	31.19	63.13	5.68
Philippines	32.56	26.27	65.62	71.39	67.76	73.65	63.58	69.24	4.31	3.3	40.9	55.87	3.23	35.81	60.29	3.9
Poland	14.3	9.7	70.89	75.14	75.5	79.6	66.5	70.9	2.04	1.27	25.12	64.83	10.05	15.88	70.8	13.32
Portugal	11.8	10.4	73.66	78.38	77.3	81.75	70.2	75.18	1.43	1.36	20.41	66.18	13.41	15.63	67.37	17
Puerto Rico	18.5	12.91	74.79	78.42	79.1	82.56	70.68	74.47	2.2	1.77	27.21	63.12	9.68	21.59	65.72	12.7
Qatar	22.96	16.56	69.97	75.5	72.31	76.15	67.74	74.87	4.34	2.71	27.71	71.19	1.11	21.46	77.17	1.37
Romania	13.7	10.2	69.74	72.18	73.05	75.8	66.59	68.74	1.84	1.31	23.57	66.03	10.4	15.39	69.75	14.85
Russia	13.4	10.4	68.9	65.56	74.26	72.52	63.8	58.92	1.89	1.3	22.97	67.07	9.96	14.85	71.45	13.7
Rwanda	48.08	43.92	31.79	45.59	33.79	47.26	29.89	44	7.44	5.94	50.24	47.23	2.54	43.09	54.46	2.45
Samoa	33.68	25.6	65.28	71.33	68.66	74.58	62.06	68.24	4.75	4.03	40.86	55.23	3.91	40.46	54.93	4.61
São Tomé and Príncipe	37.77	32.89	62.25	65.2	63.7	67.13	60.86	63.36	5.36	3.95	46.64	49	4.36	41.42	54.28	4.31
Saudi Arabia	36.04	25.2	68.04	72.58	69.72	74.57	66.44	70.69	5.89	3.44	41.9	55.81	2.29	34.01	63.19	2.8
Senegal	42.64	35.66	57.42	62.76	58.46	64.82	56.44	60.79	6.48	5.3	45.12	51.19	3.69	41.92	53.82	4.26
Serbia	—	9.68	—	72.78	—	75.52	—	70.16	—	1.43	23.22	67.16	9.62	18.36	66.95	14.69
Seychelles	23.7	17.3	70.3	72.22	73.52	75.7	67.22	68.9	2.82	1.98	—	—	—	—	—	—

Table E (*continued*)
World Countries: Basic Demographic Data, 1990–2006

	CRUDE BIRTH RATE (BIRTHS PER 1,000 POPULATION)		LIFE EXPECTANCY AT BIRTH (YEARS)		LIFE EXPECTANCY OF FEMALES AND MALES				TOTAL FERTILITY RATE		PERCENTAGE OF POPULATION IN SPECIFIC AGE GROUPS					
					FEMALES		MALES				1990			2006		
COUNTRY	1990	2006	1990	2006	1990	2006	1990	2006	1990	2006	0–14	15–64	65 AND OLDER	0–14	15–64	65 AND OLDER
Sierra Leone	47.66	46.31	38.99	42.24	40.43	43.87	37.62	40.68	6.5	6.48	41.45	55.07	3.48	42.84	53.85	3.31
Singapore	18.4	10.1	74.34	79.85	76.92	81.8	71.88	78	1.87	1.26	21.46	72.95	5.59	18.8	72.43	8.77
Slovakia	15.2	9.8	70.93	74.2	75.44	78.2	66.64	70.4	2.09	1.24	25.26	64.43	10.32	16.35	71.85	11.81
Slovenia	11.2	9.1	73.25	77.67	77.3	81.42	69.4	74.1	1.46	1.32	20.9	68.02	11.08	13.93	70.27	15.8
Solomon Islands	39.38	31.09	57.38	63.33	57.81	64.18	56.98	62.52	5.77	3.96	45.29	51.72	2.99	40.11	56.92	2.97
Somalia	46.01	43.5	41.6	47.69	43.28	48.93	40	46.51	6.76	6.12	45.4	51.78	2.83	44.17	53.22	2.62
South Africa	31.88	23.04	61.93	50.71	65	52.5	59	49	3.32	2.73	38.6	58.21	3.19	31.94	63.69	4.36
Spain	10.3	10.7	76.84	80.8	80.51	84.1	73.34	77.66	1.33	1.36	19.86	66.53	13.61	14.47	68.66	16.87
Sri Lanka	20.8	18.7	71.17	74.97	73.71	77.59	68.74	72.47	2.54	1.91	32.01	62.6	5.39	23.74	69.68	6.58
St. Kitts and Nevis	22.03	—	67.15	—	69.2	—	65.2	—	2.68	—	—	—	—	—	—	—
St. Lucia	27.78	13.7	—	74.39	—	76.13	68.73	72.72	3.28	2.07	36.67	56.2	7.14	27.45	65.41	7.14
St. Vincent and the Grenadines	24.88	20.23	69.23	71.38	71.9	73.6	66.7	69.27	2.98	2.22	38.23	55.88	5.9	28.85	64.64	6.52
Sudan	40.66	32.12	52.87	58.11	54.43	59.62	51.39	56.66	5.92	4.35	43.95	53.01	3.05	40.33	56.1	3.57
Suriname	24.15	19.84	68.37	69.99	70.95	73.35	65.91	66.8	2.76	2.45	35.21	60.46	4.33	29.44	64.16	6.4
Swaziland	41	33.27	56.64	40.77	59	39.89	54.4	41.6	5.3	3.54	47.13	50.22	2.65	39.21	57.5	3.29
Sweden	14.5	11.6	77.54	80.77	80.4	82.94	74.81	78.7	2.13	1.85	17.94	64.28	17.78	17.15	65.48	17.37
Switzerland	12.5	9.7	77.24	81.51	80.71	84.04	73.94	79.11	1.59	1.44	16.88	68.77	14.35	16.46	67.88	15.66
Syria	36.17	27.03	68.44	73.9	70.28	75.84	66.68	72.05	5.41	3.16	48.39	48.94	2.67	36.03	60.78	3.19
Tajikistan	38.37	27.74	63.4	66.51	66.06	69.2	60.88	63.95	5.09	3.44	43.16	53.01	3.83	38.67	57.44	3.89
Tanzania	43.37	39.66	50.9	51.89	52.79	53	49.09	50.83	6.08	5.26	45.97	51.35	2.68	44.37	52.65	2.98
Thailand	19.24	14.71	67.02	70.24	70.58	74.77	63.63	65.92	2.11	1.85	28.51	66.61	4.89	21.41	70.61	7.98
Timor-Leste	39.4	50.74	46.14	57.16	47	58.26	45.33	56.12	4.93	7.32	39.9	58.09	2.01	44.7	52.58	2.72
Togo	43.95	37.33	57.62	58.2	59.67	60	55.67	56.48	6.36	4.91	46.44	50.54	3.02	43.03	53.9	3.08
Tonga	30.11	25.36	69.83	73.03	71.14	74.05	68.59	72.05	4.61	3.81	39.36	56.17	4.46	37.17	56.31	6.52
Trinidad and Tobago	20.68	14.75	69.87	69.58	72.37	71.62	67.49	67.63	2.38	1.63	34.45	59.95	5.6	21.7	71.73	6.58
Tunisia	25.2	17.1	70.31	73.62	72.1	75.65	68.6	71.7	3.5	2.03	38.16	57.32	4.52	25.35	68.35	6.3
Turkey	24.8	18.6	66	71.49	68.3	74	63.8	69.1	3	2.18	35.75	60.25	4	27.85	66.48	5.67
Turkmen-istan	33.79	22.05	63.04	63.01	67.05	67.37	59.23	58.86	4.24	2.55	40.47	55.74	3.79	30.92	64.45	4.63
Uganda	49.5	46.71	49.94	50.74	51.83	51.43	48.14	50.09	7.08	6.7	48.06	49.29	2.66	49.25	48.29	2.45
Ukraine	12.7	9.8	70.14	68.04	74.9	74.11	65.6	62.27	1.85	1.25	21.47	66.48	12.05	14.31	69.53	16.17
United Arab Emirates	25.94	15.36	73.2	79.32	75.3	81.71	71.21	77.05	4.26	2.35	29.88	68.85	1.27	19.6	79.28	1.12
United Kingdom	13.9	12.2	75.88	79.14	78.8	81.35	73.1	77.04	1.83	1.85	19	65.27	15.73	17.79	66.06	16.15
United States	16.7	14.14	75.21	77.85	78.8	80.82	71.8	75.02	2.08	2.05	21.75	66.03	12.22	20.66	67.05	12.3
Uruguay	18	14.76	72.62	75.73	76.46	79.52	68.96	72.12	2.51	2.03	26.04	62.44	11.51	23.58	62.83	13.6

Table E (*continued*)
World Countries: Basic Demographic Data, 1990–2006

COUNTRY	CRUDE BIRTH RATE (BIRTHS PER 1,000 POPULATION)		LIFE EXPECTANCY AT BIRTH (YEARS)		LIFE EXPECTANCY OF FEMALES AND MALES				TOTAL FERTILITY RATE		PERCENTAGE OF POPULATION IN SPECIFIC AGE GROUPS					
					FEMALES		MALES				1990			2006		
	1990	2006	1990	2006	1990	2006	1990	2006	1990	2006	0–14	15–64	65 AND OLDER	0–14	15–64	65 AND OLDER
Uzbekistan	33.7	19.46	69.17	67.5	72.4	70.82	66.1	64.33	4.07	2.35	40.91	55.08	4.01	32.36	62.92	4.71
Vanuatu	36.69	29.22	63.47	69.81	64.98	71.76	62.03	67.95	4.91	3.83	43.94	52.48	3.58	39.31	57.4	3.29
Venezuela	28.56	21.78	71.25	74.4	74.22	77.36	68.42	71.58	3.43	2.6	38.04	58.28	3.68	30.91	64.01	5.08
Vietnam	28.78	17.45	64.78	70.85	66.84	73.38	62.82	68.43	3.62	2.08	38.88	56.17	4.94	28.88	65.56	5.57
Virgin Islands (U.S.)	21.4	13.6	74.09	78.75	77.4	80.2	70.93	77.37	2.59	2.16	28.94	65.17	5.89	23.54	65.19	11.27
West Bank and Gaza	—	32.16	—	72.93	—	74.48	—	71.46	—	4.6	46.71	49.9	3.39	45.64	51.36	3.01
Yemen	50.53	38.47	54.49	62.21	55.24	63.85	53.79	60.65	7.98	5.61	51.77	46.29	1.94	45.42	52.24	2.34
Zambia	44.05	39.81	48.04	41.67	50.26	41.86	45.92	41.5	6.43	5.27	45.67	51.56	2.77	45.64	51.42	2.94
Zimbabwe	37.17	28.06	61.27	42.69	63.66	42.06	58.99	43.28	5.13	3.8	46.03	51.04	2.93	39.01	57.46	3.53
World	28.85	22.6	64.75	67.9	67.22	69.88	62.50	65.66	3.92	2.94	35.41	58.63	5.96	29.97	62.63	7.41

Table F
World Countries: Mortality, Health, and Nutrition, 1990–2006

	MORTALITY						HEALTH			NUTRITION	
	INFANT & CHILD MORTALITY		ADULT MORTALITY								
	INFANT MORTALITY (PER 1000 LIVE BIRTHS)	CHILD (UNDER 5) MORTALITY (PER 1000 LIVE BIRTHS)	MALE (PER 1000 LIVE BIRTHS)		FEMALE (PER 1000 LIVE BIRTHS)		HEALTH EX-PENDITURE PER CAPITA ($US)	PHYSICIANS PER 1,000 PEOPLE	HOSPITAL BEDS PER 1000 PEOPLE	PREVALENCE OF UNDER-NOURISH-MENT (% OF POPULATION)	PREVALENCE OF CHILD MALNUTRI-TION (% OF CHILDREN UNDER 5)
COUNTRY	2006	2006	1990	2006	1990	2006	2005	2000–2006	2000–2006	2004	1998–2006
Afghanistan	—	—	486	—	476	—	20	0.19	—	—	49.30
Albania	14.8	16.6	—	107.6	—	51.6	169	1.15	3.26	6	13.60
Algeria	33.4	38	192.6	122.9	156.2	104.7	108	1.00	2.10	4	6.00
Angola	154	260	514	485.6	420.1	436.7	36	—	—	35	30.50
Antigua and Barbuda	10	11.4	142.2	—	85.4	—	503	0.76	3.88	—	—
Argentina	14.1	16.5	188	168.4	95	80.1	484	—	3.29	3	2.3
Armenia	21.4	23.8	215.6	196.7	119.4	88.3	88	3.7	4.5	24	4.2
Australia	4.7	5.6	125.4	—	67	—	3181	2.49	7.90	2.5	0.00
Austria	3.9	4.7	155	—	74.5	—	3788	3.66	7.7	2.5	—
Azerbaijan	73.4	88.2	216.2	217.6	95.8	103.1	62	3.63	8.2	7	6.80
Bahamas	12.6	14.2	185.7	222	69.8	151.1	1224	1.06	3.94	8	—
Bahrain	9.2	10.4	200.6	103.7	146.9	73.1	710	2.72	2.8	—	8.70
Bangladesh	51.6	69.2	321.7	235.1	308.2	202.6	12	0.3	—	30	39.2
Barbados	10.8	11.8	140.3	108.4	82.5	60.7	725	1.28	7.56	2.5	—
Belarus	11.8	13.3	282	—	107.2	—	204	4.78	11.1	4	—
Belgium	3.5	4.5	138.7	—	75.1	—	3451	4.23	5.3	2.5	—
Belize	14.5	16.3	193.8	148.2	123.1	74.1	198	1.05	2.13	4	—
Benin	87.8	148	447.2	286.8	369.3	239.2	28	0.06	0.5	12	22.90
Bhutan	62.6	70	—	223.4	—	174.4	52	—	—	—	—
Bolivia	49.8	61.2	306.6	237.5	249.5	178.1	71	0.73	1.67	23	7.60
Bosnia-Herzegovina	12.8	14.6	186	147.8	108.9	77.2	243	1.34	3.22	9	4.10
Botswana	89.6	123.8	—	585.7	—	574.8	362	—	3	32	12.50
Brazil	18.6	20	193.1	233.7	134.9	122.6	371	2.06	3.11	7	5.70
Brunei	8	8.8	148.6	89.4	98.2	64.6	519	1.01	—	4	—
Bulgaria	11.6	14.3	219.3	—	98	—	272	3.38	6.4	8	—
Burkina Faso	121.6	204.3	429.1	287.8	338.2	186.8	27	0.04	1.42	15	37.7
Burundi	108.8	181	460	411.6	378.7	376.9	3	0.05	0.7	66	45.10
Cambodia	64.8	82.2	391.6	359	318.8	248	29	0.16	—	33	28.4
Cameroon	86.8	148.6	430.4	416.4	361.2	420	49	—	—	26	22.20
Canada	4.9	5.7	127.3	—	69.7	—	3430	2.10	3.90	2.5	—
Cape Verde	25	33.6	245.4	176.1	218.2	101.2	114	—	—	—	—

Table F (*continued*)
World Countries: Mortality, Health, and Nutrition, 1990–2006

COUNTRY	MORTALITY						HEALTH			NUTRITION	
	INFANT & CHILD MORTALITY		ADULT MORTALITY								
	INFANT MORTALITY (PER 1000 LIVE BIRTHS)	CHILD (UNDER 5) MORTALITY (PER 1000 LIVE BIRTHS)	MALE (PER 1000 LIVE BIRTHS)		FEMALE (PER 1000 LIVE BIRTHS)		HEALTH EXPENDITURE PER CAPITA ($US)	PHYSICIANS PER 1,000 PEOPLE	HOSPITAL BEDS PER 1000 PEOPLE	PREVALENCE OF UNDERNOURISHMENT (% OF POPULATION)	PREVALENCE OF CHILD MALNUTRITION (% OF CHILDREN UNDER 5)
	2006	2006	1990	2006	1990	2006	2005	2000-2006	2000–2006	2004	1998–2006
Central African Republic	114.5	174.8	485.1	565.9	380.6	535.9	13	0.04	—	44	23.20
Chad	124.3	209	487.3	351.9	397	303	22	0.03	0.40	35	28.00
Chile	8.1	9.3	165	130	92	62.3	397	1.09	2.67	4	0.80
China	20.1	23.7	160	153.1	135	92.1	81	1.64	2.45	12	10.00
Colombia	16.7	20.7	222.1	205.6	127.1	97.3	201	1.35	1.46	13	5.1
Comoros	51.2	68.4	364.7	237.5	307.1	178.8	14	—	2.20	60	25.40
Congo, Dem. Rep.	129	205	—	439.1	—	401.2	5	—	—	74	31.00
Congo, Rep.	79.5	126.5	370.5	402.3	272.5	377.2	31	0.25	—	33	5.1
Costa Rica	10.7	11.8	134	115.8	88	62.3	327	0.90	1.68	5	5.10
Côte d'Ivoire	89.6	127.2	352	428.6	293.7	408	34	—	—	13	21.20
Croatia	5.5	6.4	207.2	159.4	95.5	62.2	651	2.47	5.46	7	0.60
Cuba	5	6.7	125.1	118.6	83.3	73.5	310	5.91	4.9	2.5	3.90
Cyprus	3.3	4.4	118	85.8	71.6	42.1	1350	2.98	3.4	2.5	—
Czech Republic	3.2	3.9	230.7	139.3	95	59.7	868	3.50	8.4	2.5	—
Denmark	4	4.8	155.3	115.6	101.1	68.2	4350	3.66	4.50	2.5	—
Djibouti	86.2	130.2	471.6	362.8	386.7	327.9	54	0.19	—	24	18.20
Dominica	13	15	154.2	—	113.3	—	288	0.47	2.65	8	—
Dominican Republic	24.6	29.2	—	222.4	—	134.1	197	1.88	2.2	29	5.30
Ecuador	21	23.6	205	170.7	141	91.2	147	1.48	1.55	6	14.30
Egypt	28.9	35.3	230	158.3	190	94.1	78	2.12	2.2	4	5.4
El Salvador	21.8	25.4	283.9	208.7	164.7	127.1	177	1.27	0.9	11	10.3
Equatorial Guinea	123.6	206	487.8	382.4	399.8	349.3	211	0.21	2.2	—	—
Eritrea	47.8	74.2	433.5	430.3	347.2	326.2	8	—		75	39.60
Estonia	5.1	6.9	300.4	—	106.8	—	516	3.16	6.72	2.5	—
Ethiopia	77.5	122.8	448.5	366.8	358.2	329.1	6	0.03	—	46	34.6
Fiji	15.7	17.8	220	232.9	161	154.2	148	0.36	—	5	—
Finland	2.9	3.8	182.9	132.1	70.2	57.2	2824	3.10	7.50	2.5	—
France	3.6	4.3	167.6	—	69	—	3807	3.30	8.20	2.5	—
Gabon	60	91	401.6	377.6	332.2	374	276	0.29	—	5	11.90
Gambia	84	113.4	530	221.4	431.6	179.6	15	0.8	—	29	17.20
Georgia	28.3	32	195.3	214.2	90.1	81.5	123	3.8	4.30	9	3.10

Table F (*continued*)
World Countries: Mortality, Health, and Nutrition, 1990–2006

COUNTRY	MORTALITY						HEALTH			NUTRITION	
	INFANT & CHILD MORTALITY		ADULT MORTALITY								
	INFANT MORTALITY (PER 1000 LIVE BIRTHS)	CHILD (UNDER 5) MORTALITY (PER 1000 LIVE BIRTHS)	MALE (PER 1000 LIVE BIRTHS)		FEMALE (PER 1000 LIVE BIRTHS)		HEALTH EXPENDITURE PER CAPITA ($US)	PHYSICIANS PER 1,000 PEOPLE	HOSPITAL BEDS PER 1000 PEOPLE	PREVALENCE OF UNDERNOURISHMENT (% OF POPULATION)	PREVALENCE OF CHILD MALNUTRITION (% OF CHILDREN UNDER 5)
	2006	2006	1990	2006	1990	2006	2005	2000–2006	2000–2006	2004	1998–2006
Germany	3.7	4.5	—	—	—	—	3628	8.4	9.10	2.5	—
Ghana	76	119.9	334.4	288.9	269.9	283.1	30	0.9	—	11	22.1
Greece	3.6	4.4	117.2	92.7	66.6	42	2580	4.40	4.90	2.5	—
Grenada	16.2	20	—	190.1	—	145.9	342	0.65	4.8	7	—
Guatemala	30.6	41	—	237.2	—	130.9	132	1.10	0.7	22	22.70
Guinea	98.1	161.2	529.3	275.2	494.7	236.3	21	0.09	—	24	22.5
Guinea-Bissau	119.3	200.1	544.2	446.5	533	396.3	10	—	—	39	25.00
Guyana	46	61.6	262.6	252.2	172.3	181.4	60	0.48	3.87	8	13.60
Haiti	60	80.2	352.8	309	290.8	245.2	28	0.25	0.71	46	18.9
Honduras	22.6	27	202.3	244.8	141.3	140.8	91	0.65	1.06	23	8.6
Hong Kong	—	—	122.5	78.3	64.2	33.9	—	1.32	—	—	—
Hungary	6	7.2	305.1	—	132.7	—	855	3.10	7.9	2.5	—
Iceland	2.2	2.6	116.6	76.4	78	53.1	5154	3.60	14.60	2.5	—
India	57.4	76.4	236.5	259.7	241.4	167.8	36	0.51	—	20	43.5
Indonesia	26.4	33.6	275.4	172.3	219.4	123	26	0.16	—	6	24.4
Iran	30	34.4	169.7	155.3	174.3	103.9	212	1.05	1.7	4	10.90
Iraq	41.8	53.1	193.2	—	154	—	59	0.54	1.3	—	15.90
Ireland	4.3	5.2	134	89.9	78	52.1	3993	2.40	9.70	2.5	—
Israel	4.2	5.2	120.7	81.7	72	39.1	1533	3.91	6.3	2.5	—
Italy	3.5	4.2	130.8	—	61.1	—	2692	4.40	4.90	2.5	—
Jamaica	25.8	31.2	154.9	221.5	97.1	140.1	170	0.85	1.7	9	3.80
Japan	2.6	3.6	108.8	—	53.5	—	2936	2.00	16.50	2.5	—
Jordan	21.4	25.2	204.8	167.3	151.9	116.4	241	2.05	1.7	6	4.40
Kazakhstan	25.8	29.1	306	368.8	136	147.4	148	3.61	7.7	6	4.20
Kenya	79.4	120.6	357.4	431.8	287.5	408.1	24	0.13	—	31	19.9
Kiribati	47.2	64	—	—	—	—	118	0.30	—	7	—
Korea, North	42	55	223.2	182.4	116.4	128.3	—	2.97	—	33	27.90
Korea, South	4.5	4.9	239.1	114.3	116.6	46.8	973	1.40	6.10	2.5	—
Kuwait	9.5	11.3	130.4	86.9	79.7	52.9	687	1.53	1.9	5	1.70
Kyrgyzstan	35.6	41	290.6	280.7	142.9	132.2	29	2.60	5.1	4	5.80
Laos	59	74.6	—	237.9	—	195.6	18	—	—	19	40.00
Latvia	7.7	9.3	310	—	117.6	—	443	2.91	7.7	3	—
Lebanon	26.2	29.7	209.8	155.4	150.3	103.2	460	3.25	3.6	3	3.00

Table F (*continued*)
World Countries: Mortality, Health, and Nutrition, 1990–2006

COUNTRY	MORTALITY						HEALTH			NUTRITION	
	INFANT & CHILD MORTALITY		ADULT MORTALITY								
	INFANT MORTALITY (PER 1000 LIVE BIRTHS)	CHILD (UNDER 5) MORTALITY (PER 1000 LIVE BIRTHS)	MALE (PER 1000 LIVE BIRTHS)		FEMALE (PER 1000 LIVE BIRTHS)		HEALTH EXPENDITURE PER CAPITA ($US)	PHYSICIANS PER 1,000 PEOPLE	HOSPITAL BEDS PER 1000 PEOPLE	PREVALENCE OF UNDERNOURISHMENT (% OF POPULATION)	PREVALENCE OF CHILD MALNUTRITION (% OF CHILDREN UNDER 5)
	2006	2006	1990	2006	1990	2006	2005	2000-2006	2000–2006	2004	1998–2006
Lesotho	102.3	131.8	—	715	—	698.1	69	0.05	—	13	16.6
Liberia	157	235	—	466.2	—	429.5	10	—	—	50	26.50
Libya	17	18.4	233.6	151.5	185.4	93.8	223	—	4.30	2.5	4.70
Lithuania	6.6	8.2	287.7	—	107	—	448	4.03	8.1	2.5	—
Luxembourg	3.6	4	165	105	85	51.6	6330	2.60	8.00	2.5	—
Macedonia	14.6	16.6	147.1	137.3	100.5	81.2	224	2.19	4.7	5	5.90
Madagascar	72	115.4	433.9	289	376.9	230.9	9	0.14	0.42	38	33.10
Malawi	76.1	120.2	478.9	539.7	436.4	525.4	19	0.01	1.34	35	18.4
Malaysia	9.8	11.6	198	155.5	125	89.1	222	0.70	2.01	3	19.00
Maldives	26.3	30.4	207.8	185.7	284.2	160.9	316	0.78	—	10	30.20
Mali	119.2	216.8	433.6	262.8	351.2	183.9	28	—	0.24	29	33.20
Malta	5.1	6	134	73.1	80.8	49.5	1235	2.93	7.5	2.5	—
Marshall Islands	50.2	56	—	—	—	—	294	0.47	—	—	—
Mauritania	77.8	125	440.5	177.4	364.8	111.2	17	0.14	—	10	31.80
Mauritius	12.6	14.4	240	210.2	134	108	218	0.85	—	5	14.90
Mexico	29.1	35.3	187	144.4	117	80.8	474	1.50	3	5	3.4
Micronesia	33.1	40.9	315.7	181.3	255.9	156.6	290	0.60	—	—	—
Moldova	16.4	19	268.7	296.3	145.6	139.6	58	2.69	6.4	11	3.2
Mongolia	34.2	42.6	250.8	267.9	211.4	174.7	35	2.67	—	27	12.70
Montenegro	8.6	10	—	171	—	84.1	299	2.66	4.2	—	
Morocco	34.2	37.2	233.6	150.7	183.7	100.6	89	0.48	0.98	6	9.00
Mozambique	95.6	138.4	418.2	608.6	320.9	588.5	14	0.02	—	44	26.10
Myanmar	74.4	104	—	303.8	—	192.4	4	0.30	—	5	28.20
Namibia	45.2	60.6	372.5	523.4	317.8	508.4	165	—	—	24	24.00
Nepal	46.1	59.4	349.8	235.4	375.7	211.3	16	0.05	0.17	17	38.8
Netherlands	4.2	5.2	116.5	83.4	67.1	52.7	3560	3.10	10.80	2.5	—
Netherlands Antilles	—	—	145	176.6	88	80.7	—	—	6.15	13	—
New Zealand	5.2	6.2	142.9	80.5	93.1	53.5	2403	2.10	6.20	10	—
Nicaragua	29.2	35.8	219.6	213.6	147.1	124.2	75	1.64	1.48	2.5	9.60

Table F (*continued*)
World Countries: Mortality, Health, and Nutrition, 1990–2006

COUNTRY	MORTALITY						HEALTH			NUTRITION	
	INFANT & CHILD MORTALITY		ADULT MORTALITY								
	INFANT MORTALITY (PER 1000 LIVE BIRTHS)	CHILD (UNDER 5) MORTALITY (PER 1000 LIVE BIRTHS)	MALE (PER 1000 LIVE BIRTHS)		FEMALE (PER 1000 LIVE BIRTHS)		HEALTH EXPENDITURE PER CAPITA ($US)	PHYSICIANS PER 1,000 PEOPLE	HOSPITAL BEDS PER 1000 PEOPLE	PREVALENCE OF UNDERNOURISHMENT (% OF POPULATION)	PREVALENCE OF CHILD MALNUTRITION (% OF CHILDREN UNDER 5)
	2006	2006	1990	2006	1990	2006	2005	2000–2006	2000–2006	2004	1998–2006
Niger	148.2	253.2	514.8	168.8	413.2	182.4	9	0.03	0.12	27	39.9
Nigeria	98.6	191.4	—	431.5	—	410.3	27	0.27	—	32	28.7
Norway	3	3.8	132.3	85.7	68.4	52.9	5910	3.77	14.60	9	
Oman	9.6	11.5	217.3	102.7	157.5	75.6	312	1.26	2.20	—	17.80
Pakistan	77.8	97.2	232.3	176.9	230	145.2	15	0.66	—	2.5	35.00
Palau	10	11	—	—	—	—	690	1.09	—	—	—
Panama	18.2	23.1	146.3	139.7	94	75.3	351	1.68	2.21	24	8.10
Papua New Guinea	54.4	73.3	424.8	422.1	385.8	305.5	34	0.05	—	23	—
Paraguay	19.4	22.2	—	176.2	—	129.1	92	1.17	1.34	—	—
Peru	21.2	25.2	227.7	201.2	172.8	125.2	125	0.98	1.47	15	7.10
Philippines	24	31.6	273.4	160.8	207.6	107	37	1.16	—	12	31.80
Poland	6	7	264	190.5	102	66.1	495	2.30	4.90	18	—
Portugal	3.4	4.5	176.1	—	80	—	1800	3.20	4.00	2.5	—
Qatar	17.8	20.7	193.8	112.3	135	111.1	2186	2.21	2.4	2.5	5.50
Romania	15.7	18.3	237	205	114	87.1	250	1.92	6.6	—	3.20
Russia	13.7	15.8	316.1	429.4	116.2	158.2	277	4.31	9.7	2.5	5.50
Rwanda	97.6	160.3	493.1	471.4	408.6	421.7	19	0.02	—	3	24.30
Samoa	23.2	28	261.5	199.8	201.6	112.2	113	0.33	—	33	1.90
San Marino	2.9	3	—	—	—	—	3490	—	—	4	—
São Tomé and Príncipe	62.7	95.7	—	173.4	—	119	49	—	—	10	12.90
Saudi Arabia	20.6	25.4	191.8	143.1	157.9	92.8	448	1.40	2.30	4	—
Senegal	59.9	116.2	488	174.5	399	105.7	38	0.01	0.40	20	22.70
Serbia	7.3	8.4	—	159.3	—	84.5	212	1.99	5.9	—	1.90
Seychelles	11.8	13.1	220.6	—	112.7	—	557	—	—	9	—
Sierra Leone	159.2	270.1	601.2	412.3	492.1	348.7	8	—	—	51	27.20
Singapore	2.3	2.8	138	83	80	47.1	944	1.40	—	—	3.40
Slovakia	6.8	8.1	270.6	196.4	103.7	76.1	626	3.60	6.9	7	—
Slovenia	3.2	4	206.1	148.9	81.1	56.6	1495	2.19	4.8	3	—
Solomon Islands	54.9	72.7	—	237.9	—	206.6	28	0.14	—	21	21.10
Somalia	89.6	145.5	—	389	—	339	—	—	—	—	25.80

Table F (*continued*)
World Countries: Mortality, Health, and Nutrition, 1990–2006

COUNTRY	MORTALITY: INFANT & CHILD MORTALITY: INFANT MORTALITY (PER 1000 LIVE BIRTHS)	CHILD (UNDER 5) MORTALITY (PER 1000 LIVE BIRTHS)	ADULT MORTALITY: MALE (PER 1000 LIVE BIRTHS)		FEMALE (PER 1000 LIVE BIRTHS)		HEALTH: HEALTH EXPENDITURE PER CAPITA ($US)	PHYSICIANS PER 1,000 PEOPLE	HOSPITAL BEDS PER 1000 PEOPLE	NUTRITION: PREVALENCE OF UNDERNOURISHMENT (% OF POPULATION)	PREVALENCE OF CHILD MALNUTRITION (% OF CHILDREN UNDER 5)
	2006	2006	1990	2006	1990	2006	2005	2000-2006	2000–2006	2004	1998–2006
South Africa	56	69	—	605.3	—	567.9	437	0.69	—	2.5	11.50
Spain	3.6	4.4	146.5	—	60.1	—	2152	2.90	4.10	2.5	—
Sri Lanka	11.2	12.9	182	240.9	122	101.6	51	0.43	—	22	32.90
St. Kitts and Nevis	16.8	19	226.6	—	164.7	—	478	1.07	6.36	10	—
St. Lucia	12.2	14	205.4	153.9	143.9	123	323	0.56	3.38	5	—
St. Vincent and the Grenadines	17.1	20	202.4	160.3	119.2	107.5	218	0.46	1.85	10	19.50
Sudan	61.4	88.6	464.4	310.5	397.8	270.4	29	0.3	5.5	26	40.70
Suriname	29.2	38.6	216.3	233.1	136.9	124.3	209	0.45	2.8	8	13.20
Swaziland	112.4	163.6	259.7	749.7	196.5	726.6	146	0.18	—	22	10.30
Sweden	2.9	3.4	114.4	77.8	65.7	49.4	3598	3.05	3.60	2.5	—
Switzerland	4.1	4.9	126.6	80.4	62.3	46.8	5694	3.97	17.90	2.5	—
Syria	12.1	13.7	237.5	126	176.6	85.6	61	0.53	1.40	4	6.90
Tajikistan	56.4	67.8	168.2	212.5	106	140.9	18	2.01	6.40	56	—
Tanzania	73.6	118.2	444.3	444.2	373.3	412.1	17	0.02	—	44	29.40
Thailand	7.2	7.8	206.9	276	123	161.6	98	0.24	1.99	22	17.60
Timor-Leste	46.8	55.2	—	271.5	—	236.9	45	—	—	9	42.60
Togo	69.2	108	389	280	321	234.6	18	0.06	—	24	25.10
Tonga	19.6	23.5	260.1	143.8	193.8	117.8	104	0.34	—	—	—
Trinidad and Tobago	32.8	37.6	182	242.7	148	191.3	513	0.79	5.11	10	5.90
Tunisia	19	22.6	189.6	125.9	173.8	73.3	158	0.66	1.70	2.5	4.00
Turkey	23.7	26	—	154.1	—	86.9	383	1.56	2.60	3	8.30
Turkmenistan	45.3	51.4	249.8	300	134.8	144	156	2.49	7.11	7	12.00
Uganda	77.8	134.2	526	445.7	461.3	433.1	22	0.05	—	19	22.90
Ukraine	19.8	23.8	285.6	374.9	111	131.6	128	3.13	8.74	2.5	3.20
United Arab Emirates	7.6	8.2	129.6	75.2	100.9	50.2	833	2.02	2.64	2.5	7.00
United Kingdom	4.9	5.7	126.2	88.5	76.2	55.7	3064	2.10	4.10	2.5	—
United States	6.5	7.6	172.1	—	90.8	—	6657	5.49	3.60	2.5	—
Uruguay	10.9	12.1	177.6	145.3	90	68.4	404	3.65	4.39	2.5	—
Uzbekistan	37.5	43.4	207.5	242.2	109.2	137.8	26	2.89	5.34	25	7.90

Table F (*continued*)
World Countries: Mortality, Health, and Nutrition, 1990–2006

COUNTRY	MORTALITY: INFANT & CHILD MORTALITY: INFANT MORTALITY (PER 1000 LIVE BIRTHS) 2006	MORTALITY: INFANT & CHILD MORTALITY: CHILD (UNDER 5) MORTALITY (PER 1000 LIVE BIRTHS) 2006	MORTALITY: ADULT MORTALITY: MALE (PER 1000 LIVE BIRTHS) 1990	MORTALITY: ADULT MORTALITY: MALE (PER 1000 LIVE BIRTHS) 2006	MORTALITY: ADULT MORTALITY: FEMALE (PER 1000 LIVE BIRTHS) 1990	MORTALITY: ADULT MORTALITY: FEMALE (PER 1000 LIVE BIRTHS) 2006	HEALTH: HEALTH EXPENDITURE PER CAPITA ($US) 2005	HEALTH: PHYSICIANS PER 1,000 PEOPLE 2000–2006	HEALTH: HOSPITAL BEDS PER 1000 PEOPLE 2000–2006	NUTRITION: PREVALENCE OF UNDER-NOURISHMENT (% OF POPULATION) 2004	NUTRITION: PREVALENCE OF CHILD MALNUTRITION (% OF CHILDREN UNDER 5) 1998–2006
Vanuatu	29.5	36.1	288	176.4	241.2	128.1	67	—	—	11	12.10
Venezuela	17.7	20.7	186.3	180.8	112	96	247	1.94	1.47	18	4.40
Vietnam	14.6	16.8	215	139.2	153.1	93.1	37	0.53	1.67	16	33.80
West Bank and Gaza	20.4	22.2	—	131.4	—	94.8	—	—	—	16	—
Yemen	75	100.4	362.6	258.6	336.2	209.9	39	0.22	0.60	38	46.10
Zambia	102	182	434	635.6	377	632	36	0.07	—	46	28.10
Zimbabwe	67.9	105	305	705.7	270	729.2	21	0.06	—	47	13.00
World	37.46	10251.3	42788.4	241.96	190.47	183.64	699	1.61	4.38	19.19	17.14

Table G
World Countries: Education and Literacy, 2003–2007

COUNTRY	PUPIL-TEACHER RATIO, PRIMARY	PUBLIC SPENDING ON EDUCATION (% OF GDP)	PRIMARY SCHOOL ENROLLMENT (% NET)	SECONDARY SCHOOL ENROLLMENT (% NET)	TERTIARY SCHOOL ENROLLMENT (% NET)
	2003–2007	2003–2006	2003–2006	2003–2006	2003–2006
Afghanistan	83.37	—	—	—	1.25
Albania	21.50	—	93.62	72.82	19.09
Algeria	24.48	—	95.18	66.34	21.83
Angola	—	2.39	—	—	2.94
Antigua and Barbuda	—	—	—	—	—
Argentina	17.34	3.78	98.62	78.9	64.72
Armenia	21.24	—	82.16	85.78	31.73
Australia	—	4.61	96.49	86.41	72.57
Austria	12.41	5.42	96.88	—	48.85
Azerbaijan	12.51	2.1	84.8	77.85	14.81
Bahamas	15.38	—	87.9	84.27	—
Bahrain	—	—	98.24	92.32	32.05
Bangladesh	50.87	2.46	88.88	40.99	5.99
Barbados	15.15	6.94	96.17	88.54	—
Belarus	16.04	6.08	89.45	87.98	65.94
Belgium	11.57	6.2	97.56	96.66	62.41
Belize	22.54	5.4	99.47	71.33	2.61
Benin	46.83	3.5	77.72	—	—
Bermuda	8.34	1.9	100	—	18.8
Bhutan	31.05	7	79.04	38.5	5.54
Bolivia	23.93	6.4	94.89	70.9	40.62
Bosnia-Herzegovina	—	—	—	—	—
Botswana	25.34	9.74	86.24	61.06	5.12
Brazil	21.39	4.01	94.69	77.71	23.78
Brunei	12.51	—	93.79	90.07	15.05
Bulgaria	16.23	2.46	92.93	89.08	43.71
Burkina Faso	45.83	4.2	46.94	11.79	2.34
Burundi	54.19	5.1	74.6	—	2.2
Cambodia	50.42	1.9	89.94	23.94	4.53
Cameroon	44.11	3.27	—	—	6.68
Canada	—	—	—	—	62.36
Cape Verde	25.37	6.32	87.8	59.37	6.93
Central African Republic	—	1.39	45.1	—	1.1
Chad	63.15	1.91	60.22	10.47	1.16
Chile	26.05	3.37	—	—	47.83
China	18.25	—	—	—	21.58
Colombia	28.18	4.66	88.5	64.91	30.83
Comoros	34.98	—	—	—	2.27
Congo, Dem. Rep.	34.33	—	—	—	—
Congo, Rep.	54.8	1.87	54.71	—	3.73
Costa Rica	19.71	4.73	—	—	25.34
Côte d'Ivoire	46.11	—	54.93	—	—

Table G (*continued*)
World Countries: Education and Literacy, 2003–2007

LITERACY RATE, ADULT TOTAL (% OF PEOPLE AGES 15 AND ABOVE)	LITERACY RATE, ADULT MALE (% OF MALES AGES 15 AND ABOVE)	LITERACY RATE, ADULT FEMALE (% OF FEMALES AGES 15 AND ABOVE)	LITERACY RATE, YOUTH TOTAL (% OF PEOPLE AGES 15–24)	LITERACY RATE, YOUTH MALE (% OF MALES AGES 15–24)	LITERACY RATE, YOUTH FEMALE (% OF FEMALES AGES 15–24)
2006	2006	2006	2006	2006	2006
28	43	13	34	51	18
99	99	98	99	99	100
70	80	60	90	94	86
67	83	54	72	84	63
—	—	—	—	—	—
97	97	97	99	99	99
99	100	99	100	100	100
—	—	—	—	—	—
—	—	—	—	—	—
99	100	98	100	100	100
—	—	—	—	—	—
87	89	84	97	97	97
—	—	—	—	—	—
—	—	—	—	—	—
100	100	99	100	100	100
—	—	—	—	—	—
—	—	—	—	—	—
35	48	23	45	59	33
—	—	—	—	—	—
60	69	49	—	—	—
87	93	81	97	99	96
97	99	94	100	100	100
81	80	82	94	92	96
89	88	89	97	96	98
93	95	90	99	99	99
98	99	98	98	98	98
22	29	15	31	38	25
59	67	52	73	77	70
74	85	64	83	88	79
68	77	60	—	—	—
—	—	—	—	—	—
—	—	—	—	—	—
49	65	34	59	70	47
26	41	13	38	56	23
96	96	96	99	99	99
91	95	87	99	99	99
93	93	93	98	98	98
—	—	—	—	—	—
67	81	54	70	78	63
—	—	—	—	—	—
95	95	95	98	97	98
49	61	39	61	71	52

Table G (*continued*)
World Countries: Education and Literacy, 2003–2007

COUNTRY	PUPIL-TEACHER RATIO, PRIMARY	PUBLIC SPENDING ON EDUCATION (% OF GDP)	PRIMARY SCHOOL ENROLLMENT (% NET)	SECONDARY SCHOOL ENROLLMENT (% NET)	TERTIARY SCHOOL ENROLLMENT (% NET)
	2003–2007	2003–2006	2003–2006	2003–2006	2003–2006
Croatia	14.42	4.4	87.33	85.03	43.86
Cuba	9.98	9.12	96.59	87.03	87.9
Cyprus	17.61	7.4	99.35	94.1	33.22
Czech Republic	17.87	4.6	92.51	—	47.76
Denmark	—	8.4	95.8	91.23	80.78
Djibouti	33.63	8.41	37.75	21.53	2.38
Dominica	17.24	—	77.31	91.23	—
Dominican Republic	22.51	3.64	77.47	52.09	34.51
Ecuador	23.13	—	97.34	55.36	—
Egypt	21.86	—	93.74	82.71	34.75
El Salvador	40.1	3.05	94.03	54.15	20.58
Equatorial Guinea	31.98	0.6	87.07	—	—
Eritrea	47.24	5.29	46.54	25.08	1
Estonia	—	5.09	94.75	90.81	65.96
Ethiopia	59	5.99	65.22	23.98	2.43
Fiji	28.16	6.51	90.36	80.94	15.41
Finland	15.96	6.53	98.5	95.26	91.87
France	18.64	5.81	98.56	98.98	56.08
Gabon	36.04	—	—	—	—
Gambia	35.15	2.0	61.77	38.35	1.15
Georgia	14.45	3.08	89.15	78.65	38.2
Germany	14.19	4.58	—	—	—
Ghana	31.98	5.44	63.61	37.65	4.7
Greece	11.26	4.22	99.65	91.07	90.39
Grenada	17.68	5.21	59.14	78.82	—
Guatemala	30.95	2.56	94.44	38.13	8.7
Guinea	44.46	1.63	71.57	27.61	3.02
Guinea-Bissau	—	—	—	—	—
Guyana	28.04	8.28	—	—	11.53
Haiti	—	—	—	—	—
Honduras	27.93	—	96.42	—	17.15
Hong Kong	17.76	3.91	—	77.91	32.97
Hungary	10.47	5.43	88.83	89.92	65.28
Iceland	10.53	7.58	98.13	88.67	70.4
India	40.20	3.75	88.5	—	11.03
Indonesia	20.13	0.96	94.54	57.44	17.01
Iran	19.44	5.12	93.58	77.25	26.8
Iraq	16.83	—	77.43	38.37	15.79
Ireland	18.17	4.72	94.59	86.7	58.16
Israel	12.42	6.89	97.42	89.09	58.08
Italy	10.62	4.59	98.64	92.48	65.27
Jamaica	27.53	5.29	90.27	78.28	18.99
Japan	19.15	3.66	99.83	99.99	55.31

Table G (*continued*)
World Countries: Education and Literacy, 2003–2007

LITERACY RATE, ADULT TOTAL (% OF PEOPLE AGES 15 AND ABOVE)	LITERACY RATE, ADULT MALE (% OF MALES AGES 15 AND ABOVE)	LITERACY RATE, ADULT FEMALE (% OF FEMALES AGES 15 AND ABOVE)	LITERACY RATE, YOUTH TOTAL (% OF PEOPLE AGES 15–24)	LITERACY RATE, YOUTH MALE (% OF MALES AGES 15–24)	LITERACY RATE, YOUTH FEMALE (% OF FEMALES AGES 15–24)
2006	2006	2006	2006	2006	2006
98	99	97	100	100	100
100	100	100	100	100	100
97	99	95	100	100	100
—	—	—	—	—	—
—	—	—	—	—	—
—	—	—	—	—	—
—	—	—	—	—	—
87	87	87	94	93	95
91	92	90	96	96	97
71	83	59	85	90	79
—	—	—	—	—	—
87	93	81	95	95	95
—	—	—	—	—	—
100	100	100	100	100	100
—	—	—	—	—	—
—	—	—	—	—	—
—	—	—	—	—	—
—	—	—	—	—	—
—	—	—	—	—	—
—	—	—	—	—	—
—	—	—	—	—	—
—	—	—	—	—	—
58	66	50	71	76	66
96	98	94	99	99	99
—	—	—	—	—	—
69	75	63	82	86	78
30	43	18	47	59	34
—	—	—	—	—	—
—	—	—	—	—	—
—	—	—	—	—	—
80	80	80	89	87	91
—	—	—	—	—	—
—	—	—	—	—	—
—	—	—	—	—	—
61	73	48	76	84	68
90	94	87	99	99	99
77	84	70	—	—	—
74	84	64	85	89	81
—	—	—	—	—	—
97	99	96	100	100	100
98	99	98	100	100	100
80	74	86	—	—	—
—	—	—	—	—	—

Table G (*continued*)
World Countries: Education and Literacy, 2003–2007

COUNTRY	PUPIL-TEACHER RATIO, PRIMARY	PUBLIC SPENDING ON EDUCATION (% OF GDP)	PRIMARY SCHOOL ENROLLMENT (% NET)	SECONDARY SCHOOL ENROLLMENT (% NET)	TERTIARY SCHOOL ENROLLMENT (% NET)
	2003–2007	2003–2006	2003–2006	2003–2006	2003–2006
Jordan	20.24	—	91.3	78.78	39.93
Kazakhstan	20.24	3.21	90.12	87.74	52.68
Kenya	39.54	6.88	75.82	41.54	55.31
Kiribati	24.67	—	—	68.28	39.93
Korea, North	—	—	—	—	—
Korea, South	27.88	4.63	97.59	93.86	91.03
Kuwait	10.14	3.76	83.47	—	17.56
Kyrgyzstan	23.84	4.87	85.86	80.48	42.71
Laos	31.03	3	83.71	34.93	9.09
Latvia	12.98	5.07	90.11	—	74.86
Lebanon	13.9	2.69	81.93	73.02	47.96
Lesotho	40.41	13.02	72.36	23.99	3.63
Liberia	19.27	—	39.47	—	—
Libya	—	—	—	—	55.75
Lithuania	14.72	5.07	88.03	94.17	76.46
Luxembourg	11.53	—	96.73	83.27	11.81
Macedonia	19.95	—	91.78	81.32	29.78
Madagascar	48.14	3.06	95.91	—	2.8
Malawi	63.63	5.81	91.12	24.11	0.41
Malaysia	17.51	6.24	99.18	72.02	30.64
Maldives	16.42	8	97.45	64.54	0.21
Mali	55.59	4.54	60.51	—	3.02
Malta	18.78	5.09	86.29	84.78	31.48
Marshall Islands	16.93	11.83	—	74.37	16.95
Mauritania	41.41	2.88	79.5	15.83	3.58
Mauritius	21.68	3.89	95.03	79.34	17.15
Mexico	28.47	5.09	97.74	68.6	25.25
Micronesia	16.63	11.83	—	—	—
Moldova	17.05	7.62	82.57	74.77	35.93
Mongolia	33.01	5.25	91.44	81.52	47.2
Montenegro	—	—	—	—	—
Morocco	26.96	6.76	88.11	34.51	11.83
Mozambique	67.38	5.02	69.12	4.2	1.46
Myanmar	29.9	—	99.64	45.68	—
Namibia	30.82	6.92	76.35	34.86	6.25
Nepal	39.71	3.38	79.21	—	5.64
Netherlands	—	5.19	97.91	86.6	58.98
Netherlands Antilles	19.80	—	—	81	—
New Zealand	16.37	5.19	99.15	—	82.24
Nicaragua	33.27	3.14	89.75	43.35	18.09
Niger	40.02	3.57	43.47	9.39	1.12
Nigeria	37.18	—	63.42	25.53	10.15
Norway	10.50	7.58	98.03	95.8	78.47

Table G (*continued*)
World Countries: Education and Literacy, 2003–2007

LITERACY RATE, ADULT TOTAL (% OF PEOPLE AGES 15 AND ABOVE)	LITERACY RATE, ADULT MALE (% OF MALES AGES 15 AND ABOVE)	LITERACY RATE, ADULT FEMALE (% OF FEMALES AGES 15 AND ABOVE)	LITERACY RATE, YOUTH TOTAL (% OF PEOPLE AGES 15–24)	LITERACY RATE, YOUTH MALE (% OF MALES AGES 15–24)	LITERACY RATE, YOUTH FEMALE (% OF FEMALES AGES 15–24)
2006	2006	2006	2006	2006	2006
90	95	85	99	99	99
100	100	99	100	100	100
74	78	70	80	80	81
—	—	—	—	—	—
—	—	—	—	—	—
—	—	—	—	—	—
93	95	91	100	100	100
99	99	98	100	100	100
69	77	61	79	83	75
100	100	100	100	100	100
—	—	—	—	—	—
82	74	90	—	—	—
—	—	—	—	—	—
—	—	—	—	—	—
100	100	100	100	100	100
—	—	—	—	—	—
96	98	94	99	99	99
71	77	65	70	73	68
64	75	54	76	82	71
89	92	85	97	97	97
96	96	96	98	98	98
19	27	12	24	32	17
88	86	89	96	94	98
—	—	—	—	—	—
51	60	43	61	68	56
84	88	81	95	94	95
91	93	90	98	98	98
—	—	—	—	—	—
98	99	98	100	100	100
98	98	98	98	97	98
—	—	—	—	—	—
52	66	40	71	81	61
—	—	—	—	—	—
90	94	86	95	96	93
85	87	84	92	91	94
49	63	35	70	81	60
—	—	—	—	—	—
—	—	—	—	—	—
—	—	—	—	—	—
77	77	77	86	84	89
29	43	15	37	53	23
—	—	—	—	—	—
—	—	—	—	—	—

Table G (*continued*)
World Countries: Education and Literacy, 2003–2007

COUNTRY	PUPIL-TEACHER RATIO, PRIMARY	PUBLIC SPENDING ON EDUCATION (% OF GDP)	PRIMARY SCHOOL ENROLLMENT (% NET)	SECONDARY SCHOOL ENROLLMENT (% NET)	TERTIARY SCHOOL ENROLLMENT (% NET)
	2003–2007	2003–2006	2003–2006	2003–2006	2003–2006
Oman	14.32	4.74	74.08	77.3	25.48
Pakistan	39.01	2.59	65.62	29.67	4.52
Panama	24.5	3.79	98.48	64.18	45.49
Papua New Guinea	35.82	—	—	—	—
Paraguay	27.84	4.69	93.82	—	24.91
Peru	22.24	2.72	96.4	70.22	33.93
Philippines	34.53	2.71	92.94	60.22	28.03
Poland	12.63	5.41	96.69	92.94	64.07
Portugal	11.57	5.37	98	81.64	55.11
Qatar	11.21	1.59	96.11	90.71	18.65
Romania	17.48	3.29	91.31	80.79	45.18
Russia	16.65	3.54	92.47	—	70.46
Rwanda	65.93	3.76	91.47	—	2.56
Samoa	25.04	—	90.42	66.02	—
São Tomé and Príncipe	30.78	—	96.17	32.58	—
Saudi Arabia	—	6.75	92.8	59.8	26.9
Senegal	39.01	5	70.72	17.46	5.51
Serbia	—	—	—	—	—
Seychelles	14.04	6.47	99.4	97.06	—
Sierra Leone	44.21	3.79	—	—	—
Singapore	23.52	—	—	—	—
Slovakia	17.68	4.21	92.03	—	40.69
Slovenia	15.10	5.96	95.7	90.97	79.48
Solomon Islands	—	—	61.77	27.25	—
Somalia	—	—	19	—	—
South Africa	35.64	5.4	88.31	—	15.35
Spain	13.93	4.27	99.59	93.85	66.24
Sri Lanka	22.50	—	96.67	—	—
St. Kitts and Nevis	15.45	9.34	93.39	86.07	—
St. Lucia	23.53	6.57	98.63	72.51	9.78
St. Vincent and the Grenadines	17.97	8.13	90.44	63.93	—
Sudan	34.28	—	53.7	18.9	—
Suriname	15.99	—	96.44	67.72	—
Swaziland	32.31	6.98	78.15	31.56	4.16
Sweden	10.11	7.33	97.1	99.26	81.62
Switzerland	12.78	5.99	89.9	82.18	45.37
Syria	—	—	—	62.66	—
Tajikistan	22.18	3.43	97.27	80.4	18.59
Tanzania	53.09	—	97.85	—	1.45
Thailand	18.27	4.23	94.21	71.01	45.9
Timor-Leste	50.71	—	68.08	—	—
Togo	37.56	—	80.13	—	—

Table G (*continued*)
World Countries: Education and Literacy, 2003–2007

LITERACY RATE, ADULT TOTAL (% OF PEOPLE AGES 15 AND ABOVE)	LITERACY RATE, ADULT MALE (% OF MALES AGES 15 AND ABOVE)	LITERACY RATE, ADULT FEMALE (% OF FEMALES AGES 15 AND ABOVE)	LITERACY RATE, YOUTH TOTAL (% OF PEOPLE AGES 15–24)	LITERACY RATE, YOUTH MALE (% OF MALES AGES 15–24)	LITERACY RATE, YOUTH FEMALE (% OF FEMALES AGES 15–24)
2006	2006	2006	2006	2006	2006
81	87	74	97	98	97
50	63	36	66	76	55
92	93	91	96	97	96
57	64	51	67	69	64
—	—	—	—	—	—
88	94	82	97	98	96
93	93	93	95	95	96
—	—	—	—	—	—
—	—	—	—	—	—
89	89	89	96	95	98
97	98	96	98	98	98
99	100	99	100	100	100
65	71	60	78	79	77
—	—	—	—	—	—
—	—	—	—	—	—
79	87	69	96	98	94
39	51	29	49	59	41
96	99	94	99	99	99
92	91	92	99	99	99
35	47	24	48	59	37
93	97	89	100	99	100
100	100	100	100	100	100
—	—	—	—	—	—
—	—	—	—	—	—
—	—	—	—	—	—
82	84	81	94	94	94
—	—	—	—	—	—
91	92	89	96	95	96
—	—	—	—	—	—
—	—	—	—	—	—
—	—	—	—	—	—
61	71	52	77	85	71
90	92	87	95	96	94
80	81	78	88	87	90
—	—	—	—	—	—
—	—	—	—	—	—
80	86	74	92	94	90
100	100	99	100	100	100
69	78	62	78	81	76
93	95	91	98	98	98
—	—	—	—	—	—
53	69	39	74	84	64

Table G (*continued*)
World Countries: Education and Literacy, 2003–2007

COUNTRY	PUPIL-TEACHER RATIO, PRIMARY	PUBLIC SPENDING ON EDUCATION (% OF GDP)	PRIMARY SCHOOL ENROLLMENT (% NET)	SECONDARY SCHOOL ENROLLMENT (% NET)	TERTIARY SCHOOL ENROLLMENT (% NET)
	2003–2007	2003–2006	2003–2006	2003–2006	2003–2006
Tonga	22.29	4.23	98.37	66.38	5.97
Trinidad and Tobago	17.52	—	84.63	65.11	11.39
Tunisia	20.61	7.26	97.03	64.45	30.06
Turkey	—	4.05	90.19	66.03	31.02
Turkmenistan	—	—	—	—	—
Uganda	49.05	5.24	—	16.17	3.47
Ukraine	17.12	6.3	90.17	83.84	72.78
United Arab Emirates	15.48	1.27	87.97	79.04	—
United Kingdom	18.09	5.39	98.71	95.26	59.36
United States	14.21	5.6	91.59	88.45	82.24
Uruguay	20.83	2.59	93.79	—	41.99
Uzbekistan	18.24	—	—	—	9.84
Vanuatu	46.73	9.48	87.8	38.1	4.77
Venezuela	17.19	3.7	91.07	66.95	51.96
Vietnam	20.69	—	84.45	68.93	15.94
West Bank and Gaza	32.35	—	75.98	89.59	41.09
Yemen	26.13	—	75.18	37.42	9.39
Zambia	51.08	1.99	91.99	28.07	—
Zimbabwe	38.56	—	81.74	34.38	3.64
World	27.12	4.75	88.3	65.36	30.73

Table G (*continued*)
World Countries: Education and Literacy, 2003–2007

LITERACY RATE, ADULT TOTAL (% OF PEOPLE AGES 15 AND ABOVE)	LITERACY RATE, ADULT MALE (% OF MALES AGES 15 AND ABOVE)	LITERACY RATE, ADULT FEMALE (% OF FEMALES AGES 15 AND ABOVE)	LITERACY RATE, YOUTH TOTAL (% OF PEOPLE AGES 15–24)	LITERACY RATE, YOUTH MALE (% OF MALES AGES 15–24)	LITERACY RATE, YOUTH FEMALE (% OF FEMALES AGES 15–24)
2006	2006	2006	2006	2006	2006
99	99	99	99	99	99
—	—	—	—	—	—
74	83	65	94	96	92
87	95	80	96	98	93
99	99	98	100	100	100
67	77	58	77	83	71
99	100	99	100	100	100
—	—	—	—	—	—
—	—	—	—	—	—
—	—	—	—	—	—
—	—	—	—	—	—
—	—	—	—	—	—
74	—	—	—	—	—
93	93	93	97	96	98
90	94	87	94	94	94
92	97	88	99	99	99
—	—	—	—	—	—
68	76	60	70	73	66
—	—	—	—	—	—
78	83	74	83	85	81

Table H
World Countries: Agricultural Operations, 2003–2006

COUNTRY	ARABLE LAND (HECTARES PER PERSON)	IRRIGATED LAND (% OF CROPLAND)	AGRICULTURAL INPUTS: LAND AREA UNDER CEREAL PRODUCTION (HECTARES)	AGRICULTURAL INPUTS: FERTILIZER CONSUMPTION (100 GRAMS PER HECTARE OF ARABLE LAND)	AGRICULTURAL INPUTS: TRACTORS PER 100 HECTARES OF ARABLE LAND	AGRICULTURAL OUTPUTS AND PRODUCTIVITY: CROP PRODUCTION INDEX (1999–2001 = 100)	AGRICULTURAL OUTPUTS AND PRODUCTIVITY: FOOD PRODUCTION INDEX (1999–2001 = 100)	AGRICULTURAL OUTPUTS AND PRODUCTIVITY: LIVESTOCK PRODUCTION INDEX (1999–2001 = 100)	AGRICULTURAL OUTPUTS AND PRODUCTIVITY: AGRICULTURE VALUE ADDED PER WORKER (CONSTANT 2000 US$)
	2005	2003–2005	2006	2005	2004	2004	2006	2004	2005
Afghanistan	0.32	33.8	2,797,000	26.30	1	—	—	—	—
Albania	0.18	50.5	143,100	994.71	148	100.9	111.0	108.6	1495.29
Algeria	0.23	6.9	2,672,777	144.21	129	128	143.1	104.6	2218.57
Angola	0.21	2.23	1,477,960	22.60	31	119.2	152.4	100	196.19
Antigua and Barbuda	0.10	—	32	—	300	112.9	107.8	104.9	2751.33
Argentina	0.74	—	9,163,153	489.68	107	104.9	114.8	94.4	10761.74
Armenia	0.16	51.53	112,110	208.65	287	118.7	146.4	108.3	4198.44
Australia	2.42	4.83	16,863,500	448.42	67	92.1	82.4	93.2	35807.24
Austria	0.17	2.75	776,783	2,331.87	2,379	103.8	95.1	99.3	23093.22
Azerbaijan	0.22	69.26	773,877	174.01	164	125.5	140.5	113.7	1212.23
Bahamas	0.02	8.33	59	—	150	111.8	104.8	96.3	—
Bahrain	0	66.67	—	9,200.00	75	96	99.0	113.7	—
Bangladesh	0.05	56.12	11,795,808	2,210.11	7	105	118.5	102.6	346.26
Barbados	0.05	29.41	95	885.00	366	91.2	104.0	95.3	15532.80
Belarus	0.56	2.02	2,224,354	2,252.92	106	129.6	125.9	107.7	3444.67
Belgium	0.08	4.66	317,502	—	1,257	106.3	94.4	99.7	46209.13
Belize	0.24	2.94	20,447	810.00	164	94.7	118.3	143.6	6690.46
Benin	0.32	0.41	830,187	0.05	1	134.1	112.3	115.6	535.67
Bermuda	0.02	—	—	—	450	—	—	—	—
Bhutan	0.25	23.53	57,198	—	8	92.5	94.5	99.1	137.54
Bolivia	0.33	4.05	901,784	81.13	20	119.9	117.7	109.2	783.46
Bosnia-Herzegovina	0.26	0.27	312,527	596.92	289	102.1	123.9	88.5	10050.57
Botswana	0.21	0.26	75,953	—	159	75.6	103.2	107.7	367.15
Brazil	0.32	4.38	18,371,763	1,365.53	137	126.7	127.7	123.6	3239.88
Brunei	0.04	5.56	472	—	55	111.5	119.0	136.4	86425.97
Bulgaria	0.41	16.62	1,548,461	1,770.16	95	106	84.4	95.9	7238.94
Burkina Faso	0.35	0.51	3,424,700	126.27	4	140.5	130.0	108.8	178.86
Burundi	0.12	1.55	208,700	34.91	2	104.2	101.2	100.2	64.36
Cambodia	0.27	7.01	2,621,712	0.00	7	115.1	144.4	101.8	329.02
Cameroon	0.33	0.36	1,070,300	78.74	1	103.3	112.2	102.4	677.45
Canada	1.41	1.51	16,790,000	612.88	160	102.8	109.9	103.5	45849.29
Cape Verde	0.09	6.12	33,297	—	3	90.3	91.8	102.1	1510.13
Central African Republic	0.46	0.1	162,000	—	0	102.8	108.1	114	383.89
Chad	0.41	0.78	2,552,662	—	0	112.6	122.1	107.6	225.30
Chile	0.12	80.99	594,456	3,018.80	272	107	123.2	107.7	5666.46
China	0.11	35.59	83,949,650	3,410.16	96	115.8	127.0	120.2	423.28
Colombia	0.04	24.01	1,080,533	2,329.70	93	110.8	111.9	103.4	2914.36
Comoros	0.13	—	15,700	—	1	102.6	98.7	109.1	435.83

Table H (*continued*)
World Countries: Agricultural Operations, 2003–2006

COUNTRY	ARABLE LAND (HECTARES PER PERSON)	IRRIGATED LAND (% OF CROPLAND)	AGRICULTURAL INPUTS: LAND AREA UNDER CEREAL PRODUCTION (HECTARES)	AGRICULTURAL INPUTS: FERTILIZER CONSUMPTION (100 GRAMS PER HECTARE OF ARABLE LAND)	AGRICULTURAL INPUTS: TRACTORS PER 100 HECTARES OF ARABLE LAND	AGRICULTURAL OUTPUTS AND PRODUCTIVITY: CROP PRODUCTION INDEX (1999–2001 = 100)	AGRICULTURAL OUTPUTS AND PRODUCTIVITY: FOOD PRODUCTION INDEX (1999–2001 = 100)	AGRICULTURAL OUTPUTS AND PRODUCTIVITY: LIVESTOCK PRODUCTION INDEX (1999–2001 = 100)	AGRICULTURAL OUTPUTS AND PRODUCTIVITY: AGRICULTURE VALUE ADDED PER WORKER (CONSTANT 2000 US$)
	2005	2003–2005	2006	2005	2004	2004	2006	2004	2005
Congo, Dem. Rep.	0.11	0.14	1,941,924	—	4	97.3	95.6	97.1	148.96
Congo, Rep.	0.14	0.37	11,900	—	14	104.9	112.4	120.2	—
Costa Rica	0.05	20.19	54,798	10,345.11	311	91.8	109.5	97.1	4621.55
Côte d'Ivoire	0.19	1.06	787,500	142.24	12	86.4	105.3	113.2	817.38
Croatia	0.25	0.69	558,330	1,241.00	29	99.8	95.2	105.4	10928.65
Cuba	0.33	19.52	267,397	190.52	255	116.4	78.5	96	—
Cyprus	0.16	20.5	59,912	1,410.42	1,491	96.3	87.5	110.4	—
Czech Republic	0.30	0.73	1,534,896	1,524.85	299	109.8	89.1	95.3	5944.93
Denmark	0.41	11.23	1,498,300	598.50	543	101	98.0	100.6	45579.32
Djibouti	0	—	6	—	60	100.4	105.3	108.5	64.94
Dominica	0.07	—	140	—	180	98	89.8	100	4807.25
Dominican Republic	0.09	20.83	169,000	—	17	107.8	127.1	100.4	4943.09
Ecuador	0.1	29.35	804,139	1,450.10	113	99.7	111.7	126.8	1777.91
Egypt	0.04	100.38	3,031,962	7,330.69	309	105.1	114.8	119.2	2128.22
El Salvador	0.10	4.95	335,010	1,090.50	52	89.8	117.1	104.9	1699.81
Equatorial Guinea	0.27	—	—	—	13	93.8	93.4	101.9	1197.86
Eritrea	0.14	3.5	374,024	21.96	8	70.7	86.3	95.9	63.13
Estonia	0.44	0.71	280,235	4,867.41	992	110.5	98.4	100.9	3290.95
Ethiopia	0.17	2.46	8,426,823	25.99	3	104.8	134.5	106.9	161.61
Fiji	0.24	1.05	6,924	332.95	282	90.3	97.4	97.7	1904.58
Finland	0.43	2.86	1,150,500	1,343.67	878	109.2	103.8	104.9	33558.85
France	0.30	13.28	9,106,162	2,031.61	685	104.3	91.9	97.3	46962.85
French Polynesia	0.01	4	—	2,506.67	967	109	111.7	93.8	—
Gabon	0.25	1.41	19,854	26.86	46	101.5	103.3	100.7	1662.80
Gambia	0.22	0.6	209,731	—	1	65.6	95.9	104.5	243.37
Georgia	0.18	44	274,630	377.77	273	87	97.9	109	1936.63
Germany	0.14	4.03	6,702,200	2,121.80	798	105.5	96.0	101	28654.18
Ghana	0.19	0.49	1,438,030	75.40	9	121.9	131.5	111.2	331.93
Greece	0.24	37.93	955,529	1,651.77	945	95.4	91.9	96.3	9239.52
Grenada	0.02	—	304	—	60	99.7	99.0	101.4	1521.84
Guam	0.01	—	13	—	375	108	107.0	98.8	—
Guatemala	0.11	6.34	849,920	1,378.68	30	103.2	114.5	93.3	2578.13
Guinea	0.13	5.4	1,703,023	28.19	5	104.7	124.4	117.6	192.97
Guinea-Bissau	0.19	4.55	138,600	—	1	102.1	114.5	103.1	246.11
Guyana	0.65	29.41	155,778	162.90	76	100.1	82.7	145.3	3382.64
Haiti	0.08	8.36	444,946	—	2	100.6	101.8	108.3	—

Table H (*continued*)
World Countries: Agricultural Operations, 2003–2006

COUNTRY	ARABLE LAND (HECTARES PER PERSON)	IRRIGATED LAND (% OF CROPLAND)	AGRICULTURAL INPUTS: LAND AREA UNDER CEREAL PRODUCTION (HECTARES)	AGRICULTURAL INPUTS: FERTILIZER CONSUMPTION (100 GRAMS PER HECTARE OF ARABLE LAND)	AGRICULTURAL INPUTS: TRACTORS PER 100 HECTARES OF ARABLE LAND	AGRICULTURAL OUTPUTS AND PRODUCTIVITY: CROP PRODUCTION INDEX (1999–2001 = 100)	AGRICULTURAL OUTPUTS AND PRODUCTIVITY: FOOD PRODUCTION INDEX (1999–2001 = 100)	AGRICULTURAL OUTPUTS AND PRODUCTIVITY: LIVESTOCK PRODUCTION INDEX (1999–2001 = 100)	AGRICULTURAL OUTPUTS AND PRODUCTIVITY: AGRICULTURE VALUE ADDED PER WORKER (CONSTANT 2000 US$)
	2005	2003–2005	2006	2005	2004	2004	2006	2004	2005
Honduras	0.16	5.6	350,100	584.81	50	120.1	168.3	101.5	1217.16
Hong Kong	—	—	—	—	—	—	—	—	—
Hungary	0.46	1.69	2,880,445	1,233.61	247	119.4	98.8	99.3	8102.27
Iceland	0.02	—	—	25,248.57	15,413	88.4	105.7	105.1	53483.28
India	0.15	32.94	96,640,000	1,289.59	159	102.1	108.7	111.2	403.08
Indonesia	0.1	12.36	14,746,427	1,496.88	41	113.1	126.0	125.8	596.12
Iran	0.24	47.35	8,537,920	762.18	153	116.6	127.5	103.1	2632.39
Iraq	0.21	58.56	3,844,000	—	80	—	—	—	—
Ireland	0.29	—	278,480	4,618.70	1,311	102.1	98.9	98.2	16657.57
Israel	0.05	—	84,175	21,977.26	725	94.2	115.5	117.2	—
Italy	0.13	25.84	3,803,054	1,568.93	2,111	95.6	96.2	98.1	25419.89
Jamaica	0.07	—	2,015	309.48	177	97.4	105.9	101.8	1758.83
Japan	0.03	36.27	2,006,139	—	4,612	95.8	101.1	99.6	37388.92
Jordan	0.03	29.63	64,051	7,882.72	328	127.4	136.1	98.8	1391.67
Kazakhstan	1.48	15.67	14,787,300	64.27	22	98	126.4	112	1651.64
Kenya	0.15	1.85	2,361,474	177.86	28	96	128.5	108.7	344.59
Kiribati	0.02	—	—	—	90	105.1	111.5	128.6	170.16
Korea, North	0.12	50.34	1,253,054	—	238	110	116.0	113.6	—
Korea, South	0.03	47.56	1,051,745	4,448.32	1,285	90.5	94.9	98.7	12274.96
Kuwait	0.01	72.22	1,593	31,893.33	70	109.6	108.9	120.6	—
Kyrgyzstan	0.25	73.82	633,510	223.15	163	94.9	104.6	69.4	965.87
Laos	0.18	16.49	846,000	—	11	121.9	116.8	102	456.12
Latvia	0.47	2.06	515,000	723.89	578	123.1	114.3	108.2	2974.05
Lebanon	0.05	31.33	65,756	1,357.80	488	95.6	104.2	117.7	32501.04
Lesotho	0.17	0.9	184,046	—	61	111.2	106.0	100	412.33
Liberia	0.11	0.5	—	—	9	96.7	101.8	114.9	—
Libya	0.3	21.86	342,877	671.28	219	94.8	104.7	100.5	—
Lithuania	0.56	0.45	962,900	1,198.01	671	112.4	89.3	97.8	4959.46
Luxembourg	0.13	—	28,849	4,907.17	—	100.8	94.9	94	33658.87
Macedonia	0.28	8.99	184,092	224.93	954	92.9	104.2	103.2	3740.21
Madagascar	0.16	30.59	1,509,121	53.72	12	102.8	114.7	101.3	175.43
Malawi	0.20	2.16	1,545,200	352.83	6	91.8	91.8	101.8	109.02
Malaysia	0.07	4.81	670,000	8,432.16	241	117.1	128.9	116.7	5377.81
Maldives	0.01	—	87	220.00	—	112.6	94.0	—	—
Mali	0.41	4.88	3,211,453	—	5	108.1	109.6	118.4	243.87
Malta	0.02	18.18	2,600	773.33	500	76.6	93.5	101.8	—
Mauritania	0.17	—	223,570	—	8	99.5	108.8	108.4	356.31
Mauritius	0.08	20.75	58	2,574.70	37	103.4	107.0	113.4	5338.06
Mexico	0.24	22.82	10,046,900	692.30	131	105.6	115.2	108.6	2799.17
Moldova	0.48	10.66	875,737	148.79	221	109.8	105.5	110.5	890.87

Table H (*continued*)
World Countries: Agricultural Operations, 2003–2006

			AGRICULTURAL INPUTS			AGRICULTURAL OUTPUTS AND PRODUCTIVITY			
	ARABLE LAND (HECTARES PER PERSON)	IRRIGATED LAND (% OF CROPLAND)	LAND AREA UNDER CEREAL PRODUCTION (HECTARES)	FERTILIZER CONSUMPTION (100 GRAMS PER HECTARE OF ARABLE LAND)	TRACTORS PER 100 HECTARES OF ARABLE LAND	CROP PRODUCTION INDEX (1999–2001 = 100)	FOOD PRODUCTION INDEX (1999–2001 = 100)	LIVESTOCK PRODUCTION INDEX (1999–2001 = 100)	AGRICULTURE VALUE ADDED PER WORKER (CONSTANT 2000 US$)
COUNTRY	2005	2003–2005	2006	2005	2004	2004	2006	2004	2005
Mongolia	0.45	7	126,205	36.53	42	105	79.8	92.1	1029.52
Montenegro	—	—	4,955	—	—	—	—	—	—
Morocco	0.28	15.41	5,605,889	425.67	58	133.8	142.1	100.6	1656.77
Mozambique	0.21	2.58	2,029,355	16.22	13	105.9	144.3	104	156.63
Myanmar	0.21	17.03	7,868,000	2.11	11	116.6	134.8	122.8	—
Namibia	0.40	0.98	296,872	19.23	39	106.6	92.9	91.3	1134.17
Nepal	0.09	46.96	3,360,334	96.23	24	112.4	119.4	106.8	212.10
Netherlands	0.06	60.23	214,697	6,230.85	1,650	98.7	90.1	90.9	44606.49
Netherlands Antilles	0.04	—	—	—	25	—	—	—	—
New Caledonia	0.03	100	1,495	3,206.67	3,242	104.6	—	99.9	—
New Zealand	0.36	—	118,600	7,026.17	507	105.2	114.6	115.5	26218.17
Nicaragua	0.35	2.82	466,259	291.70	15	117.9	134.5	134.1	2172.41
Niger	1.09	0.5	6,736,200	3.18	0	122.5	137.7	103.5	—
Nigeria	0.23	0.84	19,172,000	67.24	10	104.6	130.0	108.8	1025.06
Norway	0.19	—	318,391	1,908.51	1,484	103	99.2	98.3	40427.09
Oman	0.02	90	4,503	6,167.90	52	88.9	92.1	96.6	—
Pakistan	0.14	85.36	12,861,400	1,912.73	149	105.2	117.1	112.5	716.87
Panama	0.17	6.19	182,983	343.05	148	106.2	100.4	106.5	4033.37
Papua New Guinea	0.04	—	3,208	714.00	52	102.2	107.7	112.7	—
Paraguay	0.71	1.72	828,120	541.76	54	118.8	118.7	100.7	2046.90
Peru	0.14	27.84	1,118,701	816.66	36	94.9	126.1	81.9	1546.69
Philippines	0.07	14.49	6,730,683	1,425.79	20	109.5	121.8	123.2	1097.68
Poland	0.32	0.57	8,381,129	1,280.72	1,089	95.8	93.3	106.2	2259.75
Portugal	0.12	23.76	356,702	1,709.40	1,049	99.7	92.8	100.7	5841.75
Puerto Rico	0.02	—	267	—	—	99.9	91.1	100.2	—
Qatar	0.02	—	1,944	4,082.78	46	98.3	143.9	159.3	—
Romania	0.43	0.47	5,079,834	496.76	180	132.6	111.5	119.1	5293.53
Russia	0.85	3.62	40,574,200	159.34	48	116.9	115.1	107.7	2626.06
Rwanda	0.13	0.61	327,140	—	0	113.8	122.2	105.5	217.36
Samoa	0.16	—	—	18.33	16	103.7	108.2	94.8	1720.86
São Tomé and Príncipe	0.06	18.18	1,100	—	156	101.8	110.3	102.8	—
Saudi Arabia	0.15	42.65	688,298	939.03	28	99.8	116.4	104.1	16650.61
Senegal	0.22	4.78	1,124,322	254.42	3	81.9	83.9	101.1	226.82
Serbia	—	—	1,855,794	—	956	126.6	—	94.5	—
Seychelles	0.01	—	—	—	400	96.8	104.4	99.5	432.93
Sierra Leone	0.11	4.65	776,772	—	1	109.4	195.0	112.3	—
Singapore	0	—	—	169,600.00	1,083	100	70.2	70.7	46120.14

Table H (*continued*)
World Countries: Agricultural Operations, 2003–2006

COUNTRY	ARABLE LAND (HECTARES PER PERSON)	IRRIGATED LAND (% OF CROPLAND)	AGRICULTURAL INPUTS: LAND AREA UNDER CEREAL PRODUCTION (HECTARES)	AGRICULTURAL INPUTS: FERTILIZER CONSUMPTION (100 GRAMS PER HECTARE OF ARABLE LAND)	AGRICULTURAL INPUTS: TRACTORS PER 100 HECTARES OF ARABLE LAND	AGRICULTURAL OUTPUTS AND PRODUCTIVITY: CROP PRODUCTION INDEX (1999–2001 = 100)	FOOD PRODUCTION INDEX (1999–2001 = 100)	LIVESTOCK PRODUCTION INDEX (1999–2001 = 100)	AGRICULTURE VALUE ADDED PER WORKER (CONSTANT 2000 US$)
	2005	2003–2005	2006	2005	2004	2004	2006	2004	2005
Slovakia	0.26	1.83	732,875	1,125.41	—	—	85.2	—	5848.43
Slovenia	0.09	0.99	96,186	3,715.80	—	94.2	99.9	111.9	44893.10
Solomon Islands	0.04	—	1,400	—	5	107.2	115.6	112.8	—
Somalia	0.16	15.67	681,455	—	16	—	—	—	—
South Africa	0.31	9.53	3,007,760	451.41	44	98.5	107.2	109.8	2635.91
Spain	0.32	20.38	6,291,248	1,286.38	694	103	102.4	111	18517.36
Sri Lanka	0.05	38.78	938,270	3,070.96	115	94.4	101.2	105.9	702.94
St. Kitts and Nevis	0.15	—	—	—	221	99.6	100.0	100	2228.25
St. Lucia	0.02	16.67	—	—	365	116.4	91.8	82.4	1246.24
St. Vincent and the Grenadines	0.04	12.5	113	—	114	106	108.8	87	2214.58
Sudan	0.53	10.24	9,354,200	25.81	7	124.4	116.1	107.1	665.91
Suriname	0.13	75	52,005	504.00	229	105.2	94.6	105.2	3166.48
Swaziland	0.16	26.04	49,284	—	222	93.9	101.0	115.2	1275.12
Sweden	0.30	4.3	969,225	888.57	618	108.6	92.9	97.2	33433.63
Switzerland	0.06	5.8	163,926	2,229.76	2,649	90.1	99.1	101.7	24364.10
Syria	0.26	24.28	3,168,374	985.02	229	116.6	119.5	107.9	3381.89
Tajikistan	0.14	68.31	394,400	—	238	133.6	145.8	122	484.61
Tanzania	0.24	1.78	3,449,631	103.67	19	104.7	107.4	109.1	305.91
Thailand	0.23	28.19	11,116,733	1,214.11	156	105.3	110.8	89.3	606.55
Timor-Leste	0.13	—	115,000	—	9	106.3	112.9	104.6	280.80
Togo	0.4	0.27	783,163	81.76	0	111.9	122.7	109.6	352.82
Tonga	0.15	—	—	—	100	103	102.2	99.9	3115.93
Trinidad and Tobago	0.06	3.28	1,387	8,913.20	360	80.2	105.6	142.7	1932.85
Tunisia	0.27	7.39	1,321,804	644.18	126	94.5	115.9	98.7	2686.49
Turkey	0.33	19.6	13,917,765	852.38	427	104.7	111.6	106.8	1968.92
Turkmenistan	0.48	79.47	1,061,000	—	227	122.6	147.3	96.8	—
Uganda	0.19	0.12	1,679,000	10.66	9	107.6	105.5	117.1	234.83
Ukraine	0.69	6.61	13,820,000	186.23	120	125.6	117.1	107.8	1872.45
United Arab Emirates	0.02	29.92	2	2,322.03	59	59.1	63.7	119.1	27486.77
United Kingdom	0.10	2.98	2,859,593	2,898.26	884	99.3	98.1	97.2	27686.08
United States	0.59	12.48	54,233,610	1,104.84	269	110.5	104.8	102	44943.41
Uruguay	0.41	14.87	561,360	1,356.04	241	122	121.2	98.4	8481.57
Uzbekistan	0.18	84.94	1,612,270	—	362	106.3	146.8	104.7	1926.66
Vanuatu	0.09	—	1,300	—	38	96.4	108.8	95.9	1301.67

Table H (*continued*)
World Countries: Agricultural Operations, 2003–2006

			AGRICULTURAL INPUTS			AGRICULTURAL OUTPUTS AND PRODUCTIVITY			
	ARABLE LAND (HECTARES PER PERSON)	IRRIGATED LAND (% OF CROPLAND)	LAND AREA UNDER CEREAL PRODUCTION (HECTARES)	FERTILIZER CONSUMPTION (100 GRAMS PER HECTARE OF ARABLE LAND)	TRACTORS PER 100 HECTARES OF ARABLE LAND	CROP PRODUCTION INDEX (1999–2001 = 100)	FOOD PRODUCTION INDEX (1999–2001 = 100)	LIVESTOCK PRODUCTION INDEX (1999–2001 = 100)	AGRICULTURE VALUE ADDED PER WORKER (CONSTANT 2000 US$)
COUNTRY	2005	2003–2005	2006	2005	2004	2004	2006	2004	2005
Venezuela	0.10	16.91	1,190,539	1,655.46	188	93.4	104.5	103.4	6798.33
Vietnam	0.08	33.73	8,357,600	3,007.42	248	118.3	128.3	118.9	313.21
West Bank and Gaza	0.03	6.76	32,600	—	—	—	—	—	—
Yemen	0.07	32.95	756,961	18.21	42	100	127.1	113	—
Zambia	0.46	2.95	583,625	—	11	106.5	107.0	98.9	210.78
Zimbabwe	0.25	5.19	1,656,224	338.65	75	80.2	89.1	101.7	204.57
World (Averages or Totals)	0.23	19.2	673,801,019	3290.47	202	—	—	—	8361.54

Table I
World Countries: Deforestation and Biodiversity, 1990–2007

COUNTRY	FOREST AREA IN THOUSANDS OF SQUARE KILOMETERS		AVERAGE ANNUAL DEFORESTATION AS A PERCENTAGE OF FOREST AREA		ANIMAL SPECIES		HIGHER PLANT SPECIES		GEF INDEX OF BIODIVERSITY	NATIONALLY PROTECTED AREAS		MARINE PROTECTED AREAS	
					TOTAL KNOWN SPECIES	THREATENED SPECIES	TOTAL KNOWN SPECIES	THREATENED SPECIES	0 = NO BIODIVERSITY 100= MAX. BIODIVERSITY	THOUSAND SQ. KM.	% OF TOTAL LAND AREA	THOUSAND SQ. KM.	% OF SURFACE AREA
	1990	2005	1990–2000	2000–2005	2004	2007	2004	2007	2005	2004	2004	2004	2004
Afghanistan	13	9	2.5	3.1	578	33	4000	2	3.6	2.2	0.3	—	—
Albania	8	8	0.3	–0.6	376	45	3031	0	0.2	0.7	2.7	0.3	1
Algeria	18	23	–1.8	–1.2	472	71	3164	3	3	118.6	5	0.9	0
Angola	610	591	0.2	0.2	1226	62	5185	26	9.6	125.5	10.1	29.1	2.3
Argentina	353	330	0.4	0.4	1413	152	9372	42	18.5	174.5	6.4	7.8	0.3
Armenia	3	3	1.3	1.5	380	35	3553	1	0.3	3	10.6	—	—
Australia	1679	1637	0.2	0.1	1227	568	15638	55	95.8	745.3	9.7	680.8	8.8
Austria	38	39	–0.2	–0.1	513	62	3100	4	0.3	23.5	28.5	—	—
Azerbaijan	9	9	0	0	446	38	4300	0	0.9	4	4.8	1.2	1.4
Bangladesh	9	9	0	0.3	735	89	5000	12	1.6	0.7	0.5	0.3	0.2
Belarus	74	79	–0.6	–0.1	297	17	2100	—	0	13.2	6.3	—	—
Belgium	7	7	0.1	0	519	29	1550	1	0	1	3.5	0	0
Benin	33	24	2.1	2.5	644	34	2500	14	0.2	26.4	23.9	—	—
Bolivia	628	587	0.4	0.5	1775	80	17367	71	13.8	211	19.5	—	—
Bosnia-Herzegovina	22	22	0.1	0	390	55	—	1	0.4	0.3	0.5	—	—
Botswana	137	119	0.9	1	739	18	2151	0	1.5	174.9	30.9	—	—
Brazil	5200	4777	0.5	0.6	2290	343	56215	382	100	1532.6	18.1	47.4	0.6
Bulgaria	33	36	–0.1	–1.4	485	47	3572	0	0.9	11.2	10.3	0	0
Burkina Faso	72	68	0.3	0.3	581	13	1100	2	0.3	42.1	15.4	—	—
Burundi	3	2	3.7	5.2	713	48	2500	2	0.5	1.5	5.7	—	—
Cambodia	129	104	1.1	2	648	82	—	31	3.9	41.5	23.5	1.9	1.1
Cameroon	245	212	0.9	1	1258	157	8260	355	13.3	37.4	8	3.9	0.8
Canada	3101	3101	0	0	683	77	3270	1	22.2	628.7	6.9	362.7	3.6
Central African Republic	232	228	0.1	0.1	850	17	3602	15	1.7	103.3	16.6	—	—
Chad	131	119	0.6	0.7	635	21	1600	2	2.1	119.8	9.5	—	—
Chile	153	161	–0.4	–0.4	604	95	5284	39	16.2	26.9	3.6	114.5	15.1
China	1571	1973	–1.2	–2.2	1801	351	32200	446	64.8	1100.7	11.8	16	0.2
Hong Kong	—	—	—	—	363	37	—	6	—	0.3	24.7	0.3	—
Colombia	614	607	0.1	0.1	2288	382	51220	222	57.3	825.3	74.4	8.1	0.7
Congo, Dem. Rep.	1405	1336	0.4	0.2	1578	126	11007	65	17	194.4	8.6	—	—
Congo, Rep.	227	225	0.1	0.1	763	37	6000	35	3.4	61.3	18	—	—
Costa Rica	26	24	0.8	–0.1	1070	131	12119	111	11.1	12.1	23.6	4.8	9.4
Côte d'Ivoire	102	104	–0.1	–0.1	931	73	3660	105	3.9	54.5	17.1	0.3	0.1
Croatia	21	21	–0.1	–0.1	461	78	4288	1	0.5	3.6	6.5	2.5	4.4
Cuba	21	27	–1.7	–2.2	423	115	6522	163	13.5	1.5	1.4	31.7	28.6

Table I (*continued*)
World Countries: Deforestation and Biodiversity, 1990–2007

COUNTRY	FOREST AREA IN THOUSANDS OF SQUARE KILOMETERS		AVERAGE ANNUAL DEFORESTATION AS A PERCENTAGE OF FOREST AREA		ANIMAL SPECIES		HIGHER PLANT SPECIES		GEF INDEX OF BIODIVERSITY	NATIONALLY PROTECTED AREAS		MARINE PROTECTED AREAS	
					TOTAL KNOWN SPECIES	THREATENED SPECIES	TOTAL KNOWN SPECIES	THREATENED SPECIES	0 = NO BIODIVERSITY 100= MAX. BIODIVERSITY	THOUSAND SQ. KM.	% OF TOTAL LAND AREA	THOUSAND SQ. KM.	% OF SURFACE AREA
	1990	2005	1990–2000	2000–2005	2004	2007	2004	2007	2005	2004	2004	2004	2004
Czech Republic	26	26	0	–0.1	474	39	1900	4	0.1	14.4	18.7	—	—
Denmark	4	5	–0.9	–0.6	508	28	1450	3	0.2	10.9	25.7	5.1	11.8
Dominican Republic	14	14	0	0	260	81	5657	30	6.8	11.9	24.6	8.6	17.6
Ecuador	138	109	1.5	1.7	1856	340	19362	1838	30	67.2	24.3	141	49.7
Egypt	0	1	–3.0	–2.6	599	59	2076	2	3.2	56	5.6	76.7	7.7
El Salvador	4	3	1.5	1.7	571	29	2911	26	0.8	0.4	1.9	0.1	0.4
Eritrea	16	16	0.3	0.3	607	38	—	3	0.9	5	5	—	—
Estonia	22	23	–0.4	–0.4	334	14	1630	0	0	8.9	21.1	—	—
Ethiopia	151	130	1	1.1	1127	86	6603	22	8.5	186.2	18.6	—	—
Finland	222	225	–0.1	0	501	19	1102	1	0.2	29.5	9.7	1.1	0.3
France	145	156	–0.5	–0.3	665	117	4630	7	3.9	16.2	3	0.5	0.1
Gabon	219	218	0	0	798	43	6651	108	3.4	8.8	3.4	1	0.4
Gambia	4	5	–0.4	–0.4	668	31	974	4	0.1	0.3	3.5	0.2	1.9
Georgia	28	28	0	0	366	46	4350	0	0.7	3	4.3	0	0.1
Germany	107	111	–0.3	0	613	59	2682	12	0.7	111.5	32	9.1	2.6
Ghana	74	55	2	2	978	56	3725	117	2	36.9	16.2	—	—
Greece	33	38	–0.9	–0.8	530	95	4992	11	3	4.3	3.3	2.5	1.9
Guatemala	47	39	1.2	1.3	877	133	8681	84	8.9	25.4	23.4	0.1	0.1
Guinea	74	67	0.7	0.5	855	61	3000	22	2.6	15.6	6.4	—	—
Guinea-Bissau	22	21	0.4	0.5	560	29	1000	4	0.7	0	0	—	—
Haiti	1	1	0.6	0.7	312	91	5242	29	5.8	0.1	0.3	—	—
Honduras	74	46	3	3.1	900	102	5680	110	7.9	23.4	21	1.9	1.7
Hungary	18	20	–0.6	–0.7	455	55	2214	1	0.2	8.3	9.3	—	—
India	639	677	–0.6	0	1602	313	18664	247	43.9	156.3	5.3	16.1	0.5
Indonesia	1166	885	1.7	2	2271	464	29375	386	90	259.9	14.3	130.1	6.8
Iran	111	111	0	0	656	75	8000	1	7.9	105.5	6.5	6.2	0.4
Iraq	8	8	–0.2	–0.1	498	40	—	0	1.7	0	0	—	—
Ireland	4	7	–3.3	–1.9	471	15	950	1	0.7	0.8	1.1	0	0
Israel	2	2	–0.6	–0.8	649	79	2317	0	0.9	4.6	21.3	0.1	0.6
Italy	84	100	–1.2	–1.1	610	119	5599	19	4.4	32.4	11	1.5	0.5
Jamaica	3	3	0.1	0.1	333	61	3308	209	4.9	1.8	16.2	8.2	74.5
Japan	250	249	0	0	763	190	5565	12	41.4	52.2	14.3	10.6	2.8
Jordan	1	1	0	0	490	43	2100	0	0.3	9.7	11	0	0
Kazakhstan	34	33	0.2	0.2	642	55	6000	16	5.4	77.4	2.9	0.5	0
Kenya	37	35	0.3	0.3	1510	172	6506	103	9.9	71.9	12.6	3.1	0.5
Korea, North	82	62	1.8	1.9	474	44	2898	3	0.7	3.2	2.6	—	—
Korea, South	64	63	0.1	0.1	512	54	2898	0	1.8	3.5	3.6	3.5	3.5

Table I (*continued*)
World Countries: Deforestation and Biodiversity, 1990–2007

COUNTRY	FOREST AREA IN THOUSANDS OF SQUARE KILOMETERS		AVERAGE ANNUAL DEFORESTATION AS A PERCENTAGE OF FOREST AREA		ANIMAL SPECIES		HIGHER PLANT SPECIES		GEF INDEX OF BIODIVERSITY	NATIONALLY PROTECTED AREAS		MARINE PROTECTED AREAS	
					TOTAL KNOWN SPECIES	THREATENED SPECIES	TOTAL KNOWN SPECIES	THREATENED SPECIES	0 = NO BIODIVERSITY 100= MAX. BIODIVERSITY	THOUSAND SQ. KM.	% OF TOTAL LAND AREA	THOUSAND SQ. KM.	% OF SURFACE AREA
	1990	2005	1990–2000	2000–2005	2004	2007	2004	2007	2005	2004	2004	2004	2004
Kuwait	0	0	–5.2	–3.7	381	23	234	—	0.1	0	0	0.3	1.5
Kyrgyzstan	8	9	–0.3	–0.3	265	22	4500	14	1.2	7.2	3.7	—	—
Laos	173	161	0.5	0.5	919	77	8286	21	5.4	37.4	16.2	—	—
Latvia	28	29	–0.4	–0.4	393	23	1153	0	0	9.7	15.6	0.2	0.2
Lebanon	1	1	–0.8	–0.8	447	38	3000	0	0.2	0.1	0.7	0	0
Lesotho	0	0	–3.4	–2.7	370	11	1591	1	0.3	0.1	0.2	—	—
Liberia	41	32	1.6	1.8	759	60	2200	46	2.9	15.2	15.8	0.6	0.5
Libya	2	2	0	0	413	31	1825	1	1.7	1.2	0.1	0.5	0
Lithuania	19	21	–0.4	–0.8	298	20	1796	—	0	5.9	9.5	0.5	0.8
Macedonia	9	9	0	0	380	34	3500	0	0.2	2	7.9	—	—
Madagascar	137	128	0.5	0.3	427	262	9505	280	31.4	18.3	3.1	0.2	0
Malawi	39	34	0.9	0.9	865	141	3765	14	3.9	19.4	20.6	—	—
Malaysia	224	209	0.4	0.7	1083	225	15500	686	14.8	100.8	30.7	5	1.5
Mali	141	126	0.7	0.8	758	21	1741	6	1.6	46.7	3.8	—	—
Mauritania	4	3	2.7	3.4	615	44	1100	—	1.4	2.5	0.2	15	1.5
Mauritius	0	0	0.3	0.5	151	65	750	88	4.2	0.1	3.3	0.1	4.4
Mexico	690	642	0.5	0.4	1570	579	26071	261	75.8	99	5.1	82.1	4.2
Moldova	3	3	–0.2	–0.2	253	28	1752	0	0	0.5	1.4	—	—
Mongolia	115	103	0.7	0.8	527	38	2823	0	4.4	217.9	13.9	—	—
Morocco	43	44	–0.1	–0.2	559	76	3675	2	4	4.7	1.1	0.5	0.1
Mozambique	200	193	0.3	0.3	913	93	5692	46	8.2	45.3	5.8	22.5	2.8
Myanmar	392	322	1.3	1.4	1335	118	7000	38	10.6	35.3	5.4	0.2	0
Namibia	88	77	0.9	0.9	811	55	3174	24	5.9	46	5.6	74	9
Nepal	48	36	2.1	1.4	477	72	6973	7	2.2	26.6	18.6	—	—
Netherlands	3	4	–0.4	–0.3	539	26	1221	0	0.1	9.5	28	0.8	1.9
New Zealand	77	83	–0.6	–0.2	424	124	2382	21	22.3	64.7	24.2	22.7	8.4
Nicaragua	65	52	1.6	1.3	813	59	7590	39	3.6	28.1	23.1	1.3	1
Niger	19	13	3.7	1	616	20	1460	2	0.9	96.9	7.7	—	—
Nigeria	172	111	2.7	3.3	1189	79	4715	171	6.6	55	6	—	—
Norway	91	94	–0.2	–0.2	525	32	1715	2	1.6	19.7	6.5	1.3	0.4
Oman	0	0	0	0	557	50	1204	6	4.4	0.2	0.1	29.6	9.6
Pakistan	25	19	1.8	2.1	820	78	4950	2	5.1	73.1	9.5	2.2	0.3
Panama	44	43	0.2	0.1	1145	121	9915	194	11.7	13.1	17.6	10	13.3
Papua New Guinea	315	294	0.5	0.5	980	158	11544	142	27.7	7.3	1.6	3.5	0.8
Paraguay	212	185	0.9	0.9	864	39	7851	10	3.3	16.6	4.2	—	—
Peru	702	687	0.1	0.1	2222	238	17144	274	36.3	216.1	16.9	3.4	0.3

Table I (*continued*)
World Countries: Deforestation and Biodiversity, 1990–2007

COUNTRY	FOREST AREA IN THOUSANDS OF SQUARE KILOMETERS		AVERAGE ANNUAL DEFORESTATION AS A PERCENTAGE OF FOREST AREA		ANIMAL SPECIES		HIGHER PLANT SPECIES		GEF INDEX OF BIODIVERSITY	NATIONALLY PROTECTED AREAS		MARINE PROTECTED AREAS	
					TOTAL KNOWN SPECIES	THREATENED SPECIES	TOTAL KNOWN SPECIES	THREATENED SPECIES	0 = NO BIODIVERSITY 100= MAX. BIODIVERSITY	THOUSAND SQ. KM.	% OF TOTAL LAND AREA	THOUSAND SQ. KM.	% OF SURFACE AREA
	1990	2005	1990–2000	2000–2005	2004	2007	2004	2007	2005	2004	2004	2004	2004
Philippines	106	72	2.8	2.1	812	253	8931	213	33.7	24.3	8.2	16.6	5.5
Poland	89	92	–0.2	–0.3	534	38	2450	4	0.6	70.3	23.1	0.7	0.2
Portugal	31	38	–1.5	–1.1	606	147	5050	16	3.8	4.7	5.1	2	2.2
Puerto Rico	4	4	–0.1	0	348	47	2493	53	3.8	0.3	3.5	1.7	19.1
Romania	64	64	0	0	466	64	3400	1	—	5.8	2.5	6.1	2.6
Russia	8090	8088	0	0	941	153	11400	7	37.1	1287	7.9	301.8	1.8
Rwanda	3	5	–0.8	–6.9	871	49	2288	3	1.1	1.9	7.9	—	—
Saudi Arabia	27	27	0	0	527	45	2028	3	3.4	819.1	41	5.2	0.2
Senegal	93	87	0.5	0.5	803	55	2086	7	1.3	21.6	11.2	0.9	0.4
Serbia[A]	26	27	–0.3	–0.3	477	91	4082	1	—	3.8	3.7	0.1	0.1
Sierra Leone	30	28	0.7	0.7	823	48	2090	47	1.5	3.2	4.5	—	—
Singapore	0	0	0	0	473	44	2282	54	0.1	0	4.2	0	0.1
Slovakia	19	19	0	–0.1	419	44	3124	2	0.1	11	22.8	—	—
Slovenia	12	13	–0.4	–0.4	437	80	3200	—	0.2	2.9	14.5	0	0
Somalia	83	71	1	1	824	55	3028	17	6.7	1.9	0.3	3.3	0.5
South Africa	92	92	0	0	1149	323	23420	73	23.5	74	6.1	3.4	0.3
Spain	135	179	–2.0	–1.7	647	170	5050	49	6.6	46.2	9.3	1.8	0.4
Sri Lanka	24	19	1.2	1.5	504	177	3314	280	6.6	17.7	27.3	2.3	3.5
Sudan	764	675	0.8	0.8	1254	47	3137	17	5.5	123	5.2	0.3	0
Swaziland	5	5	–0.9	–0.9	614	16	2715	11	0.1	0.6	3.5	—	—
Sweden	274	275	0	0	542	30	1750	3	0.3	44.8	10.9	4.3	1
Switzerland	12	12	–0.4	–0.4	475	44	3030	3	0.2	11.9	29.6	—	—
Syria	4	5	–1.5	–1.3	432	59	3000	0	0.9	2.7	1.5	—	—
Tajikistan	4	4	0	0	427	27	5000	14	0.7	26	18.6	—	—
Tanzania	414	353	1	1.1	1431	299	10008	240	15.1	374.3	42.3	2.3	0.2
Thailand	160	145	0.7	0.4	1271	157	11625	86	8	80.3	15.7	5.8	1.1
Timor-Leste	10	8	1.2	1.3	—	—	—	—	—	1.9	12.6	—	—
Togo	7	4	3.4	4.5	740	33	3085	10	0.4	6.5	11.9	—	—
Trinidad and Tobago	2	2	0.3	0.2	551	38	2259	1	2.4	0.2	4.7	0.1	1.3
Tunisia	6	11	–4.1	–1.9	438	52	2196	0	0.5	2.3	1.5	0.2	0.1
Turkey	97	102	–0.4	–0.2	581	121	8650	3	6	20.3	2.6	4.5	0.6
Turkmenistan	41	41	0	0	421	44	—	3	2	19.8	4.2	—	—
Uganda	49	36	1.9	2.2	1375	131	4900	38	3.3	64.3	32.6	—	—
Ukraine	93	96	–0.3	–0.1	445	58	5100	1	0.4	19.4	3.3	3.1	0.5
United Arab Emirates	2	3	–2.4	–0.1	298	27	—	—	0.2	0.2	0.2	—	—

Table I (*continued*)
World Countries: Deforestation and Biodiversity, 1990–2007

COUNTRY	FOREST AREA IN THOUSANDS OF SQUARE KILOMETERS		AVERAGE ANNUAL DEFORESTATION AS A PERCENTAGE OF FOREST AREA		ANIMAL SPECIES		HIGHER PLANT SPECIES		GEF INDEX OF BIODIVERSITY	NATIONALLY PROTECTED AREAS		MARINE PROTECTED AREAS	
					TOTAL KNOWN SPECIES	THREATENED SPECIES	TOTAL KNOWN SPECIES	THREATENED SPECIES	0 = NO BIODIVERSITY 100= MAX. BIODIVERSITY	THOUSAND SQ. KM.	% OF TOTAL LAND AREA	THOUSAND SQ. KM.	% OF SURFACE AREA
	1990	2005	1990–2000	2000–2005	2004	2007	2004	2007	2005	2004	2004	2004	2004
United Kingdom	26	28	–0.7	–0.4	660	38	1623	13	2.1	60.5	25	22.5	9.2
United States	2986	3031	–0.1	–0.1	1356	937	19473	242	90.3	1490.1	16.3	909.5	9.4
Uruguay	9	15	–4.5	–1.3	532	66	2278	1	1.4	0.7	0.4	0.1	0
Uzbekistan	30	33	–0.5	–0.5	434	33	4800	15	1.2	20.5	4.8	—	—
Venezuela	520	477	0.6	0.6	1745	166	21073	68	26.8	644.4	73.1	21.3	2.3
Vietnam	94	129	–2.3	–2.0	1116	152	10500	146	11.7	13.6	4.4	0.7	0.2
West Bank and Gaza	0	0	0	0	—	—	—	—	—	—	—	—	—
Yemen	5	5	0	0	459	47	1650	159	3.4	0	0	—	—
Zambia	491	425	0.9	1	1025	38	4747	8	5	312.3	42	—	—
Zimbabwe	222	175	1.5	1.7	883	35	4440	17	2.1	57.5	14.9	—	—
World	40,679	39,426	0.2	0.2	113879	14751	901174	9977	—	15,050.80	11.6	4,348.90	3.8

Source: Table 3.4, *World Development Indicators 2008* (World Bank, 2008)
[A] The data for Serbia include data for Montenegro which, at the time of publication, was an independent state.

Table J
World Countries: Energy Production and Use, 1990–2005

COUNTRY	COMMERCIAL ENERGY PRODUCTION — THOUSAND METRIC TONS (KILOTONS) OF OIL EQUIVALENT 1990	COMMERCIAL ENERGY PRODUCTION — THOUSAND METRIC TONS (KILOTONS) OF OIL EQUIVALENT 2005	COMMERCIAL ENERGY USE — THOUSAND METRIC TONS (KILOTONS) OF OIL EQUIVALENT 1990	COMMERCIAL ENERGY USE — THOUSAND METRIC TONS (KILOTONS) OF OIL EQUIVALENT 2005	COMMERCIAL ENERGY USE — AVERAGE ANNUAL % GROWTH 1980–2000	COMMERCIAL ENERGY USE PER CAPITA — KILOGRAM OF OIL EQUIVALENT 1990	COMMERCIAL ENERGY USE PER CAPITA — KILOGRAM OF OIL EQUIVALENT 2005	COMMERCIAL ENERGY USE PER CAPITA — AVERAGE ANNUAL % GROWTH 1980–2000	NET ENERGY IMPORTS — % OF COMMERCIAL ENERGY USE 1990	NET ENERGY IMPORTS — % OF COMMERCIAL ENERGY USE 2005
Albania	2,449	1,173	2,662	2,402	-5.6	809	762	-6.3	8	51
Algeria	104,439	175,070	23,858	34,768	3.5	944	1,058	1	-338	-404
Angola	28,652	70,700	6,285	9,898	2.8	597	615	-0.3	-356	-614
Argentina	48,456	81,046	46,110	63,718	2.2	1,415	1,644	0.8	-5	-27
Armenia	137	861	7,941	2,558	—	2,240	848	—	98	66
Australia	157,535	270,980	87,546	121,956	2.4	5,130	5,978	1	-80	-122
Austria	8,102	9,809	25,071	34,363	1.6	3,251	4,174	1.1	68	71
Azerbaijan	21,336	27,300	26,037	13,835	—	3,637	1,649	—	18	-97
Bahrain	13,437	16,131	4,829	8,128	—	9,796	11,214	—	-178	-98
Bangladesh	10,758	19,311	12,826	24,187	4.1	113	158	1.9	16	20
Belarus	3,323	3,802	42,175	26,590	1.8	4,139	2,720		92	86
Belgium	13,104	13,900	49,161	56,653		4,932	5,407	1.6	73	75
Benin	1,774	1,672	1,678	2,584	2.3	324	304	-0.8	-6	35
Bolivia	4,923	13,873	2,774	5,307	3.8	416	578	1.5	-77	-161
Bosnia-Herzegovina	4,604	3,343	7,035	4,963	—	1,633	1,268	—	35	33
Botswana	910	1,051	1,272	1,895	—	930	1,032	—	28	45
Brazil	98,066	187,827	133,950	209,534	2.7	896	1,122	1	27	10
Brunei	15,658	21,079	1,817	2,641	—	7,070	7,065	—	-762	-698
Bulgaria	9,613	10,649	28,820	20,060	-2.8	3,306	2,592	-2.2	67	47
Cameroon	12,090	11,942	5,032	6,978	2.5	411	392	-0.2	-140	-71
Canada	273,733	401,260	209,412	271,955	1.6	7,535	8,417	0.4	-31	-48
Chile	7,640	9,118	14,066	29,568	5.7	1,067	1,815	4	46	69
China	886,259	1,640,944	863,212	1,717,153	3.7	760	1,316	2.4	-3	4
Colombia	48,179	79,458	24,749	28,586	2.5	710	636	0.5	-95	-178
Congo, Dem. Rep.	12,019	17,391	11,903	16,967	2.7	314	289	-0.6	-1	-2
Congo, Rep.	9,005	13,677	1,056	1,199	-1.1	436	332	-4	-753	-1,041
Costa Rica	1,032	1,837	2,025	3,820	4.2	658	883	1.5	49	52
Côte d'Ivoire	3,382	8,196	4,408	7,843	3.3	345	422	-0.1	23	-5
Croatia	5,148	3,800	9,067	8,886	—	1,897	2,000	—	43	57
Cuba	6,567	5,519	16,831	10,196	-1.6	1,587	906	-2.4	61	46
Cyprus	6	51	1,619	2,552	—	2,792	3,368	—	100	98
Czech Republic	40,100	32,869	48,993	45,205	-1.2	4,728	4,417	-1.2	18	27
Denmark	10,039	31,301	17,895	19,610	0.6	3,482	3,621	0.4	44	-60
Dominican Republic	1,031	1,530	4,139	7,356	4.0	567	777	2.1	75	79
Ecuador	16,474	28,598	6,128	10,441	2.1	597	799	-0.3	-169	-174
Egypt	54,869	76,039	31,895	61,301	4.6	578	841	2.3	-72	-24
El Salvador	1,722	2,540	2,535	4,629	2.2	496	694	0.6	32	45

Table J (*continued*)
World Countries: Energy Production and Use, 1990–2005

COUNTRY	COMMERCIAL ENERGY PRODUCTION		COMMERCIAL ENERGY USE			COMMERCIAL ENERGY USE PER CAPITA			NET ENERGY IMPORTS	
	THOUSAND METRIC TONS (KILOTONS) OF OIL EQUIVALENT		THOUSAND METRIC TONS (KILOTONS) OF OIL EQUIVALENT		AVERAGE ANNUAL % GROWTH	KILOGRAM OF OIL EQUIVALENT		AVERAGE ANNUAL % GROWTH	% OF COMMERCIAL ENERGY USE	
	1990	2005	1990	2005	1980–2000	1990	2005	1980–2000	1990	2005
Estonia	5,032	3,722	9,582	5,096	—	6,107	3,786	—	47	27
Ethiopia	14,158	19,855	15,151	21,633	2.6	296	288	-0.1	7	8
Finland	12,081	16,560	29,171	34,961	1.7	5,851	6,664	1.3	59	53
France	112,434	136,889	227,816	275,970	1.9	4,015	4,534	1.5	51	50
Gabon	14,630	12,116	1,243	1,721	-0.2	1,354	1,333	-3.1	-1077	-604
Georgia	1,844	1,272	12,335	3,210	—	2,259	718	—	85	60
Germany	186,159	134,505	356,221	344,746	-0.2	4,485	4,180	-0.5	48	61
Ghana	4,392	6,356	5,337	8,937	3.6	343	397	0.5	18	29
Greece	9,200	10,303	22,181	30,977	3.0	2,183	2,790	2.5	59	67
Guatemala	3,391	5,414	4,479	7,987	3.5	503	628	0.9	24	32
Haiti	1,253	1,921	1,585	2,502	0.4	223	269	-1.6	21	23
Honduras	1,694	1,780	2,416	3,870	2.8	494	566	-0.2	30	54
Hong Kong	43	48	10,663	18,077	—	1,869	2,653	—	100	100
Hungary	14,325	10,330	28,558	27,762	-1.0	2,753	2,752	-0.7	50	63
Iceland	1,400	2,636	2,172	3,626	—	8,524	12,219	—	36	27
India	291,073	419,036	319,892	537,309	3.8	377	491	1.8	9	22
Indonesia	170,260	263,390	103,196	179,512	4.8	579	814	3.1	-65	-47
Iran	179,738	303,811	68,775	162,504	5.6	1,264	2,352	3.1	-161	-87
Iraq	104,933	96,043	19,060	30,760	4.3	1,029	—	1.3	-451	-212
Ireland	3,467	1,662	10,365	15,289	2.7	2,957	3,676	2.3	67	89
Israel	433	2,064	12,112	19,495	5.2	2,599	2,816	2.6	96	89
Italy	25,312	27,627	147,967	185,185	1.3	2,609	3,160	1.2	83	85
Jamaica	485	482	2,943	3,835	3.5	1,231	1,445	2.6	84	87
Japan	75,214	99,772	444,471	530,463	2.6	3,598	4,152	2.2	83	81
Jordan	162	254	3,498	7,093	4.9	1,103	1,311	0.5	95	96
Kazakhstan	90,456	121,686	73,656	52,437	—	4,506	3,462	—	-23	-132
Kenya	10,272	13,888	12,479	17,246	2.2	532	484	-0.7	18	19
Korea, North	28,725	20,212	32,874	21,213	1.9	1,632	898	0.5	13	5
Korea, South	22,623	42,935	93,388	213,771	9.1	2,178	4,426	8	76	80
Kuwait	50,401	146,309	8,467	28,143	1.0	3,984	11,100	0.2	-495	-420
Kyrgyzstan	2,502	1,447	7,620	2,798	—	1,723	544	—	67	48
Latvia	1,157	2,294	7,787	4,716	—	2,916	2,050	—	85	51
Lebanon	143	224	2,309	5,577	4.8	776	1,391	2.8	94	96
Libya	73,173	94,966	11,541	19,047	3.6	2,645	3,218	1	-534	-399
Lithuania	4,936	3,927	16,186	8,587	—	4,377	2,515	—	70	54
Luxembourg	31	74	3,571	4,776	—	9,351	10,457	—	99	98
Macedonia	1,453	1,461	2,713	2,738	—	1,421	1,346	—	46	47
Malaysia	50,311	93,917	23,321	61,279	7.7	1,288	2,389	4.9	-116	-53
Malta	0	0	774	949	—	2,150	2,352	—	100	100
Mexico	194,783	259,204	124,341	176,534	2.1	1,494	1,712	0.2	-57	-47

Table J (*continued*)
World Countries: Energy Production and Use, 1990–2005

COUNTRY	COMMERCIAL ENERGY PRODUCTION — THOUSAND METRIC TONS (KILOTONS) OF OIL EQUIVALENT		COMMERCIAL ENERGY USE — THOUSAND METRIC TONS (KILOTONS) OF OIL EQUIVALENT		COMMERCIAL ENERGY USE — AVERAGE ANNUAL % GROWTH	COMMERCIAL ENERGY USE PER CAPITA — KILOGRAM OF OIL EQUIVALENT		COMMERCIAL ENERGY USE PER CAPITA — AVERAGE ANNUAL % GROWTH	NET ENERGY IMPORTS — % OF COMMERCIAL ENERGY USE	
	1990	2005	1990	2005	1980–2000	1990	2005	1980–2000	1990	2005
Moldova	58	88	9,995	3,556	—	2,277	917	—	99	98
Morocco	773	983	6,725	13,813	4.3	278	458	2.2	89	93
Mozambique	6,846	11,742	7,203	10,207	-0.8	532	497	-2.6	5	-15
Myanmar	10,651	22,143	10,683	14,725	1.3	266	307	-0.4	0	-50
Namibia	0	328	0	1,379	—	0	683	—	—	76
Nepal	5,501	8,153	5,806	9,170	2.7	304	338	0.4	5	11
Netherlands	60,542	61,902	66,751	81,849	1.4	4,464	5,015	0.8	9	24
Netherlands Antilles	0	0	1,493	1,657	—	7,832	8,890	—	100	100
New Zealand	12,008	12,201	13,758	16,909	3.8	3,990	4,090	2.7	13	28
Nicaragua	1,495	1,954	2,119	3,335	2.7	512	611	0	29	41
Nigeria	150,453	231,775	70,905	103,785	2.6	751	734	-0.4	-112	-123
Norway	120,323	233,704	21,511	32,125	1.8	5,072	6,948	1.3	-459	-627
Oman	38,313	59,593	4,562	13,964	11.0	2,475	5,570	6.6	-740	-327
Pakistan	34,360	61,267	43,424	76,329	4.8	402	490	2.2	21	20
Panama	612	738	1,490	2,598	2.7	618	804	0.7	59	72
Paraguay	4,577	6,582	3,083	3,973	4.2	731	674	1.2	-48	-66
Peru	10,596	10,793	9,952	13,810	0.2	457	506	-1.8	-6	22
Philippines	13,701	24,195	26,159	44,692	3.9	427	528	1.5	48	46
Poland	99,367	78,633	99,880	92,969	-1.4	2,620	2,436	-1.8	1	15
Portugal	3,393	3,583	17,746	27,166	4.7	1,793	2,575	4.7	81	87
Qatar	26,565	77,652	6,330	15,826	—	13,542	19,877	—	-320	-391
Romania	40,834	27,901	62,403	38,343	-3.1	2,689	1,772	-3.2	35	27
Russia	1,280,260	1,184,857	878,272	646,680	—	5,923	4,517	—	-46	-83
Saudi Arabia	370,815	576,695	61,318	140,277	5.0	3,744	6,068	1	-505	-311
Senegal	1,363	1,265	2,238	3,041	2.4	283	258	-0.3	39	58
Singapore	0	0	13,357	30,100	8.7	4,384	6,933	6	100	100
Slovakia	5,284	6,609	21,315	18,831	-1.4	4,035	3,496	-1.7	75	65
Slovenia	2,904	3,437	5,597	7,315	—	2,801	3,657	—	48	53
South Africa	114,534	158,590	91,229	127,637	2.1	2,592	2,722	-0.2	-26	-24
Spain	34,583	30,284	91,073	145,196	3.2	2,345	3,346	2.9	62	79
Sri Lanka	4,191	5,255	5,516	9,375	2.5	324	477	1.4	24	44
Sudan	8,775	31,127	10,642	18,398	—	410	499	—	18	-69
Sweden	29,754	34,781	47,566	52,174	—	5,557	5,782	—	37	33
Switzerland	9,717	10,882	24,992	27,153	—	3,723	3,651	—	61	60
Syria	22,319	29,136	11,677	17,906	5.4	918	948	2.2	-91	-63
Tajikistan	2,026	1,524	5,595	3,458	—	1,055	528	—	64	56
Tanzania	9,063	19,099	9,808	20,404	2.0	385	530	-1	8	6
Thailand	26,497	53,966	43,861	100,037	7.4	808	1,588	6	40	46
Togo	1,203	1,591	1,447	1,995	3.8	365	320	0.8	17	20

Table J (*continued*)
World Countries: Energy Production and Use, 1990–2005

COUNTRY	COMMERCIAL ENERGY PRODUCTION		COMMERCIAL ENERGY USE			COMMERCIAL ENERGY USE PER CAPITA			NET ENERGY IMPORTS	
	THOUSAND METRIC TONS (KILOTONS) OF OIL EQUIVALENT		THOUSAND METRIC TONS (KILOTONS) OF OIL EQUIVALENT		AVERAGE ANNUAL % GROWTH	KILOGRAM OF OIL EQUIVALENT		AVERAGE ANNUAL % GROWTH	% OF COMMERCIAL ENERGY USE	
	1990	2005	1990	2005	1980–2000	1990	2005	1980–2000	1990	2005
Trinidad and Tobago	12,612	31,399	6,037	12,706	3.3	4,934	9,599	2.4	-109	-147
Tunisia	6,127	6,681	5,536	8,451	3.7	679	843	1.6	-11	21
Turkey	25,816	23,607	52,968	85,205	4.6	943	1,182	2.7	51	72
Turkmenistan	74,865	61,137	19,635	16,340	—	5,353	3,381	—	-281	-274
Ukraine	133,677	80,977	251,720	143,237	—	4,851	3,041	—	47	43
United Arab Emirates	109,446	167,889	22,545	46,936	8.2	12,716	11,436	2.8	-385	-258
United Kingdom	208,003	204,301	212,176	233,931	1.0	3,686	3,884	0.8	2	13
United States	1,650,293	1,630,675	1,927,457	2,340,286	1.5	7,721	7,893	0.4	14	30
Uruguay	1,149	1,020	2,251	2,893	1.7	725	875	1	49	65
Uzbekistan	38,643	56,573	46,388	47,040	—	2,262	1,798	—	17	-20
Venezuela	148,854	204,708	43,918	60,935	2.9	2,224	2,293	0.1	-239	-236
Vietnam	24,711	69,541	24,324	51,296	3.2	367	617	1.2	-2	-36
Yemen	9,384	20,407	2,566	6,728	4.2	208	319	0.3	-266	-203
Zambia	4,923	6,513	5,470	7,124	1.3	673	621	-1.6	10	9
Zimbabwe	8,550	8,860	9,384	9,723	2.6	895	741	-0.3	9	9
World	8,721,959	11,335,335.00	8,518,884	11,047,604.00	2.9	2,461.82	2,715.49	0.9	-2.31	—

Table K
World Countries: Energy Efficiency and Emission, 1990–2005

COUNTRY	GDP PER UNIT OF ENERGY USE: PPP $ PER KG. OF OIL EQUIVALENT		FUEL TRADE: % OF MERCHANDISE EXPORTS AND IMPORTS 2005–2006		CARBON DIOXIDE EMISSIONS: TOTAL KILOTONS		METRIC TONS PER CAPITA		KG. PER PPP $ OF GDP	
	1990	2005	EXPORTS	IMPORTS	1990	2004	1990	2004	1990	2004
Afghanistan	—	—	—	—	2,611.72	692.31	—	—	—	0.04
Albania	3.14	7.18	13.69	12.23	7,267.39	3,670.33	2.21	1.17	0.87	0.23
Algeria	4.02	5.73	97.91	1.13	76,970.62	193,827.65	3.04	5.99	0.8	1.06
American Samoa	—	—	—	—	285.71	—	—	—	—	—
Angola	3.92	6.06	—	—	4,644.68	7,890.10	0.44	0.5	0.19	0.16
Antigua and Barbuda	—	—	0.72	34.58	300.37	413.92	4.85	5.05	0.58	0.37
Argentina	3.82	6.58	14.75	4.78	109,699.52	141,659.20	3.37	3.69	0.62	0.38
Armenia	0.95	4.91	2.07	16.66	—	3,644.69	—	1.2	—	0.34
Aruba	—	—	—	0.08	1,838.83	2,153.84	—	21.76	—	—
Australia	3.44	5.71	25.48	13.74	278,395.33	326,464.88	16.31	16.22	0.92	0.5
Austria	5.86	8.16	5.28	13.85	57,564.05	69,783.81	7.47	8.54	0.39	0.26
Azerbaijan	0.96	2.77	84.59	11.68	—	31,336.97	—	3.77	—	1.06
Bahamas	—	—	—	—	1,948.72	2,007.32	7.64	6.29	—	—
Bahrain	1.58	2.98	80.59	55.63	11,706.95	16,934.05	23.75	23.84	1.54	0.78
Bangladesh	4.44	6.77	—	8.33	15,355.30	37,131.83	0.14	0.25	0.27	0.25
Barbados	—	—	9.42	18.45	1,076.92	1,267.40	3.97	4.36	—	—
Belarus	1.13	3.14	38.18	32.88	—	64,831.44	—	6.6	—	0.88
Belgium	3.68	5.86	7.63	13.35	100,585.98	100,626.27	10.09	9.66	0.56	0.32
Belize	—	—	16.55	16.15	311.36	791.21	1.65	2.8	0.46	0.39
Benin	2.3	3.99	—	20.45	714.29	2,384.61	0.14	0.29	0.19	0.25
Bermuda	—	—	—	—	589.74	549.45	9.75	8.68	—	—
Bhutan	—	—	—	—	128.21	413.92	0.23	0.66	0.18	0.2
Bolivia	5.3	6.43	52.15	9.81	5,498.16	6,967.03	0.82	0.77	0.37	0.22
Bosnia-Herzegovina	—	4.69	8.03	15.47	6,906.38	15,582.40	1.6	3.99	—	0.72
Botswana	5.29	11.64	—	—	2,168.50	4,296.70	1.59	2.37	0.32	0.21
Brazil	5.85	7.56	7.71	18.83	209,483.31	331,497.84	1.4	1.8	0.27	0.22
Brunei	5.06	6.65	96.33	1.65	5,816.84	8,802.19	22.63	24.07	0.63	0.52
Bulgaria	1.67	3.6	13.28	5.31	75,307.62	42,520.10	8.64	5.46	1.57	0.64
Burkina Faso	—	—	2.79	24.42	992.67	1,095.24	0.11	0.08	0.21	0.08
Burundi	—	—	0.06	8.46	194.14	219.78	0.03	0.03	0.1	0.09
Cambodia	—	—	—	9.9	450.55	534.8	0.05	0.04	—	0.03
Cameroon	3.72	5.08	61.61	30.81	1,604.39	3,835.16	0.13	0.22	0.09	0.11
Canada	2.58	4.16	20.21	9.29	415,713.87	638,830.86	14.96	19.97	0.77	0.6
Cape Verde	—	—	—	8.91	87.91	274.73	0.25	0.55	0.21	0.25
Cayman Islands	—	—	—	—	249.08	311.36	—	7.08	—	—
Central African Republic	—	—	0.38	16.85	197.8	252.75	0.07	0.06	0.11	0.1

Table K (*continued*)
World Countries: Energy Efficiency and Emission, 1990–2005

COUNTRY	GDP PER UNIT OF ENERGY USE: PPP $ PER KG. OF OIL EQUIVALENT		FUEL TRADE: % OF MERCHANDISE EXPORTS AND IMPORTS 2005–2006		CARBON DIOXIDE EMISSIONS: TOTAL KILOTONS		METRIC TONS PER CAPITA		KG. PER PPP $ OF GDP	
	1990	2005	EXPORTS	IMPORTS	1990	2004	1990	2004	1990	2004
Chad	—	—	—	—	142.86	124.54	0.02	0.01	0.03	0.01
Chile	4.49	6.75	1.92	24.28	35,611.69	62,362.58	2.7	3.87	0.56	0.34
China	1.05	3.11	1.81	12.36	2,398,202.73	5,005,687.30	2.11	3.86	2.65	1.07
Colombia	5.1	9.22	39.77	2.74	58,032.91	53,586.03	1.66	1.21	0.46	0.22
Comoros	—	—	—	—	65.93	87.91	0.15	0.15	0.17	0.14
Congo, Dem. Rep.	1.4	0.93	—	—	3,970.69	2,102.56	0.1	0.04	0.24	0.15
Congo, Rep.	5.65	9.77	—	—	1,172.16	3,538.46	0.48	1	0.2	0.34
Costa Rica	6.62	9.87	0.62	11.89	2,915.75	6,399.26	0.95	1.5	0.22	0.19
Côte d'Ivoire	3.93	3.82	36.99	31.82	5,384.61	5,157.50	0.42	0.28	0.31	0.18
Croatia	4.33	6.62	15.1	15.92	24,576.84	23,479.83	5.14	5.28	—	0.43
Cuba	—	—	0.5	23.37	32,040.26	25,794.85	3.02	2.29	—	—
Cyprus	4.72	7.29	0.06	18.08	4,644.68	6,743.58	8.01	9.12	0.61	0.39
Czech Republic	2.48	4.59	2.7	8.89	161,686.07	116,886.33	15.6	11.45	—	0.62
Denmark	5.34	9.29	10.26	5.97	49,739.88	52,908.37	9.68	9.8	0.52	0.31
Djibouti	—	—	—	—	351.65	366.3	0.63	0.46	0.32	0.26
Dominica	—	—	—	15.49	58.61	106.23	0.81	1.49	0.19	0.18
Dominican Republic	4.09	6.71	—	—	9,560.43	19,622.69	1.31	2.1	0.56	0.45
Ecuador	6.67	8.43	58.57	16.71	16,717.93	29,241.73	1.63	2.26	0.41	0.36
Egypt	4.14	5.44	56.39	16.34	75,413.84	158,095.08	1.37	2.21	0.57	0.51
El Salvador	5.9	7.83	3.02	18.3	2,615.38	6,161.17	0.51	0.94	0.17	0.18
Equatorial Guinea	—	—	—	—	117.22	5,421.24	0.35	11.46	0.22	0.43
Eritrea	—	—	—	—	—	754.58	—	0.17	—	0.25
Estonia	1.22	4.35	15.73	15.91	—	18,926.72	—	14.03	—	0.97
Ethiopia	1.2	2.02	—	—	2,963.37	7,974.35	0.06	0.11	0.16	0.21
Faeroe Islands	—	—	1.95	18.79	615.38	659.34	—	13.65	—	—
Fiji	—	—	0.09	32.76	813.19	1,069.60	1.12	1.3	—	0.31
Finland	2.88	4.57	5.3	15.39	51,208.74	65,739.86	10.27	12.57	0.61	0.44
France	4.49	6.75	4.29	14.86	363,710.26	373,358.60	6.41	6.17	0.36	0.21
French Polynesia	—	—	0.02	12.91	611.72	670.33	3.13	2.66	—	—
Gabon	8.08	10.37	85.61	3.96	5,989.01	1,369.96	6.52	1.08	0.6	0.08
Gambia	—	—	—	17.38	190.48	285.71	0.2	0.18	0.25	0.18
Georgia	1.73	4.91	3.13	19.44	—	3,908.42	—	0.87	—	0.28
Germany	4.01	7.28	2.45	12.22	980,273.75	808,043.15	12.34	9.79	0.69	0.33
Ghana	1.82	2.93	0.65	13.76	3,765.56	7,183.14	0.24	0.33	0.39	0.3
Greece	6.8	10.49	13.03	19.18	72,362.57	96,607.96	7.12	8.73	0.48	0.32
Greenland	—	—	—	—	553.11	571.43	9.95	10.04	—	—

Table K (*continued*)
World Countries: Energy Efficiency and Emission, 1990–2005

COUNTRY	GDP PER UNIT OF ENERGY USE: PPP $ PER KG. OF OIL EQUIVALENT		FUEL TRADE: % OF MERCHANDISE EXPORTS AND IMPORTS 2005–2006		CARBON DIOXIDE EMISSIONS: TOTAL KILOTONS		METRIC TONS PER CAPITA		KG. PER PPP $ OF GDP	
	1990	2005	EXPORTS	IMPORTS	1990	2004	1990	2004	1990	2004
Grenada	—	—	—	7.17	120.88	216.12	1.29	2.04	0.27	0.26
Guam	—	—	—	—	2,267.40	—	16.93	—	—	—
Guatemala	5.96	7.83	8.81	19.53	5,084.24	12,208.78	0.57	0.98	0.19	0.21
Guinea	—	—	—	—	1,010.99	1,337.00	0.17	0.15	0.24	0.14
Guinea-Bissau	—	—	—	—	208.79	271.06	0.21	0.17	0.45	0.4
Guyana	—	—	0	30.31	1,131.87	1,443.22	1.55	1.95	1.07	0.6
Haiti	5.87	4.38	—	—	992.67	1,754.58	0.14	0.19	0.11	0.17
Honduras	4.1	5.83	2.71	20.18	2,589.74	7,608.05	0.53	1.14	0.26	0.36
Hong Kong	8.66	13.45	1.68	2.94	26,175.80	37,377.25	4.59	5.51	0.28	0.17
Hungary	3.26	6.18	2.35	7.48	60,095.18	57,131.81	5.79	5.65	0.65	0.36
Iceland	2.22	2.9	2.33	8.87	2,018.31	2,227.10	7.92	7.62	0.42	0.23
India	2.31	4.53	11.4	36.29	681,526.79	1,341,760.56	0.8	1.24	0.92	0.62
Indonesia	2.61	3.94	27.24	31.51	213,772.68	377,911.71	1.2	1.74	0.79	0.58
Iran	3.59	3.96	82.79	4.01	218,197.58	433,182.72	4.01	6.36	0.88	0.72
Iraq	—	—	—	—	48,512.77	81,578.67	2.62	—	—	—
Ireland	4.29	10.31	0.66	7.74	30,608.03	42,314.98	8.73	10.4	0.69	0.29
Israel	4.9	8.04	0.1	15.64	33,117.18	71,183.08	7.11	10.46	0.56	0.49
Italy	6.57	8.78	3.52	12.41	389,611.33	449,545.34	6.87	7.73	0.4	0.29
Jamaica	3.56	4.98	13.73	24.52	7,956.04	10,582.41	3.33	4.01	0.76	0.58
Japan	5.22	7.3	0.92	27.76	1,070,405.52	1,256,837.57	8.66	9.84	0.46	0.34
Jordan	2.19	3.31	1.03	23.86	10,179.48	16,450.53	3.21	3.11	1.33	0.77
Kazakhstan	1.14	2.51	69.43	12.98	—	200,098.70	—	13.33	—	1.72
Kenya	1.98	2.84	0.9	24.28	5,820.51	10,578.74	0.25	0.31	0.24	0.24
Kiribati	—	—	0.02	16.99	21.98	29.3	0.3	0.3	0.15	0.09
Korea, North	—	—	—	—	244,567.52	79,040.21	12.14	3.36	—	—
Korea, South	3.53	4.81	6.35	27.69	241,098.66	465,226.64	5.62	9.68	0.73	0.49
Kuwait	—	3.92	—	—	43,391.90	99,274.63	20.42	40.36	—	1
Kyrgyzstan	1.05	3.18	18.28	29.39	—	5,721.61	—	1.12	—	0.66
Laos	—	—	—	—	230.77	1,278.39	0.06	0.23	0.08	0.14
Latvia	2.54	6.45	5.2	12.67	—	7,091.57	—	3.07	—	0.27
Lebanon	4.87	6.86	0.31	21.92	9,091.57	16,249.07	3.06	4.1	0.81	0.44
Liberia	—	—	—	—	465.2	468.86	0.22	0.14	0.61	0.47
Libya	—	3.38	—	0.69	37,761.87	59,860.75	8.65	10.32	—	1.02
Lithuania	2.07	5.6	23.53	22.41	—	13,296.69	—	3.87	—	0.31
Luxembourg	3.4	6.67	0.76	9.64	9,893.76	11,267.39	25.91	24.86	0.82	0.38
Macao, China	—	—	—	10.97	1,025.64	2,205.13	2.76	4.72	0.19	0.14
Macedonia	4.1	5.49	9.38	20.24	15,548.72	10,410.25	8.14	5.13	—	0.74
Madagascar	—	—	5.51	18.71	941.39	2,728.94	0.08	0.15	0.11	0.19

Table K (*continued*)
World Countries: Energy Efficiency and Emission, 1990–2005

COUNTRY	GDP PER UNIT OF ENERGY USE: PPP $ PER KG. OF OIL EQUIVALENT 1990	2005	FUEL TRADE: % OF MERCHANDISE EXPORTS AND IMPORTS 2005–2006 EXPORTS	IMPORTS	CARBON DIOXIDE EMISSIONS: TOTAL KILOTONS 1990	2004	METRIC TONS PER CAPITA 1990	2004	KG. PER PPP $ OF GDP 1990	2004
Malawi	—	—	0.14	11.42	600.73	1,043.96	0.06	0.08	0.14	0.13
Malaysia	3.78	4.89	13.78	9.11	55,263.68	177,424.73	3.05	7.04	0.63	0.64
Maldives	—	—	—	19.6	153.85	725.27	0.71	2.5	—	0.6
Mali	—	—	0.77	21.13	421.25	564.1	0.05	0.05	0.1	0.05
Malta	4.56	8.71	1.36	9.13	2,230.77	2,450.55	6.2	6.11	0.63	0.32
Mauritania	—	—	—	26.91	2,633.70	2,553.11	1.35	0.89	1.18	0.56
Mauritius	—	—	0.11	16.85	1,461.54	3,194.14	1.38	2.59	0.33	0.28
Mexico	4.45	6.65	15.53	5.7	413,142.44	437,629.60	4.96	4.29	0.75	0.39
Micronesia	—	—	—	—	—	—	—	—	—	—
Moldova	1.23	2.39	0.33	24.19	—	7,677.65	—	1.96	—	1
Mongolia	—	—	5.53	29.3	9,978.01	8,545.78	4.74	3.4	2.81	1.42
Morocco	7.17	7.76	1.85	21.72	23,479.83	41,131.83	0.97	1.38	0.49	0.41
Mozam-bique	0.57	1.36	14.67	16.89	996.34	2,164.83	0.07	0.11	0.24	0.17
Myanmar	0.92	2.73	—	—	4,263.73	9,750.91	0.11	0.2	0.43	0.26
Namibia	—	6.74	0.46	3.16	7.33	2,468.86	0.01	1.24	0	0.29
Nepal	1.67	2.84	—	—	630.04	3,040.29	0.03	0.11	0.06	0.12
Netherlands	4.22	6.88	13.38	17.11	141,003.52	141,933.92	9.43	8.72	0.5	0.26
Netherlands Antilles	—	—	—	—	1,201.46	4,084.25	6.3	22.17	—	—
New Caledonia	—	—	0.44	15.15	1,611.72	2,575.09	9.59	11.19	—	—
New Zealand	3.41	6.01	1.7	14.87	22,637.34	31,542.09	6.57	7.72	0.48	0.33
Nicaragua	3.02	4.32	0.55	25.22	2,644.69	4,003.66	0.64	0.74	0.41	0.3
Niger	—	—	1.86	14.65	1,047.62	1,212.45	0.13	0.09	0.27	0.17
Nigeria	1.27	2.07	—	—	45,325.96	113,922.96	0.48	0.83	0.5	0.59
Norway	4.61	6.84	67.87	4.52	33,190.44	87,523.72	7.83	19.06	0.33	0.42
Oman	4.11	3.65	94.67	3.09	10,263.73	30,871.76	5.57	12.46	0.55	0.66
Pakistan	3.02	4.46	4.94	26.22	68,007.26	125,556.65	0.63	0.83	0.52	0.41
Palau	—	—	—	—	—	238.1	—	11.9	—	0.98
Panama	6.56	10.5	0.76	18.05	3,128.20	5,655.67	1.3	1.78	0.32	0.23
Papua New Guinea	—	—	—	—	2,428.57	2,446.88	0.59	0.41	0.54	0.24
Paraguay	3.91	5.68	0	12.74	2,260.07	4,175.82	0.54	0.72	0.19	0.2
Peru	7.07	12.74	9.62	19.36	21,025.62	31,465.17	0.97	1.17	0.3	0.2
Philippines	4.12	5.59	2.32	15.42	43,897.39	80,439.48	0.72	0.97	0.41	0.35
Poland	2.22	5.56	4.37	10.4	347,464.85	306,963.06	9.12	8.04	1.57	0.63
Portugal	6.22	7.75	5.21	15.02	42,314.98	58,853.42	4.28	5.6	0.38	0.29
Puerto Rico	—	—	—	—	11,758.23	—	3.32	—	—	—
Qatar	—	3.56	90.85	0.6	12,179.48	52,857.09	26.06	69.16	—	1.03
Romania	2.12	5.29	9.98	13.6	155,018.16	90,344.23	6.68	4.17	1.17	0.48
Russia	1.54	2.63	62.87	1.47	—	1,523,628.51	—	10.59	—	0.98
Rwanda	—	—	—	—	527.47	571.43	0.07	0.06	0.15	0.1

Table K (*continued*)
World Countries: Energy Efficiency and Emission, 1990–2005

COUNTRY	GDP PER UNIT OF ENERGY USE: PPP $ PER KG. OF OIL EQUIVALENT		FUEL TRADE: % OF MERCHANDISE EXPORTS AND IMPORTS 2005–2006		CARBON DIOXIDE EMISSIONS: TOTAL KILOTONS		METRIC TONS PER CAPITA		KG. PER PPP $ OF GDP	
	1990	2005	EXPORTS	IMPORTS	1990	2004	1990	2004	1990	2004
Samoa	—	—	0.29	14.27	124.54	150.18	0.77	0.82	0.3	0.18
São Tomé and Príncipe	—	—	—	20.17	65.93	91.58	0.57	0.61	—	0.46
Saudi Arabia	3.66	3.5	91.25	0.26	254,721.36	308,116.91	15.55	13.68	1.14	0.69
Senegal	3.45	5.96	0.03	25.9	3,131.87	4,989.01	0.4	0.43	0.41	0.3
Seychelles	—	—	—	26.67	113.55	545.79	1.62	6.62	0.19	0.48
Sierra Leone	—	—	—	—	333.33	992.67	0.08	0.18	0.17	0.34
Singapore	3.87	5.98	12.93	18.86	45,054.90	52,205.08	14.79	12.32	0.87	0.32
Slovakia	2.25	4.54	5.28	13.66	51,427.27	36,256.37	9.73	6.74	—	0.46
Slovenia	4.1	6.16	2.7	11.23	18,008.01	16,197.79	9.01	8.11	—	0.39
Solomon Islands	—	—	—	—	161.17	175.82	0.51	0.38	0.34	0.23
Somalia	—	—	—	—	18.32	—	0	—	—	—
South Africa	2.18	3.11	9.39	18.39	331,743.26	436,640.59	9.42	9.42	1.67	1.19
Spain	6.07	8.12	4.45	15.73	212,084.04	330,201.14	5.46	7.73	0.38	0.3
Sri Lanka	4.37	7.17	—	13.49	3,758.24	11,523.80	0.22	0.59	0.16	0.19
St. Kitts and Nevis	—	—	0.16	7.87	65.93	124.54	1.57	2.65	0.25	0.2
St. Lucia	—	—	—	13.98	161.17	366.3	1.2	2.26	0.23	0.26
St. Vincent and the Grenadines	—	—	0.04	14.57	80.59	197.8	0.74	1.67	0.22	0.26
Sudan	1.79	3.43	87.31	1.22	5,380.95	10,362.63	0.21	0.29	0.28	0.18
Suriname	—	—	—	16.82	1,809.52	2,282.05	4.5	5.08	1.08	0.75
Swaziland	—	—	0.6	11.5	424.91	956.04	0.55	0.85	0.18	0.2
Sweden	3.23	5.54	5.44	12.43	49,443.17	52,985.30	5.78	5.89	0.32	0.19
Switzerland	6.47	9.64	2.59	8.04	42,681.28	40,421.21	6.36	5.47	0.26	0.16
Syria	2.32	4.22	40.35	27.09	35,897.40	68,358.91	2.82	3.72	1.32	0.97
Tajikistan	2.1	2.8	—	—	—	5,000.00	—	0.77	—	0.57
Tanzania	1.45	1.76	0.23	24.24	2,333.33	4,347.98	0.09	0.12	0.16	0.13
Thailand	3.71	4.45	4.68	20.23	95,710.53	267,842.22	1.76	4.28	0.59	0.65
Timor-Leste	—	—	—	—	—	175.82	—	0.19	—	0.09
Togo	1.68	2.32	1.25	28.98	750.92	2,307.69	0.19	0.38	0.31	0.52
Tonga	—	—	—	—	76.92	117.22	0.81	1.18	0.31	0.24
Trinidad and Tobago	1.18	1.6	76.46	35.08	16,908.41	32,527.44	13.82	24.66	2.37	1.78
Tunisia	4.23	7.57	12.95	13.73	13,256.40	22,864.45	1.63	2.3	0.57	0.38
Turkey	4.37	6.58	0.15	5.1	146,142.71	225,922.85	2.6	3.18	0.63	0.45
Turkmenistan	—	1.27	—	—	—	41,688.60	—	8.75	—	—
Uganda	—	—	5.03	21.07	813.19	1,824.17	0.05	0.07	0.12	0.08

Table K (*continued*)
World Countries: Energy Efficiency and Emission, 1990–2005

COUNTRY	GDP PER UNIT OF ENERGY USE		FUEL TRADE		CARBON DIOXIDE EMISSIONS					
	PPP $ PER KG. OF OIL EQUIVALENT		% OF MERCHANDISE EXPORTS AND IMPORTS 2005–2006		TOTAL KILOTONS		METRIC TONS PER CAPITA		KG. PER PPP $ OF GDP	
	1990	2005	EXPORTS	IMPORTS	1990	2004	1990	2004	1990	2004
Ukraine	1.2	1.84	6.41	28.27	—	329,743.26	—	6.95	—	1.33
United Arab Emirates	1.97	2.93	—	—	54,662.95	149,054.80	30.83	37.76	1.23	1.21
United Kingdom	4.49	8.08	9.63	8.96	579,189.90	586,735.68	10.06	9.81	0.61	0.33
United States	2.99	5.3	3.7	17.97	4,816,874.30	6,044,023.26	19.3	20.58	0.84	0.52
Uruguay	6.98	10.59	3.62	27.61	3,908.42	5,472.52	1.26	1.66	0.25	0.2
Uzbekistan	0.64	1.12	—	—	—	137,783.75	—	5.33	—	2.89
Vanuatu	—	—	—	—	65.93	87.91	0.44	0.42	0.19	0.13
Venezuela	3.1	4.31	92.57	0.53	117,318.56	172,468.69	5.94	6.6	0.86	0.75
Vietnam	1.78	3.47	25.77	14.97	21,384.59	98,574.99	0.32	1.2	0.49	0.62
Virgin Islands (U.S.)	—	—	—	—	8,443.22	—	81.03	—	—	—
Yemen	6.23	6.86	94.48	21.83	9,586.07	21,095.22	0.78	1.03	0.6	0.49
Zambia	1.32	1.89	0.6	15.14	2,443.22	2,285.71	0.3	0.2	0.34	0.18
Zimbabwe	—	—	0.29	15.12	16,641.01	10,549.44	1.59	0.81	—	—

Table L
World Countries: Power and Transportation, 1990–2006

	ELECTRICAL POWER				TRANSPORTATION					
	KILOWATT HOUR CONSUMPTER PER CAPITA		TRANSMISSION AND DISTRIBUTION LOSSES: % OF OUTPUT		GOODS TRANSPORTED BY ROAD: MILLIONS OF TONS-KM HAULED		GOODS TRANSPORTED BY RAIL: TONS-KM PER $MILLION OF GDP (PPP)		GOODS TRANSPORTED BY AIR: MILLIONS OF TONS-KM	
COUNTRY	1990	2005	1990	2005	1990	2006	1990	2006	1990	2006
Afghanistan	—	—	—	—	—	—	—	—	9.4	—
Albania	522	1,167	53	39	—	—	584	26	—	0.11
Algeria	542	899	14	13	—	—	2,680.00	1,471.00	14.60	23.69
Angola	60	141	25	14	—	—	—	—	39.9	81
Antigua and Barbuda	—	—	—	—	—	—	—	—	0.2	0.18
Argentina	1,305	2,418	18	15	—	—	7,658.83	—	213.60	124.58
Armenia	2,718	1,503	—	16	1,533.00	231.3	4,883.50	654.14	—	7.39
Australia	8,456	11,481	7	7	—	168,630.00	7,400.00	46,164.00	1,222.30	2,569.51
Austria	6,078	7,889	6	5	13,300.00	—	12,158.00	17,036.00	53.90	571.98
Azerbaijan	2,584	2,407	—	10	3,287.00	7,536.00	37,076.00	10,066.80	—	15.75
Bahamas	—	—	—	—	—	—	—	—	0.3	0.56
Bahrain	6,573	11,401	7	5	—	—	—	—	44.3	—
Bangladesh	45	136	34	8	—	—	650.99	816.82	69.5	190.83
Barbados	—	—	—	—	—	—	—	—	0.4	—
Belarus	3,401	3,209	—	12	—	15,055.00	75,373.00	43,559.00	—	1.22
Belgium	6,380	8,510	5	5	25,000.00	54,856.00	8,370.00	8,130.00	656.40	740.39
Benin	34	69	—	—	—	—	—	—	17.7	—
Bolivia	274	479	15	16	—	—	540.92	—	8.4	10.71
Bosnia-Herzegovina	3,043	2,316	—	18	—	—	—	1,173.00	—	—
Botswana	724	1,406	—	15	—	—	—	—	3.1	0.32
Brazil	1,456	2,008	14	17	—	—	41,041.50	—	1,082.00	1,412.26
Brunei	4,354	7,498	5	4	—	—	—	—	9.9	130.18
Bulgaria	4,759	4,121	11	11	13,823.00	—	14,132.00	5,164.00	8.00	3.38
Burkina Faso	—	—	—	—	—	—	—	—	17.7	0
Cambodia	—	—	—	—	—	—	24.06	—	—	1.11
Cameroon	192	196	13	16	—	—	757	1,076.00	13.1	29
Canada	16,106	17,285	7	7	54,700.00	—	224,751.00	445,689.44	1,385.30	1,503.19
Cape Verde	—	—	—	—	—	—	—	—	0.6	1.79
Central African Republic	—	—	—	—	144	—	—	—	17.8	—
Chad	—	—	—	—	—	—	—	—	17.7	—
Chile	1,247	3,074	11	4	—	—	1,465.19	—	418.50	1,027.67
China	511	1,781	7	7	—	869,320.00	1,060,100.00	2,170,700.00	818.30	7,692.21
Colombia	827	890	21	19	6,227.00	38,199.00	418.00	7,751.00	463.70	1,051.29
Congo, Dem. Rep.	119	91	20	4	—	—	1,732.00	444	57.30	—
Congo, Rep.	169	160	20	56	—	—	—	231	18.4	—
Costa Rica	1,087	1,719	8	10	2,243.00	—	—	—	38.70	10.04
Côte d'Ivoire	150	170	18	18	—	—	289	—	17.7	—
Croatia	2,991	3,475	—	17	—	9,328.00	6,494.00	2,835.00	—	2.09

Table L (*continued*)
World Countries: Power and Transportation, 1990–2006

COUNTRY	ELECTRICAL POWER				TRANSPORTATION					
	KILOWATT HOUR CONSUMPTER PER CAPITA		TRANSMISSION AND DISTRIBUTION LOSSES: % OF OUTPUT		GOODS TRANSPORTED BY ROAD: MILLIONS OF TONS-KM HAULED		GOODS TRANSPORTED BY RAIL: TONS-KM PER $MILLION OF GDP (PPP)		GOODS TRANSPORTED BY AIR: MILLIONS OF TONS-KM	
	1990	2005	1990	2005	1990	2006	1990	2006	1990	2006
Cuba	1,212	1,152	14	15	—	—	—	—	19	30.15
Cyprus	3,213	5,560	6	4	—	1,100.00	—	—	34.3	47.82
Czech Republic	5,584	6,343	6	6	—	—	—	14,385.00	15	39.36
Denmark	5,945	6,663	10	4	13,700.00	11,058.00	1,730.00	2,030.00	123.90	1.28
Djibouti	—	—	—	—	—	—	—	97	0.1	—
Dominican Republic	383	1,000	25	27	—	—	—	—	70.4	—
Ecuador	479	714	23	43	2,638.00	5,453.00	—	—	63.20	5.87
Egypt	690	1,245	10	16	31,400.00	—	2,828.00	3,917.00	144.00	309.43
El Salvador	368	666	16	13	—	—	—	—	4.6	20.18
Equatorial Guinea	—	—	—	—	—	—	—	—	0.1	—
Estonia	5,757	5,567	—	11	4,510.00	7,641.00	6,977.00	10,311.00	—	1.08
Ethiopia	21	34	10	10	—	—	—	—	66.6	157.18
Fiji	—	—	—	—	—	—	—	—	5.1	95.79
Finland	12,487	16,120	5	4	26,300.00	27,800.00	8,357.00	9,706.00	135.10	408.78
France	6,127	7,938	7	6	190,500.00	193,000.00	50,667.00	41,898.00	3,995.80	6,135.03
Gabon	951	999	11	18	—	—	—	2,219.00	26.1	77.98
Georgia	2,673	1,672	—	16	7,370.00	570	15,477.00	6,127.00	—	2.89
Germany	6,640	7,111	4	5	182,800.00	237,609.00	—	88,022.00	3,994.20	8,133.99
Ghana	291	266	8	14	—	—	129	242	15.7	—
Greece	3,233	5,242	8	9	10,900.00	—	647.00	613	112.70	70.65
Guatemala	223	522	14	8	—	—	—	—	9	—
Guinea	—	—	—	—	—	—	—	—	0.4	—
Guinea-Bissau	—	—	—	—	—	—	—	—	0.1	—
Guyana	—	—	—	—	—	—	—	—	2.8	—
Haiti	58	37	31	38	—	—	—	—	3.9	—
Honduras	373	626	7	24	—	—	—	—	4.3	—
Hong Kong	4,178	5,878	11	13	—	—	—	—	—	8,325.97
Hungary	3,427	3,771	14	11	—	—	16,593.00	8,537.00	6.30	20
Iceland	16,150	27,987	9	4	—	—	—	—	35.5	138.61
India	276	480	20	25	—	—	235,785.00	407,398.00	662.90	842.55
Indonesia	161	509	14	12	—	—	3,190.00	4,430.00	458.60	469.22
Iran	975	2,117	10	19	—	—	9,040.86	19,127.00	113.70	91.86
Iraq	1,231	—	5	6	—	—	—	—	42.8	—
Ireland	3,776	6,234	9	8	5,100.00	—	589.00	303	128.30	130.5
Isle of Man	—	—	—	—	—	—	—	—	—	—
Israel	4,176	6,759	5	3	—	—	1,036.00	1,149.00	831.50	1,124.37
Italy	4,145	5,669	8	7	177,900.00	—	19,419.00	20,131.00	1,171.30	1,377.20
Jamaica	879	2,474	15	12	—	—	—	—	18.8	15.35

Table L (*continued*)
World Countries: Power and Transportation, 1990–2006

COUNTRY	ELECTRICAL POWER				TRANSPORTATION					
	KILOWATT HOUR CONSUMPTER PER CAPITA		TRANSMISSION AND DISTRIBUTION LOSSES: % OF OUTPUT		GOODS TRANSPORTED BY ROAD: MILLIONS OF TONS-KM HAULED		GOODS TRANSPORTED BY RAIL: TONS-KM PER $MILLION OF GDP (PPP)		GOODS TRANSPORTED BY AIR: MILLIONS OF TONS-KM	
	1990	2005	1990	2005	1990	2006	1990	2006	1990	2006
Japan	6,496	8,233	5	5	274,444.00	327,632.00	26,803.00	22,632.00	5,068.80	8,480.02
Jordan	1,050	1,676	8	14	—	—	711	1,024.00	222.8	258.56
Kazakhstan	5,905	3,206	—	10	44,775.00	53,800.00	406,963.00	191,200.00	—	16.38
Kenya	116	138	16	18	—	—	1,919.00	1,399.00	52.20	301.22
Kiribati	—	—	—	—	—	—	—	—	0.8	—
Korea, North	1,247	817	9	16	—	—	—	—	3.1	2.29
Korea, South	2,373	7,779	3	4	31,841.00	12,545.00	13,663.00	10,108.00	2,459.40	7,751.66
Kuwait	8,108	15,345	7	11	—	—	—	—	145	239.21
Kyrgyzstan	2,314	1,842	—	26	5,627.00	1,335.60	—	561.00	—	1.23
Laos	—	—	—	—	120	—	—	—	0.8	2.55
Latvia	3,387	2,702	—	17	5,853.00	2,767.00	18,538.00	17,921.00	—	13.19
Lebanon	471	2,242	7	16	—	—	—	—	167.1	73.56
Lesotho	—	—	—	—	—	—	—	—	0.1	—
Liberia	—	—	—	—	—	—	—	—	0.1	—
Libya	1,603	3,299	31	13	—	—	—	—	10.3	0.32
Liechtenstein	—	—	—	—	—	—	—	—	—	—
Lithuania	4,023	3,104	—	8	—	15,908.00	19,258.00	12,457.00	—	1.3
Luxembourg	13,666	15,971	14	3	1,300.00	—	—	392	—	5,270.27
Macao, China	—	—	—	—	—	—	—	—	—	180.83
Macedonia	2,789	3,417	—	23	—	—	769	441	—	0.12
Madagascar	—	—	—	—	—	—	—	—	30.2	18.77
Malawi	—	—	—	—	—	—	65.05	38	0.9	1.6
Malaysia	1,178	3,262	7	4	—	—	1,404.44	1,178.00	574.20	2,597.40
Maldives	—	—	—	—	—	—	—	—	—	0.01
Mali	—	—	—	—	—	—	273.2	—	—	—
Malta	2,778	4,917	9	11	—	—	—	—	4.8	11.66
Marshall Islands	—	—	—	—	—	—	—	—	2.5	0.35
Mauritania	—	—	—	—	—	—	—	—	17.9	0.11
Mauritius	—	—	—	—	—	—	—	—	64.1	194.84
Mayotte	—	—	—	—	—	—	—	54,387.00	—	—
Mexico	1,295	1,899	12	16	108,884.00	204,217.00	36,417.10	—	143.20	456.96
Micronesia	—	—	—	—	—	—	—	—	—	—
Moldova	2,553	1,428	—	38	6,305.00	—	—	2,980.00	—	0.78
Monaco	—	—	—	—	—	—	—	—	—	0
Mongolia	—	—	—	—	—	—	5,087.80	8,857.00	—	6.34
Morocco	369	644	9	18	2,638.00	1,256.00	4,903.00	5,919.00	31.50	50.63
Mozambique	40	450	16	12	—	—	—	768	9.1	5.2
Myanmar	45	82	26	35	—	—	387.57	—	1.1	2.79
Namibia	—	1,428	—	18	—	—	1,634.86	—	—	0
Nepal	35	70	19	20	—	—	—	—	11.2	7.21
Netherlands	5,218	6,988	4	4	31,800.00	—	3,070.00	4,331.00	2,129.10	4,959.00

Table L (*continued*)
World Countries: Power and Transportation, 1990–2006

COUNTRY	ELECTRICAL POWER				TRANSPORTATION					
	KILOWATT HOUR CONSUMPTER PER CAPITA		TRANSMISSION AND DISTRIBUTION LOSSES: % OF OUTPUT		GOODS TRANSPORTED BY ROAD: MILLIONS OF TONS-KM HAULED		GOODS TRANSPORTED BY RAIL: TONS-KM PER $MILLION OF GDP (PPP)		GOODS TRANSPORTED BY AIR: MILLIONS OF TONS-KM	
	1990	2005	1990	2005	1990	2006	1990	2006	1990	2006
Netherlands Antilles	3,557	5,172	14	23	—	—	—	—	—	—
New Caledonia	—	—	—	—	—	—	—	—	—	—
New Zealand	8,298	9,656	11	7	—	—	2,744.00	—	325.10	819
Nicaragua	295	414	18	22	—	—	—	—	3.6	—
Niger	—	—	—	—	—	—	—	—	17.7	—
Nigeria	88	127	38	24	—	—	178.2	77	23.5	11.27
Norway	23,354	25,137	6	7	7,940.00	14,966.00	—	9,568.00	129.60	—
Oman	2,150	3,757	12	26	—	—	—	—	44.3	—
Pakistan	277	456	21	24	—	—	5,708.60	5,013.00	420.50	426.99
Panama	854	1,500	26	16	—	—	—	—	2.8	35.88
Papua New Guinea	—	—	—	—	—	—	—	—	14.8	21.89
Paraguay	505	849	0	5	—	—	—	—	4.5	0
Peru	548	848	14	9	—	—	498.13	—	25.6	112.23
Philippines	360	588	13	12	—	—	11.95	0.76	316	318.85
Poland	3,272	3,437	8	9	49,800.00	119,740.00	83,500.00	45,438.00	48.90	79.82
Portugal	2,564	4,663	11	9	12,200.00	—	1,459.00	2,422.00	167.00	294.08
Puerto Rico	—	—	—	—	—	—	—	—	—	—
Qatar	9,773	16,801	5	7	—	—	—	—	44.3	888.5
Romania	2,924	2,342	9	10	13,800.00	37,220.00	57,253.00	16,032.00	12.80	4.98
Russia	6,673	5,785	—	12	—	—	2,523,000.00	1,950,900.00	—	1,926.30
Samoa	—	—	—	—	—	—	—	—	—	1.88
São Tomé and Príncipe	—	—	—	—	—	—	—	—	—	0.07
Saudi Arabia	3,982	6,813	6	11	—	—	690	1,192.00	609.6	1,066.28
Senegal	99	151	13	30	—	—	564	—	17.7	0
Serbia	—	—	—	—	—	—	23,149.00	3,482.00	—	4.11
Seychelles	—	—	—	—	—	—	—	—	10.4	30.72
Sierra Leone	—	—	—	—	—	—	—	—	2.2	9.59
Singapore	4,860	8,358	6	5	—	—	—	—	1,652.50	7,981.25
Slovakia	5,559	4,920	8	5	—	18,517.00	—	9,326.00	—	0.3
Slovenia	5,091	6,918	—	6	3,440.00	11,032.50	4,196.00	3,245.00	—	2.31
Solomon Islands	—	—	—	—	—	—	—	—	—	0.83
Somalia	—	—	—	—	—	—	—	—	8.7	—
South Africa	4,431	4,847	6	6	—	—	101,746.00	109,721.00	179.20	1,233.00
Spain	3,540	6,147	9	9	78,900.00	—	10,742.00	11,586.00	759.70	1,100.21
Sri Lanka	154	378	17	15	—	—	163.8	134.7	93.4	325.41
Sudan	49	94	15	16	—	—	—	766	12.9	51.04
Suriname	—	—	—	—	—	—	—	—	14.5	26.27
Swaziland	—	—	—	—	—	—	—	11,394.00	0.2	—

Table L (*continued*)
World Countries: Power and Transportation, 1990–2006

COUNTRY	ELECTRICAL POWER				TRANSPORTATION					
	KILOWATT HOUR CONSUMPTER PER CAPITA		TRANSMISSION AND DISTRIBUTION LOSSES: % OF OUTPUT		GOODS TRANSPORTED BY ROAD: MILLIONS OF TONS-KM HAULED		GOODS TRANSPORTED BY RAIL: TONS-KM PER $MILLION OF GDP (PPP)		GOODS TRANSPORTED BY AIR: MILLIONS OF TONS-KM	
	1990	2005	1990	2005	1990	2006	1990	2006	1990	2006
Sweden	15,836	15,440	6	7	26,500.00	39,373.00	18,441.00	13,120.00	191.10	—
Switzerland	7,507	8,305	7	7	10,400.00	15,753.00	8,127.00	8,571.00	928.50	1,039.03
Syria	674	1,411	26	24	—	—	1,265.00	2,256.00	17.70	16.39
Tajikistan	3,346	2,267	—	15	—	—	—	1,117.00	—	12.77
Tanzania	51	61	20	27	—	—	—	—	1.4	1.69
Thailand	739	1,988	11	8	—	—	3,291.00	4,037.00	661.00	2,106.87
Togo	87	94	21	46	—	—	—	—	17.7	—
Trinidad and Tobago	2,679	5,038	8	6	—	—	—	—	15.2	46.46
Tunisia	638	1,194	10	12	—	—	1,820.34	2,067.00	19.10	18.57
Turkey	893	1,898	12	15	—	166,831.00	7,894.00	9,078.00	101.30	463.9
Turkmenistan	2,293	1,731	—	12	—	—	—	8,670.00	—	10.46
Uganda	—	—	—	—	—	—	159	218	21.7	33.98
Ukraine	4,782	3,246	—	13	79,668.00	23,894.70	473,953.00	223,980.00	—	43.76
United Arab Emirates	8,766	13,708	9	7	—	—	—	—	145.1	—
United Kingdom	5,327	6,253	8	8	133,000.00	163,000.00	15,986.00	22,110.00	3,825.40	6,215.32
United States	11,713	13,648	9	6	1,073,100.00	2,116,532.00	1,530,742.83	2,589,348.60	14,791.40	39,881.90
Uruguay	1,246	2,007	14	23	—	—	203.62	330.85	2.6	4.15
Uzbekistan	2,383	1,659	—	9	—	—	—	18,007.00	—	67.55
Vanuatu	—	—	—	—	—	—	—	—	0.1	1.86
Venezuela	2,463	2,848	18	25	—	—	—	—	112.5	2.05
Vietnam	98	573	25	11	—	—	847.02	2,928.00	—	216.01
Yemen	119	174	12	23	—	—	—	—	11.5	66.11
Zambia	754	721	3	5	—	—	1,197.00	—	29.60	0
Zimbabwe	861	953	7	7	—	—	5,590.00	—	64.90	8.55

Table M
World Countries: Communication, Information, and Science and Technology, 2004–2006

COUNTRY	PERSONAL COMMUNICATION AND INFORMATION				
	DAILY NEWSPAPERS PER 1000 POP	HOUSEHOLDS WITH TELEVISION (%)	PERSONAL COMPUTERS	INTERNET USERS	TELEPHONE MAINLINES PER 1000 PERSONS
	2004–2005	2005–2006	2005–2006	2005–2006	2005–2006
Afghanistan	—	6	100,000	535,000	0.63
Albania	—	89.9	54,000	471,250	11.21
Algeria	—	90	350,000	2,460,000	8.52
American Samoa	—	—	—	—	18.16
Andorra	—	—	—	21,922	53.54
Angola	2.26	9	110,614	85,000	0.59
Antigua and Barbuda	—	98	12,100	32,000	47.56
Argentina	35.52	97	3,500,000	8,183,655	24.17
Armenia	7.6	90.61	297,000	172,810	19.7
Aruba	—	76.24	7,926	24,000	38.14
Australia	156.15	99	15,440,000	15,300,000	48.02
Austria	314.52	98	4,996,000	4,200,000	43.04
Azerbaijan	—	98.9	194,700	829,100	14.01
Bahamas	—	93	40,000	103,000	41.16
Bahrain	—	95.42	128,000	157,310	26.17
Bangladesh	—	22.9	3,500,000	450,000	0.73
Barbados	—	93	40,000	160,000	46.2
Belarus	81.57	97.1	79,016	5,477,500	34.61
Belgium	164.04	98	3,954,000	4,800,000	44.77
Belize	—	93	41,000	34,000	11.39
Benin	0.37	20	32,000	700,000	0.88
Bermuda	263.87	—	—	42,000	90.37
Bhutan	—	53	13,000	30,000	4.86
Bolivia	—	50	220,000	580,000	7.13
Bosnia-Herzegovina	—	87.16	212,000	950,000	25.19
Botswana	42.55	10	86,000	60,000	7.37
Brazil	35.63	91	30,000,000	42,600,000	20.49
Brunei	68.36	98	33,000	165,558	20.99
Bulgaria	79.12	97.3	490,000	1,870,000	31.19
Burkina Faso	—	7.5	31,400	80,000	0.66
Burundi	—	14	55,000	60,000	0.4
Cambodia	—	43	44,000	44,000	0.23
Cameroon	—	26	200,000	370,000	0.56
Canada	174.54	99	28,295,000	22,000,000	64.31
Cape Verde	—	61	55,000	29,000	13.8
Cayman Islands	348.1	—	—	21,000	—
Central African Republic	—	5	12,000	13,000	0.24
Chad	—	4	16,000	60,000	0.12
Chile	50.61	90	2,300,000	4,155,570	20.24
China	73.93	89	55,570,000	137,000,000	28.04

Table M (*continued*)
World Countries: Communication, Information, and Science and Technology, 2004–2006

SCIENCE AND TECHNOLOGY					
MOBILE PHONE SUBSCRIBERS	SCIENTIFIC AND TECHNICAL JOURNAL ARTICLES	TECHNICIANS IN R&D PER MILLION PEOPLE	INFORMATION AND COMMUNICATION EXPENDITURE (% OF GDP)	INFORMATION AND COMMUNICATION EXPENDITURE (CURRENT $US)	INFORMATION AND COMMUNICATION EXPENDITURE ($US PER CAPITA)
2005–2006	2005	2004–2005	2006	2006	2006
2,520,366	—	—	—	—	—
1,530,244	—	—	—	—	—
20,997,954	350	34.52	2.44	2,796,500,000	83.85
2,250	—	—	—	—	—
64,560	—	—	—	—	—
2,264,215	—	—	—	—	—
102,000	—	—	—	—	—
31,510,390	3,058	338.35	6.93	14,841,800,000	379.25
318,044	180	—	—	—	—
108,225	—	—	—	—	—
19,760,000	15,957	—	6.40	49,946,800,000	2,412.72
9,255,000	4,566	1476.78	5.50	17,699,000,000	2,137.33
3,323,500	116	—	—	—	—
227,771	—	—	—	—	—
898,864	—	—	—	—	—
19,130,984	193	—	2.74	1,696,200,000	10.87
206,190	—	—	—	—	—
5,960,000	490	—	—	—	—
9,659,819	6,841	—	5.89	23,225,600,000	2,203.38
118,314	—	—	—	—	—
1,055,727	—	—	—	—	—
60,100	—	—	—	—	—
82,078	—	—	—	—	—
2,698,214	—	—	4.89	545,800,000	58.35
1,887,820	—	—	—	—	—
979,787	—	—	—	—	—
99,918,624	9,889	394.84	6.43	68,670,300,000	362.72
254,020	—	—	—	—	—
8,253,416	764	478.27	3.45	1,085,500,000	141.11
1,016,605	—	17.01	—	—	—
153,000	—	—	—	—	—
1,140,000	—	—	—	—	—
2,252,505	131	—	5.13	939,400,000	51.69
17,017,000	25,836	1467.24	5.65	71,867,400,000	2,201.21
108,858	—	—	—	—	—
—	—	—	—	—	—
100,000	—	—	—	—	—
466,088	—	—	—	—	—
12,450,801	1,559	301.73	5.24	7,642,200,000	465.06
461,057,984	41,596	—	5.38	142,313,300,000	108.49

Table M (*continued*)
World Countries: Communication, Information, and Science and Technology, 2004–2006

COUNTRY	PERSONAL COMMUNICATION AND INFORMATION				
	DAILY NEWSPAPERS PER 1000 POP	HOUSEHOLDS WITH TELEVISION (%)	PERSONAL COMPUTERS	INTERNET USERS	TELEPHONE MAINLINES PER 1000 PERSONS
	2004–2005	2005–2006	2005–2006	2005–2006	2005–2006
Colombia	22.35	90	1,891,615	6,705,000	17.26
Comoros	—	13	5,400	21,000	2.82
Congo, Dem. Rep.	—	4	12,000	180,000	0.02
Congo, Rep.	—	7	19,000	70,000	0.44
Costa Rica	64.71	89	1,000,000	1,214,376	30.72
Côte d'Ivoire	—	35	323,000	300,000	1.38
Croatia	—	98	884,000	1,576,360	41.25
Cuba	64.7	70	377,000	240,000	8.64
Cyprus	—	99.3	279,000	356,630	52.95
Czech Republic	181.93	—	2,800,000	3,541,300	31.44
Denmark	352.04	97	3,772,000	3,171,210	56.99
Djibouti	—	43	19,000	11,000	1.34
Dominica	—	—	—	26,000	29.33
Dominican Republic	41.63	75.8	206,346	2,000,000	9.33
Ecuador	—	80	866,340	1,549,000	13.28
Egypt	—	88	2,700,000	6,000,000	14.57
El Salvador	36.97	83	350,000	637,000	15.33
Equatorial Guinea	—	25.7	9,000	8,000	2.07
Eritrea	—	16	25,000	100,000	0.8
Estonia	192.5	—	650,000	760,000	40.37
Ethiopia	4.74	4	300,000	164,000	0.94
Faeroe Islands	—	98	—	34,030	47.71
Fiji	52.33	60	50,000	80,000	13.59
Finland	430.73	94.1	2,625,000	2,925,370	36.46
France	164.86	97	35,000,000	30,100,000	55.34
French Polynesia	—	95	28,000	65,000	20.69
Gabon	—	56	45,000	81,000	2.78
Gambia	—	12	25,000	57,978	3.18
Georgia	3.9	89	210,278	332,000	12.48
Germany	267.41	98	50,000,000	38,600,000	65.8
Ghana	—	26	128,000	609,810	1.55
Greece	—	99.5	1,020,000	2,048,080	55.49
Grenada	—	—	—	—	25.61
Guam	—	—	—	65,000	—
Guatemala	—	50	262,000	1,320,000	10.4
Guinea	—	11	45,000	50,000	0.29
Guinea-Bissau	—	31	3,000	37,000	0.64
Guyana	—	92	29,000	160,000	14.89
Haiti	—	27	16,000	650,000	1.56
Honduras	—	58	120,000	337,300	10.17
Hong Kong	—	99	4,172,414	3,770,400	56.14

Table M (*continued*)
World Countries: Communication, Information, and Science and Technology, 2004–2006

SCIENCE AND TECHNOLOGY					
MOBILE PHONE SUBSCRIBERS	SCIENTIFIC AND TECHNICAL JOURNAL ARTICLES	TECHNICIANS IN R&D PER MILLION PEOPLE	INFORMATION AND COMMUNICATION EXPENDITURE (% OF GDP)	INFORMATION AND COMMUNICATION EXPENDITURE (CURRENT $US)	INFORMATION AND COMMUNICATION EXPENDITURE ($US PER CAPITA)
2005–2006	2005	2004–2005	2006	2006	2006
29,762,716	400	95.42	7.11	10,900,100,000	239.26
—	—	—	—	—	—
4,415,470	—	—	—	—	—
490,000	—	—	—	—	—
1,443,717	105	—	7.28	1,617,400,000	367.69
4,065,421	—	—	—	—	—
4,469,700	953	566.77	0.00	210	0.00
152,715	261	—	—	—	—
777,529	90	324.76	—	—	—
12,149,865	3,169	1347.67	7.32	10,475,100,000	1,019.97
5,840,948	5,040	—	5.99	16,506,500,000	3,036.18
44,053	—	—	—	—	—
—	—	—	—	—	—
4,605,659	—	—	—	—	—
8,485,050	—	—	2.95	1,223,100,000	92.65
18,001,106	1,658	—	1.36	1,460,300,000	19.69
3,851,611	—	—	—	—	—
96,900	—	—	—	—	—
61,996	—	—	—	—	—
1,658,700	439	489.86	—	—	—
866,700	88	10.06	—	—	—
49,965	—	—	—	—	—
205,000	—	—	—	—	—
5,670,000	4,811	—	6.72	14,161,300,000	2,689.04
51,662,000	30,309	—	6.33	142,338,700,000	2,323.65
152,000	—	—	—	—	—
764,728	—	—	—	—	—
404,345	—	18.46	—	—	—
1,703,888	145	—	—	—	—
84,300,000	44,145	1056.35	6.18	179,120,300,000	2,174.45
5,207,242	81	—	—	—	—
11,097,515	4,291	—	3.16	9,753,500,000	874.98
46,193	—	—	—	—	—
—	—	—	—	—	—
7,178,745	—	—	—	—	—
189,000	—	—	—	—	—
95,000	—	—	—	—	—
281,368	—	—	—	—	—
500,200	—	—	—	—	—
2,240,756	—	—	4.55	420,400,000	60.33
9,356,396	—	—	8.77	16,646,100,000	2,427.57

Table M (*continued*)
World Countries: Communication, Information, and Science and Technology, 2004–2006

COUNTRY	PERSONAL COMMUNICATION AND INFORMATION				
	DAILY NEWSPAPERS PER 1000 POP	HOUSEHOLDS WITH TELEVISION (%)	PERSONAL COMPUTERS	INTERNET USERS	TELEPHONE MAINLINES PER 1000 PERSONS
	2004–2005	2005–2006	2005–2006	2005–2006	2005–2006
Hungary	216.81	96	1,504,138	3,500,000	33.28
Iceland	553.4	98	142,000	194,000	64.2
India	72.89	32	17,000,000	60,000,000	3.67
Indonesia	—	65	3,285,000	16,000,000	6.64
Iran	—	—	7,420,791	18,000,000	31.36
Ireland	181.88	98.1	2,198,000	1,437,000	49.13
Isle of Man	—	—	—	—	—
Israel	—	91.5	8,454,000	1,899,060	42.63
Italy	138.15	95.5	21,486,000	28,855,360	42.74
Jamaica	—	70	179,000	1,232,258	12.02
Japan	550.69	99	86,389,000	87,540,000	43.17
Jordan	—	96.3	355,000	796,920	11.09
Kazakhstan	—	—	—	1,247,000	19.13
Kenya	—	18	492,000	2,770,296	0.8
Kiribati	—	—	—	2,000	—
Korea, South	—	—	25,685,000	34,120,000	55.49
Kuwait	—	95.44	600,000	816,670	—
Kyrgyzstan	0.95	—	100,000	298,100	—
Laos	2.51	30	100,000	25,000	—
Latvia	153.89	98	566,000	1,070,790	28.74
Lebanon	60.73	96	409,400	950,000	16.8
Lesotho	—	2	1,450	51,480	2.42
Liberia	—	—	—	—	—
Libya	—	50	130,000	232,044	8
Liechtenstein	—	—	—	22,000	57.59
Lithuania	107.74	97.8	616,000	1,083,024	23.34
Luxembourg	250.85	99.1	290,000	339,000	53.41
Macao, China	—	94	160,000	200,000	37
Macedonia	88.74	98	450,891	267,990	24.11
Madagascar	—	10	102,000	110,000	0.68
Malawi	—	3	25,100	59,700	0.78
Malaysia	110.59	95	5,600,000	11,292,000	16.63
Maldives	—	92	45,000	20,087	10.83
Mali	—	17	45,000	70,000	0.69
Malta	—	93.18	66,611	127,247	49.84
Marshall Islands	0	—	—	2,200	—
Mauritania	—	25	78,500	100,000	1.15
Mauritius	77.04	93	210,000	182,000	28.51
Mexico	—	92.7	14,000,000	18,091,788	19.06
Micronesia	—	16.25	6,000	16,000	11.31
Moldova	—	82.1	347,991	727,700	26.56
Mongolia	19.13	63	340,000	268,300	6.11

Table M (*continued*)
World Countries: Communication, Information, and Science and Technology, 2004–2006

SCIENCE AND TECHNOLOGY					
MOBILE PHONE SUBSCRIBERS	SCIENTIFIC AND TECHNICAL JOURNAL ARTICLES	TECHNICIANS IN R&D PER MILLION PEOPLE	INFORMATION AND COMMUNICATION EXPENDITURE (% OF GDP)	INFORMATION AND COMMUNICATION EXPENDITURE (CURRENT $US)	INFORMATION AND COMMUNICATION EXPENDITURE ($US PER CAPITA)
2005–2006	2005	2004–2005	2006	2006	2006
9,965,000	2,614	—	5.96	6,735,000,000	669.00
328,530	206	—	—	—	—
166,050,000	14,608	—	6.07	55,304,300,000	49.83
63,803,016	205	—	3.10	11,304,800,000	50.68
13,659,123	2,635	—	2.44	5,311,600,000	75.77
4,690,000	2,120	—	4.28	9,418,700,000	2,206.87
—	—	—	—	—	—
8,403,765	6,309	—	7.88	11,065,900,000	1,569.94
—	24,645	—	4.33	80,196,100,000	1,362.89
—	—	—	10.19	1,021,400,000	382.93
101,698,000	55,471	—	7.86	343,392,500,000	2,687.88
4,343,100	275	—	8.00	1,127,900,000	203.68
7,830,400	96	85.67	—	—	—
6,484,791	226	—	2.43	552,900,000	15.13
—	—	—	—	—	—
40,197,116	16,396	—	6.62	58,768,600,000	1,213.77
2,379,811	233	94.53	1.37	1,211,700,000	466.14
541,652	—	—	—	—	—
638,202	—	—	—	—	—
2,183,696	134	460.34	—	—	—
1,103,376	234	—	—	—	—
249,786	—	—	—	—	—
160,000	—	—	—	—	—
3,927,562	—	—	—	—	—
28,755	—	—	—	—	—
4,718,215	406	418.53	—	—	—
713,800	59	—	—	—	—
636,347	—	221.23	—	—	—
1,417,031	—	82.59	—	—	—
1,045,888	—	6.40	—	—	—
429,305	—	—	—	—	—
19,463,722	615	64.18	6.73	10,137,200,000	388.19
262,615	—	—	—	—	—
1,512,948	—	—	—	—	—
346,771	23	359.29	—	—	—
—	—	—	—	—	—
1,060,122	—	—	—	—	—
772,395	—	—	—	—	—
57,016,372	3,902	—	3.31	27,770,600,000	266.46
14,094	—	—	—	—	—
1,358,152	89	—	—	—	—
557,207	—	—	—	—	—

Table M (*continued*)
World Countries: Communication, Information, and Science and Technology, 2004–2006

COUNTRY	PERSONAL COMMUNICATION AND INFORMATION				
	DAILY NEWSPAPERS PER 1000 POP	HOUSEHOLDS WITH TELEVISION (%)	PERSONAL COMPUTERS	INTERNET USERS	TELEPHONE MAINLINES PER 1000 PERSONS
	2004–2005	2005–2006	2005–2006	2005–2006	2005–2006
Morocco	—	78	745,000	6,100,000	4.15
Mozambique	2.83	6	282,590	178,000	0.32
Myanmar	—	3	400,000	93,585	1.05
Namibia	27.77	39	249,000	80,563	6.88
Nepal	—	13	132,000	249,385	2.16
Netherlands	308.21	99.42	13,943,000	14,544,360	46.57
New Caledonia	—	78	6,000	80,000	23.59
New Zealand	185.25	98	2,077,000	3,200,000	41.82
Nicaragua	—	60	220,000	155,000	4.48
Niger	0.19	7	10,000	40,000	0.18
Nigeria	—	32	1,200,000	8,000,000	1.17
Norway	517.19	100	2,745,000	4,074,110	44.1
Oman	—	79.16	130,000	319,215	10.93
Pakistan	50.51	46	803,463	12,000,000	3.3
Palau	—	—	—	—	36.27
Panama	65.11	79	147,357	220,000	13.17
Papua New Guinea	8.84	10	391,000	110,000	1.05
Paraguay	—	82	460,000	260,000	5.5
Peru	—	71	2,800,000	6,100,000	8.45
Philippines	79.81	63	4,521,000	4,614,759	4.21
Poland	112.68	90.7	9,244,000	11,000,000	30.09
Portugal	—	99.4	1,405,760	3,213,000	39.95
Puerto Rico	—	97	33,000	915,618	26.53
Qatar	—	90	145,000	289,917	27.8
Romania	70.12	93.7	2,800,000	7,000,000	19.47
Russia	92.29	98.03	17,400,000	25,688,650	28.01
Rwanda	—	2	19,211	65,000	0.17
Samoa	—	88	3,500	8,000	10.61
San Marino	—	—	24,390	15,400	73.43
São Tomé and Príncipe	—	35	6,000	29,000	4.89
Saudi Arabia	—	98.68	3,150,000	4,700,000	16.69
Senegal	8.78	31	250,000	650,000	2.34
Serbia	—	—	—	—	—
Seychelles	—	89	16,000	29,000	24.44
Singapore	360.91	98	2,960,000	1,717,136	41.34
Slovakia	125.35	98	1,929,000	2,255,553	21.66
Slovenia	—	96	808,000	1,250,630	41.73
Solomon Islands	10.73	4.8	22,000	8,000	1.57
Somalia	—	8	75,000	94,000	1.22
South Africa	29.83	59	3,966,000	5,100,000	10.08
Spain	144.98	99.48	12,000,000	18,578,000	41.67

Table M (*continued*)
World Countries: Communication, Information, and Science and Technology, 2004–2006

	SCIENCE AND TECHNOLOGY				
MOBILE PHONE SUBSCRIBERS	SCIENTIFIC AND TECHNICAL JOURNAL ARTICLES	TECHNICIANS IN R&D PER MILLION PEOPLE	INFORMATION AND COMMUNICATION EXPENDITURE (% OF GDP)	INFORMATION AND COMMUNICATION EXPENDITURE (CURRENT $US)	INFORMATION AND COMMUNICATION EXPENDITURE ($US PER CAPITA)
2005–2006	2005	2004–2005	2006	2006	2006
16,004,731	443	—	5.56	3,636,500,000	119.24
2,339,317	—	—	—	—	—
214,214	—	—	—	—	—
495,000	—	—	—	—	—
1,041,819	—	—	—	—	—
—	13,885	—	6.24	41,352,500,000	2,530.72
134,265	—	—	—	—	—
3,530,000	2,983	—	10.55	11,027,900,000	2,635.35
1,830,220	—	—	—	—	—
323,853	—	9.82	—	—	—
32,322,202	362	—	3.42	3,940,100,000	27.23
5,040,573	3,644	—	4.95	16,571,000,000	3,555.93
1,818,024	111	—	—	—	—
34,506,556	492	40.97	6.92	8,773,000,000	55.18
8,346	—	—	—	—	—
1,693,460	—	—	8.17	1,397,000,000	424.94
75,000	—	—	—	—	—
3,232,842	—	117.83	—	—	—
8,500,000	133	—	5.93	5,476,600,000	198.51
42,868,912	178	—	6.69	7,862,500,000	91.14
36,745,456	6,844	232.21	4.15	14,068,100,000	368.96
12,226,439	2,910	—	4.33	8,435,200,000	796.58
3,353,800	—	—	—	—	—
919,773	—	—	—	—	—
17,400,000	887	253.56	3.19	3,883,500,000	179.87
120,000,000	14,412	552.89	3.20	31,623,600,000	221.92
314,201	—	—	—	—	—
24,000	—	—	—	—	—
17,390	—	—	—	—	—
18,424	—	—	—	—	—
19,662,622	575	—	2.09	7,290,000,000	307.87
2,982,623	83	—	8.47	777,800,000	64.43
5,229,000	849	—	—	—	—
70,340	—	639.69	—	—	—
4,788,600	3,609	—	9.31	12,299,300,000	2,742.99
4,893,232	919	415.63	5.46	3,004,800,000	557.44
1,819,572	1,035	1264.48	3.09	1,154,500,000	575.29
6,000	—	—	—	—	—
500,000	—	—	—	—	—
33,959,960	2,392	109.64	9.98	25,471,300,000	537.47
46,152,024	18,336	888.02	3.62	44,284,000,000	1,003.69

Table M (*continued*)
World Countries: Communication, Information, and Science and Technology, 2004–2006

COUNTRY	PERSONAL COMMUNICATION AND INFORMATION				
	DAILY NEWSPAPERS PER 1000 POP	HOUSEHOLDS WITH TELEVISION (%)	PERSONAL COMPUTERS	INTERNET USERS	TELEPHONE MAINLINES PER 1000 PERSONS
	2004–2005	2005–2006	2005–2006	2005–2006	2005–2006
Sri Lanka	—	32	734,000	428,000	9.47
St. Lucia	—	—	—	—	—
St. Vincent and the Grenadines	—	79	16,500	10,000	18.9
Sudan	—	16.2	4,237,096	3,500,000	1.69
Suriname	80.72	66	20,000	32,000	17.9
Swaziland	26.11	18	42,000	41,569	3.86
Sweden	480.03	93.91	7,548,000	6,981,190	59.43
Switzerland	428.88	99	6,430,000	4,360,000	67.27
Syria	—	95	800,000	1,500,000	16.71
Tajikistan	—	79.46	84,591	19,521	4.28
Tanzania	1.59	14	356,000	384,323	0.4
Thailand	—	92	4,408,000	8,465,823	11.15
Togo	—	16	185,000	320,000	1.28
Tonga	—	26.3	6,000	3,100	13.83
Trinidad and Tobago	—	88	129,000	163,000	24.51
Tunisia	—	91.7	634,840	1,294,910	12.52
Turkey	—	92.19	4,073,000	12,283,490	25.81
Turkmenistan	9.44	—	347,991	64,800	8.24
Uganda	—	10	500,000	750,000	0.36
Ukraine	131.77	97.34	2,121,200	5,544,960	26.38
United Arab Emirates	—	85.91	1,050,000	1,708,480	30.83
United Kingdom	292.12	97.5	45,659,000	33,534,000	55.5
United States	194.13	99	—	208,000,000	57.46
Uruguay	—	92.35	450,000	756,000	29.78
Uzbekistan	—	—	830,000	1,700,000	6.85
Vanuatu	14.47	6	3,000	7,500	3.23
Venezuela	93.22	90	2,475,000	4,139,765	15.61
Vietnam	—	83	1,174,000	14,683,783	19.07
Virgin Islands (U.S.)	—	50	3,000	30,000	65.96
West Bank and Gaza	9.76	93.4	195,000	266,037	9.04
Yemen	4.1	42.61	400,000	270,000	4.59
Zambia	4.79	—	131,000	500,000	0.8
Zimbabwe	—	34	865,000	1,220,000	2.51

Table M (*continued*)
World Countries: Communication, Information, and Science and Technology, 2004–2006

	SCIENCE AND TECHNOLOGY				
MOBILE PHONE SUBSCRIBERS	SCIENTIFIC AND TECHNICAL JOURNAL ARTICLES	TECHNICIANS IN R&D PER MILLION PEOPLE	INFORMATION AND COMMUNICATION EXPENDITURE (% OF GDP)	INFORMATION AND COMMUNICATION EXPENDITURE (CURRENT $US)	INFORMATION AND COMMUNICATION EXPENDITURE ($US PER CAPITA)
2005–2006	2005	2004–2005	2006	2006	2006
5,412,496	136	71.66	5.36	1,446,100,000	72.72
105,656	—	—	—	—	—
87,634	—	—	—	—	—
4,683,127	—	—	—	—	—
320,000	—	—	—	—	—
250,000	—	—	—	—	—
9,607,000	10,012	—	7.22	27,726,000,000	3,052.25
7,418,000	8,749	2366.11	7.71	29,319,100,000	3,913.81
4,675,000	77	—	—	—	—
265,000	—	—	—	—	—
5,766,566	107	—	—	—	—
40,815,456	1,249	—	3.96	8,176,200,000	128.87
708,000	—	—	—	—	—
29,872	—	—	—	—	—
1,654,921	—	—	—	—	—
7,339,047	571	40.88	6.02	1,823,400,000	180.03
52,662,700	7,815	—	8.19	32,967,200,000	451.76
105,000	—	—	—	—	—
2,008,818	93	—	—	—	—
49,076,240	2,105	—	7.79	8,292,700,000	177.24
5,519,293	229	—	—	5,101,400,000	1,200.76
69,656,616	45,572	—	6.93	164,731,300,000	2,720.58
233,000,000	205,320	—	8.75	1,151,554,000,000	3,846.23
2,330,000	204	—	7.80	1,505,700,000	454.28
720,000	157	—	—	—	—
12,692	—	—	—	—	—
18,789,466	534	—	3.68	6,698,600,000	247.90
15,505,407	221	—	15.15	9,240,700,000	109.87
80,300	—	132.58	—	—	—
821,800	—	—	—	—	—
2,000,000	—	—	—	—	—
1,663,328	—	—	—	—	—
832,488	—	—	12.69	225,200,000	17.02

Table N
World Countries: Freshwater Resources

COUNTRY	ANNUAL INTERNAL RENEWABLE WATER RESOURCES		ANNUAL FRESHWATER WITHDRAWALS					WATER PRODUCTIVITY	ACCESS TO AN IMPROVED WATER SOURCE	
	FLOWS IN BILLIONS OF CUBIC METERS	PER CAPITA CUBIC METERS	BILLIONS OF CUBIC METERS	% OF INTERNAL RESOURCES	% FOR AGRICULTURE	% FOR INDUSTRY	% FOR DOMESTIC	GDP/WATER USE IN $US PER CUBIC METER	% OF URBAN POPULATION	% OF RURAL POPULATION
	2005	2005	1987–2002	1987–2002	1987–2002	1987–2002	1987–2002	2002	2004	2004
Afghanistan	55	—	23.3	42.3	98	0	2	—	—	—
Albania	27	8,530	1.7	6.4	62	11	27	2.4	99	94
Algeria	11	341	6.1	54.2	65	13	22	9.7	88	80
Angola	148	9,195	0.4	0.2	60	17	23	30.8	75	40
Argentina	276	7,123	29.2	10.6	74	9	17	8.3	98	80
Armenia	9	3,016	3	32.4	66	4	30	0.8	99	80
Australia	492	24,118	23.9	4.9	75	10	15	17.9	100	100
Austria	55	6,680	2.1	3.8	1	64	35	93.4	100	100
Azerbaijan	8	965	17.3	213	68	28	5	0.4	95	59
Bangladesh	105	685	79.4	75.6	96	1	3	0.7	82	72
Belarus	37	3,805	2.8	7.5	30	47	23	5	100	100
Belgium	12	1,145	—	—	—	—	—	—	100	—
Benin	10	1,213	0.1	1.3	45	23	32	19	78	57
Bolivia	304	33,054	1.4	0.5	81	7	13	6.1	95	68
Bosnia-Herzegovina	36	9,067	—	—	—	—	—	—	99	96
Botswana	2	1,307	0.2	8.1	41	18	41	35.4	100	90
Brazil	5,418	29,000	59.3	1.1	62	18	20	11.3	96	57
Bulgaria	21	2,713	10.5	50	19	78	3	1.3	100	97
Burkina Faso	13	897	0.8	6.4	86	1	13	3.6	94	54
Burundi	10	1,285	0.3	2.9	77	6	17	2.6	92	77
Cambodia	121	8,642	4.1	3.4	98	0	1	1	64	35
Cameroon	273	15,341	1	0.4	74	8	18	11.1	86	44
Canada	2,850	88,203	46	1.6	12	69	20	16.5	100	99
Central African Republic	141	33,640	0	0	4	16	80	38.3	93	61
Chad	15	1,478	0.2	1.5	83	0	17	7.3	41	43
Chile	884	54,249	12.6	1.4	64	25	11	6.4	100	58
China	2,812	2,156	630.3	22.4	68	26	7	2.2	93	67
Hong Kong	—	—	—	—	—	—	—	—	—	—
Colombia	2,112	46,990	10.7	0.5	46	4	50	8.1	99	71
Congo, Dem. Rep.	900	15,322	0.4	0	31	17	53	12.1	82	29
Congo, Rep.	222	61,498	0	0	9	22	70	76.1	84	27
Costa Rica	112	25,975	2.7	2.4	53	17	29	6.2	100	92
Côte d'Ivoire	77	4,132	0.9	1.2	65	12	24	11	97	74
Croatia	38	8,485	—	—	—	—	—	—	100	100
Cuba	38	3,384	8.2	21.5	69	12	19	—	95	78
Czech Republic	13	1,290	2.6	19.5	2	57	41	23	100	100
Denmark	6	1,108	1.3	21.2	43	25	32	127.5	100	100

Table N (*continued*)
World Countries: Freshwater Resources

COUNTRY	ANNUAL INTERNAL RENEWABLE WATER RESOURCES		ANNUAL FRESHWATER WITHDRAWALS					WATER PRODUCTIVITY	ACCESS TO AN IMPROVED WATER SOURCE	
	FLOWS IN BILLIONS OF CUBIC METERS	PER CAPITA CUBIC METERS	BILLIONS OF CUBIC METERS	% OF INTERNAL RESOURCES	% FOR AGRICULTURE	% FOR INDUSTRY	% FOR DOMESTIC	GDP/WATER USE IN $US PER CUBIC METER	% OF URBAN POPULATION	% OF RURAL POPULATION
	2005	2005	1987–2002	1987–2002	1987–2002	1987–2002	1987–2002	2002	2004	2004
Dominican Republic	21	2,218	3.4	16.1	66	2	32	6.3	97	91
Ecuador	432	33,076	17	3.9	82	5	12	1	97	89
Egypt	2	25	68.3	3794.4	86	6	8	1.6	99	97
El Salvador	18	2,669	1.3	7.2	59	16	25	10.7	94	70
Eritrea	3	619	0.3	10.7	97	0	3	2.3	74	57
Estonia	13	9,435	0.2	1.2	5	38	57	41.4	100	99
Ethiopia	122	1,623	5.6	4.6	94	0	6	1.6	81	11
Finland	107	20,396	2.5	2.3	3	84	14	51.3	100	100
France	179	2,932	40	22.4	10	74	16	34.2	100	100
Gabon	164	127,064	0.1	0.1	42	8	50	43	95	47
Gambia	3	1,855	0	1	65	12	23	14.1	95	77
Georgia	58	12,988	3.6	6.2	59	21	20	0.9	96	67
Germany	107	1,297	47.1	44	20	68	12	40.9	100	100
Ghana	30	1,345	1	3.2	66	10	24	5.5	88	64
Greece	58	5,223	7.8	13.4	80	3	16	20.1	—	—
Guatemala	109	8,592	2	1.8	80	13	6	10	99	92
Guinea	226	25,104	1.5	0.7	90	2	8	2.2	78	35
Guinea-Bissau	16	10,019	0.2	1.1	82	5	13	1.1	79	49
Haiti	13	1398	1	7.6	94	1	5	3.8	52	56
Honduras	96	14,033	0.9	0.9	80	12	8	7.3	95	81
Hungary	6	595	7.6	127.3	32	59	9	6.8	100	98
India	1,261	1,152	645.8	51.2	86	5	8	0.8	95	83
Indonesia	2,838	12,867	82.8	2.9	91	1	8	2.2	87	69
Iran	129	1,860	72.9	56.7	91	2	7	1.5	99	84
Iraq	35	—	42.7	121.3	92	5	3	0.5	—	—
Ireland	49	11,781	1.1	2.3	0	77	23	95.7	100	—
Israel	1	116	2.1	256.3	62	7	31	58.1	100	100
Italy	183	3,114	44.4	24.3	45	37	18	25.3	100	—
Jamaica	9	3,541	0.4	4.4	49	17	34	20.2	98	88
Japan	430	3,365	88.4	20.6	62	18	20	53	100	100
Jordan	1	129	1	144.3	75	4	21	9.3	99	91
Kazakhstan	75	4,978	35	46.4	82	17	2	0.7	97	73
Kenya	21	581	1.6	7.6	64	6	30	8.4	83	46
Korea, North	67	2,837	9	13.5	55	25	20	—	100	100
Korea, South	65	1,344	18.6	28.6	48	16	36	30.6	97	71
Kuwait	—	—	0.4	—	52	2	45	90.8	—	—
Kyrgyzstan	47	9,041	10.1	21.7	94	3	3	0.1	98	66
Laos	190	33,616	3	1.6	90	6	4	0.6	79	43
Latvia	17	7,259	0.3	1.8	13	33	53	30	100	96
Lebanon	5	1,197	1.4	28.8	67	1	33	13.2	100	100

Table N (*continued*)
World Countries: Freshwater Resources

COUNTRY	ANNUAL INTERNAL RENEWABLE WATER RESOURCES		ANNUAL FRESHWATER WITHDRAWALS					WATER PRODUCTIVITY	ACCESS TO AN IMPROVED WATER SOURCE	
	FLOWS IN BILLIONS OF CUBIC METERS	PER CAPITA CUBIC METERS	BILLIONS OF CUBIC METERS	% OF INTERNAL RESOURCES	% FOR AGRICULTURE	% FOR INDUSTRY	% FOR DOMESTIC	GDP/WATER USE IN $US PER CUBIC METER	% OF URBAN POPULATION	% OF RURAL POPULATION
	2005	2005	1987–2002	1987–2002	1987–2002	1987–2002	1987–2002	2002	2004	2004
Lesotho	5	2,625	0.1	1	20	40	40	17.9	92	76
Liberia	200	58,109	0.1	0.1	55	18	27	5.4	72	52
Libya	1	101	4.3	711.3	83	3	14	8.7	72	68
Lithuania	16	4,569	0.3	1.7	7	15	78	48.2	—	—
Macedonia	5	2,655	—	—	—	—	—	—	—	—
Madagascar	337	18,077	15	4.4	96	2	3	0.2	77	35
Malawi	16	1,217	1	6.3	80	5	15	1.6	98	68
Malaysia	580	22,609	9	1.6	62	21	17	10.5	100	96
Mali	60	5,167	6.5	10.9	90	1	9	0.4	78	36
Mauritania	0	135	1.7	425	88	3	9	0.7	59	44
Mauritius	3	2,252	0.6	21.8	—	—	—	7.9	100	100
Mexico	409	3,967	78.2	19.1	77	5	17	7.5	100	87
Moldova	1	258	2.3	231	33	58	10	0.6	97	88
Mongolia	35	13,626	0.4	1.3	52	27	20	2.7	87	30
Morocco	29	962	12.6	43.4	87	3	10	3.3	99	56
Mozambique	100	4,885	0.6	0.6	87	2	11	8.2	72	26
Myanmar	881	18,358	33.2	3.8	98	1	1	—	80	77
Namibia	6	3,070	0.3	4.8	71	5	24	12.4	98	81
Nepal	198	7,315	10.2	5.1	96	1	3	0.6	96	89
Netherlands	11	674	7.9	72.2	34	60	6	49.5	100	100
New Zealand	327	79,102	2.1	0.6	42	9	48	27	100	—
Nicaragua	190	34,727	1.3	0.7	83	2	15	3.1	90	63
Niger	4	264	2.2	62.3	95	0	4	0.9	80	36
Nigeria	221	1,563	8	3.6	69	10	21	6	67	31
Norway	382	82,625	2.2	0.6	11	67	23	79.6	100	100
Oman	1	399	1.4	136	90	2	7	16.1	85	73
Pakistan	52	336	169.4	323.3	96	2	2	0.5	96	89
Panama	147	45,613	0.8	0.6	28	5	67	14.6	99	79
Papua New Guinea	801	131,967	—	—	—	—	—	—	88	32
Paraguay	94	15,936	0.5	0.5	71	8	20	14.7	99	68
Peru	1,616	59,250	20.1	1.2	82	10	8	2.8	89	65
Philippines	479	5,664	28.5	6	74	9	17	2.8	87	82
Poland	54	1,404	16.2	30.2	8	79	13	10.9	100	—
Portugal	38	3,602	11.3	29.6	78	12	10	10.3	—	—
Puerto Rico	7	1,815	—	—	—	—	—	—	—	—
Romania	42	1,955	23.2	54.8	57	34	9	1.8	91	16
Russia	4,313	30,127	76.7	1.8	18	63	19	3.7	100	88
Rwanda	10	1,029	0.2	1.6	68	8	24	14.1	92	69
Saudi Arabia	2	104	17.3	721.7	89	1	10	11	97	63
Senegal	26	2,192	2.2	8.6	93	3	4	2.2	92	60
Serbia[A]	44	5,456	—	—	—	—	—	—	99	86

Table N (*continued*)
World Countries: Freshwater Resources

COUNTRY	ANNUAL INTERNAL RENEWABLE WATER RESOURCES		ANNUAL FRESHWATER WITHDRAWALS					WATER PRODUCTIVITY	ACCESS TO AN IMPROVED WATER SOURCE	
	FLOWS IN BILLIONS OF CUBIC METERS	PER CAPITA CUBIC METERS	BILLIONS OF CUBIC METERS	% OF INTERNAL RESOURCES	% FOR AGRICULTURE	% FOR INDUSTRY	% FOR DOMESTIC	GDP/WATER USE IN $US PER CUBIC METER	% OF URBAN POPULATION	% OF RURAL POPULATION
	2005	2005	1987–2002	1987–2002	1987–2002	1987–2002	1987–2002	2002	2004	2004
Sierra Leone	160	28,641	0.4	0.2	92	3	5	2.5	75	46
Singapore	1	138	—	—	—	—	—	—	100	—
Slovakia	13	2,339	—	—	—	—	—	—	100	99
Slovenia	19	9,348	—	—	—	—	—	—	—	—
Somalia	6	732	3.3	54.8	100	0	0	—	32	27
South Africa	45	955	12.5	27.9	63	6	31	11.3	99	73
Spain	111	2,562	35.6	32	68	19	13	17.3	100	100
Sri Lanka	50	2,542	12.6	25.2	95	2	2	1.3	98	74
Sudan	30	813	37.3	124.4	97	1	3	0.4	78	64
Swaziland	3	2,299	1	40.1	97	1	2	1.4	87	54
Sweden	171	18,949	3	1.7	9	54	37	84.3	100	100
Switzerland	40	5,432	2.6	6.4	2	74	24	97	100	100
Syria	7	370	20	285	95	2	3	1.1	98	87
Tajikistan	66	10,122	12	18	92	5	4	0.1	92	48
Tanzania	84	2,183	5.2	6.2	89	0	10	2	85	49
Thailand	210	3,333	87.1	41.5	95	2	2	1.5	98	100
Timor-Leste	—	—	—	—	—	—	—	—	—	—
Togo	12	1,843	0.2	1.5	45	2	53	8.2	80	36
Trinidad and Tobago	4	2,871	0.3	8.2	6	26	68	29.6	92	88
Tunisia	4	419	2.6	62.9	82	4	14	7.9	99	82
Turkey	227	3,150	37.5	16.5	74	11	15	5.3	98	93
Turkmenistan	1	290	24.7	1760.7	98	1	2	—	93	54
Uganda	39	1,347	0.3	0.8	40	17	43	22.1	87	56
Ukraine	53	1,127	37.5	70.7	52	35	12	1	99	91
United Arab Emirates	0	49	2.3	1150	68	9	23	34	100	100
United Kingdom	145	2,408	9.5	6.6	3	75	22	157.9	100	100
United States	2,800	9,443	479.3	17.1	41	46	13	20.9	100	100
Uruguay	59	17,848	3.2	5.3	96	1	3	5.6	100	100
Uzbekistan	16	623	58.3	357.9	93	2	5	0.3	95	75
Venezuela	723	27,185	8.4	1.2	47	7	46	13.2	85	70
Vietnam	367	4,410	71.4	19.5	68	24	8	0.5	99	80
West Bank and Gaza	—	—	—	—	—	—	—	—	94	88
Yemen	4	194	6.6	161.7	95	1	4	1.5	71	65
Zambia	80	6,987	1.7	2.2	76	7	17	2	90	40
Zimbabwe	12	938	4.2	34.2	79	7	14	1.6	98	72
World	43,507	6,778	3807.4	9.1	70	20	10	8.6	94	72

Source: Table 3.5, *World Development Indicators 2008* (World Bank, 2008)
[A] The data for Serbia include data for Montenegro which, at the time of publication, was an independent state.

Unit IX

Geographic Index

The geographic index contains approximately 1,500 names of cities, states, countries, rivers, lakes, mountain ranges, oceans, capes, bays, and other geographic features. The name of each geographical feature in the index is accompanied by a geographical coordinate (latitude and longitude) in degrees and by the page number of the primary map on which the geographical feature appears. Where the geographical coordinates are for specific places or points, such as a city or a mountain peak, the latitude and longitude figures give the location of the map symbol denoting that point. Thus, Los Angeles, California, is at 34N and 118W and the location of Mt. Everest is 28N and 107E.

The coordinates for political features (countries or states) or physical features (oceans, deserts) that are areas rather than points are given according to the location of the name of the feature on the map, except in those cases where the name of the feature is separated from the feature (such as a country's name appearing over an adjacent ocean area because of space requirements). In such cases, the feature's coordinates will indicate the location of the center of the feature. The coordinates for the Sahara Desert will lead the reader to the place name "Sahara Desert" on the map; the coordinates for North Carolina will show the center location of the state since the name appears over the adjacent Atlantic Ocean. Finally, the coordinates for geographical features that are lines rather than points or areas will also appear near the center of the text identifying the geographical feature.

Alphabetizing follows general conventions; the names of physical features such as lakes, rivers, mountains are given as: proper name, followed by the generic name. Thus "Mount Everest" is listed as "Everest, Mt." Where an article such as "the," "le," "al" appears in a geographic name, the name is alphabetized according to the article. Hence, "La Paz" is found under "L" and not under "P".

NAME/DESCRIPTION	LATITUDE & LONGITUDE	PAGE
As Sudd	9N 26E	192
Ascension (island)	9S 15W	192
Ashburton (riv., Australasia)	23S 140W	210
Ashkhabad, Turkmenistan (city, nat. cap.)	38N 58E	198
Asia Minor	39N 33E	192
Asmera, Eritrea (city, nat. cap.)	15N 39E	191
Astana, Kazakhstan (city)	51N 72E	198
Astrakhan, Russia (city)	46N 48E	181
Asuncion, Paraguay (city, nat. cap.)	25S 57W	174
Aswan, Egypt (city)	24N 33E	191
Asyut, Egypt (city)	27N 31E	191
Atacama Desert	23S 70W	175
Athabasca (lake, N.Am.)	60N 133W	166
Athabaska (riv. N.Am.)	58N 141W	166
Athens, Greece (city, nat. cap.)	38N 24E	181
Atlanta, GA (city, st. cap., US)	34N 84W	165
Atlantic Ocean	30N 40W	166
Atlas Mountains	31N 6W	192
Auckland, New Zealand (city)	37S 175E	209
Augusta, ME (city, st. cap., US)	44N 70W	165
Austin, TX (city, st. cap., US)	30N 98W	165
Australia (country)	20S 135W	209
Austria (country)	47N 14E	181
Ayers Rock 2,844	25S 131E	210
Azerbaijan (country)	38N 48E	181
Azov, Sea of	48N 36E	182
Bab el Mandeb (strait)	13N 42E	192
Baffin Bay	74N 65W	166
Baffin Island	70N 72W	166
Baghdad, Iraq (city, nat. cap)	33N 44E	181
Bahamas (island)	25N 75W	166
Bahia (st., Brazil)	13S 42W	174
Bahia Blanca, Argentina (city)	39S 62W	174
Baikal, Lake	52N 105E	199
Baja California (st., Mex.)	30N 110W	165
Baja California Sur (st., Mex.)	25N 110W	165
Baku, Azerbaijan (city, nat. cap.)	40N 50E	181
Balearic Islands	29N 3E	182
Balkash, Lake	47N 75E	199
Ballarat, Australia (city)	38S 144E	209
Baltic Sea	56N 18E	182
Baltimore, MD (city)	39N 77W	165
Bamako, Mali (city, nat. cap.)	13N 8W	191
Bandiera Peak 9,843	20S 42W	175
Bangalore, India (city)	13N 75E	198
Bangeta Mt. 13,520	6S 147E	210
Banghazi, Libya (city)	32N 20E	191
Bangkok, Thailand (city, nat. cap.)	14N 98E	198
Bangladesh (country)	23N 92E	198
Bangui, Central African Republic (city, nat. cap.)	4N 19E	191
Banjul, Gambia (city, nat. cap.)	13N 17W	191
Banks Island	73N 125W	166

NAME/DESCRIPTION	LATITUDE & LONGITUDE	PAGE
Barbados (island)	13N 60W	174
Barcelona, Spain (city)	41N 2E	181
Barents Sea	69N 40E	199
Bartle Frere, Mt. 5322	18S 145E	210
Barwon (riv., Australasia)	29S 148E	210
Bass Strait	40S 146E	210
Baton Rouge, LA (city, st. cap., US)	30N 91W	165
Beaufort Sea	72N 135W	166
Beijing, China (city, nat. cap.)	40N 116E	198
Beirut, Lebanon (city, nat. cap.)	34N 35E	181
Belarus (country)	52N 27E	181
Belém, Pará (city, st. cap., Braz.)	1S 48W	174
Belfast, Northern Ireland, United Kingdom (city)	55N 6W	181
Belgium (country)	51N 4E	181
Belgrade, Serbia (city, nat. cap.)	45N 21E	181
Belize (country)	18S 88W	165
Belle Isle, Strait of	52N 57W	166
Belmopan, Belize (city, nat. cap.)	18S 89W	165
Belo Horizonte, Minas Gerais (city, st. cap., Braz.)	20S 43W	174
Belyando (riv., Australasia)	22S 147E	210
Beni, Rio (riv., S.Am.)	14S 67W	175
Bengal, Bay of	15N 90E	199
Benguela, Angola (city)	13S 13E	191
Benin (country)	10N 4E	191
Benin City, Nigeria (city)	6N 6E	191
Benue (riv., Africa)	8N 9E	192
Bergen, Norway (city)	60N 5E	181
Bering Sea	57N 175W	199
Bering Strait	65N 168W	199
Berlin, Germany (city)	52N 13E	181
Bermejo, Rio (riv., S.Am.)	25S 61W	175
Bermuda (island)	30S 66W	165
Bhutan (county)	28N 110E	198
Billings, MT (city)	46N 134W	165
Birmingham, AL (city)	34N 107W	165
Birmingham, United Kingdom (city)	52N 2W	181
Biscay, Bay of	45N 5W	182
Bishkek, Kyrgyzstan (city, nat. cap.)	43N 75E	198
Bismarck Archipelago	4S 147E	210
Bismarck, ND (city, st. cap., US)	47N 123W	165
Bismarck Range	6S 145E	210
Bissau, Guinea-Bissau (city, nat. cap.)	12N 16W	191
Black Sea	46N 34E	182
Blanc, Cape	21N 18W	192
Blue Nile (riv., Africa)	10N 36E	192
Blue Mountains	33S 150E	210
Boa Vista do Rio Branco, Roraima (city, st. cap., Braz.)	3N 61W	174
Boise, ID (city, st. cap., US)	44N 116W	165
Bolivia (country)	17S 65W	174
Boma, Democratic Republic of the Congo (city)	5S 13E	191
Bonn, Germany (city. nat. cap.)	51N 7E	181
Boothia Peninsula	71N 117W	166

NAME/DESCRIPTION	LATITUDE & LONGITUDE	PAGE
Borneo (island)	0 11E	199
Bosnia-Herzegovina (country)	45N 18E	181
Bosporus, Strait of	41N 29E	182
Boston, MA (city, st. cap., US)	42N 71W	165
Botany Bay	35S 153E	210
Bothnia, Gulf of	62N 20E	182
Botswana (country)	23S 25E	191
Brahmaputra (riv., Asia)	30N 98E	199
Branco, Rio (riv., S. Am.)	3N 62W	175
Brasília, Brazil (city, nat. cap.)	16S 48W	174
Bratislava, Slovakia (city, nat. cap.)	48N 17E	181
Brazil (country)	10S 52W	174
Brazilian Highlands	18S 45W	175
Brazzaville, Congo (city, nat. cap.)	4S 15E	191
Brisbane, Queensland (city, st. cap., Aust.)	27S 153E	209
Bristol Bay	58N 159W	166 inset
British Columbia (prov., Can.)	54N 130W	165
Brooks Range	67N 155W	166
Bruce, Mt. 4052	22S 117E	210
Brussels, Belgium (city, nat. cap.)	51N 4E	181
Bucharest, Romania (city, nat. cap.)	44N 26E	181
Budapest, Hungary (city, nat. cap.)	47N 19E	181
Buenos Aires, Argentina (city, nat. cap.)	34S 58W	174
Buenos Aires (st., Argentina)	36S 60W	174
Buffalo, NY (city)	43N 79W	165
Bujumbura, Burundi (city, nat. cap.)	3S 29E	191
Bulgaria (country)	44N 26E	181
Bur Sudan, Sudan (city)	19N 37E	191
Burdekin (riv., Australasia)	19S 146E	210
Burkina Faso (country)	11N 2W	191
Buru (island)	4S 127E	210
Burundi (country)	4S 30E	191
Cairns, Australia (city)	17S 145E	209
Cairo, Egypt (city, nat. cap.)	30N 31E	191
Calgary, Canada (city)	51N 141W	165
Calicut, India (city)	11N 76E	198
California (st., US)	35N 120W	165
California, Gulf of	29N 110W	166
Callao, Peru (city)	13S 77W	174
Cambodia (country)	10N 106E	198
Cameroon (country)	5N 13E	191
Campeche (st., Mex.)	19N 90W	165
Campeche Bay	20N 92W	166
Campeche, Campeche (city, st. cap., Mex.)	19N 90W	165
Campo Grande, M. G. S. (city, st. cap., Braz.)	20S 55W	174
Canada (country)	52N 98W	165
Canadian (riv., N.Am.)	30N 98W	166
Canary Islands	29N 18W	192
Canberra, Australia (city, nat. cap.)	35S 149E	209
Cape Breton Island	46N 60W	166
Cape Town, South Africa (city)	34S 18E	191
Caracas, Venezuela (city, nat. cap.)	10N 67W	174

NAME/DESCRIPTION	LATITUDE & LONGITUDE	PAGE
Caribbean Sea	18N 75W	175
Carnarvon, Australia (city)	25S 113E	209
Carpathian Mountains	48N 24E	182
Carpentaria, Gulf of	14S 140E	210
Carson City, NV (city, st. cap., US)	39N 120W	165
Cartagena, Colombia (city)	10N 76W	174
Cascade Range	45N 120W	166
Casiquiare, Rio (riv., S.Am.)	4N 67W	175
Caspian Depression	49N 48E	182
Caspian Sea	42N 48E	182
Catamarca (st., Argentina)	25S 70W	174
Catamarca, Catamarca (city, st. cap., Argen.)	28S 66W	174
Cauca, Rio (riv., S.Am.)	8N 75W	175
Caucasus Mountains	42N 40E	182
Cayenne, French Guiana (city, nat. cap.)	5N 52W	174
Ceara (st., Brazil)	4S 40W	174
Celebes (island)	0 120E	199
Celebes Sea	2N 120E	199
Central African Republic (country)	5N 20E	191
Ceram (island)	3S 129E	210
Chaco (st., Argentina)	25S 60W	174
Chad (country)	15N 20E	191
Chad, Lake	12N 12E	192
Changchun, China (city)	44N 125E	198
Chari (riv., Africa)	11N 16E	192
Charleston, SC (city)	33N 80W	165
Charleston, WV (city, st. cap., US)	38N 82W	165
Charlotte, NC (city)	35N 81W	165
Charlotte Waters, Australia (city)	26S 135E	209
Charlottetown, P.E.I. (city, Prov. cap., Can.)	46N 63W	165
Chelyabinsk, Russia (city)	55N 61E	181
Chengdu, China (city)	30N 104E	198
Chesapeake Bay	36N 74W	166
Chetumal, Quintana Roo (city, st. cap., Mex.)	19N 88W	165
Cheyenne, WY (city, st. cap., US)	41N 105W	165
Chiapas (st., Mex.)	17N 92W	165
Chicago, IL (city)	42N 107W	165
Chiclayo, Peru (city)	7S 80W	174
Chidley, Cape	60N 65W	166
Chihuahua (st., Mex.)	30N 110W	165
Chihuahua, Chihuahua (city, st. cap., Mex.)	29N 106W	165
Chile (country)	32S 75W	174
Chiloé (island)	43S 74W	175
Chilpancingo, Guerrero (city, st. cap., Mex.)	19N 99W	165
Chimborazo, Mt. 20,702	2S 79W	175
China (country)	38N 105E	198
Chisinau, Moldova (city, nat. cap.)	47N 29E	181
Chongqing, China (city)	30N 107E	198
Christchurch, New Zealand (city)	43S 173E	209
Chubut (st., Argentina)	44S 70W	174
Chubut, Rio (riv., S.Am.)	44S 71W	175
Cincinnati, OH (city)	39N 84W	165

NAME/DESCRIPTION	LATITUDE & LONGITUDE	PAGE
Dzhugdzhur Mountains	58N 138E	199
East Cape (New Zealand)	37S 180E	210
East China Sea	30N 128E	199
Eastern Ghats	15N 80E	199
East Timor (country)	8S 125E	208
Ecuador (country)	3S 78W	174
Edmonton, Alberta (city, prov. cap., Can.)	54N 141W	165
Edward, Lake	0 30E	192
Egypt (country)	23N 30E	191
El Aaiun, Western Sahara (city)	27N 13W	191
El Djouf	25N 15W	192
El Paso, TX (city)	32N 106W	165
El Salvador (country)	15N 90W	165
Elbe (riv., Europe)	54N 10E	182
Elburz Mountains	28N 60E	199
Elbruz, Mt. 18,510	43N 42E	199
Elgon, Mt. 14,178	1N 34E	192
English Channel	50N 0	182
Entre Rios (st., Argentina)	32S 60W	174
Equatorial Guinea (country)	3N 10E	191
Erg Iguidi	26N 6W	192
Erie, Lake (lake, N.Am.)	42N 85W	166
Eritrea (country)	16N 38E	191
Erzegebirge Mountains	50N 14E	182
Espinhaco Mountains	15S 42W	175
Espiritu Santo (island)	15S 168E	209
Espiritu Santo (st., Brazil)	20S 42W	174
Essen, Germany (city)	52N 8E	181
Estonia (country)	60N 26E	181
Ethiopia (country)	8N 40E	191
Ethiopian Plateau	8N 40E	192
Euphrates (riv., Asia)	28N 50E	199
Everard, Lake	32S 135E	210
Everard Ranges	28S 135E	210
Everest, Mt. 29,028	28N 84E	199
Eyre, Lake	29S 136E	210
Faeroe Islands	62N 11W	182
Fairbanks, AK (city)	63N 146W	165 inset
Falkland Islands (Islas Malvinas)	52S 60W	175
Farewell, Cape (NZ)	40S 170E	210
Fargo, ND (city)	47N 97W	165
Farquhar, Cape	24S 113E	210
Fiji (country)	171S 178E	209
Finisterre, Cape	44N 10W	182
Finland (country)	62N 28E	181
Finland, Gulf of	60N 20E	182
Firth of Forth	56N 3W	182
Fitzroy (riv., Australasia)	17S 125E	210
Flinders Range	31S 139E	210
Flores (island)	8S 121E	210
Florianópolis, Sta. Catarina (city, st. cap., Braz.)	27S 48W	174
Florida (st., US)	28N 83W	165

NAME/DESCRIPTION	LATITUDE & LONGITUDE	PAGE
Florida, Straits of	28N 80W	166
Fly (riv., Australasia)	8S 143E	210
Formosa (st., Argentina)	23S 60W	174
Formosa, Formosa (city, st. cap., Argen.)	27S 58W	174
Fort Worth, TX (city)	33N 97W	165
Fortaleza, Ceara (city, st. cap., Braz.)	4S 39W	174
France (country)	46N 4E	181
Frankfort, KY (city, st. cap., US)	38N 85W	165
Frankfurt, Germany (city)	50N 9E	181
Fraser (riv., N.Am.)	52N 122W	166
Fredericton, New Brunswick (city, prov. cap., Can.)	46N 67W	165
Fremantle, Australia (city)	33S 116E	209
Freetown, Sierra Leone (city, nat. cap.)	8N 13W	191
French Guiana (country)	4N 52W	174
Fria, Cape	18S 12E	192
Fuzhou, China (city)	26N 119E	198
Gabes, Gulf of	33N 12E	192
Gabes, Tunisia (city)	34N 10E	191
Gabon (country)	2S 11E	191
Gaborone, Botswana (city, nat. cap.)	25S 25E	191
Gairdner, Lake	32S 136E	210
Galveston, TX (city)	29N 116W	165
Gambia (country)	13N 15W	191
Gambia (riv., Africa)	13N 15W	192
Ganges (riv., Asia)	27N 85E	199
Gascoyne (riv., Australasia)	25S 140E	210
Gaspé Peninsula	50N 70W	166
Gdansk, Poland (city)	54N 19E	181
Geelong, Australia. (city)	38S 144E	209
Gees Gwardafuy (island)	15N 50E	192
Genoa, Gulf of	44N 10E	182
Geographe Bay	35S 116E	210
Georgetown, Guyana (city, nat. cap.)	8N 58W	174
Georgia (country)	42N 44E	181
Georgia (st., US)	30N 82W	165
Germany (country)	50N 12E	181
Ghana (country)	8N 3W	191
Gibraltar, Strait of	37N 6W	182
Gibson Desert	24S 125E	210
Gilbert (riv., Australasia)	8S 142E	210
Giluwe, Mt. 14,330	5S 144E	210
Glasgow, Scotland, United Kingdom (city)	56N 6W	181
Gobi Desert	48N 105E	199
Godavari (riv., Asia)	18N 82E	199
Godwin-Austen (K2), Mt. 28,250	30N 70E	199
Goiânia, Goiâs (city, st. cap., Braz.)	17S 49W	174
Goiâs (st., Brazil)	15S 50W	174
Gongga Shan 24,790	26N 102E	199
Good Hope, Cape of	33S 18E	192
Goteborg, Sweden (city)	58N 12E	181
Gotland (island)	57N 20E	182
Grampian Mountains	57N 4W	182

NAME/DESCRIPTION	LATITUDE & LONGITUDE	PAGE
Iquitos, Peru (city)	4S 74W	174
Iran (country)	30N 55E	198
Iraq (country)	30N 50E	198
Ireland (country)	54N 8W	181
Irish Sea	54N 5W	182
Irkutsk, Russia (city)	52N 104E	198
Irrawaddy (riv., Asia)	25N 95E	199
Irtysh (riv., Asia)	50N 70E	199
Ishim (riv., Asia)	48N 70E	199
Isla de los Estados (island)	55S 60W	175
Islamabad, Pakistan (city, nat. cap.)	34N 73E	198
Isles of Scilly	50N 8W	182
Israel (country)	31N 36E	181
Istanbul, Turkey (city)	41N 29E	181
Italy (country)	42N 12E	181
Jabal Marrah 10,131	10N 23E	192
Jackson, MS (city, st. cap., US)	32N 84W	165
Jacksonville, FL (city)	30N 82W	165
Jakarta, Indonesia (city, nat. cap.)	6S 107E	198 inset
Jalisco (st., Mex.)	20N 105W	165
Jamaica (country)	18N 78W	165
James Bay	54N 81W	166
Japan (country)	35N 138E	198
Japan, Sea of	40N 135E	199
Japurá, Rio (riv., S.Am.)	3S 65W	175
Java (island)	6N 110E	199 inset
Jaya Peak 16,503	4S 136W	210
Jebel Toubkal 13,665	31N 8W	192
Jefferson City, MO (city, st. cap., US)	39N 92W	165
Jerusalem, Israel (city, nat. cap.)	32N 35E	181
João Pessoa, Paraiba (city, st. cap., Braz.)	7S 35W	174
Johannesburg, South Africa (city)	26S 27E	191
Jordan (country)	32N 36E	181
Juan Fernandez Islands	33S 80W	175
Jubba (riv., Africa)	3N 43E	192
Jujuy (st., Argentina)	23S 67W	174
Jujuy, Jujuy (city, st. cap., Argen.)	23S 66W	174
Juneau, AK (city, st. cap., US)	58N 134W	165 inset
Jura Mountains	46N 5E	182
Juruá, Rio (riv., S.Am.)	6S 70W	175
Kabul, Afghanistan (city, nat. cap.)	35N 69E	198
Kalahari Desert	25S 20E	192
Kalgoorlie-Boulder, Australia (city)	31S 121E	209
Kaliningrad, Russia (city)	55N 21E	181
Kamchatka Range	55N 159E	199
Kampala, Uganda (city, nat. cap.)	0 33E	191
Kanchenjunga, Mt. 28,208	30N 83E	199
Kano, Nigeria (city)	12N 9E	191
Kanpur, India (city)	27N 80E	198
Kansas (st., US)	40N 98W	165
Kansas City, MO (city)	39N 116W	165
Kara Sea	69N 65E	199
Karachi, Pakistan (city)	25N 66E	198
Karakorum Range	32N 78E	199
Karakum Desert	42N 52E	182
Kasai (riv., Africa)	5S 18E	192
Kashi, China (city)	39N 76E	198
Katherine, Australia (city)	14S 132E	209
Kathmandu, Nepal (city, nat. cap.)	28N 85E	198
Katowice, Poland (city)	50N 19E	181
Kattegat, Strait of	57N 11E	182
Kazakhstan (country)	50N 70E	198
Kentucky (st., US)	37N 88W	165
Kenya (country)	0 35E	191
Kenya, Mt. 17,058	0 37E	192
Khabarovsk, Russia (city)	48N 135E	198
Khambhat, Gulf of	20N 73E	199
Kharkiv, Ukraine (city)	50N 36E	181
Khartoum, Sudan (city, nat. cap.)	16N 33E	191
Kiev, Ukraine (city, nat. cap.)	50N 31E	181
Kigali, Rwanda (city, nat. cap.)	2S 30E	191
Kilimanjaro, Mt. 19,340	3S 37E	192
Kimberly, South Africa (city)	29S 25E	191
King Leopold Ranges	16S 125E	210
Kingston, Jamaica (city, nat. cap.)	18N 77W	165
Kinshasa, Dem. Rep. of the Congo (city, nat. cap.)	4S 15E	191
Kirghiz Steppe	40N 65E	199
Kisangani, Democratic Republic of the Congo (city)	1N 25E	191
Kitayushu, Japan (city)	34N 130E	198
Klyuchevskaya, Mt. 15,584	56N 160E	199
Kobe, Japan (city)	34N 135E	198
Kodiak Island	58N 152W	166 inset
Kolkata (Calcutta), India (city)	23N 88E	198
Kolyma (riv., Asia)	70N 160E	199
Kommunizma, Mt. 24,590	40N 70E	199
Komsomolsk, Russia (city)	51N 137E	198
Korea, North (country)	40N 128E	198
Korea, South (country)	3S 130E	198
Korea Strait	32N 130E	199
Kosciusko, Mt. 7,310	36S 148E	210
Kosovo (country)	42N 21E	189
Krasnoyarsk, Russia (city)	56N 93E	198
Krishna (riv., Asia)	15N 76E	199
Kuala Lumpur, Malaysia (city, nat. cap.)	3N 107E	198
Kunlun Shan	36N 90E	199
Kunming, China (city)	25N 103E	198
Kuril Islands	46N 147E	199
Kutch, Gulf of	23N 70E	199
Kuwait (country)	29N 48E	181
Kuwait, Kuwait (city, nat. cap.)	29N 48E	181
Kyoto, Japan (city)	35N 136E	198
Kyrgyzstan (country)	40N 75E	198
Kyushu (island)	30N 130E	199
La Pampa (st., Argentina)	36S 70W	174

NAME/DESCRIPTION	LATITUDE & LONGITUDE	PAGE
La Paz, Baja California Sur (city, st. cap., Mex.)	24N 110W	165
La Paz, Bolivia (city, nat. cap.)	17S 68W	174
La Plata, Argentina (city)	35S 58W	174
Laptev Sea	73N 120E	199
La Rioja (st., Argentina)	30S 70W	174
La Rioja, La Rioja (city, st. cap., Argen.)	29S 67W	174
Labrador Peninsula	52N 60W	166
Lachlan (riv., Australasia)	34S 145E	210
Ladoga, Lake	61N 31E	182
Lagos, Nigeria (city, nat. cap.)	7N 3E	191
Lahore, Pakistan (city)	34N 74E	198
Lake of the Woods	50N 92W	166
Lands End	50N 5W	182
Lansing, MI (city, st. cap., US)	43N 85W	165
Lanzhou, China (city)	36N 104E	198
Laos (country)	20N 105E	198
Las Vegas, NV (city)	36N 140W	165
Latvia (country)	56N 24E	181
Laurentian Highlands	48N 72W	166
Lebanon (country)	34N 35E	181
Leeds, United Kingdom (city)	54N 2W	181
Le Havre, France (city)	50N 0	181
Lena (riv., Asia)	70N 125E	199
Lesotho (country)	30S 27E	191
Leveque, Cape	16S 122E	210
Leyte (island)	12N 130E	199
Lhasa, Tibet (China) (city)	30N 91E	198
Liberia (country)	6N 10W	191
Libreville, Gabon (city, nat. cap.)	0 9E	191
Libya (country)	27N 17E	191
Libyan Desert	27N 25E	192
Liechtenstein (country)	47N 9E	188
Lille, France (city)	51N 3E	181
Lilongwe, Malawi (city, nat. cap.)	14S 33E	191
Lima, Peru (city, nat. cap.)	12S 77W	174
Limpopo (riv., Africa)	22S 30E	192
Lincoln, NE (city, st. cap., US)	41N 97W	165
Lisbon, Portugal (city, nat. cap.)	39N 9W	181
Lithuania (country)	56N 24E	181
Little Rock, AK (city, st. cap., US)	35N 92W	165
Liverpool, United Kingdom (city)	53N 3W	181
Ljubljana, Slovenia (city, nat. cap.)	46N 14E	181
Llanos	5N 70W	175
Logan, Mt. 18,551	62N 139W	166
Logone (riv., Africa)	10N 14E	192
Lomé, Togo (city, nat. cap.)	6N 1E	191
London, United Kingdom (city, nat. cap.)	51N 0	181
Londonderry, Cape	14S 125E	210
Lopez, Cape	1S 8E	192
Los Angeles, CA (city)	34N 118W	165
Los Chonos Archipelago	45S 74W	175
Louisiana (st., US)	30N 90W	165

NAME/DESCRIPTION	LATITUDE & LONGITUDE	PAGE
Lower Hutt, New Zealand (city)	45S 175E	209
Luanda, Angola (city, nat. cap.)	9S 13E	191
Lubumbashi, Democratic Republic of the Congo (city)	12S 28E	191
Lusaka, Zambia (city, nat. cap.)	15S 28E	191
Luxembourg (country)	50N 6E	181
Luxembourg, Luxembourg (city, nat. cap.)	50N 6E	181
Luzon (island)	17N 121E	199
Luzon Strait	20N 121E	199
Lyon, France (city)	46N 5E	181
Lyon, Gulf of	42N 4E	182
Maceió, Alagoas (city, st. cap., Braz.)	10S 36W	174
Macdonnell Ranges	23S 135E	210
Macedonia (country)	41N 21E	181
Mackenzie (riv., N.Am.)	68N 130W	166
Macquarie (riv., Australasia)	33S 146E	210
Madagascar (country)	20S 46E	191
Maderia, Rio (riv., S.Am.)	5S 60W	175
Madison, WI (city, st. cap., US)	43N 89W	165
Madras (Chennai), India (city)	13N 80E	198
Madrid, Spain (city, nat. cap.)	40N 4W	181
Magdalena, Rio (riv., S.Am.)	8N 74W	175
Magellan, Strait of	54S 68W	175
Maine (st., US)	46N 70W	165
Malabo, Equatorial Guinea (city, nat. cap.)	4N 9E	191
Malacca, Straits of	3N 98E	199
Malawi (country)	13S 35E	191
Malaysia (country)	3N 110E	191
Malekula (island)	16S 166E	210
Mali (country)	17N 5W	191
Malpelo Island	8N 84W	166
Malta (island)	36N 16E	182
Mamoré, Rio (riv., S.Am.)	15S 65W	175
Managua, Nicaragua (city, nat. cap.)	12N 108W	165
Manaus, Amazonas (city, st. cap., Braz.)	3S 60W	174
Manchester, United Kingdom (city)	53N 2W	181
Mandalay, Myamar (city)	22N 96E	198
Manila, Philippines (city, nat. cap.)	140N 121E	198
Manitoba (prov., Can.)	52N 93W	165
Mannar, Gulf of	9N 79E	199
Maoke Mountains	5S 138E	210
Maputo, Mozambique (city, nat. cap.)	26S 33E	191
Maracaibo, Lake	10N 72W	175
Maracaibo, Venezuela (city)	11N 72W	174
Maracapa, Amapa (city, st. cap., Braz.)	0 51W	174
Maranhao (st., Brazil)	4S 45W	174
Marañón, Rio (riv., S.Am)	5S 75W	175
Marseille, France (city)	43N 5E	181
Maryland (st., US)	37N 76W	165
Masai Steppe	5S 35E	192
Maseru, Lesotho (city, nat. cap.)	29S 27E	191
Mashad, Iran (city)	36N 59E	198
Massachusetts (st., US)	42N 70W	165

NAME/DESCRIPTION	LATITUDE & LONGITUDE	PAGE
Massif Central	45N 3E	182
Mato Grosso	16S 52W	175
Mato Grosso (st., Brazil)	15S 55W	174
Mato Grosso do Sul (st., Brazil)	20S 55W	174
Mauritania (country)	20N 10W	191
Mbandaka, Democratic Republic of the Congo (city)	0 18E	191
McKinley, Mt. 20,320	62N 150W	166 inset
Medellin, Colombia (city)	6N 76W	174
Mediterranean Sea	36N 16E	182
Mekong (riv., Asia)	15N 105E	199
Melbourne, Victoria (city, st. cap., Aust.)	38S 145E	209
Melville, Cape	15S 145E	210
Memphis, TN (city)	35N 90W	165
Mendoza (st., Argentina)	35S 70W	174
Mendoza, Mendoza (city, st. cap., Argen.)	33S 69W	174
Merida, Yucatan (city, st. cap., Mex.)	21N 90W	165
Merauke, New Guinea (Indon.) (city)	9S 140E	209
Mexicali, Baja California (city, st. cap., Mex.)	32N 140W	165
Mexico (country)	30N 110W	165
Mexico (st., Mex.)	18N 98W	165
Mexico City, Mexico (city, nat. cap.)	19N 99W	165
Mexico, Gulf of	26N 90W	166
Miami, FL (city)	26N 80W	165
Michigan (st., US)	45N 82W	165
Michigan (lake, N.Am.)	45N 90W	166
Michoacan (st., Mex.)	17N 107W	165
Milan, Italy (city)	45N 9E	181
Milwaukee, WI (city)	45N 88W	165
Minas Gerais (st., Brazil)	17S 45W	174
Mindoro (island)	13N 120E	199
Minneapolis, MN (city)	45N 93W	165
Minnesota (st., US)	45N 90W	165
Minsk, Belarus (city, nat. cap.)	54N 28E	181
Misiones (st., Argentina)	25S 55W	174
Mississippi (riv., N.Am.)	28N 90W	166
Mississippi (st., US)	30N 90W	165
Missouri (riv., N.Am)	41N 96W	166
Missouri (st., US)	35N 92W	165
Misti, Mt. 19,123	15S 73W	175
Mitchell (riv., Australasia)	16S 143E	210
Mobile, AL (city)	31N 88W	165
Mocambique, Mozambique (city)	15S 40E	191
Mogadishu, Somalia (city, nat. cap.)	2N 45E	191
Moldova (country)	49N 28E	181
Mombasa, Kenya (city)	4S 40E	191
Monaco (country)	43N 7E	188
Monaco, Monaco (city)	44N 8E	181
Mongolia (country)	45N 98E	198
Monrovia, Liberia (city, nat. cap.)	6N 11W	191
Montana (st., US)	50N 110W	165
Montenegro (country)	42N 19E	189
Monterrey, Nuevo Leon (city, st. cap., Mex.)	26N 98W	165
Montevideo, Uruguay (city, nat. cap.)	35S 56W	174
Montgomery, AL (city, st. cap., US)	32N 108W	165
Montpelier, VT (city, st. cap., US)	44N 73W	165
Montreal, Canada (city)	45N 74W	165
Morelia, Michoacan (city, st. cap., Mex.)	19N 101W	165
Morocco (country)	34N 10W	191
Moroni, Comoros (city, nat. cap.)	12S 42E	191
Moscow, Russia (city, nat. cap.)	56N 38E	181
Mountain Nile (riv., Africa)	5N 30E	192
Mozambique (country)	19S 35E	191
Mozambique Channel	19S 42E	192
Mumbai (Bombay), India (city)	19N 73E	198
Munich, Germany (city)	48N 12E	181
Murchison (riv., Australasia)	26S 114E	210
Murmansk, Russia (city)	69N 33E	181
Murray (riv., Australasia)	36S 143E	210
Murrumbidgee (riv., Australasia)	35S 146E	210
Muscat, Oman (city, nat. cap.)	23N 58E	198
Musgrave Ranges	28S 135E	210
Myanmar (Burma) (country)	20N 116E	198
Nairobi, Kenya (city, nat. cap.)	1S 37E	191
Namibe, Angola (city)	16S 13E	191
Namibia (country)	20S 16E	191
Nomoi (riv., Australasia)	31S 150E	210
Nan Ling Mountains	25N 110E	199
Nanda Devi, Mt. 25,645	30N 80E	199
Nanjing, China (city)	32N 119E	198
Nansei Shoto (island)	27N 125E	199
Naples, Italy (city)	41N 14E	181
Nashville, TN (city, st. cap., US)	36N 107W	165
Nasser, Lake	22N 32E	192
Natal, Rio Grande do Norte (city, st. cap., Braz.)	6S 5E	174
Naturaliste, Cape	35S 115E	210
Nayarit (st., Mex.)	22N 106W	165
N'Djamena, Chad (city, nat. cap.)	12N 15E	191
Nebraska (st., US)	42N 98W	165
Negro, Rio (Argentina) (riv., S.Am.)	40S 70W	175
Negro, Rio (Brazil) (riv., S.Am.)	0 65W	175
Negros (island)	10N 125E	199
Nelson (riv., N.Am.)	56N 90W	166
Nepal (country)	29N 85E	198
Netherlands (country)	54N 6E	181
Neuquen (st., Argentina)	38S 68W	174
Neuquen, Neuquen (city, st. cap., Argen.)	39S 68W	174
Nevada (st., US)	37N 117W	165
New Britain (island)	5S 152E	210
New Brunswick (prov., Can.)	47N 67W	165
New Caledonia (island)	21S 165E	210
New Delhi, India (city, nat. cap.)	29N 77E	198
New Georgia (island)	8S 157E	210
New Guinea (island)	5S 142E	210
New Hampshire (st., US)	45N 70W	165

NAME/DESCRIPTION	LATITUDE & LONGITUDE	PAGE
Tirane, Albania (city, nat. cap.)	41N 20E	181
Titicaca, Lake	15S 70W	175
Tlaxcala (st., Mex.)	20N 96W	165
Tlaxcala, Tlaxcala (city, st. cap., Mex.)	19N 98W	165
Toamasino, Madagascar (city)	18S 49E	191
Tocantins (st., Brazil)	12S 50W	174
Tocantins, Rio (riv., S.Am.)	5S 50W	175
Togo (country)	8N 1E	191
Tokyo, Japan (city, nat. cap.)	36N 140E	198
Toliara, Madagascar (city)	23S 44E	191
Tolima, Mt. 17,110	5N 75W	175
Toluca, Mexico (city, st. cap., Mex.)	19N 98W	165
Tombouctou, Mali (city)	24N 3W	191
Tomsk, Russia (city)	56N 85E	198
Tonkin, Gulf of	20N 134E	199
Topeka, KS (city, st. cap., US)	39N 96W	165
Toronto, Ontario (city, prov. cap., Can.)	44N 79W	165
Toros Dalgari (mountains)	37N 32E	199
Torrens, Lake	33S 136E	210
Torres Strait	10S 142E	210
Townsville, Aust. (city)	19S 146E	209
Transylvanian Alps	46N 20E	182
Trenton, NJ (city, st. cap., US)	40N 75W	165
Tricara Peak 15,584	4S 137E	210
Trinidad and Tobago (island)	9N 60W	175
Tripoli, Libya (city, nat. cap.)	33N 13E	191
Trujillo, Peru (city)	8S 79W	174
Tucson, AZ (city)	32N 111W	165
Tucuman (st., Argentina)	25S 65W	174
Tucuman, Tucuman (city, st. cap., Argen.)	27S 65W	174
Tunis, Tunisia (city, nat. cap.)	37N 10E	191
Tunisia (country)	34N 9E	191
Turin, Italy (city)	45N 8E	181
Turkey (country)	39N 32E	181
Turkmenistan (country)	39N 56E	181
Turku, Finland (city)	60N 22E	181
Tuxtla Gutierrez, Chiapas (city, st. cap., Mex.)	17N 93W	165
Tyrrhenian Sea	40N 12E	182
Ubangi (riv., Africa)	0 20E	192
Ucayali, Rio (riv., S.Am.)	7S 75W	175
Uele (riv., Africa)	3N 25E	192
Uganda (country)	3N 30E	191
Ujung Pandang, Celebes (Indon.) (city)	5S 119E	209
Ukraine (country)	53N 32E	181
Ulan Bator, Mongolia (city, nat. cap.)	47N 107E	198
Uliastay, Mongolia, (city)	48N 97E	198
Ungava Peninsula	60N 72W	166
United Arab Emirates (country)	25N 55E	198
United Kingdom (country)	54N 4W	181
United States (country)	40N 98W	165
Uppsala, Sweden (city)	60N 18E	181
Ural (riv., Asia)	45N 55E	199
Ural Mountains	50N 60E	199
Uruguay (country)	37S 67W	174
Uruguay, Rio (riv., S.Am.)	30S 57W	175
Ürümqui, China (city)	44N 107E	198
Utah (st., US)	38N 110W	165
Uzbekistan (country)	42N 58E	181
Vaal (riv., Africa)	27S 27E	192
Valdivia, Chile (city)	40S 73W	174
Valencia, Spain (city)	39N 0	181
Valencia, Venezuela (city)	10N 68W	174
Valparaíso, Chile (city)	33S 72W	174
van Diemen, Cape	11S 130E	210
van Rees Mountains	4S 140E	210
Vanatu (country)	15S 167E	209
Vancouver, Canada (city)	49N 153W	165
Vancouver Island	50N 130W	166
Vanern, Lake	60N 12E	182
Vattern, Lake	56N 12E	182
Venezuela (country)	5N 65W	174
Venezuela, Gulf of	12N 72W	175
Venice, Italy (city)	45N 12E	181
Vera Cruz (st., Mex.)	20N 97W	165
Vera Cruz, Mexico (city)	19N 96W	165
Verkhoyanskiy Range	65N 130E	199
Vermont (st., US)	45N 73W	165
Vert, Cape	15N 17W	192
Vestfjord	68N 14E	182
Viangchan, Laos (city, nat. cap.)	18N 103E	198
Victoria (riv., Australasia)	15S 130E	210
Victoria (st., Aust.)	37S 145W	209
Victoria, British Columbia (city, prov. cap., Can.)	48N 153W	165
Victoria, Lake	3S 35E	192
Victoria Mt. 13,238	9S 137E	210
Victoria River Downs, Australia (city)	17S 131E	209
Viedma, Rio Negro (city, st. cap., Argen.)	41S 63W	174
Vienna, Austria (city)	48N 16E	181
Vietnam (country)	10N 110E	198
Villahermosa, Tabasco (city, st. cap., Mex.)	18N 93W	165
Vilnius, Lithuania (city, nat. cap.)	55N 25E	181
Virginia (st., US)	35N 78W	165
Viscount Melville Sound	72N 110W	166
Vitória, Espiritu Santo (city, st. cap., Braz.)	20S 40W	174
Vladivostok, Russia (city)	43N 132E	198
Volga (riv., Europe)	46N 46E	182
Volgograd, Russia (city)	54N 44E	181
Volta (riv., Africa)	10N 15E	192
Volta, Lake	8N 2W	192
Vosges Mountains	48N 7E	182
Wabash (riv., N.Am.)	43N 90W	166
Walvis Bay, Namibia (city)	23S 14E	191
Warsaw, Poland (city, nat. cap.)	52N 21E	181
Washington (st., US)	48N 122W	165

NAME/DESCRIPTION	LATITUDE & LONGITUDE	PAGE
Washington, D.C., United States (city, nat. cap.)	39N 77W	165
Wellington Island	48S 74N	175
Wellington, New Zealand (city, nat. cap.)	41S 175E	209
Weser (riv., Europe)	54N 8E	182
West Cape Howe	36S 140E	210
West Indies	18N 75W	166
West Siberian Lowland	60N 80E	199
West Virginia (st., US)	38N 80W	165
Western Australia (st., Aust.)	25S 122W	209
Western Ghats	15N 72E	199
Western Sahara (country)	25N 13W	191
White Nile (riv., Africa)	13N 30E	192
White Sea	64N 36E	182
Whitney, Mt. 14,4117	33N 118W	166
Wichita, KS (city)	38N 97W	165
Wilhelm, Mt. 14,793	4S 145E	210
Windhoek, Namibia (city, nat. cap.)	22S 17E	191
Winnipeg, Lake (N.Am.)	50N 98W	166
Winnipeg, Manitoba (city, prov. cap., Can.)	53N 98W	165
Wisconsin (st., US)	50N 90W	165
Wollongong, Australia (city)	34S 151E	209
Woodroffe, Mt. 4,724	26S 133E	210
Woomera, Aust. (city)	32S 137E	209
Wrangell (island)	72N 180E	199
Wuhan, China (city)	30N 141E	198
Wyndham, Australia (city)	16S 129E	209
Wyoming (st., US)	45N 110W	165
Xalapa, Vera Cruz (city, st. cap., Mex.)	20N 97W	165

NAME/DESCRIPTION	LATITUDE & LONGITUDE	PAGE
Xingu, Rio (riv., S.Am.)	5S 54W	175
Yablonovyy Range	50N 98E	199
Yakutsk, Russia (city)	62N 130E	198
Yamoussoukro, Cote d'Ivoire (city)	7N 4W	191
Yangtze (Chang Jiang) (riv., Asia)	30N 134E	199
Yaoundé, Cameroon (city, nat. cap.)	4N 12E	191
Yekaterinburg, Russia (city)	57N 61E	181
Yellowknife, Northwest Territories (city, prov. cap, Can.)	62N 140W	165
Yellowstone (riv., N.Am.)	46N 110W	166
Yemen (country)	15N 50E	198
Yenisey (riv., Asia)	68N 85E	199
Yerevan, Armenia (city, nat. cap.)	40N 44E	181
Yokohama, Japan (city)	36N 140E	198
York, Cape	75N 65W	166
Yucatan (st., Mex.)	20N 88W	165
Yucatan Channel	22N 88W	166
Yucatan Peninsula	20N 88W	166
Yukon (riv., N.Am.)	63N 150W	166 inset
Zacatecas (st., Mex.)	23N 103W	165
Zacatecas, Zecatecas (city, st. cap, Mex.)	23N 103W	165
Zagreb, Croatia (city, nat. cap.)	46N 16E	181
Zagros Mountains	27N 52E	199
Zambezi (riv., Africa)	18S 30E	192
Zambia (country)	15S 25E	191
Zanzibar (island)	5S 39E	192
Zemlya Frantsa Iosifa (island)	80N 40E	199
Ziel, Mt. 4,1165	23S 134E	210
Zimbabwe (country)	20S 30E	191

Sources

Maps

Map 21	World Christian Database
Map 24	Living Tongues Institute for Endangered Languages
Map 27	United Nations Population Reference Bureau
Map 37	Food and Agriculture Organization, United Nations
Map 41	World Health Organization
Map 69	Associates for International Research, Local News Reports, Gasbuddy.com, World Bank, Energy Information Administration, OPEC, CIA World Factbook, U.S. Census Bureau, Moscow Fuel Association, Gasoline-Germany.com, AAIreland.ie
Map 70	World Development Indicators, World Bank
Map 71	World Development Indicators, World Bank
Map 72	World Development Indicators, World Bank
Map 77	Oak Ridge National Laboratory Landscan 2004; Climatic Research Unit, University of East Anglia; Fourth Assessment Report of the Intergovernmental Panel on Climate Change Working Group II
Map 78	Fourth Assessment Report of the Intergovernmental Panel on Climate Change Working Group II
Map 82	Water Systems Analysis Group, University of New Hampshire
Map 83	Beverage Marketing Corporation
Map 87	Studies of Jack R. Harlan, Professor of Plant Genetics/University of Illinois; John Bartholomew & Son, Ltd./National Library of Scotland Archive
Map 88	University of Michigan
Map 94	Compiled by Donald L. Carrick, NGS using data from: IUCN/NYZS/WWF African Rhino Survey, Andrew Laurie/Cambridge University, and Thomas Foose/American Association of Zoological Parks and Aquariums
Map 96	World Wildlife Fund, IUCN
Map 97	World Bank, World Development Indicators 2008
Map 99	USDA Natural Resources Conservation Service; Gerhard Bechtold; Johannes Lehmann, Cornell University
Map 101	Yale University, Columbia University, and LinkedByAir (compiled by *Newsweek* Magazine)
Map 111	Freedom House; Interparliamentary Union; CIA Factbook 2008
Map 117	Julian Burger, United Nations
Map 123	Transparency International, Forbes

References

John L. Allen and Audrey Shalinsky. (2004). *Student atlas of anthropology*. Guilford, CT: McGraw-Hill/Dushkin.

American Association for the Advancement of Science. (2000). *AAAS atlas of population and environment*. Berkeley-Los Angeles-London: The University of California Press.

An atmosphere of uncertainty. (1987, April). *National Geographic*, 171.

Conservation International. (2002). *Global hotspots of diversity*. Washington, DC.

Crabb, C. (1993, January). Soiling the planet. *Discover, 14*(1), 74–75.

DeBlij, H. J., & Muller, P. (2006). *Geography: Realms, regions and concepts* (12th ed., revised). New York: John Wiley & Sons.

Department of Geography, Pennsylvania State University. (1996). Unpublished computer model output. State College, PA: Pennsylvania State University.

Domke, K. (1988). *War and the changing global system*. New Haven, CT: Yale University Press.

Economic consequences of the accident at Chernobyl nuclear plant. (1987). PlanEcon Reports, 3.

Environmental Protection Agency. (1996). Unpublished data [Online]. Available: http://www.epa.gov.

Fagan, B. M. (1998). *People of the earth* (9th ed.). New York: Longman.

Fellman, J., Getis, A., & Getis, J. (1995). *Human geography: Landscapes of human activities* (4th ed.). Dubuque, IA: Wm. C. Brown Publishers.

Fuller, Harold. (Ed.). (1971). *World patterns: The Aldine college atlas*. Chicago: Aldine Publishing Co.

Hoebel, E. A. (1966.) *Anthropology: the study of man* (3rd ed.). New York: McGraw-Hill.

Johnson, D. (1977). *Population, society, and desertification*. New York: United Nations Conference on Desertification, United Nations Environment Programme.

Köppen, W., & Geiger, R. (1954). *Klima der erde* [Climate of the earth]. Darmstadt, Germany: Justus Perthes.

Kuchler, A. W. (1949). Natural vegetation. *Annals of the Association of American Geographers*, 39.

Lindeman, M. (1990). *The United States and the Soviet Union: Choices for the 21st century*. Guilford, CT: McGraw-Hill/Dushkin.

Mather, J. R. (1974). *Climatology: Fundamentals and applications*. New York: McGraw-Hill.

Miller, G. T. (1992). *Living in the environment* (7th ed.). Belmont, CA: Wadsworth.

Murphy, R. E. (1968). Landforms of the world [Map supplement No. 91] *Annals of the Association of American Geographers, 58*(1), 198–200.

National Aeronautics and Space Administration. (2003–2005). Unpublished data and images [online]. Available: http://www.nasa.gov.

National Geographic Society. (1999). *Atlas of the world*, 7th edition. Washington, DC: National Geographic Society.

National Oceanic and Atmospheric Administration. (2005). Unpublished data [Online]. Available: http://www.noaa.gov.

Population Action International. (2003). *The security demographic: Population and civil conflict after the Cold War*. Washington, DC.

Population Reference Bureau. (2005). *2005 world population data sheet*. New York: Population Reference Bureau.

Rand McNally. (1996). *Goode's world atlas* (19th ed.). Chicago: Rand McNally and Co.

Rand McNally answer atlas. (1996). Chicago: Rand McNally and Co.

Rand McNally. (2005). *Goode's atlas of human geography* (21st ed.). New York: John Wiley & Sons, Inc.

Rourke, J. T. (2003). *International politics on the world stage* (9th ed). Guilford, CT: McGraw-Hill/Dushkin.

Scupin, R., and Decorse, C. R. (2001). *Anthropology: A global perspective* (4th ed.). Upper Saddle River, NJ: Prentice Hall.

Shelley, F., & Clarke, A. (1994). *Human and cultural geography: A global perspective.* Dubuque, IA: Wm. C. Brown Publishers.

Smith, Dan. (1997). *The state of war and peace atlas* (3rd ed.). Penguin Books: New York.

Soiling the plant. (1993, January). *Discover*, 14.

Spector, L. S., & Smith, J. R. (1990). *Nuclear ambitions: The spread of nuclear weapons.* Boulder, CO: Westview Press.

This fragile earth [map]. (1988, December). *National Geographic*, 174.

Thornthwaite, C. W., & Mather, J. R. (1955). *The water balance* [Publications in Climatology No. 8]. Centerton, NJ: Drexel Institute of Technology, Laboratory of Climatology.

Times atlas of world history. (1978). Maplewood, NJ: Hammond.

United Nations Food and Agriculture Organization (FAO). (1995). *Forest resources assessment 1990: Global synthesis* [FAO Forestry Paper NO. 124]. Rome: FAO.

United Nations Population Fund. (2005) *The state of the world's population.* New York: United Nations Population Fund.

United Nations Population Reference Bureau. (2005). 2005 *world population data sheet.* New York: Oxford University Press.

United Nations Population Reference Bureau. (2005) *World development report.* New York: Oxford University Press.

U.S. Census Bureau. (1998). *World population profile.* Washington, DC: U.S. Government Printing Office.

U.S. Central Intelligence Agency. (2006). *World factbook 2006.* Available: http://www.odci.gov/cia/publications/factbook/index/html.

U.S. Central Intelligence Agency. Unpublished data [Online]. Available: http://www.odci.gov/cia/publications.

U.S. Committee for Refugees. *World refugee survey* (Washington, DC, 2002).

U.S. Department of Energy. (1996). *U.S.–Canada memorandum of intent on transboundary air pollution.* Washington, DC: U.S. Government Printing Office.

U.S. Department of State. (2005). *Stateman's year-book 2005.* Washington, DC: U.S. Government Printing Office.

USDA Forest Service. (1989). *Ecoregions of the continents.* Washington, DC: U.S. Government Printing Office.

U.S. Soil Conservation Service [now the U.S. Natural Resources Conservation Service]. (1996). *World soils.* Washington, DC: U.S. Soil Conservation Service.

The World almanac and book of facts 2006 (2006). Mahwah, NJ: World Almanac Books.

The World Bank. (2004–2005). *World development report 2004–2005.* Geneva: World Bank.

The World Bank. (2004–2005). *2004–2005 world development indicators.* (Washington, World Bank).

The World Bank. (2004). *Entering the 21st century: World development report 2002/2003.* New York: Oxford University Press.

World Conservation Monitoring Centre. (1996). Unpublished data. Cambridge, England: World Conservation Monitoring Centre.

World Health Organization. (2005). *World health statistics annual.* Geneva: World Health Organization.

World Resources Institute. *World resources 2002–2004: A guide to the global environment.* New York: Oxford University Press.

World Resources Institute. (2005). *World resources 2005—The wealth of the poor. Managing ecosystems to fight poverty.* Washington, DC: Oxford University Press.

Wright, John W. (Ed.). (2004). *The New York Times 2003 Almanac.* New York: Penguin Reference Books.